Pharmaceutical Quality by Design Using JMP®

Solving Product Development and Manufacturing Problems

Rob Lievense

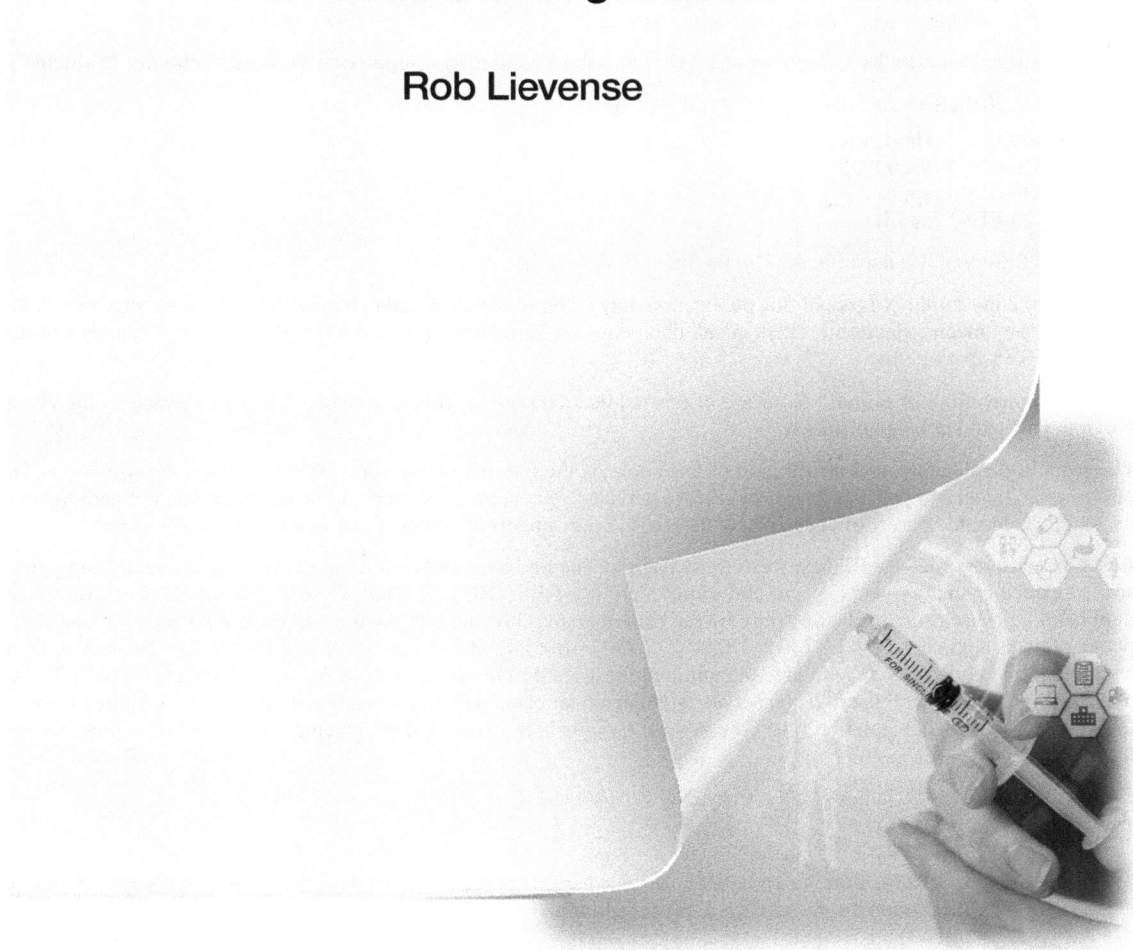

sas.com/books

Contents

About This Book

What Does This Book Cover?

Regulatory agencies for the pharmaceutical and medical device industries have released several guidelines to promote the use of elements of Quality by Design (QbD). Technical professionals have great interest in QbD, but many are unsure of where to start. This book is a guide for using data visualization and statistical analyses as elements of QbD to solve problems and support improvement throughout the product life cycle.

The book includes three areas of general focus for the topics contained. The first several chapters focus on the type of data that is available for current commercial production of healthcare products. The book then focuses on the tools and techniques that are useful for product and process development. The final chapters are more specialized and deal with utilizing data visualization to solve complex problems, as well as special topics that are unique to healthcare products.

In chapters 1 through 5, technical professionals learn how to use JMP to obtain visualizations of their data by using the Distribution platform and the Graph Builder. The powerful, dynamic nature of the data visualizations is highlighted so that readers can easily extract meaningful information quickly. Techniques for including a time element for effective visualization and identification of trends is covered as well. Methods for comparing trends in the data to specification limits are covered, enabling you to diagnose the performance of a process and effectively communicate the findings to the stakeholders of an improvement project. The stream of topics moves on to the utilization of data from a random sample to make precise estimates (via statistical inference) on an entire population of units produced. Statistical inference is expanded to analyze for relationships and differences between two variables, utilizing the rich set of techniques available in the Fit Y by X platform.

Chapters 6 through 12 begin with applications that help the reader justify why structured, multivariate, experimental designs must be used to develop robust products and processes. Comparing designs created through the Design of Experiments (DOE) platform to the typical approach that uses one factor at a time (OFAT) clearly shows the advantages of structured, multivariate, experimental designs, especially in QbD era. Examples focus on effective techniques for analyzing measurement systems and quantifying how measurement variability may affect analysis results. Various modeling techniques are covered so that you know how to utilize available historical data to use resources efficiently in experimental designs. The DOE platform is extensively utilized to teach you how to create effective experimental designs for both materials and processes. The section is rounded out with analysis techniques for completed experiments as well as simulation tools that you can use to include known process variation and simulate likely results. Simulation can save a development team time, money, and increased credibility due to the potential to mitigate future mistakes.

The context in chapters 13 through 17 expand on the predictive modeling techniques presented in section two by including predictive models that can detect inputs that have subtle influences on outputs. Basic mixture designs are covered to help you effectively deal with three-component proportional mixes of materials. Many aspects of pharmaceutical products show trends in outputs that include rates of change as a function and that cannot be studied with typical linear modeling. Examples of non-linear modeling help you gain understanding about such applications. Analyses of measurement systems from the second section is expanded on with an example of how you can use a structured, multivariate, experimental design to support analytical method development. The section wraps up with the specialized topic of stability analysis vis a tool provided in the Degradation platform. The stability analysis techniques follow

International Conference on Harmonisation of Technical Requirements for Registration of Pharmaceuticals for Human Use (ICH) guidelines regarding how to identify the likely shelf life of products. Using these techniques to dig deeper into the modeling details provides insight that is unparalleled.

The book does not offer a deep, theoretical understanding of the concepts or detail about the computational methods used by JMP to create the output. There are several references to the Help menu in JMP throughout the chapters so that you can find this detail if you are interested.

Is This Book for You?

I have read many instructional texts for data visualization and statistics. Most begin with the identification and discussion of a statistical topic or technique, followed by examples intended for readers to use to add practical ability. The typical flow of such textbooks creates barriers to technical professionals who want to efficiently apply the knowledge to solve problems involving data. They are often under time pressures and struggle most with trying to find the statistical technique that will work to extract the information they need from data. This book is written from a technical professional point of view to match the flow of work that occurs in the real world of the pharmaceutical and medical device industries. Each chapter involves a technical professional facing a problem that could benefit from the use of JMP.

Each chapter describes the problem at hand, followed by hands-on work in JMP. Examples include relevant screen shots of the JMP interface, along with figures, notes, and explanations of results. The data sets are based on actual problems in an attempt to make the examples reflect the real world. Many of the problems involve data preparation steps and table manipulation before analysis can be done, which is another issue that technical professionals encounter in the real world. Chapters culminate with practical conclusions that help the reader summarize the key points of the analysis. Most chapters include exercises for additional hands-on practice.

Scientists, engineers, and technicians involved throughout the pharmaceutical and medical device product lifecycles will find this book useful. The reliance upon principal science and professional experience for product development can combine to yield a batch that passes requirements. The use of JMP to apply data visualization and statistical modeling will create a product that robustly meets requirements for the entire life cycle. The trends in the inputs and outputs of processes are easy to explore from the creation of simple graphs to model analysis with simulations used to estimate the defect rate of a future product. The analysis completed in JMP provides a great foundation for regulatory submissions of products and processes. Submissions supported with robust statistics tend to have fewer deficiencies. Regulatory deficiencies that occur can be better answered with data visualizations and statistics, which tend to also increase the speed of product approvals.

JMP includes the versatility to be used to solve problems throughout the life cycle of a product. Quality control can monitor and assess processes through the use of control charting and capability studies. Filling processes can be optimized through the dynamic function of the distribution platform as well as predictive modeling, Stability studies are easy to create in JMP and offer the insight needed to predict the expected shelf life for multiple packaging configurations. Physical features of medical devices can be studied and optimized to ensure that variation in products is mitigated and customers are likely to enjoy consistency in the use of a product. The measurement systems used to quantify a physical or chemical attribute can be studied using JMP to ensure the highest levels of accuracy and precision in data obtained.

Products developed through the use of JMP DOE tools can reach the market in half the time required for development using principal science and experience alone. The resources required to get a product to market are greatly reduced as models are utilized to find the optimum input settings to meet all product requirements simultaneously. Fewer developmental batches need to be run and the potential for making costly mistakes is greatly reduced. This book offers more than instruction on the use of JMP; it is also a guide for saving time and money.

What Are the Prerequisites for This Book?

This book makes a few assumptions about its readers. It is assumed that you possess a general understanding of the relevant scientific and technical concepts for the pharmaceutical or medical device industries. By following the examples, you will be able to fill in any details that you are not already familiar with. Some initial familiarity with JMP is helpful. You can use the JMP website to become familiar with JMP: https://www.jmp.com.

What Should You Know about the Examples?

This book includes relevant examples from the target industries for you to follow in order to gain hands-on experience with JMP.

Software Used to Develop the Book's Content

The book uses JMP 14.0 for the majority of content and JMP Pro 14.0 for a few high-level concepts. The screen shots used to demonstrate navigating the JMP menus are captured using JMP Pro, and most have the same look as what is seen with JMP. Other versions of JMP might not have the same options or have slightly different menu options.

Example Code and Data

It is intended that you work on the examples as you read through each chapter. The exercises at the end of most chapters provide an extension of this work by either expanding on the chapter examples or by using new data sets with similar problems. A set of additional materials including the data sets used throughout the book is available for download. You can access the example code and data for this book by visiting the author page at https://support.sas.com/lievense.

Where Are the Exercise Solutions?

A full set of solutions for the end-of-chapter exercises is available on the author page at https://support.sas.com/lievense.

About the Author

Rob Lievense is a Research Fellow of Global Statistics at Perrigo, as well as an active professor of statistics at Grand Valley State University (GVSU), located in Allendale, Michigan. At Perrigo, he leads a group that supports the consumer health care research and development department with statistical analysis, data visualization, advanced modeling, data-driven Quality by Design for product development, and structured experimental design planning. Rob has more than 20 years of experience in the applied statistics industry and 10 years of experience in the use of JMP. He has presented at major conferences including JMP Discovery Summit, where he served on the Steering Committee in 2017, and the annual conference of the American Association of Pharmaceutical Scientists. Rob has a BS in Applied Statistics and an MS in Biostatistics from GVSU. He currently serves as a member of the Biostatistics Curriculum Development Committee for GVSU and has his Six Sigma Black Belt Certification..

Learn more about this author by visiting his author page at support.sas.com/lievense. There you can download free book excerpts, access example code and data, read the latest reviews, get updates, and more.

We Want to Hear from You

SAS Press books are written *by* SAS Users *for* SAS Users. We welcome your participation in their development and your feedback on SAS Press books that you are using. Please visit sas.com/books to do the following:

- Sign up to review a book
- Recommend a topic
- Request information on how to become a SAS Press author
- Provide feedback on a book

Do you have questions about a SAS Press book that you are reading? Contact the author through saspress@sas.com or https://support.sas.com/author_feedback.

SAS has many resources to help you find answers and expand your knowledge. If you need additional help, see our list of resources: sas.com/books.

Acknowledgments

I first and foremost must thank my wife Kate and my kids Ben, Sam, and Viv for being understanding and tolerant as I have been immersed in this book project. I was the quintessential optimist as I proposed this book and now have the deepest respect for all authors who invest so much time in their craft. The folks at SAS Press and JMP have offered excellent support and feedback as this book took shape, and I am truly grateful. The technical reviewers include Catherine Connolly, Senior Associate Development Editor at SAS; Hector A. Alfonso PhD, Principal Formulation Scientist *retired*; Michelle Ricci, Senior Biostatistician at Sanofi Genzyme; Richard Zink, Director of Statistics at TARGET PharmaSolutions; Tonya Mauldin, Sr Analytics Software Tester for JMP; and Sarah Seligman, Analytical Technical Support Engineer for JMP. A personal note of thanks to Dr. Bob Carver for his valued friendship, encouragement, and precise feedback in his reviews of my (often rough) work. Your advice has added greatly to the quality of the book. Thanks goes out to Bruce D, Johnson, Carlos Paz, Perry Truitt, and Inderdeep Bhatia of Perrigo for their support and encouragement of my work and this book project. Max Wettlaufer of Perrigo has been part of the "statistics underground" for years and continues to provide practical feedback to ensure that analysis efforts have business value. Tanya Davis is a valued friend and colleague who was part of the initial "Army of Two" when R&D Statistics was formed at Perrigo and continues to positively influence the culture of analytics. Thanks to Joshua Lust and Kevin Yost of Ranir and others for providing information about medical devices to include in this book. The mentorship and advice of Louis Valente of JMP is appreciated very much, especially the sharing of the OFAT and multivariate comparison used in the DOE chapter. Others whom I must thank by name for their support and encouragement include many of the extraordinary people at JMP including Erich Gundlach, JMP Systems Engineer (Midwest); Mike Pritchard, Senior Account Executive (Midwest); and Gail Massari, JMP Customer Care.

Chapter 1: Preparing Data for Analysis

Overview

Pharmaceutical product and medical device manufacturing are complex subjects that involve a significant amount of data on a multitude of subjects. Leaders in such organizations deal with a seemingly endless stream of challenges that must be dealt with quickly and effectively. Technical professionals contend with a constant flow of data that must be converted to information so that the best possible decisions are made. The idea of using statistical analysis to deal with regular problems might not be popular due to concerns over the assumed amount of time and resources required. Professionals need a tool that can efficiently handle many types of data with the ability to easily visualize a problem and identify the best course of action. JMP and JMP PRO include powerful data visualization tools that are extremely easy for non-statisticians to master. The best decisions result from data that is analyzed at a simple, high-level view, with more complex analyses completed as more information is needed. In many cases, the visualization of a single variable can offer significant amounts of information. This first chapter deals with two common problems involving the visualization and analysis of a single random variable. A problem involving measurable data from a pharmaceutical manufacturing setting is analyzed as well as a problem involving discrete data from a medical device manufacturing facility.

The Problem: Overfilling of Bulk Product Containers

The story opens with Suzanne, a manager of a facility that produces containers of a bulk, dry pharmaceutical product. Suzanne has been under increasing pressure to continue to maintain the highest standards of quality while finding ways to reduce costs. The pharmaceutical industry is becoming increasingly competitive, and the profit margins that have been enjoyed are taking some hits. Suzanne is faced with the reality of needing to make improvements as soon as possible to ensure that her facility remains viable.

Suzanne knows that her fill lines have demonstrated a robust ability to meet the label claim for product in the containers. Containers that come off the line must have an average fill that is no less than the claimed weight printed on the product label. The quality team has been very satisfied with the fill crews who do their best to make sure that each container has plenty of product. The teams' only known upper limit for fills is to make sure that the tops of the containers can be applied. The new focus must be on increased precision as the fill lines are required to robustly meet quality standards while performing consistently to maintain the least possible amount of overfilling. Suzanne knows that she will need to collect data on the fill process in order to measure the extent of the fill range, which can lead to the identification of possible improvements by the operations and engineering teams.

Suzanne is a brilliant manager and has a few advantages up her sleeve that she can use to ensure success in improvement projects. She has JMP software licenses among the tools available for the team, and she is resourceful in researching best practices for data visualization and analytics. Suzanne assembles key members of the fill process, which will enable her to plan and execute the most effective improvement process.

Collect the Data

The team knows that first they need to capture the current state of the fill process as a baseline. The fill lines have an accurate and precise digital scale used to weigh in-process samples for regular quality checks. Suzanne works with the team to pick a target product fill line to represent the process. The team determines that a sample of 50 in-process checks will be chosen from the process records. Each in-process check event involves collecting weights for 5 tubs of product; therefore, the data sheet includes 250 individual weights. A team member is chosen as a project lead for producing the data for analysis. Everything is in place and Suzanne is optimistic because her planning has enabled a good start on the project.

The data is in Figure 1.1, which has been compiled into a Microsoft Excel worksheet that is highly formatted. Suzanne is impressed by the time and effort that was put into the data sheet. However, she is unable to get much more out of it than what was already known. The line is consistently filling containers to more than the 500-gram fill weight claimed on the product label. Suzanne is not sure how to proceed. However, she will easily be able to extract valuable information from her data with JMP.

Figure 1.1: Data Sheet Provided in a Formatted Excel Spreadsheet

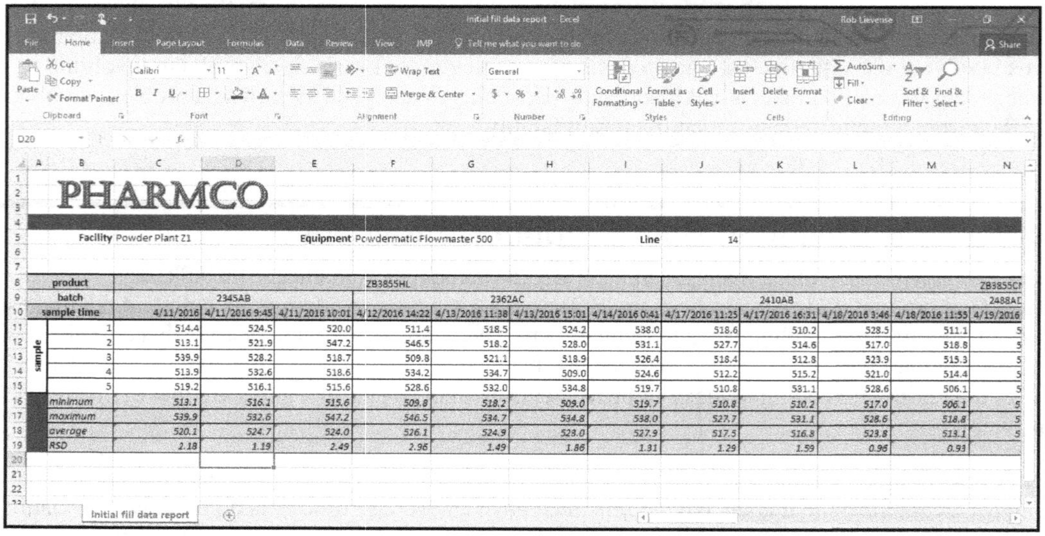

PHARMCO

Facility Powder Plant Z1 | Equipment Powdermatic Flowmaster 500 | Line 14

product		ZB3855HL							ZB3855C				
batch		2345AB			2362AC				2410AB			2488AD	
sample time		4/11/2016	4/11/2016 9:45	4/11/2016 10:01	4/12/2016 14:22	4/13/2016 11:38	4/13/2016 15:01	4/14/2016 0:41	4/17/2016 11:25	4/17/2016 16:31	4/18/2016 3:46	4/18/2016 11:55	4/19/2016
sample	1	514.4	524.5	520.0	511.4	518.5	524.2	538.0	518.6	510.2	528.5	511.1	5
	2	513.1	521.9	547.2	546.5	518.2	528.0	531.1	527.7	514.6	517.0	518.8	5
	3	539.9	528.2	518.7	509.8	521.1	518.9	526.4	518.4	512.8	523.9	515.3	5
	4	513.9	532.6	518.6	534.2	534.7	509.0	524.6	512.2	515.2	521.0	514.4	5
	5	519.2	516.1	515.6	528.6	532.0	534.8	519.7	510.8	531.1	528.6	506.1	5
	minimum	513.1	516.1	515.6	509.8	518.2	509.0	519.7	510.8	510.2	517.0	506.1	5
	maximum	539.9	532.6	547.2	546.5	534.7	534.8	538.0	527.7	531.1	528.6	518.8	5
	average	520.1	524.7	524.0	526.1	524.9	528.0	527.9	517.5	516.8	523.8	513.1	5
	RSD	2.18	1.19	2.49	2.96	1.49	1.86	1.31	1.29	1.59	0.96	0.93	

Initial fill data report

A great deal of information for getting started on a project is available through the JMP website. Suzanne uses the JMP website (https://www.jmp.com/en_us/home.html) to explore the information available on the *Learn JMP* tab, including an on-demand webcast focused on importing Excel data into JMP.

Import Data into JMP

You can easily import data from Excel sheets into JMP by using the Excel Import Wizard. The process that Suzanne used to import data from Excel is described in the following steps. With JMP open, select *File ▶ Open* and choose the *Excel Files* option (Figure 1.2) to choose the file location for the *initial fill data report.xls* file. Leave all other options to the defaults, and then click *Open* to open the file in JMP.

Figure 1.2: Open Data File Dialog Box Window

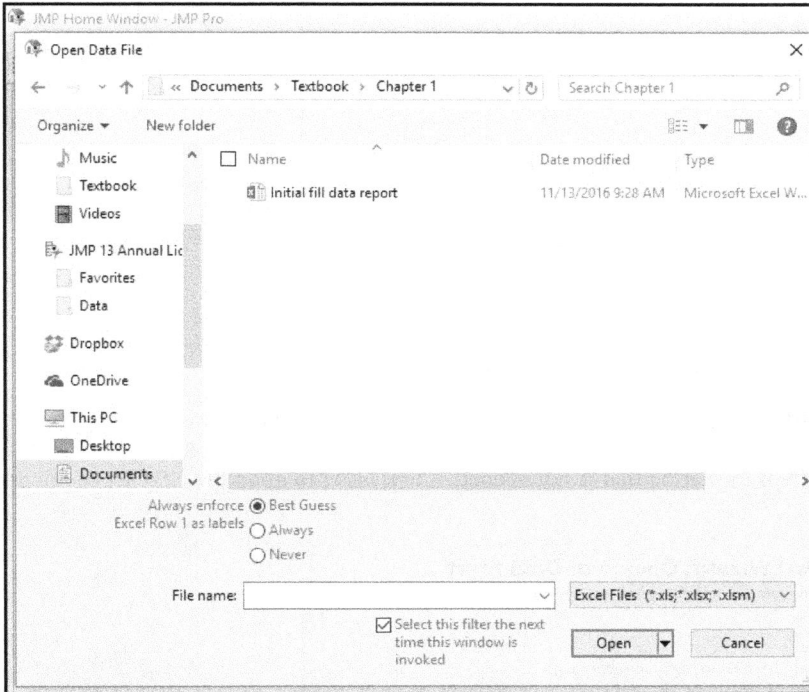

Figure 1.1 is an Excel sheet with highly formatted data. Title information, product and batch information, and group summaries are also present in the sheet. Suzanne is interested in starting her data visualization as simply as possible. She is not interested in the products or lots, or in looking at the data by the date and time of the in-process checks. The good thing is that the Excel Import Wizard, which enables the user to select the data of interest to extract into a JMP data file. Figure 1.3 displays the initial page of the wizard. The wizard can handle an Excel file with multiple worksheets. However, this file does not contain multiple worksheets. The *Data Preview* shows the entire Excel sheet initially, which will not work for our purposes.

Figure 1.3: JMP Excel Import Wizard

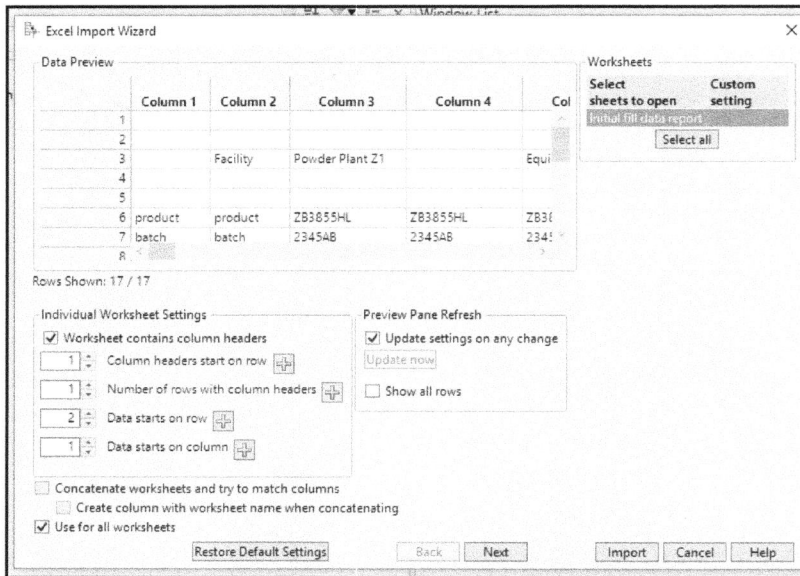

The options within the wizard enable you to selectively focus on the rows in which the actual data values begin. For this example, the *Column headers start on row* value is 9 and the *Data starts on column* value is 2 in order to eliminate information that is not needed. Click *Next* to go to the next set of options for importing the data.

Figure 1.4: JMP Excel Import Wizard: Choice of Data Start

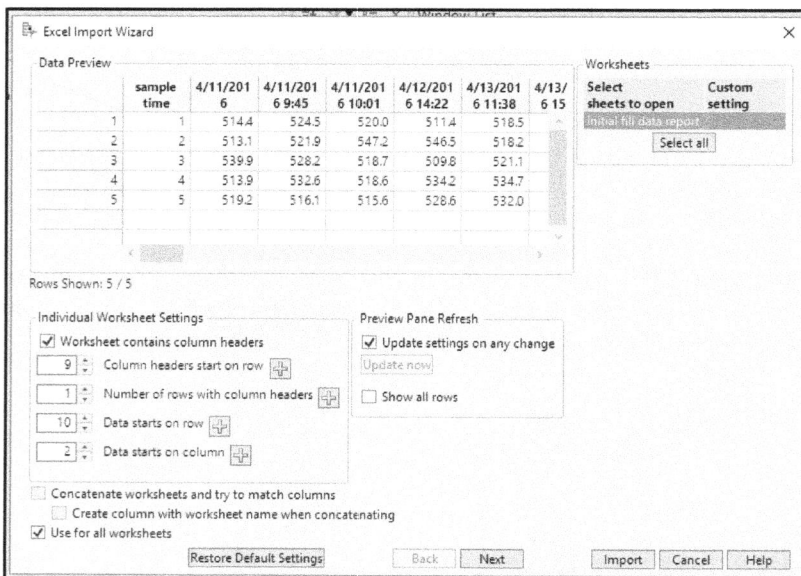

The summary statistics for each of the 50 in-process checks are not needed for this project. Figure 1.5 shows that the *Data ends with row* value is 14, which cuts off the summary statistics from the data set. No other options are required. Click *Import* to complete the process of importing the data to JMP.

Figure 1.5: JMP Excel Import Wizard: Choice of Data End

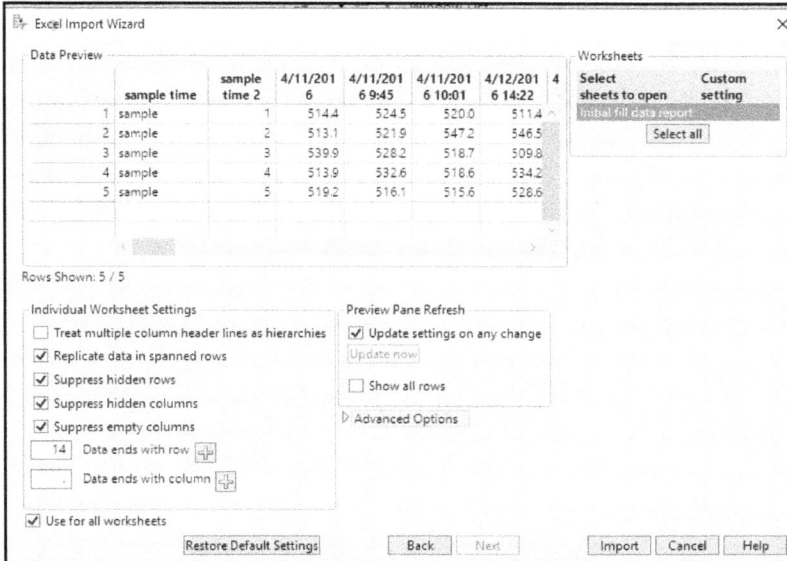

Change the Format of a JMP Table

Suzanne is impressed with how quickly she has been able to convert the highly formatted Excel sheet to a JMP data set using the Excel Import Wizard (Figure 1.6). The data is now in an unstacked format; the sampling groups (times of checks) are in individual columns with each of the five observations presented in rows. There is a bit more work needed to get the data into its most useable form.

To start, the first column *sample time* should be changed to *sample* by clicking on the column header and changing the column name. The weight information now must be converted into a single column, which is a stacked data set.

Figure 1.6: Initial Fill of JMP Data Table

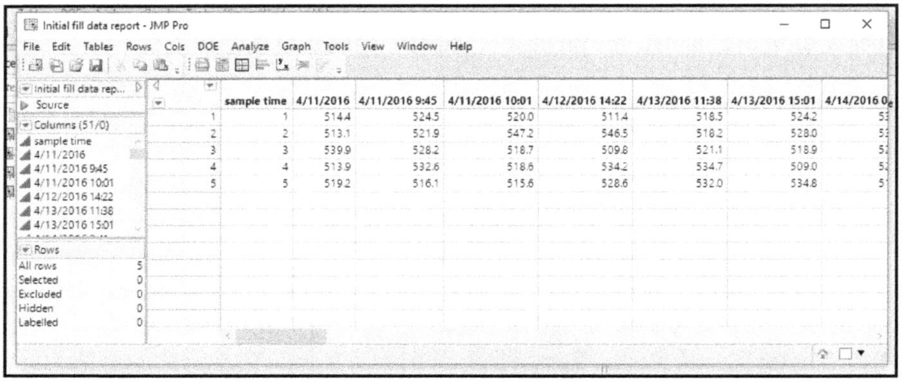

The *Tables* menu includes all of the tools needed to manipulate the data table into the format that works best. The following steps reformat the data sheet:

1. Select *Tables ▶ Stack*. The window shown in Figure 1.7 appears.
2. Select all of the time columns, and move them to the *Stack Columns* section.
3. Deselect the *Stack by row* check box, and type *stacked initial sample weight* in the *Output table name* field.
4. Deselect the *Stack By Row* check box. The default setting is to stack the observations across the columns in row order. This default option would take the data out of the date groups, which is not acceptable for the subject analysis.
5. Enter *weight* in the *Stacked Data Column* field and *sample time* in the *Source Label Column* field.
6. Click *OK* to execute the stacked data table.

Figure 1.7: Stacked Tables Window

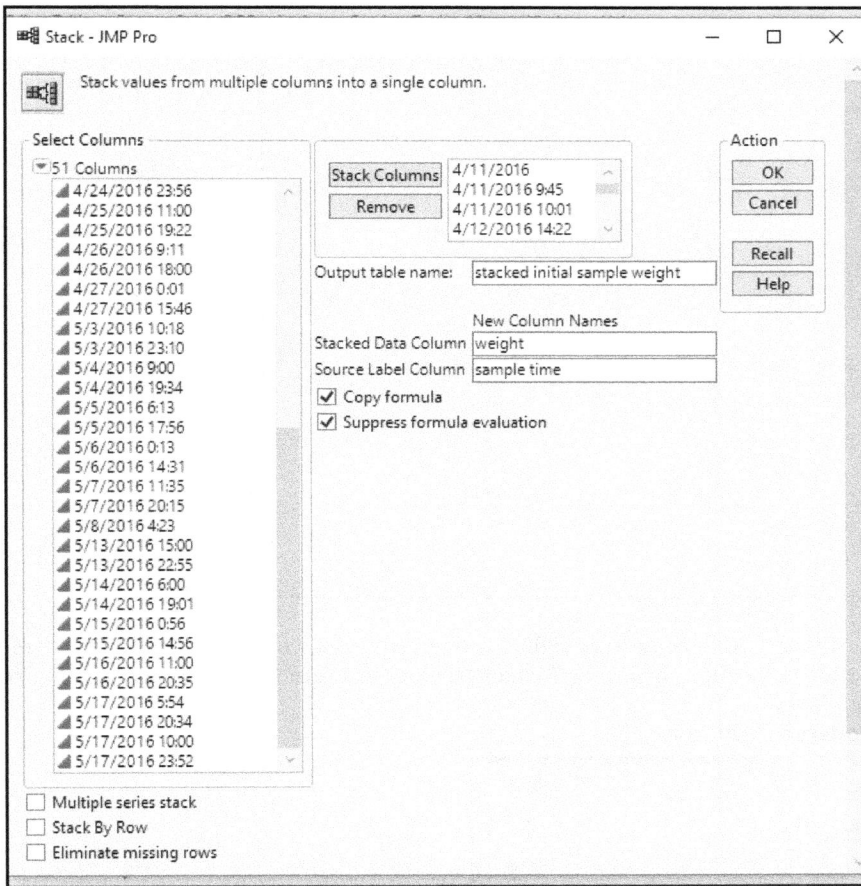

The stacked data is shown in Figure 1.8, and is almost ready for analysis. There is one more thing that is needed to maintain the organizational structure of the data since the sample time is not of interest at this time. A new column must be added to create a numbered sample group for each of the 50 process checks chosen at random for the analysis.

Figure 1.8: Initial Fill of Stacked Data Table

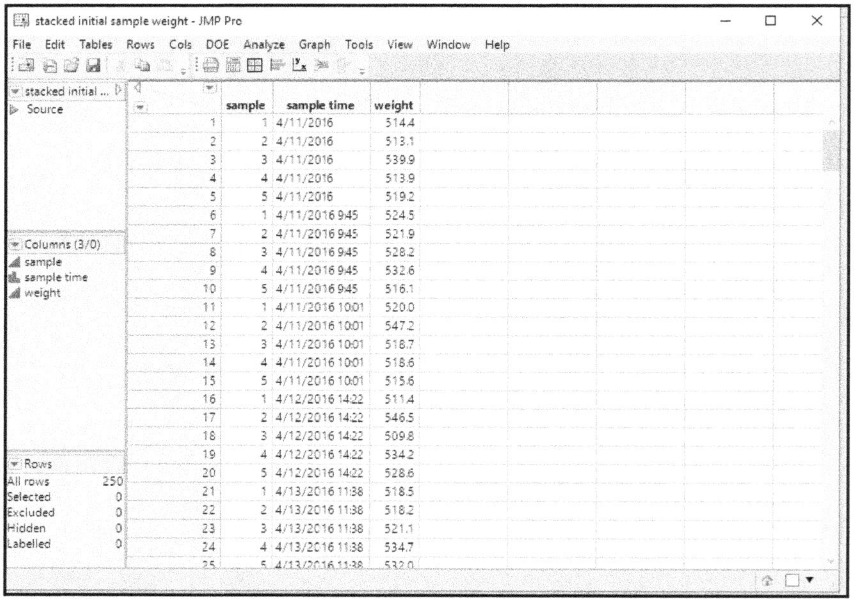

Start a new column by using the *Cols* menu or by right clicking on the open column to the right of *weight*. Name the new column *sample group*. Then, click *Missing/Empty* and select the *Sequence Data* option.

Figure 1.9: Column Properties Window

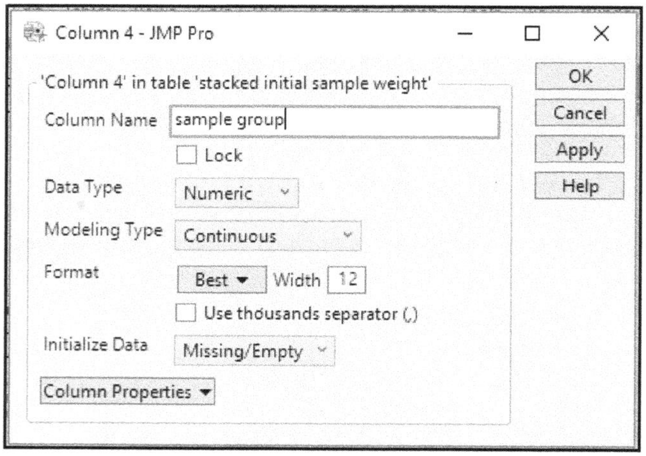

In Figure 1.10, the value for *Repeat each value N times* is 5, which causes each group number to be repeated for the five weight observations of the group. Click *OK* to complete the table. Figure 1.11 shows the resulting table.

Figure 1.10: Column Properties Window with Initialize Data

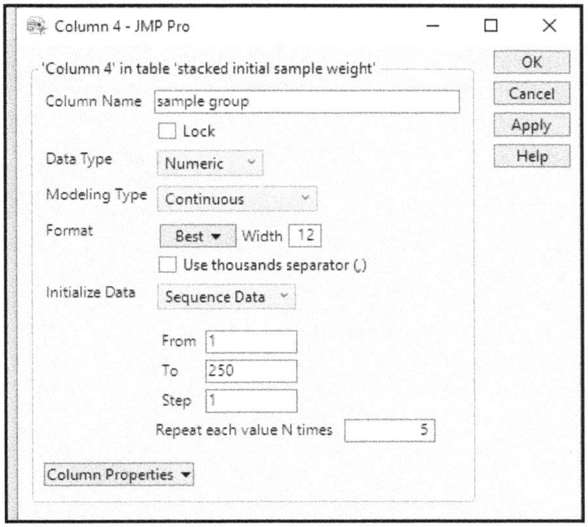

Figure 1.11: Initial Fill of Stacked Data Table Complete for Analysis

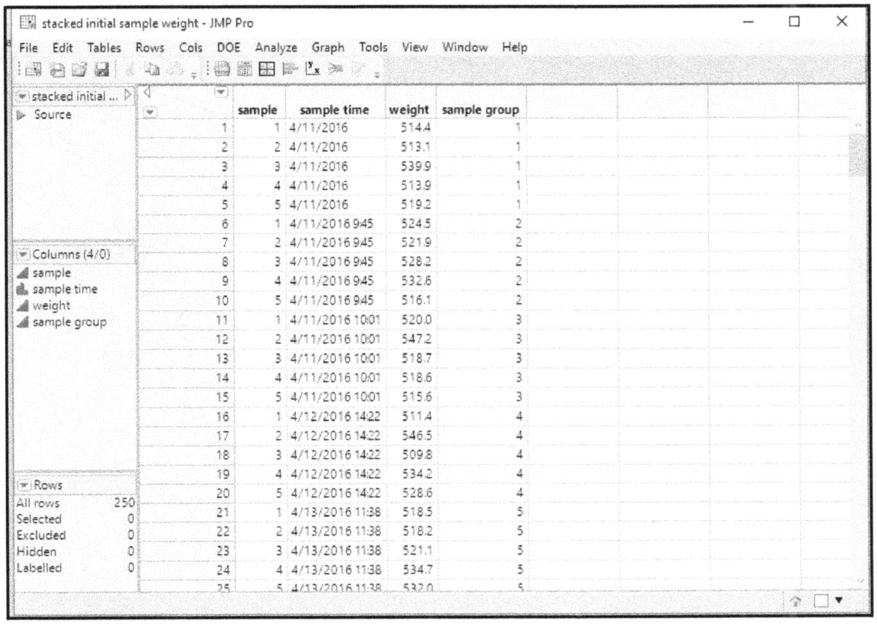

Explore Data with Distributions

The data set is formatted and ready for analysis. It is best practice to start with a basic look at the data in order to understand where the data set is located on the infinite scale of values, the extent to which the data is spread out, and the shape of the data spread. JMP enables you to easily gain a great deal of information by selecting *Analyze ▶ Distribution*, as shown in Figure 1.12.

Note: When you hold your pointer over your selection, information describing the analysis choice appears. Such help is another useful hidden feature offered by JMP to make it easy for novice users to choose the most appropriate menu options.

Figure 1.12: Create a Distribution

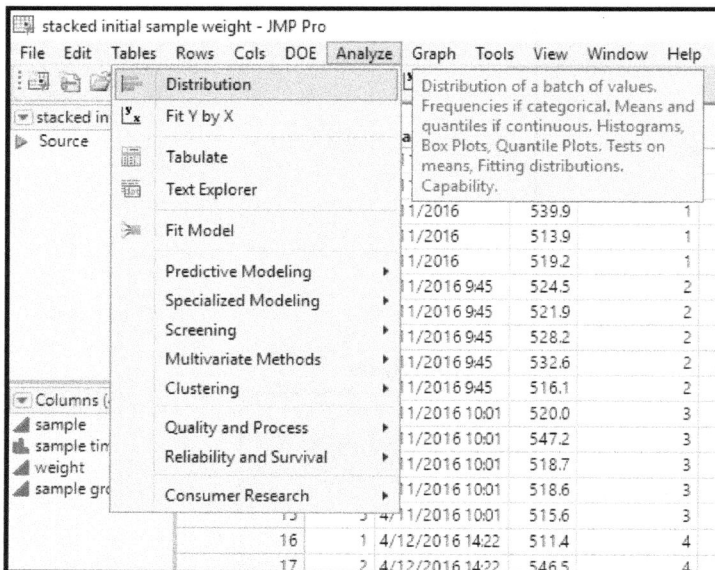

The *Distribution* window appears, as shown in Figure 1.13. All of the variables of the data sheet are listed in the *Select Columns* section. Move the **weight** variable to the *Y, Columns* box for the analysis.

Options are available to provide weighting for variable groups, add a variable that includes frequency counts, and for the ability to split distributions by a grouping variable. These options are not needed for the initial analysis and are explored in later chapters.

Figure 1.13: Distribution Window

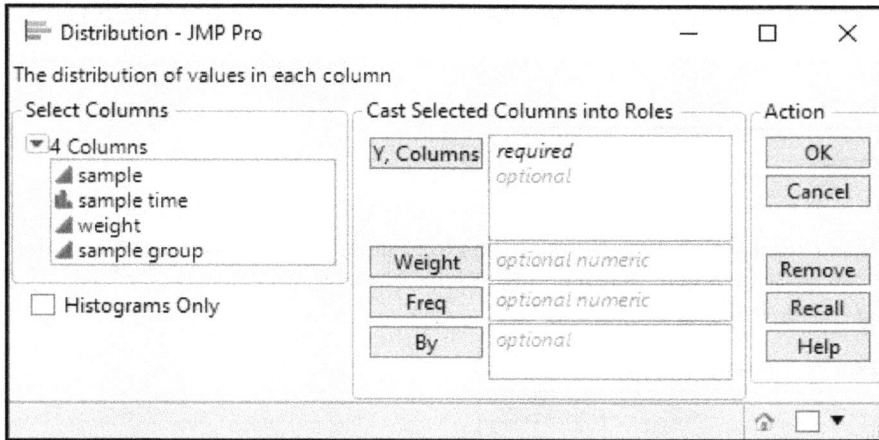

Click *OK* to display the Distribution Output window (Figure 1.14). The initial output includes a histogram, outlier box plot, Quantiles table, and Summary Statistics table. JMP output typically initiates in a stacked format. You can change this format to a view that offers optimum usability.

Figure 1.14: Distribution Output

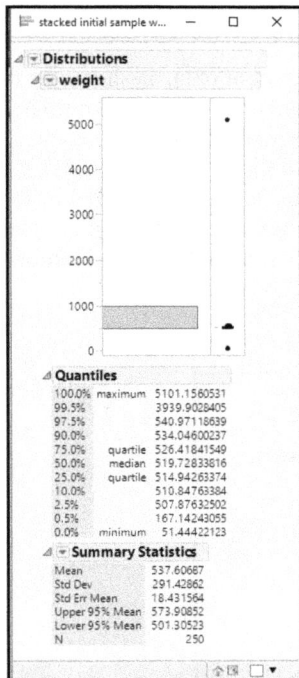

The red triangle menu located to the left of each analysis heading, shown in Figure 1.15, provides you with many custom options for extracting the maximum amount of information from the data. The examples in this book use the red triangle menu to add detail to plots and analyses.

Figure 1.15: JMP Hotspot

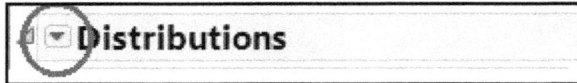

Click the red triangle menu beside the *Distributions* heading to change the output so that it is organized across the screen. Select the *Stack* option, shown in Figure 1.16. The result can improve the usability of the output for a single variable distribution.

Figure 1.16: Distribution Red Triangle Menu

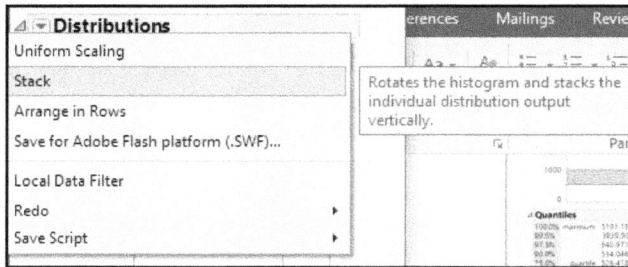

The distribution output in Figure 1.17 reveals some significant outliers in the set of data, as shown by the black dots in the outlier box plot above the histogram. JMP uses the Tukey method to illustrate outliers. The method uses the inner quartile range (IQR), which is the distance between the 25th percentile and 75th percentile of the data, and is shown as the box of a box plot. The IQR is multiplied by 1.5 because it is expected that random variation includes observations that are within 1.5 times IQR above and below the median. Any observation that is beyond this range of expected random variation is identified on the plot as a black dot. To select the two outliers, above and below the high frequency bar, hold the control button and click the outliers in question.

Figure 1.17: Distributions Output

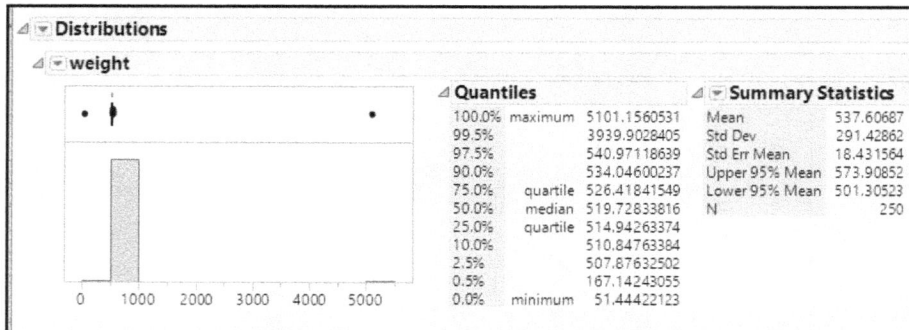

The data table shown in Figure 1.18 illustrates the dynamic features of JMP. Each of the rows with outlier values is shaded in blue, and the number of rows indicated as Selected appears in the Rows panel at the bottom left of the table. The left side of the Home window in JMP includes three panels. The top panel includes table information, the middle includes columns information, and the bottom panel includes row information.

Figure 1.18: Initial Fill Stacked Data Table with Outliers Selected

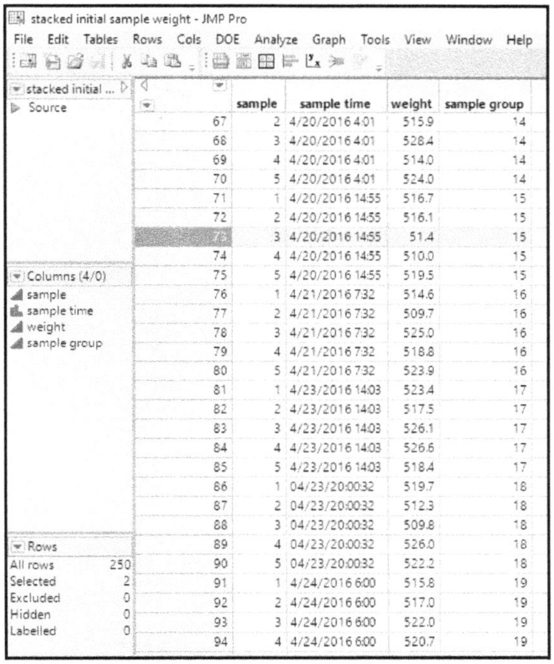

Right-click *Selected* in the Rows panel of the data table and choose *Data View* to create a new data table with the selected outliers, as shown in Figure 1.19.

Figure 1.19: Creating a Data View from a Selection

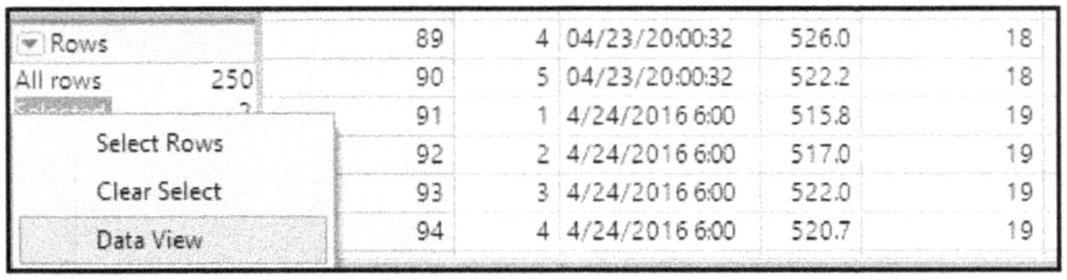

Figure 1.20 shows the outliers, which were found to be typographical errors due to incorrect decimal placement. The selected values in the original data table are corrected in the stacked data table to be 514.0 and 510.12 respectively. Close the outlier data table after the corrections have been made.

Figure 1.20: Outlier Data Table

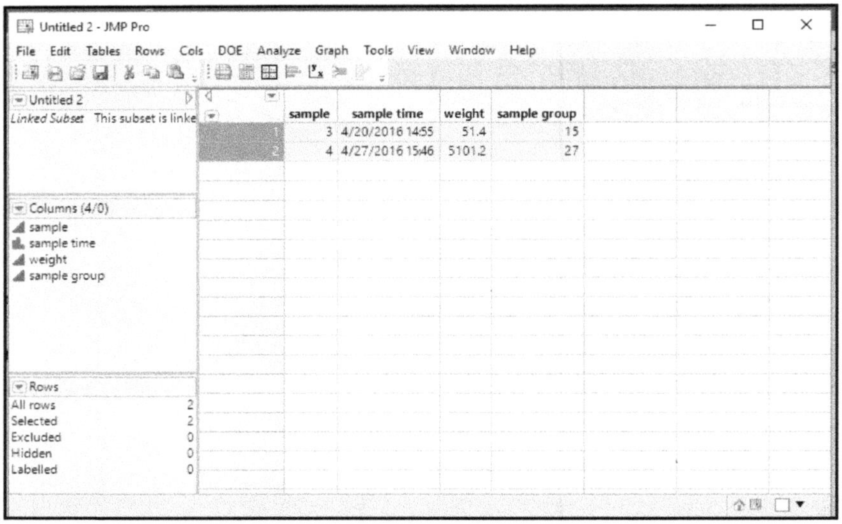

Many time saving features are embedded in JMP that might not be immediately evident. The red triangle menu options beside the *Distributions* header enable you to choose from the *Redo* options. The *Redo Analysis* option works best to quickly repeat the *Distributions* for the corrected data, as shown in Figure 1.21.

Figure 1.21: Redo Analysis of Corrected Data in Distributions

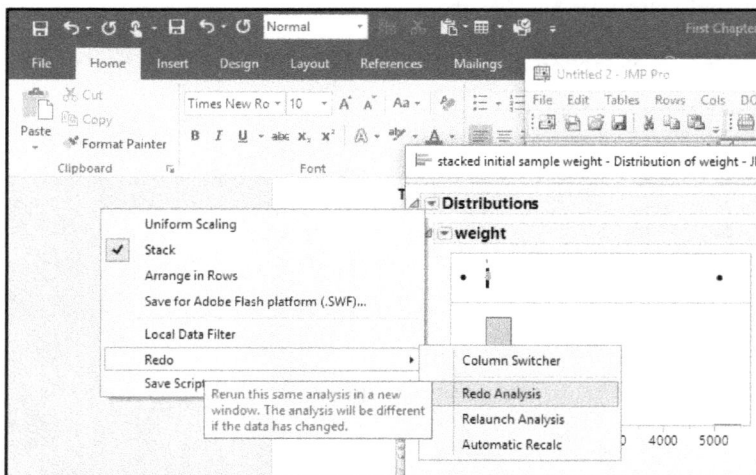

The *Distributions* plot from the corrected data shown in Figure 1.22 includes a limited number of minor high outliers. The values were matched with actual entries in the source data. Therefore, the extreme data values should not be discarded.

Figure 1.22: Distributions of Corrected Output

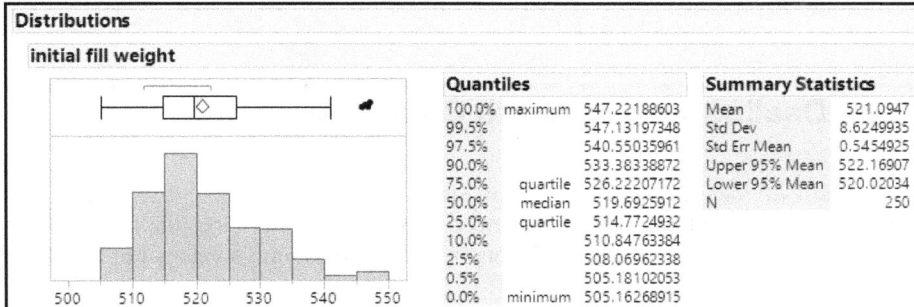

Distributions

initial fill weight

Quantiles			Summary Statistics	
100.0%	maximum	547.22188603	Mean	521.0947
99.5%		547.13197348	Std Dev	8.6249935
97.5%		540.55035961	Std Err Mean	0.5454925
90.0%		533.38338872	Upper 95% Mean	522.16907
75.0%	quartile	526.22207172	Lower 95% Mean	520.02034
50.0%	median	519.6925912	N	250
25.0%	quartile	514.7724932		
10.0%		510.84763384		
2.5%		508.06962338		
0.5%		505.18102053		
0.0%	minimum	505.16268915		

The interpretation of the Distributions analysis provides a great deal of information about the fill process. The default plots available in JMP enable the user to find anomalies more quickly than is possible by studying an Excel spreadsheet full of numbers. The minimum value of just over 505 grams provides evidence that none of the containers studied is at risk for not meeting the minimum label claim fill of 500 grams. Containers can be significantly overfilled, as identified by the maximum fill of 547 grams of product. The median value tells us that 50% of the containers include 519.7 grams or more material.

Research was completed by Suzanne's team into the production control system parameters of the product. The enterprise resource planning (ERP) system was set up with the expectation of a typical 3% overage. This means that commercial production plans for containers to be over the 500-gram label claim by 3%, which is 515 grams. The quantiles from the plotted sample distribution indicate that the current fill process exceeds the expected fill roughly 75% of the time. The practical implications of this mismatch are a cascading waterfall of system adjustments that must be completed to manage product output, caused by the following issues:

- The inventory of empty containers will continue to grow as product output does not use the expected volume of containers.
- The customer planning schedule also becomes a complex nightmare. Drop in production orders will take place regularly as the volume of completed product is regularly less than what the system expects.
- Raw materials ordering will be off, resulting in potential shortages and the need for regular inventory and adjustments.

An organization invests a significant amount of money to implement ERP systems with the expectation of saving more money through automated resource planning. The manual adjustments to the system needed to correct the overfilling problem create added costs due to lost efficiency. These costs are typically even more than the cost of the extra product provided in each container and are a significant problem.

The summary statistics provide additional information about the general trends of the fill process. The average for the random sample is a container fill of just over 521 grams, with a standard deviation of 8.6 grams. Nearly all the individual results for a distribution are contained by +/- 3 standard deviations of the mean, which is the empirical rule for a Normal distribution. The random sample includes a staggeringly wide amount of variability—the range of fills is over 50 grams, more than 10% of the label claim target! Suzanne now knows that with the level of variation present in the fill process, it will be impossible to reduce the target fill of the equipment and maintain the minimum label claim of 500 grams. JMP quickly pinpointed the extreme need to reduce variation in the fill process. Suzanne will share the results of the data visualization in JMP, justifying to leadership why it is important to provide resources in support of an improvement project for the fill process.

A Second Problem: Dealing with Discrete Characteristics of Dental Implants

Data comes in many forms. JMP identifies each variable by data type and modeling type to best represent the data. Data type refers to the general structure of the information, which determines the format in the data grid, how the column's values are saved internally, and whether the column's values can be used in calculations. The types are described briefly as follows:

- Numbers are numeric.
- Text is character.
- Row state describes attributes of the data, such as if a row is selected, excluded, hidden, or labeled, as well as graph marker type, color, shape, or hue.
- Expression is used for pictures, graphics, and functions. The variables can be identified as characters, numerical values (continuous or discrete), or expressions.

The initial container fill problem involved data that is measurable and can be meaningfully divided, which is a numeric data type with a continuous modeling type. The column properties of each variable (column) can be manipulated to properly identify the data. This problem involves data with discrete categories. However, JMP can analyze the different data types with similar tools.

The modeling type of a column indicates to JMP the type of anaylsis that can be done on the information. Data that is either entered or imported into JMP is categorized as a modeling type by default. For instance, a column of numbers defaults to continuous (numeric) data and can be analyzed with statistically appropriate techniques. A user cannot create a bar chart (appropriate for discrete data) with a continuous modeling type. Additional information about the many modeling types is easily available through the Help menu.

This section describes a problem that includes variables that are discrete to use for data visualization in JMP.

Ngong is a process engineer for a facility that manufactures dental implants. The dental implants are made up of various components, including a threaded implant (inserted into the bone), an abutment (essentially a machine screw with a flat vertical projection at the top), and a permanent crown (to be attached to the flat surface of the abutment). The components are illustrated in Figure 1.23.

Figure 1.23: Implant Components

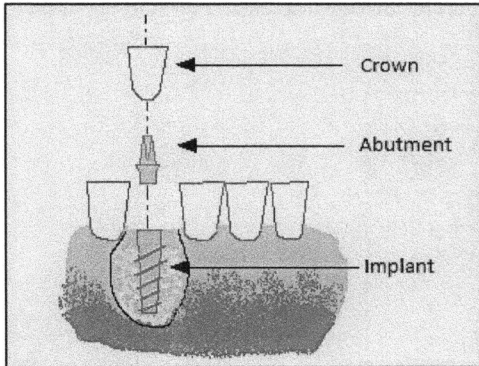

Figure 1.24: Implant Cross Section

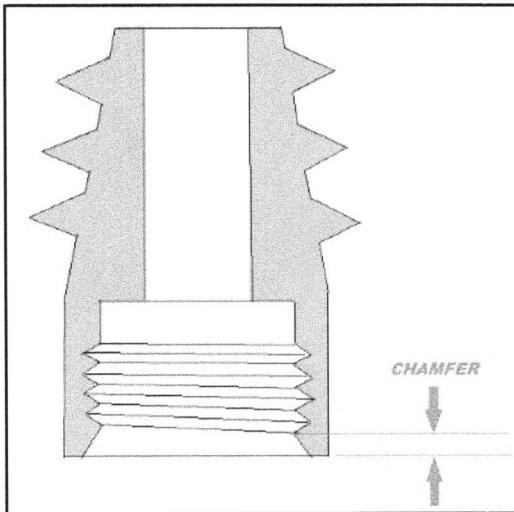

Ngong has received information from the customer services group regarding complaints from dentists who have been experiencing difficulties in starting the threading of the abutment into the implant on some procedures. Their records indicate no significant complaints for this problem until the last 14-18 months. Additional information has come from the field representatives who have narrowed down the cause of the threading difficulty to a minimal chamfer on the implant. Implants are manufactured with a machine that cuts the internal threads of the implant. The technical specifications require that "a chamfer is present" at the top of the threaded hole, as shown in the cross-section view shown in Figure 1.24. Information from the field identifies that chamfers of less than 0.75mm in depth can be problematic for starting the threads of the abutment.

Ngong holds a meeting of the operations team, and a plan is put together for measuring random samples of implants from the facility. There are four machining centers, so the data collection protocol requires that at least 400 samples from each machine be collected at random and sent to the quality team for measurement. Any implant that has a chamfer of less than 0.75mm is to be considered "minimal chamfer", otherwise the sample is to be identified as "good chamfer". The data was collected and placed into a stacked format, as shown in Figure 1.25. A good first step is to use the graph builder to view the data.

Figure 1.25: Stacked Implant Data Table

	machine	inspection result	count
1	C	good chamfer	1
2	C	good chamfer	2
3	C	good chamfer	3
4	C	good chamfer	4
5	C	good chamfer	5
6	C	good chamfer	6
7	C	good chamfer	7
8	C	marginal chamfer	8
9	C	good chamfer	9
10	C	good chamfer	10
11	C	good chamfer	11
12	C	good chamfer	12
13	C	good chamfer	13
14	C	good chamfer	14
15	C	good chamfer	15
16	C	good chamfer	16
17	C	good chamfer	17
18	C	good chamfer	18
19	C	good chamfer	19
20	C	good chamfer	20
21	C	good chamfer	21
22	C	good chamfer	22
23	C	good chamfer	23
24	C	good chamfer	24
25	C	good chamfer	25
26	C	good chamfer	26
27	C	good chamfer	27
28	C	good chamfer	28

dental implant data - JMP Pro — File Edit Tables Rows Cols DOE Analyze Graph Tools View Window Help

Columns (3/0): machine, inspection result, count

Rows: All rows 1,660 / Selected 0 / Excluded 0 / Hidden 0 / Labelled 0

Open *dental implant data.jmp*. Select *Graph ▶ Graph Builder* and drag *inspection result* from the list in the upper left of the window to the *X* drop zone in the graph to visualize the data, as shown in Figure 1.26.

Figure 1.26: Graph Builder with Discrete Data

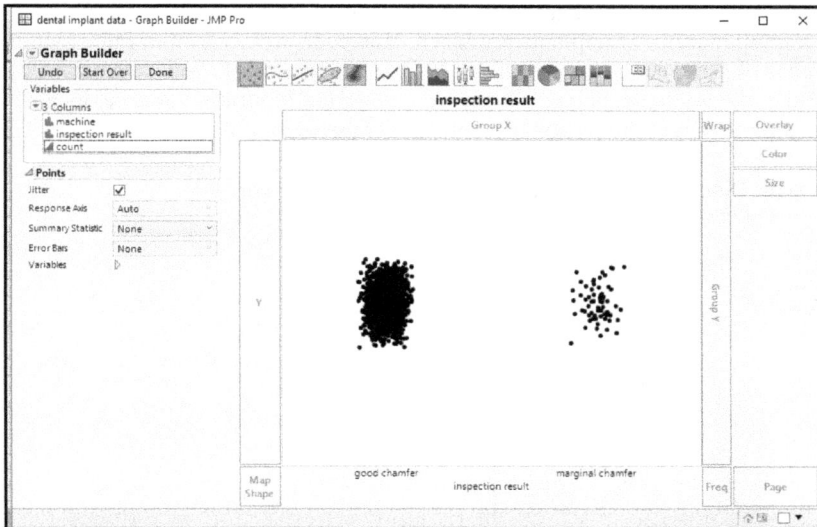

The initial view is a dot plot of the observations for each of the two categories. The default setting in JMP is to show the points jittered to better illustrate the density of observations. Most implants have a good chamfer as the mass of points is dense and black.

A better summary view can be had by clicking the bar chart icon that is roughly in the middle of chart style icons displayed across the top of the window. The control panel in the lower left of the graph builder adapts to the style of plot chosen. In the control panel, change the Label choice to *Label by Value* to show the counts for each of the categories, as shown in Figure 1.27.

Figure 1.27: Graph Builder with Discrete Data

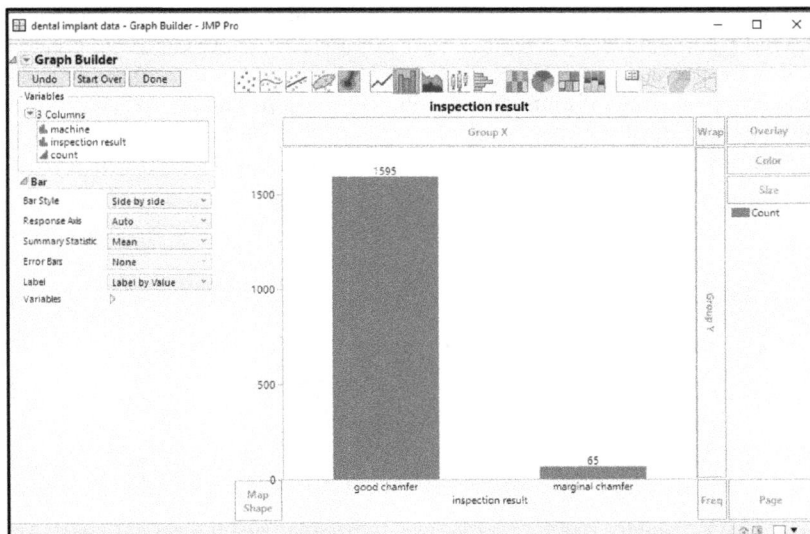

The plot shows information that identifies the clear majority (n_{gc}=1595) of implants as having a good chamfer and a small number as having minimal chamfer (n_{mc}=65). Through the analysis, the problem seems to be limited to only a small number of implants produced. However, we need more information to help narrow the focus. The control panel of the graph builder is open, and the results are categorized by each of the four machines that make implants. Ngong decides to choose the machine variable, located in the upper right of the graph, and move it to the wrap drop zone to get the final plot shown in Figure 1.28.

Figure 1.28: Graph Builder with Discrete Data Wrapped with the Machine Variable

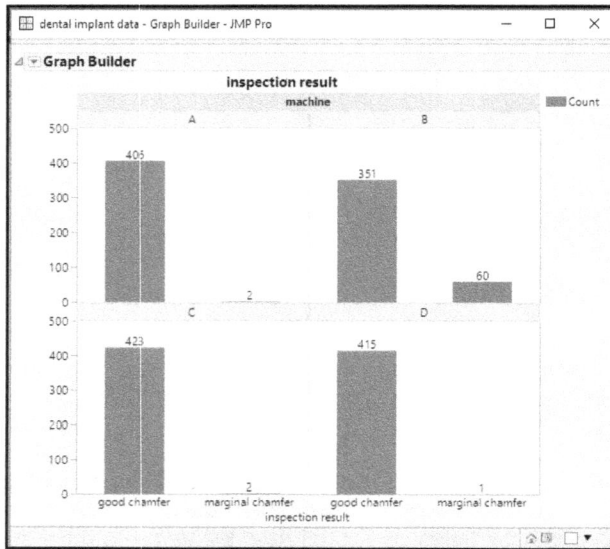

The bar charts of inspection results, wrapped by *machine*, adds an important dimension to the visualization of the data. It is very clear that machine B has a much higher count of implants with a minimal chamfer than the other three machines combined. Ngong is interested in using this chart format throughout the improvement project and does not want to have to remember all the options he had to choose to create it. JMP provides the efficient ability to save each analysis as a script, which can be run later to produce the exact same chart format. The script even works if more data is added to the table and an update is needed. Click the red triangle menu next to the *Graph Builder* header and choose the *Save Script>To Data Table...* option, as shown in Figure 1.29. There are many other options for saving a script, which are explored later in this book.

Figure 1.29: Creating a Script from a Plot

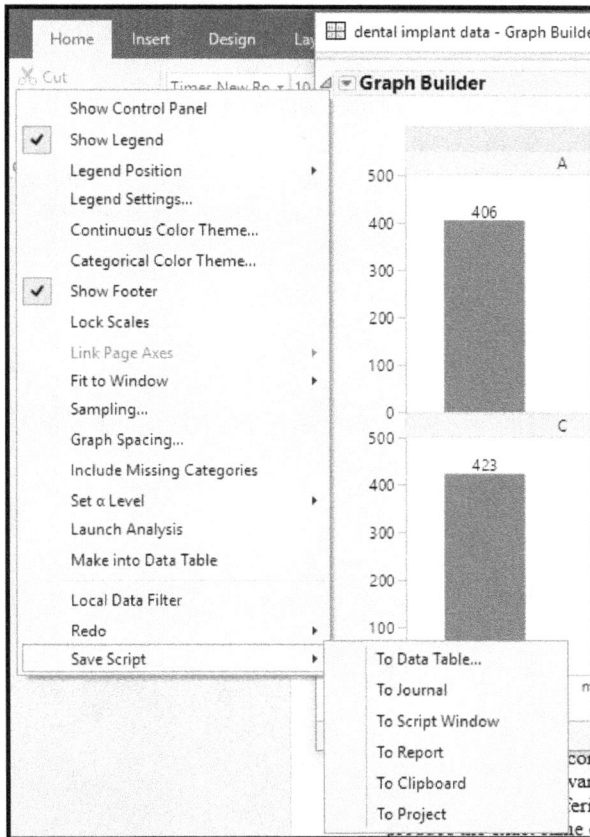

In Figure 1.30, the green triangle to the left of the name shows that the new script named "inspection result wrapped bar chart" is now available to run.

Figure 1.30: Script for Plot Saved to Data Table

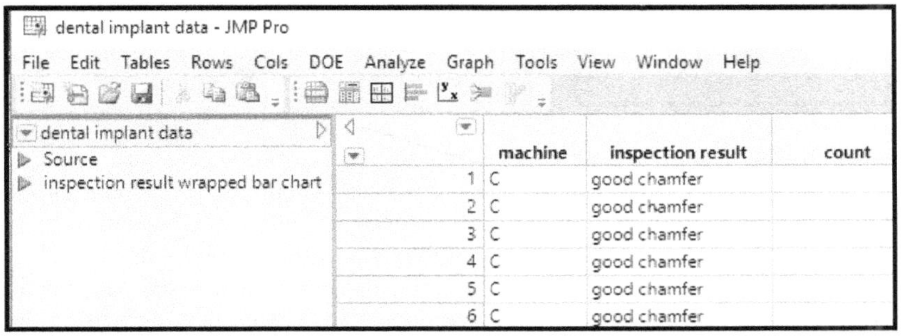

You can create a more detailed look at the inspection results data by looking at *Distributions* using the following steps:

1. Select *Analyze ▸ Distribution*.
2. Drag *inspection result* over to the *Y, Columns* box in the *Distribution* window.
3. Drag *machine* to the *By* box.
4. Once the *Distributions* output is created, choose the red triangle menu next to the *Distributions machine=A* header while pressing the control key to display all plots shown in the *Stacked* format, as seen in Figure 1.31.

Figure 1.31: Distributions of Inspection Results by Machine

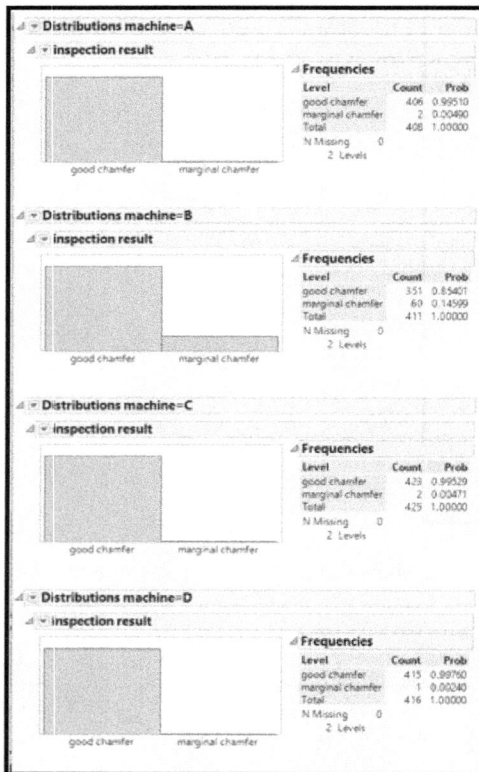

Machine B produced implants that have a 14.5 % probability of including a marginal chamfer, and conversely an 85.5% probability of making implants with a good chamfer. Machines A, C, and D produce marginally chamfered implants at a rate of between 0.2% to 0.5%. Ngong has enough information from the data to narrow the team's focus to the study of Machine B so that they can determine what is different compared with the other three machines. Operations and quality leadership are very pleased because the chances of reducing complaints from their dentist customers have improved greatly with the help of the data visualization results.

Get More Out of Simple Analysis with Column Formulas

Suzanne needs to persuade leadership with the information about overfilling occurring within the fill process to ensure that resources are available to make improvements. The data is compelling. However, work must be done to define the financial implications of the overfilling. The product cost is known to be $ 0.08 per gram. It is also known that the annual volume for the product is 50,000 dozen containers, which is 600,000 individual containers. JMP allows for calculated variables that can quickly illustrate the fixed materials cost of overfilling.

She creates a new column (variable) by selecting *Cols* ▶ *New Column* or by right-clicking on the header of the next unnamed column, as shown in Figure 1.32. This new column will be used to calculate the difference between the actual fill weight and the 515-gram baseline used for planning purposes, named "difference from baseline" shown in Figure 1.33.

Figure 1.32: Create a New Column (variable)

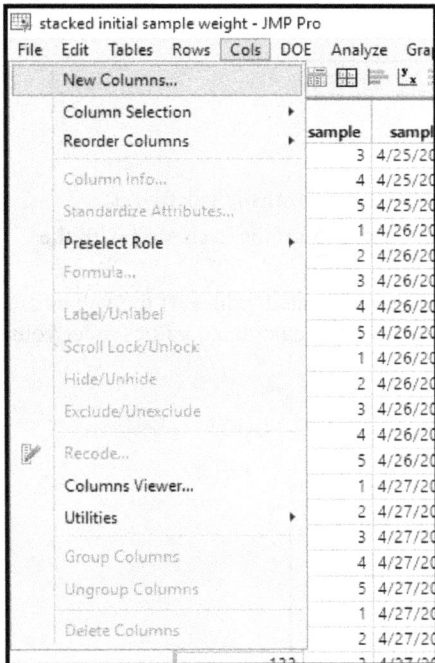

Figure 1.33: New Column Window

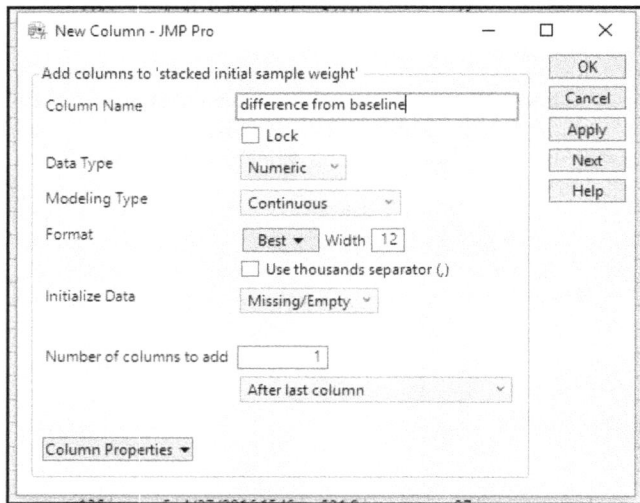

The following steps creates a calculated variable with a formula:

1. Select the *Column Properties* options and choose *Formula* to define the calculation.
2. The formula for the difference between the actual weight and 515 grams is created with the formula window shown in Figure 1.34.
3. Click *Apply* to activate the formula. The column shows the calculated values. If the values are not correct, you can change the formula and and click *Apply* until the calculated values meet your needs.
4. Click *OK* to complete the calculated column values.

Figure 1.34: Formula Window

A second new column is created for the "annual cost of difference". Use the formula editor to multiply the difference at baseline by the $0.08 cost per gram of product and by the 600,000 unit annual volume, as shown in Figure 1.35. It might be helpful to change the *Format* value *Currency* to emphasize to the observer that the data is illustrating financial costs.

Figure 1.35: New Column Window

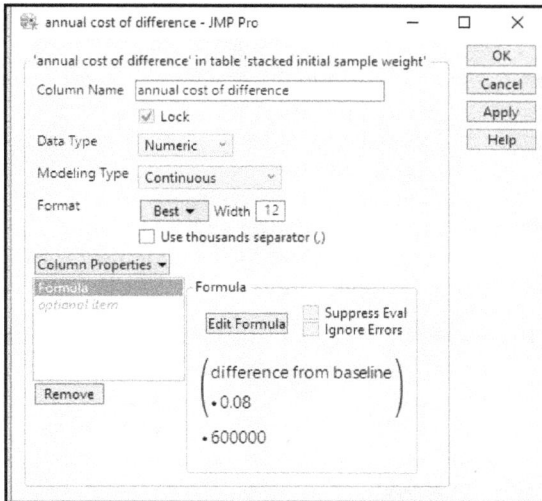

The final step is to create a distribution for the annual cost of product, as shown in Figure 1.36.

Figure 1.36: Distribution Window for Annual Cost of Difference

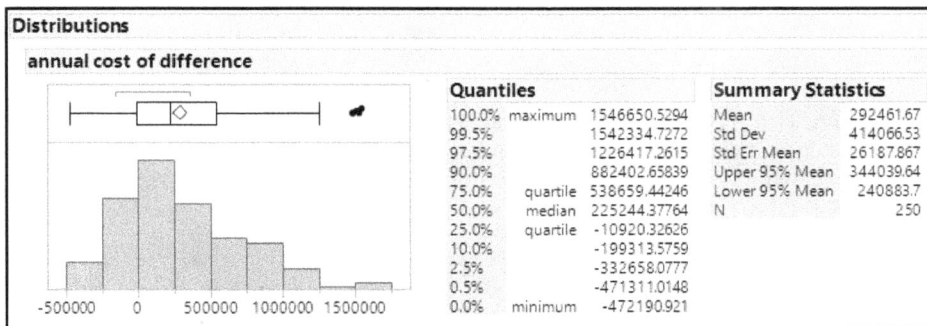

Practical Conclusions

The default settings of the *Distributions* output provide a great deal of information about the annual cost of excess materials that result from overfilling of containers. The summary statistics indicate that excess materials cost an average of more than $292,000 per year. The pattern of the annual excess costs can be used to explain the average overage more precisely than the point estimate for the average. Suzanne can confidently explain (with 95% confidence to be exact) to her leadership that the team is shipping at least $241,000 on average of "free" product per year, which may be costing as much as $344,000!

Why can Suzanne be "confident"?

The concept of confidence is one of the least understood by "consumers" of statistical analysis. It is also one that is not so easy to explain. However, it is important to be precise as to the likely population summary measure from a sample summary value.

Random variation is present among subjects for all but a uniform distribution (all values are the same). Samples of the same size taken from a population yield sample averages that differ at random. Eventually, once enough samples have been taken, the sample summary values form a bell-shaped distribution. The distribution of sample summary values is known to have the population average for the summary value, known as the grand average. The bell-shaped curve that forms this sampling distribution varies above and below the population average in a known and controlled pattern. The analyst needs to choose the level of precision desired so an estimate can be made for the range of values that are likely to contain the actual population average.

Resources are always limited and it is highly impractical to assume that leadership will support the expensive endeavor of taking many samples to create a sampling distribution. In general, one sample of subjects is taken (at random) with summaries made from the data. Statisticians have been basing estimates on the properties of sampling distributions for over 100 years, and the process is known to be quite robust. The "confidence" we have is in the process of using a sampling distribution to make the interval estimate of the summary value. The example deals with an average cost of the difference in container fill. By default, the summary statistics give the 95% confidence estimates (low and high) for the population average cost difference. Suzanne has confidence that if she were to have collected 100 samples of the same size, 95 of the intervals calculated would contain the real population average value, and 5 will not.

The hardest part of understanding a confidence interval is that there is no way to tell if the one interval made from the one sample is from the 95 that contain the population average, or if it is within the 5 intervals that do not. All values between the high and low limits have the same likelihood of being the population mean. Therefore, the interval is treated as if it is a single value. There is no way to calculate the probability of any value being the true population average. You can, however, be confident in the process of making an interval estimate of where that value is likely to be located in the distribution.

The cost of product is only one aspect of increased costs that result from overfilling of containers. Other areas of increased costs are likely to include but are not limited to: inventory adjustments required for materials and containers; added overtime as the team has immediate drop-in orders due to low yields of filled container batches; opportunity costs as the line cannot be used to make additional products; and especially the loss of customer credibility as product shipments are delayed or quantities are reallocated due to the fill process not meeting the ERP expectations. When data can be aquired for each cost, JMP can be used to visualize the information.

Suzanne has been successful using JMP to measure the amount of overfilling that is occuring in the process, as well as quantifying the financial impact to the organization. The leadership can bank on an annual savings of at least $240,000 in materials as well as all other quantifiable costs that result from overfilling the containers. Quality leadership will also be concerned about the excess variability present in the fill process. The result of Suzanne's work with JMP is that she has garnered the support of leadership to provide the resources necessary to execute an improvement project for the process. The improvement stage of her work is covered in chapter 14 as an evolutionary process study is completed.

Ngong has been able to utilize simple data visualizations with the Graph Builder to narrow the focus of the marginal chamfers to one of the four machining centers. The Distribution plots add detail that is used to quantify the extent of the problem of malformed chamfers within the implant. The time spent to quickly visualize the implant data provides great value as planning for continuous improvement resources can be focused to the machine that is the most likely source of the issue rather than working to improve the process of all four machines.

Exercises

E1.1—A manufacturing facility for surgical tubing contacts you to help them justify an improvement to their process that allows for a faster feed rate for tubing extrusion. They have data on two different tubing sizes from the current process and from the new process. The quality team requires that a validation run is completed and analyzed because they have voiced concerns over an increased rate of tube defects (tubes with an outside diameter that is outside of the internal limits). When internal limits are surpassed, technicians must stop the process and re-adjust parameters until tubes are acceptable.

1. Open *surgical tubes.jmp.*
2. Stack the tables as a multiple series stack of 3 that is contiguous and not stacked by row. This will stack the data for the new and old process for each of the three tube sizes.
3. Rename each of the data columns to *1.5mm OD*, *3mm OD*, and *5mm OD* respectively.
4. Select the *Label* column. Use *Cols* ▶ *Recode* to identify the groups as *current* and *new* to create a new column. Name the new column *process.*
5. Delete the three remaining label columns.
6. Create a new column named OOL 3mm. Create a formula with the conditional IF and comparison functions (shown in Figure 1.37) to identify each OD value outside the +/- 0.1mm internal limit as "pass" and "OOL" otherwise.

Figure 1.37: Formula Editor

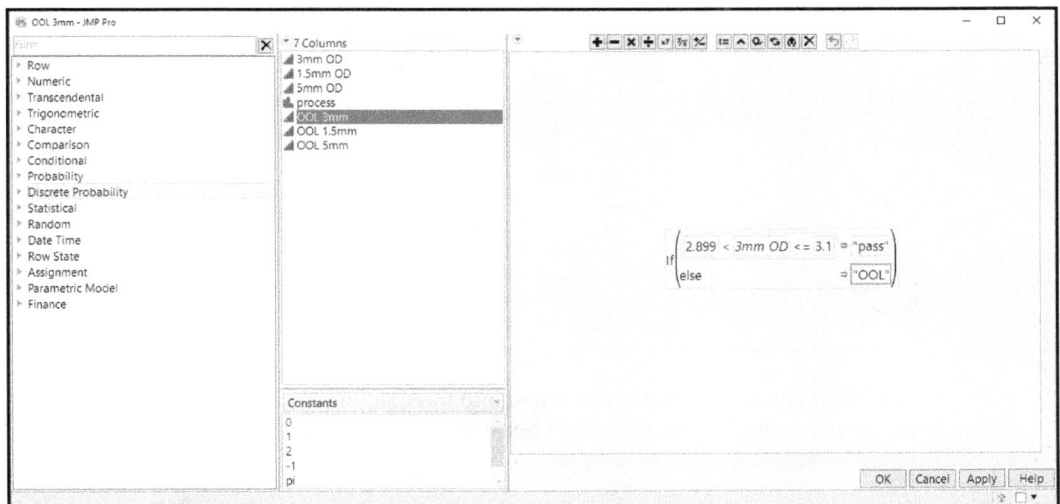

7. Repeat by creating the new columns OOL 1.5mm and OOL 5mm for the respective internal limits of +/- 0.05 and +/-0.15. Hint: Use 1.4499 and 4.8499 for the low values of the comparison formula.
8. Use the distributions function to visualize OOL 3mm, OOL 1.5mm, and OOL 5mm by process to determine whether the new process is adding risk.
9. How would you summarize the process validation run to the quality team and management?

E1.2—You are working with a materials team of a pharmaceutical research and development group. They are involved with a drug formula that has a specification on the d(0.9) of the particle size distribution for an active pharmaceutical (API) ingredient. The producer of the API is located on the other side of the world from the facility that receives lots of material. This material is sampled and tested for acceptable particle size. The material typically travels by ship. However, there have been several air-shipped containers of API due to high demand for the drug product in the first campaigns of production. The quality team has noticed that there have been an increased number of lots that are very close to not being accepted due to small d(0.9) values. The producer of the API has noticed no difference in the typical average and variation in particle size. The goal is to visualize the data to determine if any difference exists in the d(0.9) particle size values.

1. Open the *API lot data.jmp* data set. Use the *stack tables* function and drag the four columns other than *lot* to the *stack columns* window.
2. Deselect the *Stack by row* check box, and click *OK* to get the table shown in Figure 1.38.

Figure 1.38: Stacked Data Table

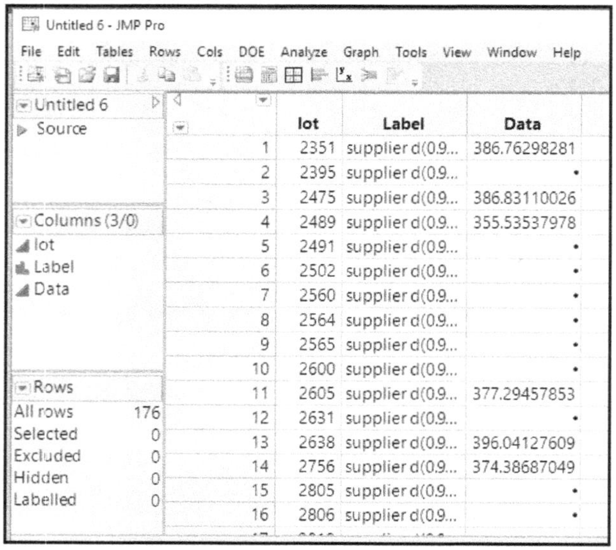

	lot	Label	Data
1	2351	supplier d(0.9...	386.76298281
2	2395	supplier d(0.9...	•
3	2475	supplier d(0.9...	386.83110026
4	2489	supplier d(0.9...	355.53537978
5	2491	supplier d(0.9...	•
6	2502	supplier d(0.9...	•
7	2560	supplier d(0.9...	•
8	2564	supplier d(0.9...	•
9	2565	supplier d(0.9...	•
10	2600	supplier d(0.9...	•
11	2605	supplier d(0.9...	377.29457853
12	2631	supplier d(0.9...	•
13	2638	supplier d(0.9...	396.04127609
14	2756	supplier d(0.9...	374.38687049
15	2805	supplier d(0.9...	•
16	2806	supplier d(0.9...	•

The information in *Label* is combined into three important elements, which need to be separated for useful analysis. JMP includes a text-to-columns option to easily create three new variables.

3. Click on the *Label* column header. Select *Cols* ▶ *Utilities* ▶ *Text to Column* to open the *Text to Columns* window. Type " " as shown in the following figure to use a blank space as the delimiter to break the information into three columns.

   Figure 1.39: Text to Columns Dialog Box

   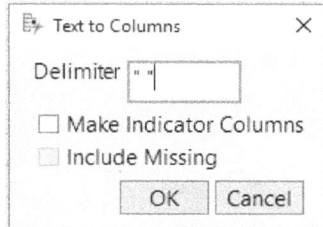

4. Change the column names as follows:
   a. Label 1 to lab.
   b. Change Label 3 to shipping mode.
   c. Change Data to psd d(0.9).

5. Create a distribution of psd d(0.9) to visualize the overall pattern of the data.

6. With the psd d(0.9) distribution plot open, click the red triangle next to *Distributions* and select *Redo* ▶ *Relaunch*.

7. Add *lab* as a *By* variable. Does the *By* variable help visualize and compare the data?

8. Click the red triangle next to *Distributions* to select *Redo* ▶ *Relaunch* to change the *By* variable to *shipping mode*.

9. How would you summarize this information to the stakeholders of the project?

E1.3—A quality control laboratory contacts you about a test method with results that are concerning. The team believes that variation in results may occur due to the analyst involved. A data set of test results has been compiled along with meta data including the sample set identification, analyst, and system name. The first task that needs to be completed is to visualize the data to see if any trends are present.

1. Open the *Analytical Data.xlsx* file directly into JMP by choosing the file type *Excel Files (*.xls, *.xlsx, *.xlsm)*. Use the Excel Import Wizard to include the appropriate row for column headers and to start the data on the row immdiately underneath the headers. The imported data creates the following JMP data sheet:

Figure 1.40: Downloaded Analytical Data Table

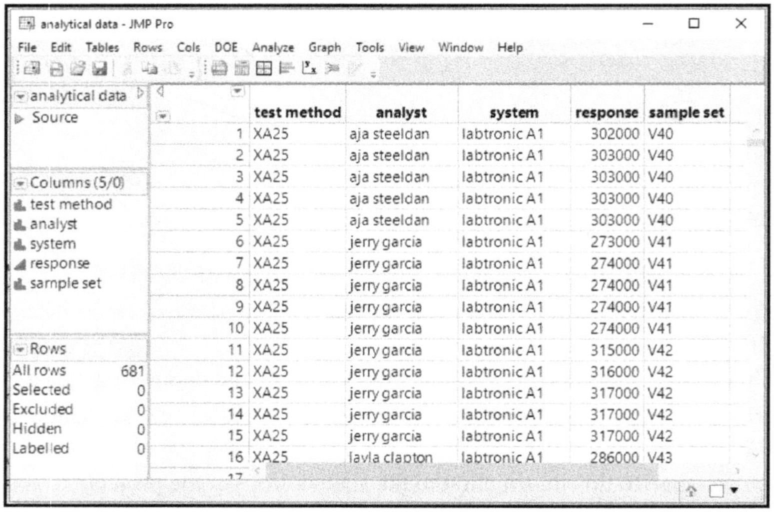

	test method	analyst	system	response	sample set
1	XA25	aja steeldan	labtronic A1	302000	V40
2	XA25	aja steeldan	labtronic A1	303000	V40
3	XA25	aja steeldan	labtronic A1	303000	V40
4	XA25	aja steeldan	labtronic A1	303000	V40
5	XA25	aja steeldan	labtronic A1	303000	V40
6	XA25	jerry garcia	labtronic A1	273000	V41
7	XA25	jerry garcia	labtronic A1	274000	V41
8	XA25	jerry garcia	labtronic A1	274000	V41
9	XA25	jerry garcia	labtronic A1	274000	V41
10	XA25	jerry garcia	labtronic A1	274000	V41
11	XA25	jerry garcia	labtronic A1	315000	V42
12	XA25	jerry garcia	labtronic A1	316000	V42
13	XA25	jerry garcia	labtronic A1	317000	V42
14	XA25	jerry garcia	labtronic A1	317000	V42
15	XA25	jerry garcia	labtronic A1	317000	V42
16	XA25	layla clapton	labtronic A1	286000	V43

2. Use the dynamic functionality of the distribution plots to look for any visible trends that might suggest relationships between the high or low end of the response range and the metadata (plot by *test method*, *analyst*, and *system*).
3. The two test methods were developed with the expectation that similar results will be produced regardless of the method used. Use the graph builder to look at the response data by the *test method* and determine whether the visualized data matches what is expected.
4. The methods should be robust to the analyst who is testing. Use the graph builder to determine whether there are any analysts who have results that seem to be very diferent from others.
5. Use the appropriate analysis to determine the 95% confidence limits for what can be expected for the average response of the population of results.
6. How would you summarize this information to the stakeholders of the project?

Chapter 2: Investigating Trends in Data over Time

Overview

Pharmaceutical organizations have rigid controls on information to ensure that it can be tracked by date and time. The traceability of information provides evidence of data integrity. Date and time coupled with the entries of the subject data provide evidence of extemporaneous documentation. A project team can evaluate how the data changes over time with time-based plots. JMP offers an array of options to investigate trends over time including simple plots, progressing to more complex quality control charting techniques. The ability to identify and separate changes in data that occur over time from random variation present in all data provides a great deal of helpful information.

The Problem: Fill Amounts Vary throughout Processing

The fill team collects fill weight data during in-process checks of five containers every 15 minutes of operating the equipment. Suzanne's improvement project team was able to provide visual evidence of the overfilling problem by utilizing plots of summary data in chapter 1. The team has the sampling date and time within the data set of random checks, and the data is already in chronological order. One step to a better understanding of the overfilling problem is to determine whether there are any patterns in fill weight variability that occur over time. Suzanne needs to provide an update to the stakeholders on how fill weight is changing over time in order to convey a better picture of the filling process.

Visualize Trends over Time with Simple Plots in the Graph Builder

Data was collected over several days including the two shifts of operational crews. JMP includes many tools to use summary analysis data and create a new table of data that is in chronological order, allowing for the ability to explore trends over a period of time. The initial fill weight data is already sorted according to the time that each group of containers was collected from the fill line. If this was not the case, the data could be easily sorted by using the tables menu options. Later chapters describe sorting data tables.

The ability to quickly create excellent graphs and visualize data is a compelling reason to use JMP. The Graph Builder is an excellent tool that gives the user a dynamic interface in which to create a graphic from data. Open *stacked initial sample weight.jmp* and select the *Graph ▶ Graph Builder*, as shown in Figure 2.1.

Figure 2.1: Graph Builder Initiation

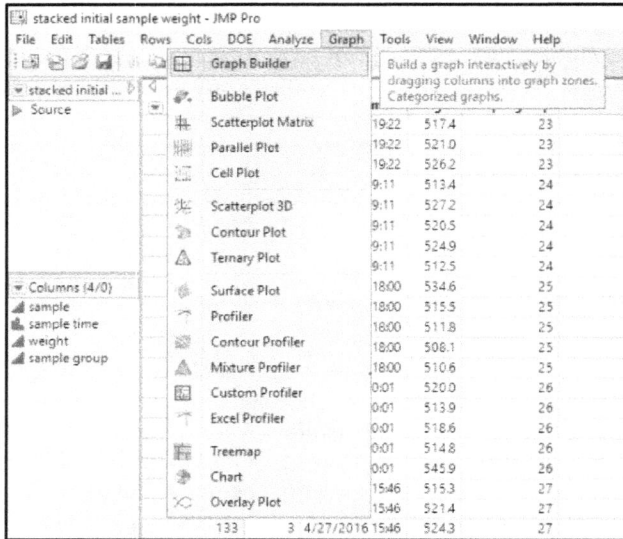

The Graph Builder window shown in Figure 2.2 is an open canvas where the user paints a picture of visualized data. Recall that Graph Builder was first used in chapter 1 to create a basic view of the dental implant data.

Figure 2.2: Graph Builder Window

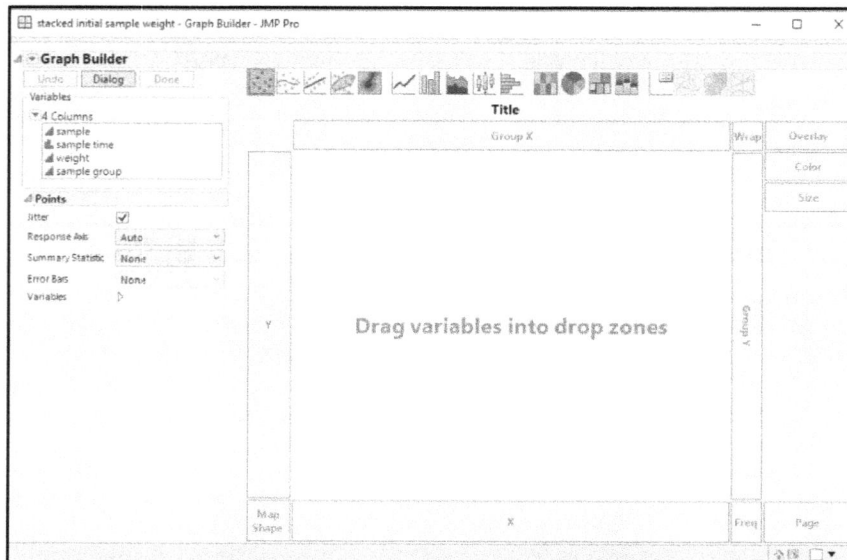

The quickest way to select variables for graphing is to click on a variable and drag it onto the Graph Builder drop zones. Notice that the Graph Builder window includes the *Dialog* button in the top left. Click *Dialog* to graph multiple variables at the same time in a graph matrix. A graph matrix is further explored in the exercises at the end of this chapter. The example utilizes the drag and drop functionality of **Graph Builder**. Select *weight* and drag it into the drop zone in the middle of the space for a quick look at response data over time. While in the Graph Builder window, select *sample group* and drag it into the *X* drop zone to organize the results by sampling order.

In Figure 2.3, the general trend of container weights from the samples taken over time is represented by the default blue smoother line. The black dots represent individual container weights. Vertical distances among the dots at each sample illustrate the amount of variability in the fill weights. Explore the graphical options at the top of the window to get different views of the data and find the best visualization for the intended audience. There are additional options underneath the *Points* header, as well as the Smoother that enables you to adjust the view of the data.

Figure 2.3: Graph Builder Window with Data

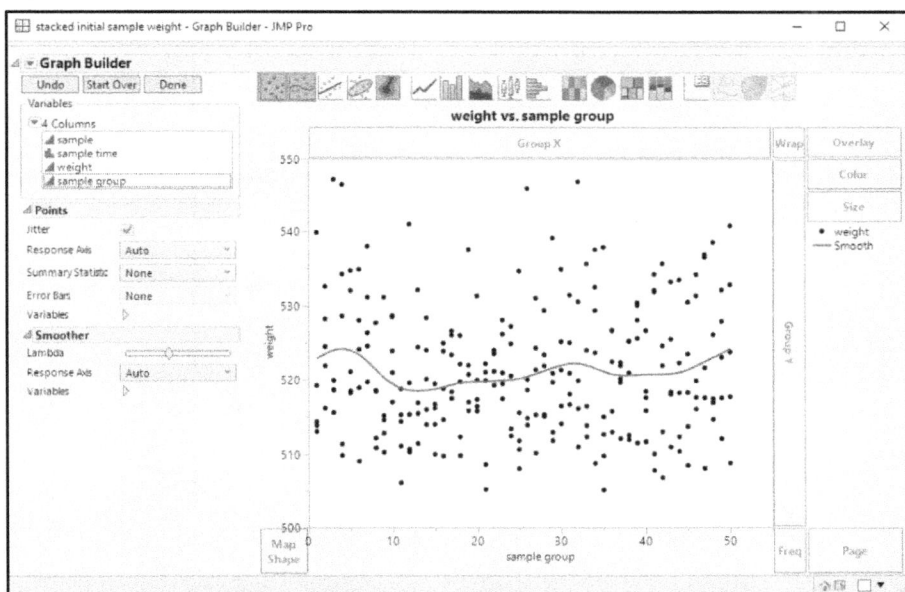

The Smoother can be adjusted to make the line more or less sensitive to variation in the data. The simple plot shows that the fill process seems to be stable over time because the smoother line is close to horizontal and the spread among samples looks to vary at random. Figures 2.4 and 2.5 illustrate the extremes of smoother sensitivity by adjusting the Lambda slider. The lambda value is a tuning parameter in the spline formula used to create the smoother line. As the value of λ decreases, the error term of the spline model has more weight and the fit becomes more flexible and curved. As the value of λ increases, the fit becomes stiff (less curved), approaching a straight line.

Figure 2.4: Smother Adjusted to High Sensitivity

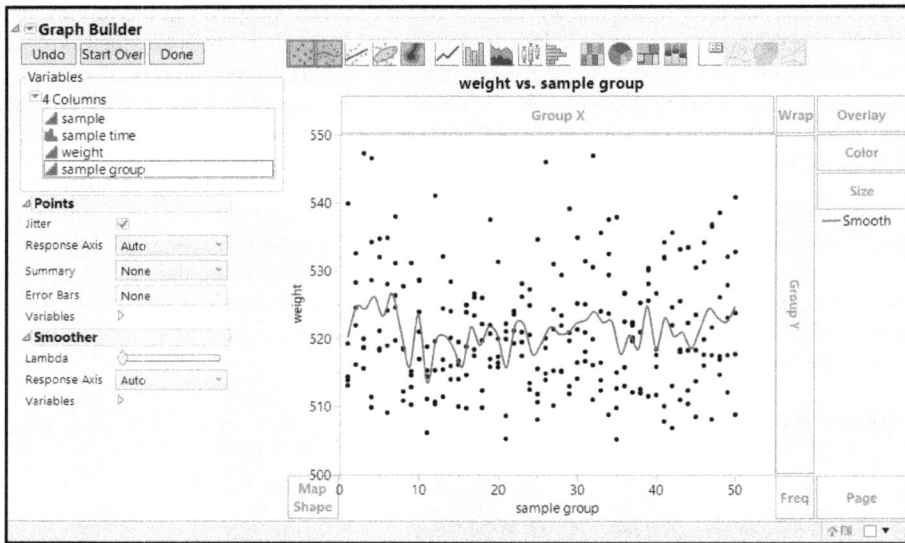

Figure 2.5: Smother Adjusted to Low Sensitivity

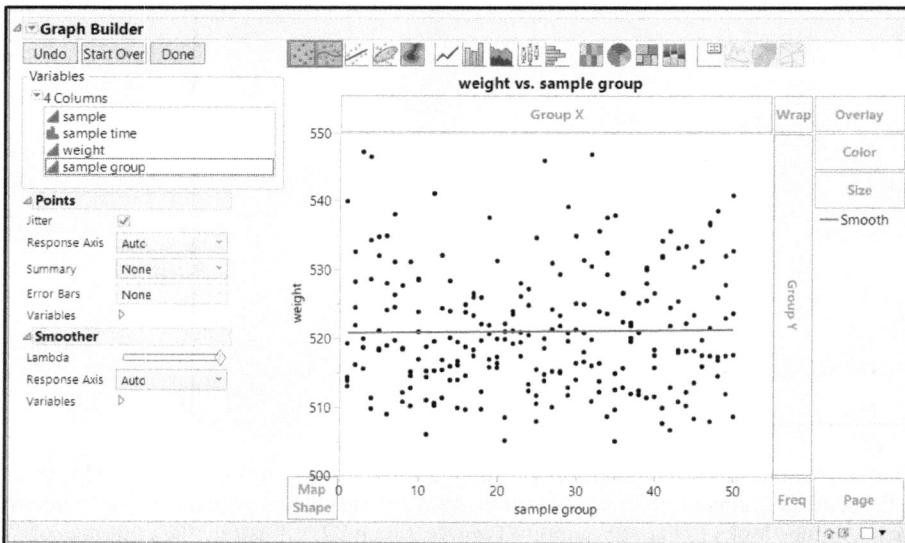

More Detail for Time-Based Trends with the Control Chart Builder

The simple plot created with Graph Builder provides a glimpse of any possible trends over time. Large-scale trends were highlighted by the dots and the blue smoother line. However, shifts that occur over shorter periods of time and shifts in averages do not stand out.

The purpose for gathering the fill data is to measure the extent of the overfilling problem. Suzanne needs to use tools that allow the fill process to "talk" by explaining trends in the average and variation within results as samples are taken over time; this is known as the voice of the process. The stakeholders of the process are likely interested in the amount of random variation expected, as well as observations that might exceed the bounds of randomness. Observations that are more extreme than what can be expected as random variation might be due to a special cause and are worth further exploration. For instance, one might identify extreme results occurring at regular intervals that might correlate with a change in shifts, material lots, or other identifiable factors. A trend in variation that occurs over time as big shifts from one time point to another can also suggest the presence of special causes. Quality control charts offer illustrations of trends over time, which include statistically calculated limits set on the edges of the zone of expected random variation. The overall average is also included on the charts to illustrate how trends over time relate to the center expectation of the process. JMP includes popular quality control charting techniques that assess the voice of the process. You can create a control chart for the fill process by selecting *Analyze ▶ Quality and Process ▶ Control Chart Builder*, as shown in Figure 2.6.

Figure 2.6: Selecting Control Chart Builder

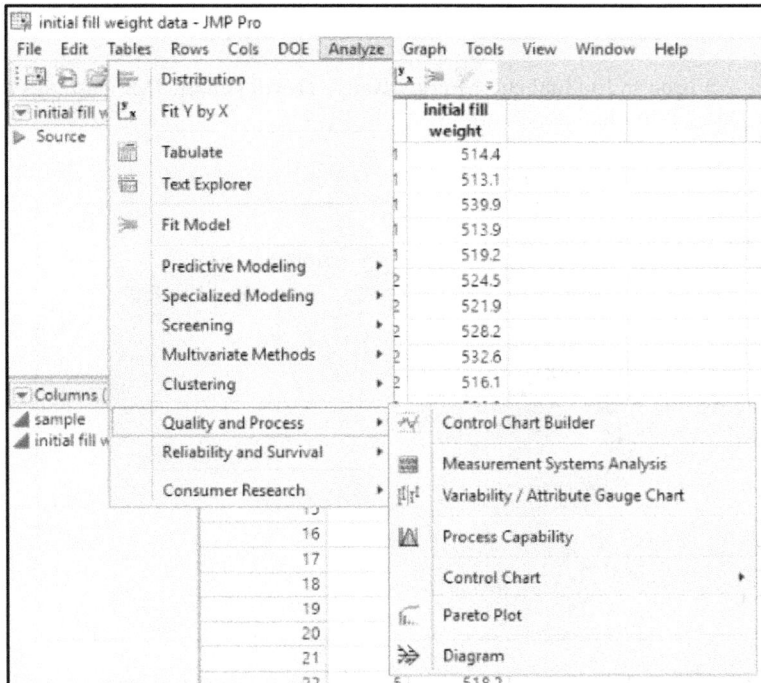

The control chart builder window shown in Figure 2.7 looks similar to Graph Builder. In the Control Chart Builder window, select *weight* and drag it into the Y drop zone.

Figure 2.7: Control Chart Builder Window

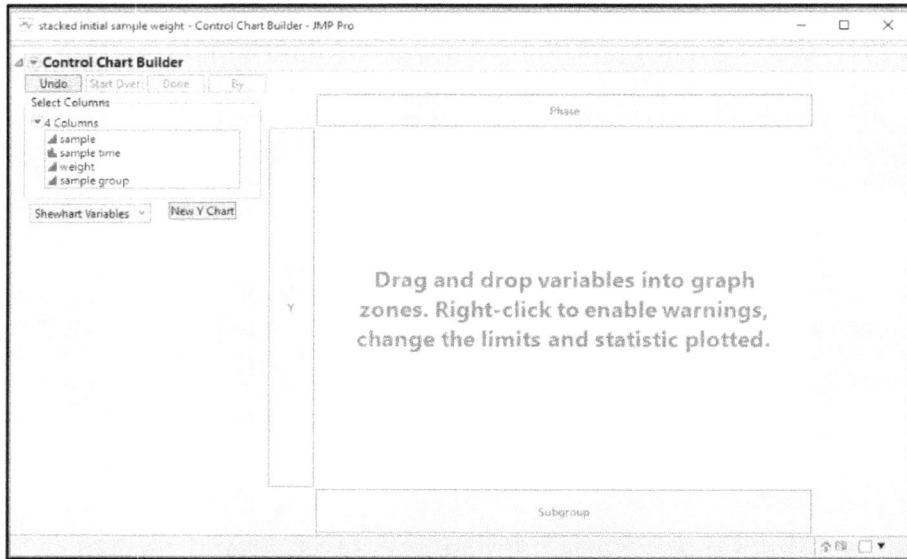

The default chart shown in Figure 2.8 is an Individual and Moving Range (I-MR) chart. JMP assumes that the samples were taken in row order as individual observations.

Figure 2.8: Control Chart Builder Window with Data

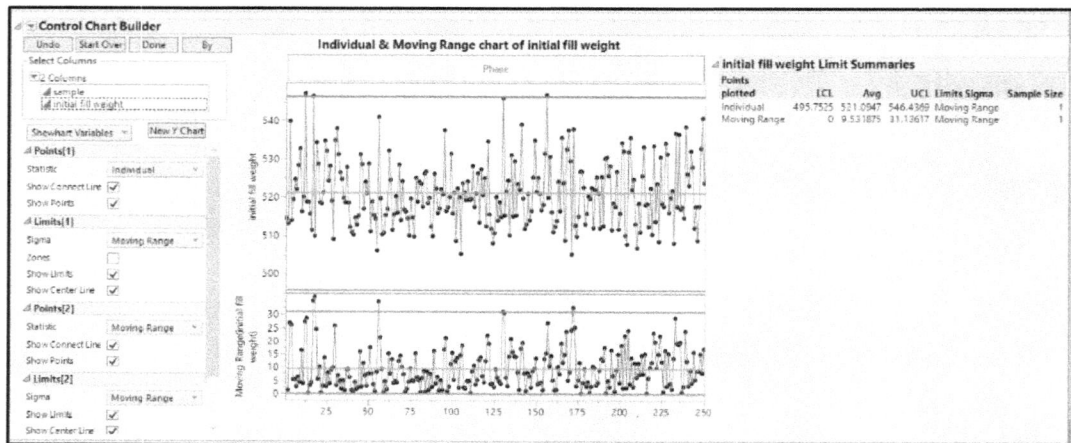

Control charts provide information regarding trends of data over time. The lower, moving range chart in Figure 2.8, known as a dispersion chart, illustrates the amount of change between a previous observation and the next. Notice that the first plotted point of moving range is at the second time point. The trend in the moving range of the process over time is a highly sensitive initial indicator of possible trends. The upper chart illustrates changes in the observations over time, providing evidence of the expected span of values from the process over the time period. A green line indicates the overall average for the period studied; on the upper chart, it represents fill weight average, and on the lower chart it represents the average moving range. The red lines provide information about the limits to random variability and are referred to as control limits. Control limits are calculated statistically, which is very different from the specifications set for the attribute measured. (Details about the formulas are available in the book *Quality and Process Methods*, accessible from the Help menu.) The moving range chart contains four points that are above the upper control limit, which might be due to a special cause. Each point that is beyond the limit represents a difference between two observations that exceeds the amount of difference that can be expected due to random variability. It is noteworthy that the extremes of the individual observations in the individuals chart correlate with the extremes of the moving range chart. The control charting technique is powerful since the observer focus on a few observations rather than research the entire set of 250 observations.

Because of the large amount of variation present among individual observations, using I-MR control charts for trend analysis is not always optimal. Extreme values might be evident but other trends over time might be difficult to interpret because the trends look so busy. Operational stakeholders would like to know whether the process is stable over time. A stable process includes a pattern of points that do not have a discernable slope, with data increasing or decreasing over time. The individual fill weights seem to follow a flat line over time, but the excessive variability make it difficult to really assess the stability of the process.

Additionally, the in-process checks did not involve just one container pulled from the line at each time point; checks were completed by pulling five containers and noting the average weight for each of the sampling points. Control Chart Builder can easily accommodate a chart based on averages of subgroups of observations taken at each time point. An X-bar and R chart includes group averages and ranges of values within subgroups. It is a popular control-charting technique for subgrouped data. Choose **Sample group** as the subgroup field and drag it onto the **X axis** of the plot to reflect the sample averages in an X-Bar and R chart shown, as shown in Figure 2.9.

Figure 2.9: Control Chart Builder Window with Averages and Ranges

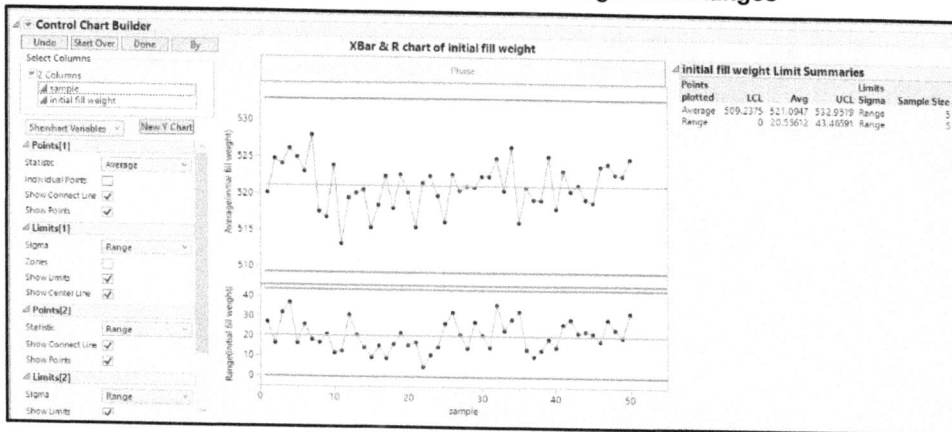

The lower chart in Figure 2.9 shows trends in the range of values within the subgroups of the process over time, which is a sensitive indicator of process trends. A green line indicates the average for the entire period studied; note that the process average is no different in the X-bar chart as compared to the previous I-MR chart. The green line in the lower chart represents the average range of values within the subgroups, which is almost double the previous average moving range chart. There are no out of control points evident in the X-bar and R charts, which indicates an average level of control for the process. The chart contains no definable trends among successive points on either chart; the variation is random. The X-bar and R charts made from the subgrouped data suggest that the process is functioning with only random variation. Improvements to the process are not likely to come from addressing a special cause situation. Reduction in the variability of the fill process must occur for improvements to be made.

Interpretation of control charts can be a bit of an art form. Different types of trendsindicate non-random patterns of interest, and various alert levels can be included to test for such patterns. The pattern of the averages over time looks to be relatively flat so the team need not be concerned that the process is not stable over time. You can select details in control panel of the Chart Builder as well as by clicking the red triangle menu at the top of the chart output header. (Several excellent books have been written regarding effective interpretation of control charts, with *Juran's Quality Handbook* one of the most classic of references.)

The control chart of the average fill weights includes a random pattern of variation with no definable trends over time. The expected population average fill is 521.1 grams but the statistical control limits specify that sample averages are expected to vary between 509.2 grams and 533.0 grams. The range of fill weights within each group of five containers is expected to be 20.6 grams on average and could be up to 43.5 grams. The information extracted from the data using JMP clearly illustrates that the fill process is highly variable.

Dynamically Selecting Data from JMP Plots

The data were collected randomly from the in-process check weight records. Suzanne decides to dig into the details of variability to compare the extreme high sample average weights with the extreme low sample average weights. She can easily accomplish this by using the dynamic features of JMP graphics.

Press the Ctrl key while selecting the points that represent the five highest weight averages. Figure 2.10 shows the selected points with black dots. All other points and the trend line fade to gray to show that they are not part of the selection. The individual observations that make up the selected average weights show as blue highlighted rows in the data table shown in Figure 2.11. The Rows panel in the lower left of the data table specifies that 25 rows of data have been selected because each point represents the average of five weights due to the choice of the 5 average points on the control chart plot.

Figure 2.10: Control Chart Builder Window with (High Weight) Selected Points

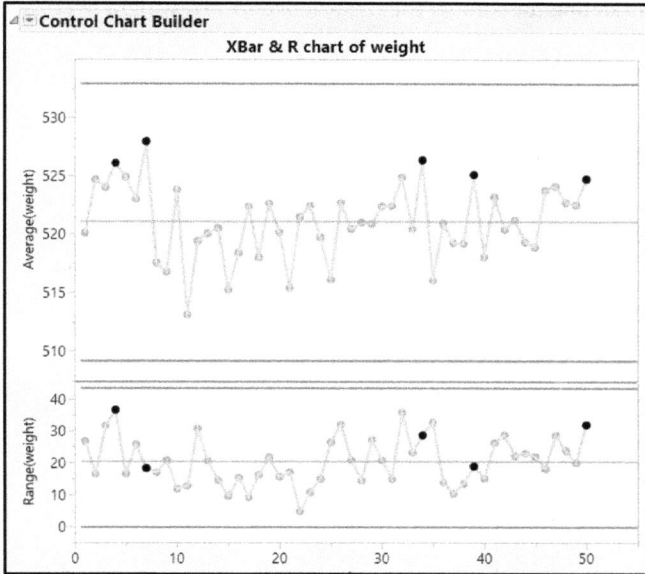

Figure 2.11: JMP Data Table With (High Weight) Selected Rows

	sample	sample time	weight	sample group
25	5	4/13/2016 11:38	532.0	5
26	1	4/13/2016 15:01	524.2	6
27	2	4/13/2016 15:01	528.0	6
28	3	4/13/2016 15:01	518.9	6
29	4	4/13/2016 15:01	509.0	6
30	5	4/13/2016 15:01	534.8	6
31	1	4/14/2016 11:41	538.0	7
32	2	4/14/2016 11:41	531.1	7
33	3	4/14/2016 11:41	526.4	7
34	4	4/14/2016 11:41	524.6	7
35	5	4/14/2016 11:41	519.7	7
36	1	4/17/2016 11:25	518.6	8
37	2	4/17/2016 11:25	527.7	8
38	3	4/17/2016 11:25	518.4	8
39	4	4/17/2016 11:25	512.2	8
40	5	4/17/2016 11:25	510.8	8
41	1	4/17/2016 16:31	510.2	9
42	2	4/17/2016 16:31	514.6	9
43	3	4/17/2016 16:31	512.8	9
44	4	4/17/2016 16:31	515.2	9
45	5	4/17/2016 16:31	531.1	9
46	1	4/18/2016 3:46	528.5	10
47	2	4/18/2016 3:46	517.0	10
48	3	4/18/2016 3:46	523.9	10

Now, press the Ctrl key and select the lowest average weights, as shown in Figure 2.12. Figure 2.13 indicates that 50 rows have been selected, reflecting the total of the highest and lowest weight averages.

Figure 2.12: Control Chart Builder Window with High and Low Weight Selected Points

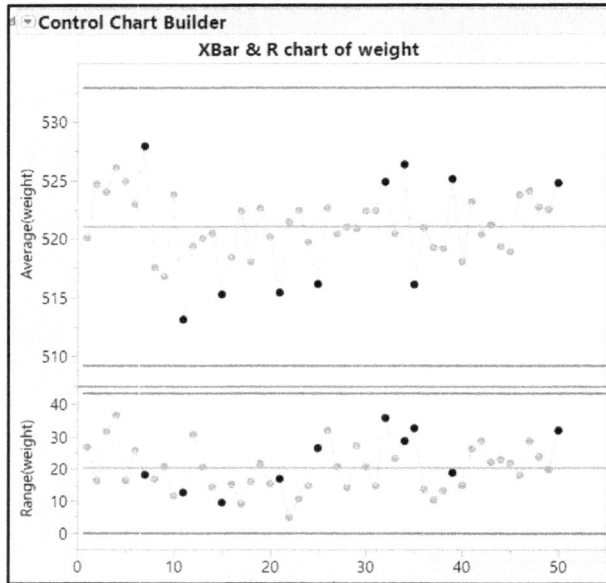

Figure 2.13: JMP Data Table with High and Low Weight Selected Rows

	sample	sample time	weight	sample group
1	1	4/11/2016	514.4	1
2	2	4/11/2016	513.1	1
3	3	4/11/2016	539.9	1
4	4	4/11/2016	513.9	1
5	5	4/11/2016	519.2	1
6	1	4/11/2016 9:45	524.5	2
7	2	4/11/2016 9:45	521.9	2
8	3	4/11/2016 9:45	528.2	2
9	4	4/11/2016 9:45	532.6	2
10	5	4/11/2016 9:45	516.1	2
11	1	4/11/2016 10:01	520.0	3
12	2	4/11/2016 10:01	547.2	3
13	3	4/11/2016 10:01	518.7	3
14	4	4/11/2016 10:01	518.6	3
15	5	4/11/2016 10:01	515.6	3
16	1	4/12/2016 14:22	511.4	4
17	2	4/12/2016 14:22	546.5	4
18	3	4/12/2016 14:22	509.8	4
19	4	4/12/2016 14:22	534.2	4
20	5	4/12/2016 14:22	528.6	4
21	1	4/13/2016 11:38	518.5	5
22	2	4/13/2016 11:38	518.2	5
23	3	4/13/2016 11:38	521.1	5
24	4	4/13/2016 11:38	534.7	5

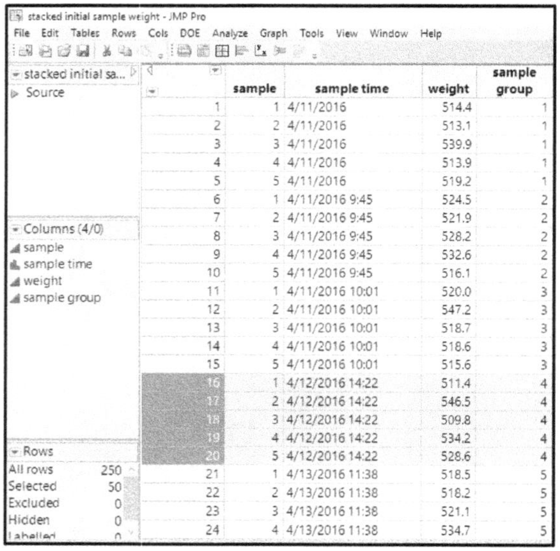

Place your pointer over *Selection* in the *Rows* panel, and then right-click to get a list of detailed options. Select *Data View*. JMP creates a subset table including only the 50 selected rows shown in Figure 2.15.

Figure 2.14: Row Selection Options

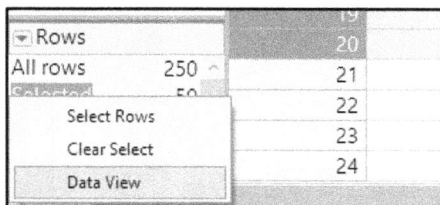

Figure 2.15: Data Table of Selected Rows

	sample	sample time	weight	sample group
1	1	4/12/2016 08:22	511.4	4
2	2	4/12/2016 08:22	546.5	4
3	3	4/12/2016 08:22	509.8	4
4	4	4/12/2016 08:22	534.2	4
5	5	4/12/2016 08:22	528.6	4
6	1	4/14/2016 11:41	538.0	7
7	2	4/14/2016 11:41	531.1	7
8	3	4/14/2016 11:41	526.4	7
9	4	4/14/2016 11:41	524.6	7
10	5	4/14/2016 11:41	519.7	7
11	1	4/18/2016 18:55	511.1	11
12	2	4/18/2016 18:55	518.8	11
13	3	4/18/2016 18:55	515.3	11
14	4	4/18/2016 18:55	514.4	11
15	5	4/18/2016 18:55	506.1	11
16	1	4/20/2016 20:55	516.7	15
17	2	4/20/2016 20:55	516.1	15
18	3	4/20/2016 20:55	514.0	15
19	4	4/20/2016 20:55	510.0	15
20	5	4/20/2016 20:55	519.5	15
21	1	4/24/2016 23:56	521.1	21
22	2	4/24/2016 23:56	508.6	21
23	3	4/24/2016 23:56	522.2	21
24	4	4/24/2016 23:56	519.9	21
25	5	4/24/2016 23:56	505.2	21
26	1	4/26/2016 18:00	534.6	25
27	2	4/26/2016 18:00	515.5	25

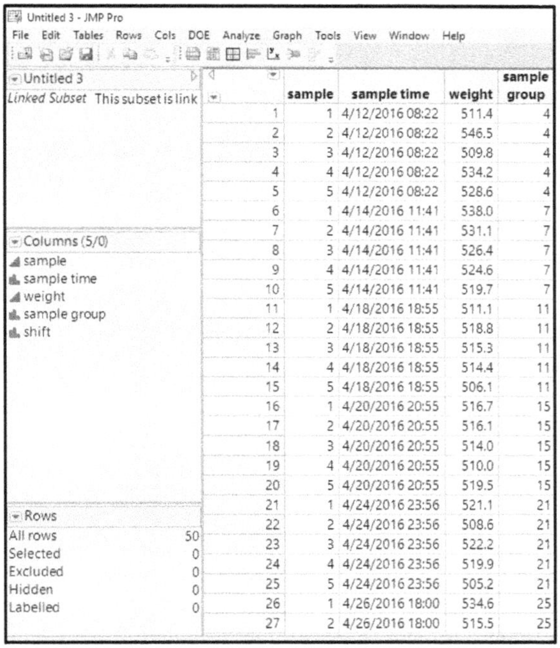

Creating Subset Tables

JMP users quickly realize that many options are available to get desired results. This is true for narrowing focus from a large table of data to a selected number of observations that are of specific interest to the user. An alternate way to create a data table of selected data uses the Tables menu options. Select your rows of interest, and select *Tables* ► *Subset*,as shown in Figure 2.16. By default, the *Subset* window specifies *Selected Rows*, as shown in Figure 2.17. Click *OK* to create a subset table of the 50 selected rows. The table is shown in Figure 2.18.

Figure 2.16: Tables Menu

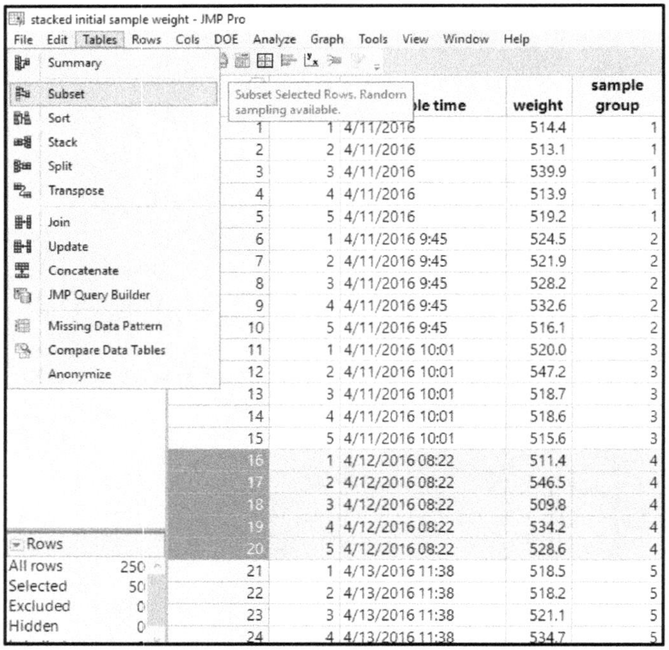

Figure 2.17: Subset Window

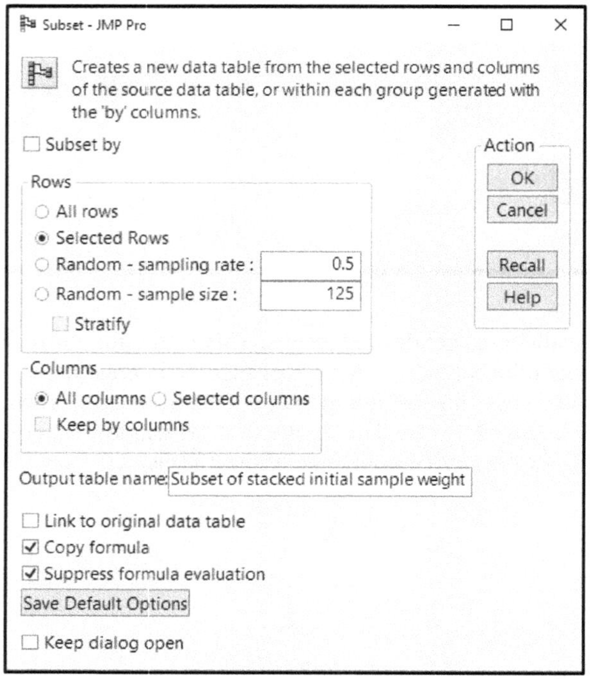

Figure 2.18: Subset Table of Selected Rows

		sample	sample time	weight	sample group
	1	1	4/12/2016 08:22	511.4	4
	2	2	4/12/2016 08:22	546.5	4
	3	3	4/12/2016 08:22	509.8	4
	4	4	4/12/2016 08:22	534.2	4
	5	5	4/12/2016 08:22	528.6	4
	6	1	4/14/2016 11:41	538.0	7
	7	2	4/14/2016 11:41	531.1	7
	8	3	4/14/2016 11:41	526.4	7
	9	4	4/14/2016 11:41	524.6	7
	10	5	4/14/2016 11:41	519.7	7
	11	1	4/18/2016 18:55	511.1	11
	12	2	4/18/2016 18:55	518.8	11
	13	3	4/18/2016 18:55	515.3	11
	14	4	4/18/2016 18:55	514.4	11
	15	5	4/18/2016 18:55	506.1	11
	16	1	4/20/2016 20:55	516.7	15

Columns (4/0): sample, sample time, weight, sample group

Rows: All rows 50, Selected 0, Excluded 0, Hidden 0, Labelled 0

The subset of data with the five highest and five lowest average fill weights has been created from the original set of random data. Sample time is a variable in the data. However, categorizing the sample groups by work shift helps determine whether the extremes have a specific shift in common. You could do this by creating a new variable named "shift" and manually typing the shift that correlates with sample time, but that would be tedious for the set of 50 observations. A more efficient technique uses the columns recode feature. Each time point includes five replicates so it will be easier to change 10 values verses the 50 observations in the data table. Complete the following steps to create a shift variable.

1. Select *Cols* ▶ *Recode* to open a window that enables you to change the old values to new values.

2. Utilizing each of the two shift start and end times (shift 1: 6:00 to 17:59; shift 2: 18:00 to 23:59), enter the appropriate shift as a new value for each observation.

3. Enter "shift" as the name for the new column, as shown in Figure 2.19. (By default, in JMP 14 creates a new column for the recoded data.)

4. Click *Recode* to execute the new column of recoded values.

Figure 2.19: Subset Table of Selected Rows with shift Variable

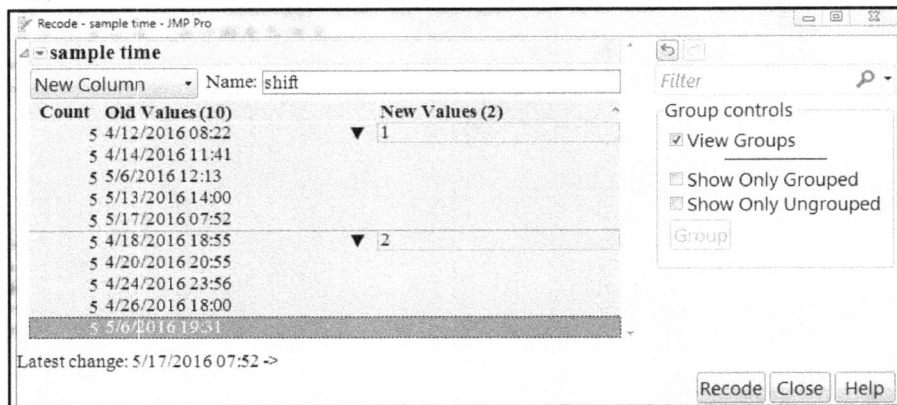

The extreme sample averages for fill weights categorized by shift can be studied with data visualization to explore for potential trends. It is typical for data visualization to start at the highest level, focusing on interesting aspects of the data. The large set of tools available in JMP make the practice of digging into data easy and efficient.

Using Graph Builder to View Trends in Selected Data

The versatility of Graph Builder allows for efficient visualization of data to look for various trends. Select *Graph ▶ Graph Builder*. In the *Graph Builder* window, select *weight* and move it to the *Y* drop zone, move *sample group* to the *X* drop zone and move *shift* to the *Group X* drop zone. The smoother line is not needed, so click the graph icon above the plot to show only the points.

Figure 2.20 shows the individual product weights. The interpretation of the plot seems to suggest that the first shift might have more variability within the sample groups and that the weights seem to be higher. Select *Mean* in the *Points/Summary* options located in the lower left of Graph Builder to convert the plot to show average weights, as shown in Figure 2.21.

Figure 2.20: Plot of Individual Product Weight Values by Sample Group and Shift

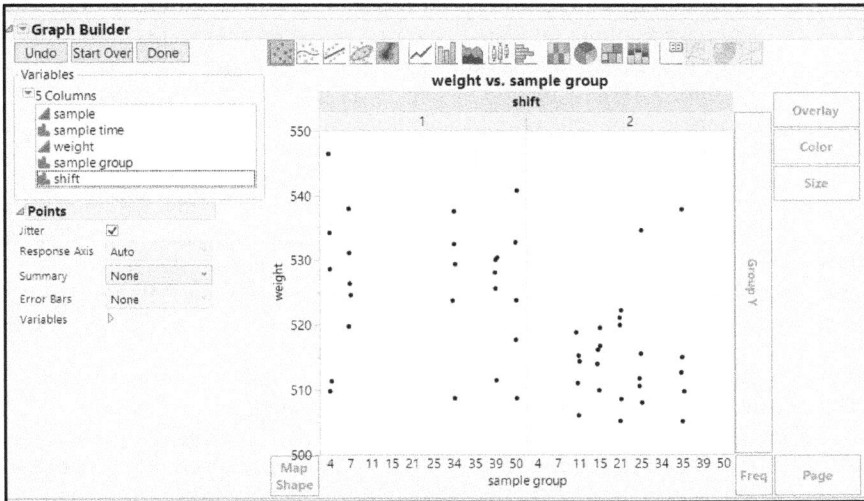

The interpretation of the plot in Figure 2.19 is more clear. It shows that the first shift tends to have the highest average weights and the second shift tends to have the lowest. The fill weights from shift 1 also seem to have more variability because they are more spread out than the averages from shift 2.

Figure 2.21: Plot of Average Product Weight Values by Sample Group and Shift

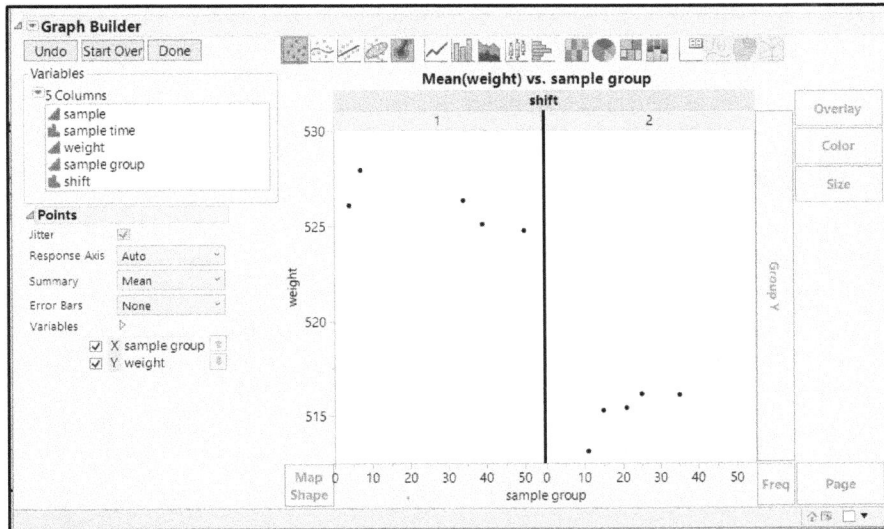

Practical Conclusions

Analysis of data over time is a common technique for visualizing trends in a process. The Graph Builder provides a flexible format for looking at variables over a time axis. The central trend is shown by the smoother line, which can be adjusted for a desired amount of sensitivity to shifts. An understanding of trends in variation is shwon by the spread in points.. Often, using the plot enables teams to uncover changes in process data that can be traced back to events that occured at specific times.

Control charting of data over time includes the dimension of statistical control limits that define the zone of random variation for changes between samples, as well as the variability within samples. Points that are extreme can be selected for further study in subset tables so that the project team can use them to determine the cause of the variation in fill weights. The information provided with the JMP plots is crucial for the team as they investigate relationships between operational information and variability in fill weight.

Suzanne's improvement project team has been able to focus effort from the confusing entirety of the process data to the trends of extreme fill weights by shift with a minimal amount of work. A more comprehensive set of data labeled by shift might help the team to better understand variation in weights over time by shift. Investigation into time-based trends is of great value for many sources of data obtained from manufacturing operations. Time-based plots can offer many interesting clues to relationships among operational aspects and to variation in the data. Dynamic features that are built into JMP plots allow for a great deal of exploration into trends to extract useable information.

Exercises

E2.1—Uniformity of dosage requirements depends on processes that remain stable over time. Compressed tablets must be made out of a blend of materials that is consistent, with no increasing or decreasing trends of the amount of API over time. You have been asked to use JMP to evaluate a set of individual tablet assay data from a confirmation batch of tablets to determine whether a trend over time can be detected in the data.

1. Open the data table *individual tablet assay.jmp*.
2. Stack the table so that there is a location column (numeric, continuous) and an assay column (numeric, continuous).
3. Use Graph Builder to visualize the assay data by location. Is a trend present in the data?
4. Create a control chart of the assay averages by location. What are the upper and lower control limits? Are there any extreme values outside of the control limits? Is the process stable (no large scale increasing or decreasing trends over time)?
5. Create an executive summary of the information for the project stakeholders, listing the three most important aspects of the plots.

E2.2—A medical device manufacturer creates sterile surgical kits that are hermetically sealed by a plasticized foil induction that is sealed to a plastic tray. The process is monitored by five kits pulled off the line every hour and pressure tested to check the strength of the seal. The seal must survive exposure of up to 36,000 ft above sea level, which is a seal strength that resists at least 23 inches of mercury. The testing device slowly increases pressure inside the sealed tray and records the point at which seal failure occurs, in inches of mercury.

1. Open the data table *burst testing.jmp*.
2. Use Graph Builder to visualize the burst test results by date and time. Are any trends present in the data?
3. Create an X-bar and R control chart of the burst test results by date and time. What are the upper and lower control limits? Is there any risk of not meeting the minimum seal strength of a burst test result of 23 inches of mercury?
4. Use the red triangle menu in the *Variables Control* chart header, and select the *Redo* and *Relaunch* options to create an additional chart.
5. Use the *shift* variable in the *By* box to get control charts for each shift. Do the control limits indicate a potential trend?
6. How would you summarize the information into a report to the project stakeholders?

E2.3—Pharmaceutical products go through an annual review process for all of the critical quality attributes (CQA). The data for the first year of production of a new capsule drug product has been compiled, and you are to look at the CQA's to determine whether there are any non-random trends over the span of the year. The table includes a total of 62 commercial batches that have been made in the calendar year.

1. Open the data table *capsule APR data.jmp.* Each row of the data table includes results for four of the five CQAs. Use Graph Builder to plot the five CQAs on a graph matrix.

2. The *Dialog* button in Graph Builder was mentioned previously as a way to create a graph matrix. Figure 2.22 shows the window with the choices needed to obtain a graph matrix for all five CQAs by batch.

Figure 2.22: Using the Dialog Button in Graph Builder

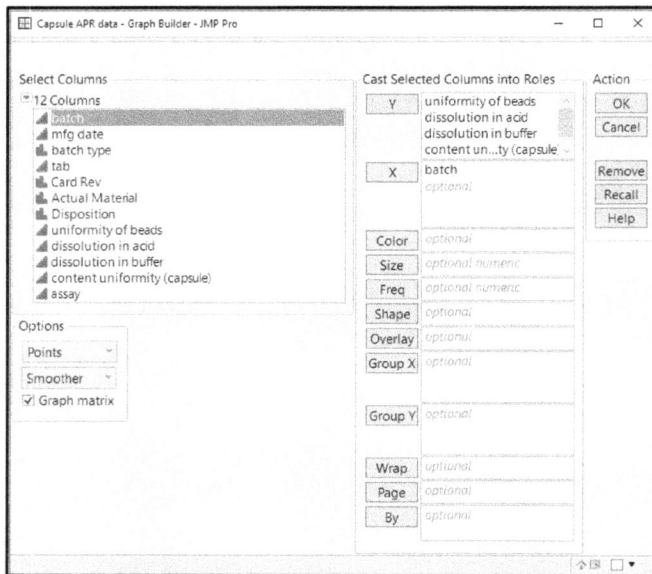

3. Click *Dialog* in Graph Builder, hold the Shift key, and select all five CQAs. Drag the selection into the *Y* box, and drag batch to the *X* box.

4. Select the *Graph Matrix* check box, and click *OK* to create the plot.

5. Change the Points() detail in the lower left of the control panel of the plot for *content uniformity (capsule), dissolution in buffer, dissolution in acid*, and *uniformity of beads* CQAs. Summarize each CQA for the **mean**, with **confidence interval** error bars.

6. Adjust the smoother of the five CQAs as appropriate to highlight any non-random patterns if present.

7. Create appropriate control charts for the five CQAs with batch on the X axis.

How would you summarize the information into a report to the project stakeholders?

Chapter 3: Assessing How Well a Process Performs to Specifications with Capability Analyses

Overview

Manufacturers of pharmaceutical and medical device products must plan in order to produce products that comply with specifications and ensure high levels of quality. Descriptive statistics and time-based plotting provide useful information about the processes, but neither can adequately summarize the performance of the process to specifications and estimate robustness. The previous chapter discussed analysis of a process to determine whether outputs are stable over time, since only stable processes should be analyzed for capability. Capability studies are a priority for the manufacture of high-quality products to quantify the performance of the process with detailed summaries. Stakeholders utilize the results to determine the risk of producing products that are outside of specifications. The studies can also help differentiate the sources of risks that originate from excessive variability, improper targeting, or both. This section uses sample data with capability studies to provide a comprehensive understanding of the robustness of a process and the products made from it.

The Problems: Assessing the Capability of the Fill Process and the Dental Implant Manufacturing Processes

Until now, the fill weight team has used data visualization techniques to describe the overfilling problem in great detail. The previous chapter focused on using time-based graphics to illustrate the "voice of the process". The average results and the spread in results over time explain the process as though it is talking to the analyst. Process data over time yields statistical limits for fill weight, which tell us the spread of random variation for the in-process averages that can be expected throughout commercial production. Suzanne is asked by leadership to identify how well the fill process can meet the label claim of 500 grams, which is the lowest average amount allowed. The team needs to demonstrate how capable the fill process is at meeting the label claim quality requirement in order to answer the question.

Ngong's improvement team determined that a specific machining center is producing implants that result in complaints from the dental customers. They collected a random sample of 200 implants with measurements of two physical attributes documented in a data set. The team needs to quantify the robustness of the process and gain information about where improvements are needed. Capability studies for the measured attributes will provide the needed information in a concise analysis summary.

One-Sided Capability Analysis for Fill Weight

The study of the capability of a process involves tools that compare the summary of the process measurements with the specification limits. With respect to the label claim for the amount of product customers can expect in a container, the fill process involves a single lower specification. The expectation is that the distribution of results meets or exceeds the label claim minimum fill weight. A sample with a distribution of weights with an appreciable density above the lower specification represents a population that can be expected to be increasingly capable of meeting the specification.

One way to visualize this problem is to think in terms of playing tennis. Figure 3.1 shows three different tennis players and the distributions of their forehand shots. Player A is very consistent, as shown in the tight spread of shots in Figure 3.1 A. The average of the shots is located at a distance from the net that ensures that the spread of all shots made will go over the net; the capability of player A is very good. Player B also has a consistent spread of shots, but the location of the average is precisely on the net limit. The capability is not very good because roughly half of the shots will not make it over the low specification (the net). It could be argued that player C has good capability for playing tennis because all of the shots make it over the net. The problem is that the consistency of player C is not as good as A or B, as can be seen by the wide distribution of shots. It is clear that player C is more capable than player B, but slight reductions in the average height could result in balls that go into the net.

Figure 3.1: Spread and Location in Tennis

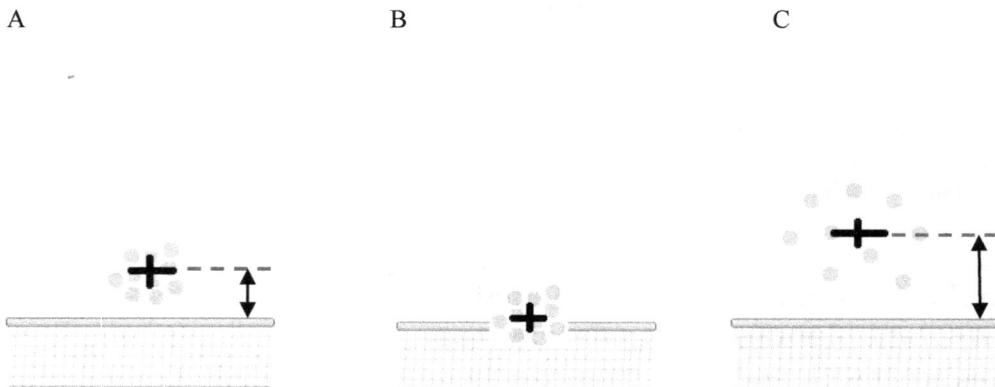

There are practical costs for the high capability of Player C, as she has had to compensate for the lack of consistency by adjusting the average higher on the net than the other two players. As a player hits the ball in higher arcs over the net, the speed of the ball is reduced giving, the opponent more time to set up for a shot. The cost to player C for the high target is a reduction in competitive advantage. If you have ever played tennis against a power player delivering consistently fast and flat shots, you will agree with the superior results of player A. The need for an adjusted target above a minimum specification as shown by these players illustrates the overfilling issue perfectly.

The best first step in capability analysis is to use simple, descriptive statistics to visualize and summarize the distribution of results. Open *initial fill weight data.jmp* and select *Analyze ▶ Distribution*, as shown in Figure 3.2, to get the initial visualization of the weight distribution.

Figure 3.2: Fill Weight Distribution

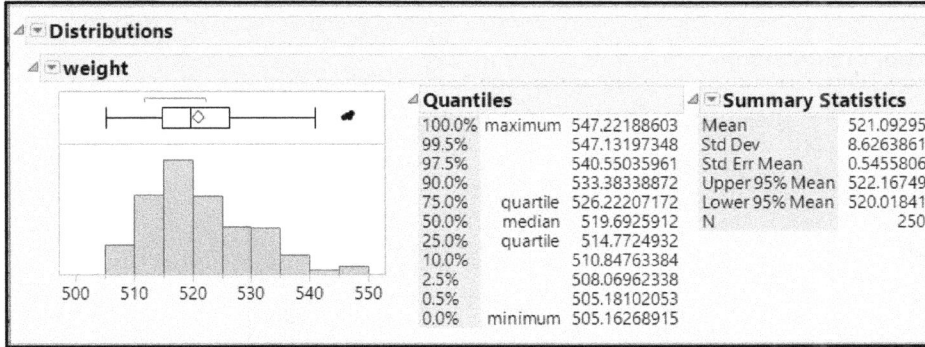

Suzanne knows that visualizing the distribution of a variable is always a great first step for deeper analysis. The weight data looks to be relatively normal with the exception of a skew toward higher values. Further diagnostics are suggested to determine whether assumptions are met for the use of the default capability tools.

Checking Assumptions for Fill Weight Data

The first thing to consider is whether the distribution of results can be studied with the default capability analysis based on parametric statistics, which are based on a normal distribution. Distributions with skew might not be good candidates for parametric statistics, especially if the sample size is small. A distribution illustrates skew when there are extreme values either to the lower or upper end of the range of values, which creates a long tail. The direction that the tail is pointing is the direction of skew, shown in Figure 3.3. (The histogram in Figure 3.2 illustrates a skew toward high fill values.)

Figure 3.3: Distribution Skewness

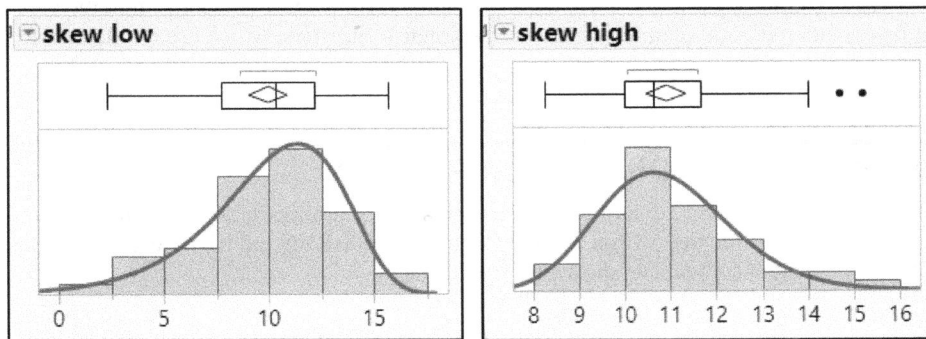

The sample of fill events includes 250 observations. Through the central limit theorem, we know that large sample sizes (n>50) tend to create sampling distributions that are of a normal shape, even when the distribution of the sample is not exactly normal and symmetric. The large number of observations can somewhat mitigate the effect of the error that comes from skewed data when using parametric statistics. However, additional diagnostics can help determine whether the skew might be detrimental.

Click the red triangle meu next to the *weight* header and select *Normal Quantile Plot* for an additional test of the normality of the distribution, shown in Figure 3.4.

Figure 3.4: Fill Weight Distribution

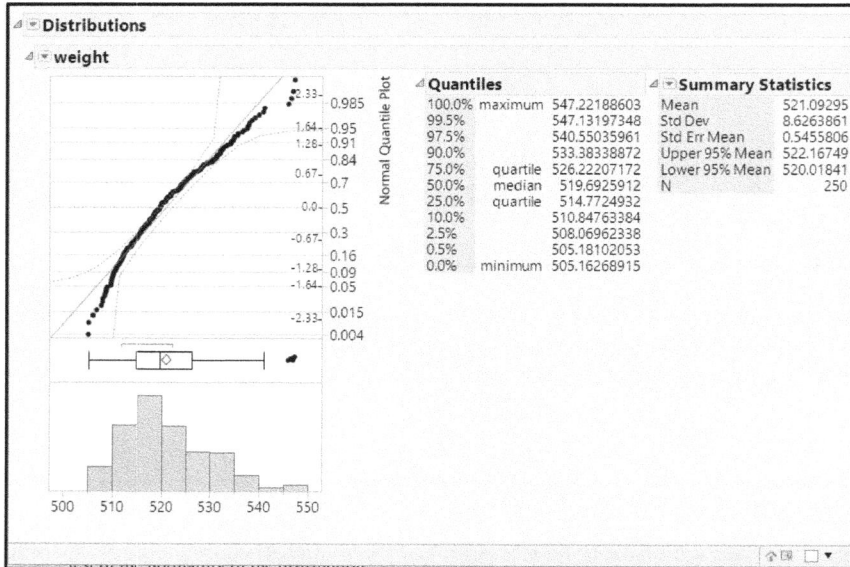

The normal quantile plot displays the trend of a normal distribution as a solid red upwardly diagonal line. The 95% confidence interval about the normal model line is reflected by red segmented lines on either side of the solid diagonal line, forming an hourglass pattern. The user looks for points that fall within the confidence interval to visually assess the normality of a distribution. The plot shows that a large number of observations follow the normal model line very closely with a few observations on the extremes that stray marginally away from the line. All of the observations fall within a 95% confidence interval for a normal distribution. The large sample size and the normal quantile plot correlate with evidence that the weight distribution can be considered good data for the use of parametric statistics, which are the default for capability studies.

Capability Studies from the Distribution Platform

Chapter 1 described some of the many dynamic analysis options that are available in the JMP red triangle menu. The red triangle menu is located within an analysis header. Click the red triangle next to the *weight* header, and select *Capability Analysis*, as shown in Figure 3.5.

Figure 3.5: Distribution Red Triangle Menu Capability Analysis Option

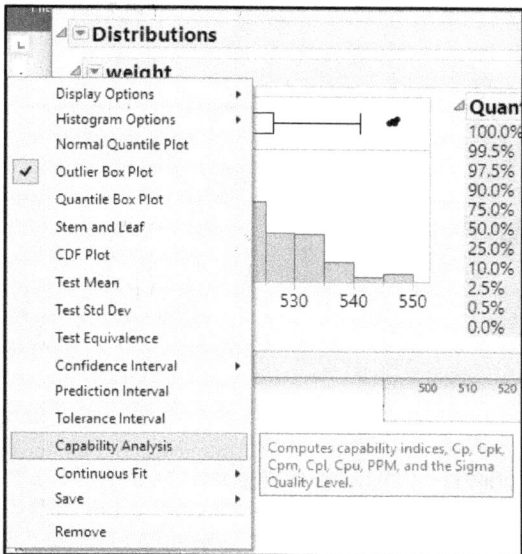

To create a capability study, complete the following steps:

1. Enter the lower specification limit for the 500-gram label claim in the *Capability Analysis Setting Specification Limits 'weight'* window, shown in Figure 3.6.

   Figure 3.6: Entering Specification Limit

   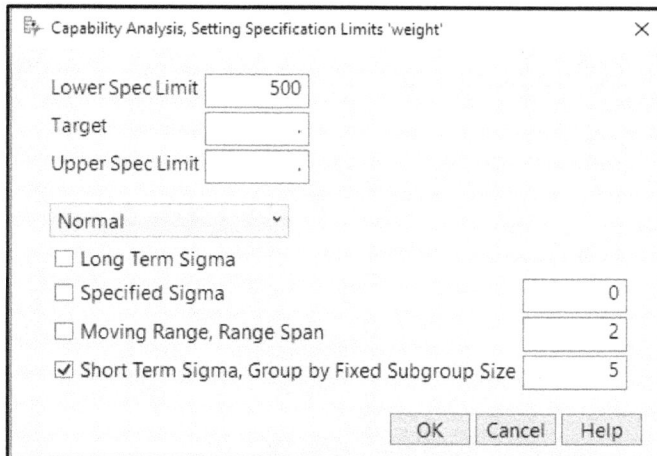

2. Deselect the *Long Term Sigma* check box since the sample represents a short-term pull of the fill process.

3. Select the *Short Term Sigma, Grouped by Fixed Subgroup Size* check box, and ensure that the number of observations is 5, representing the five units measured at each in-process check.

4. Leave the other default values.

5. Click *OK* to get the results added to the *Distributions* window, shown in Figure 3.7.

Figure 3.7: Distributions with Capability Analysis

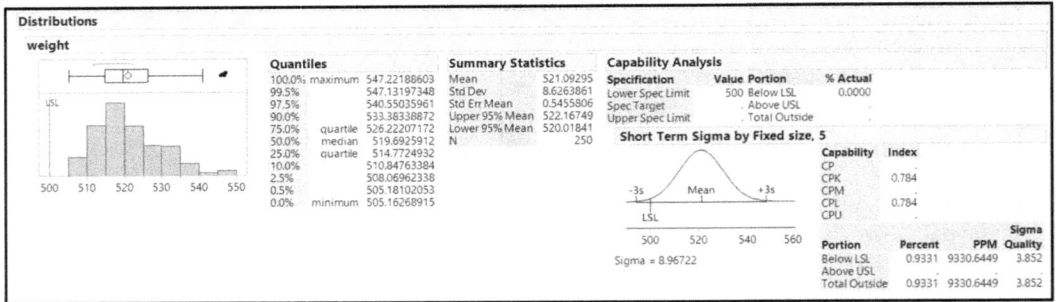

The capability analysis option is added to the Distributions window. The normal quantile plot has been removed for clarity. Fill weight has the one-sided minimum label claim used to assess the quality performance of the process. Many options are available to customize the analysis, including the ability to specify a sampling distribution used for making estimates of the population. This example uses the default values for simplicity since the assumptions of the fill data being a normal distribution have been met.

Figure 3.8 provides a larger view of the analysis details. Practical interpretations of the analysis are typically of great interest to stakeholders and are covered here first. The percent of actual observations that are outside of the minimum specification (% Actual) gives the JMP user with the good news that the process determined that 0% of the actual in-process checks are below the label claim specification limit of 500 grams. The remainder of the capability study information now involves estimates of the trends in the population of fill results.

Figure 3.8: Capability Analysis Details

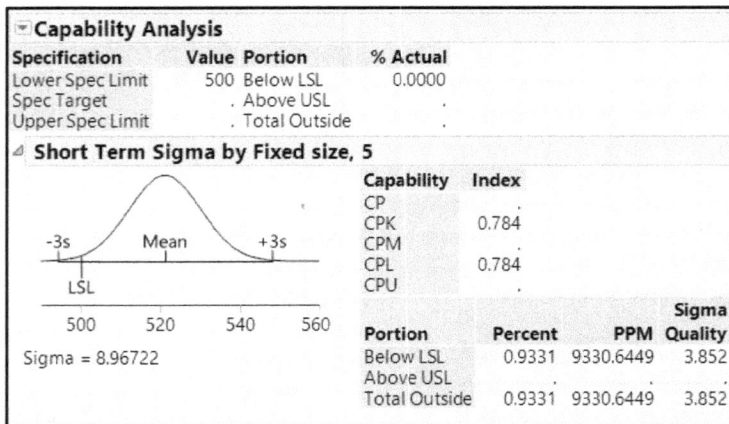

Capability indices are a collection of values that describe how well the distribution of results fits within specifications. In our fill weight case, we are dealing with a single, minimum specification and working to determine the portion of the fill weight distribution that exceeds it. JMP uses "within" capability as the default. The capability index is Cpk, which is a single value that reflects the proportion of the distribution values that are likely to exceed the minimum specification. The fill weight data involves five observations at each check interval. The within estimate of standard deviation refers to the variability of weights for the five observations taken at each subgroup. Targeted capability (CpK) = the difference between the process average and the label claim ÷ the statistically expected amount of spread within the in-process checks.

Through the empirical rule, Suzanne can estimate that almost all values of a distribution are likely to fall within an interval that is +/- 3 units of the sample standard deviation from the mean. Targeted capability is typically applied to an outcome that includes both a higher and lower specification. The targeted capability index that reflects the worst case is used as a summary. The fill weight process is subject to only a minimum specification, which makes things a little easier for the team. Basically, the distance of the sample mean from the minimum specification is divided by 3 units of within standard deviation to estimate the number of distributions of results that will exceed the low specification. The statistical interpretation of the results gives more detail on the performance of the process.

$$\overline{x} = \text{sample average of in-process weight checks (521.09)}$$
$$s_w = \text{sample std. deviation within subgroups (8.967)}$$

$$CpK = \frac{\overline{x} - label\_claim}{3 * s_w}$$

$$CpK = \frac{521.09 - 500}{(3 * 8.967)}$$

$$CpK = 21.09/26.901$$

$$CpK = 0.784$$

The fill process has achieved a CpK of 0.784, explaining that a little less than the one distribution of fill weight results (78.4%) is likely to exceed the low specification. The statistic assumes that the fill weights are behaving as a normal distribution with a single high frequency peak (the mean), and symmetric tailing of frequencies above and below the mean. The fill data has a slightly higher potential to contain a few results that are very high as opposed to extremely low results. The capability analysis assumes an equal likelihood for a few extreme high and low results. Therefore, the capability statistics based on a normal distribution will be conservative.

The output provides the quantification of potential risk for the population of results relative to the minimum specified fill weight (or label claim). The fill process might contribute up to 0.9% of average weights that are less than 500 grams, as seen in the lower rows of the output. The estimate of just under 1% of fill weight averages not meeting the label claim is practically higher than the actual risk due to the shape of the fill weight distribution. This outcome is to the result of the conservative nature of the parametric estimate. It is reasonable to expect that for every million in-process checks executed, there is a chance than just over 9,000 might have an average weight than is less than 500 grams. The distribution includes nearly 4 units of standard deviation that will exceed the label claim, noted by the sigma quality level of the process. The higher the sigma level, the more likely it is that the results will meet and exceed the specification.

The improvement team is utilizing the capability analysis to create a baseline for comparison. JMP provides additional tools to find a continuous distribution that has a better fit to the fill weight data and that obtain a statistically precise result. The downside of the added precision is that the model becomes more complex, which is more difficult to explain to the stakeholders of the project. Suzanne decides to postpone the more precise analysis and keep the improvement team focused on the goals of reducing the variability in fills and reducing the average fill target if possible.

Two-Sided (Bilateral) Capability Analysis for Implant Dimensions

Process monitoring often involves using subgrouping to track average results from intervals involving multiple individual values. Chapter 1 analyzed discrete results of dental implants collected from dental customers. Recall that the manufacturer of the implants is responding to customer complaints regarding difficulties in getting the abutment threaded into the implant. The initial analysis resulted in tracking a chamfering problem back to one of the several machining centers used for manufacturing. The improvement project team discovered that quality checks for the manufacturing process are limited to the ability of the operators to start a test abutment into the threaded implant. The current check method is not adequate because it does not mimic the angles dentists deal with for the installation of an abutment into an implant when it is within a patient's mouth. More data is needed to adequately study the problem.

The project leader, Ngong, worked with the engineering team and learned that an inadequate depth of the chamfer is the likely cause of the threading problem. A chamfer is a lead-in angle that is machined at the top of the threaded portion of the implant. The engineers design a fixture to be able to accurately gather data on the actual depth (in mm) of the chamfer as well as the overall depth of the threaded hole of the implant. Measurement systems analysis was completed prior to the initiation of the new in-process checks for the machining center to ensure the highest level of accuracy and precision in the measurements. (Details on how to run measurement systems analyses are included in chapter 7.) A sample of in-process checks involves the operator grabbing five implants at a specified process interval and measuring them; the results are shown in Figure 3.9. The sample variable illustrates the grouping of in-process checks as a series of five repeat values as the sample interval value increases.

Figure 3.9: Implant Dimensional Data

	sample	chamfer depth	threading depth
1	1	0.67	2.73
2	1	0.67	2.51
3	1	0.64	2.47
4	1	0.58	2.51
5	1	0.58	2.52
6	2	0.62	2.57
7	2	0.67	2.61
8	2	0.63	2.70
9	2	0.54	2.45
10	2	0.66	2.42
11	3	0.69	2.73
12	3	0.56	2.71
13	3	0.54	2.46
14	3	0.57	2.61
15	3	0.65	2.60
16	4	0.57	2.55

A design review of the components was completed and a specification limit for adequate chamfer depth is defined as between 0.60 mm and 0.80 mm. The threading depth of the hole must be between 2.0 mm and 2.8 mm. Ngong is interested in using JMP to run capability studies on the data so that the team can better understand how well implants from the subject machining center meet the specifications for chamfer depth and threading depth.

Get started by opening *dental implant dimensional checks with speeds.jmp.* Select *Analyze ▶ Distribution* to view the pooled data for *chamfer depth* and *threading depth*, as shown in Figure 3.10.

Figure 3.10: Distributions of Implant Measurements

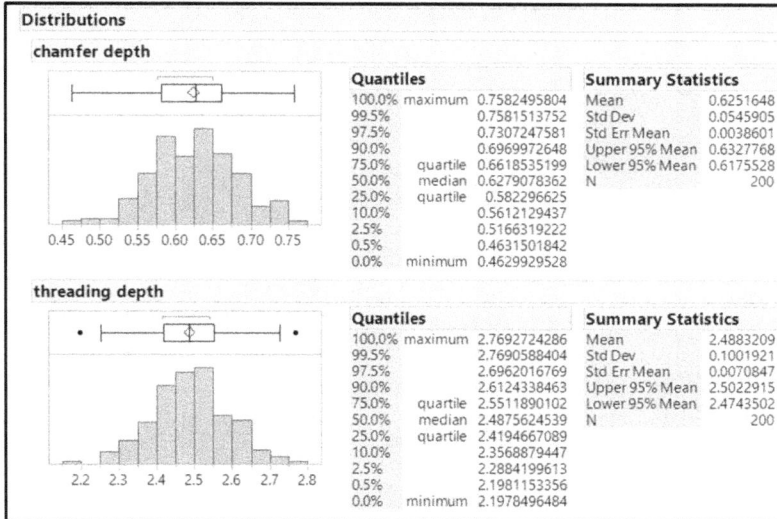

Checking Assumptions for Implant Measures Data

The histograms show that the data for each of the measurements is normally distributed (symmetric). A quick check of normality involves the comparison of the means and medians, which are similar in value confirming the symmetry of the data. If skew were present, the mean of the distribution would follow the tail of the distribution and differ from the median. Since the data are normally distributed, the typical parametric statistics are appopriate to analyze capability. The chamfer depth minimum of 0.46 mm is clearly less than the 0.60 mm minimum specification, and the maximum depth of 0.76 mm is well within the upper specification of 0.80 mm. The threading depth looks to be within the specifications, although it is difficult to determine how well the process performs to maintain proper threading depth by studying the distribution summary alone.

Capability Analysis from the Quality and Process Options

Ngong could use the Distributions hot spot for capability study of each measurement, but the team would like to get a comprehensive view of dimensional capability for the implant with both measurements. Dental implant data is known to be stacked, with each group of five measurements in the order referenced by **sample**; knowledge of this data structure is important because it is the basis for how the data will be subgrouped. The analyze menu includes a capability studies option within the quality tools of the analysis menu. The following example provides the steps necessary to get the needed output and the interpretation of results.

Select *Analyze ▶ Quality and Process ▶ Process Capability* to initiate the analysis, as shown in Figure 3.11.

Figure 3.11: Initiating a Process Capability Study

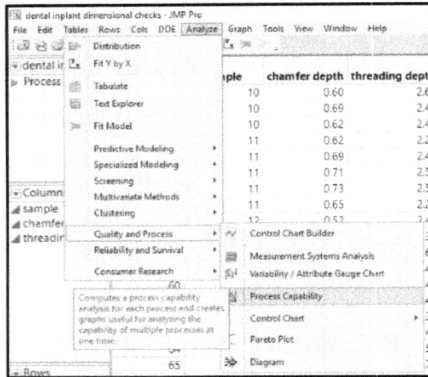

1. In the *Process Capability* window, select *chamfer depth* and *threading depth*, and move them to the *Cast Selected Columns into Roles* box for the analysis.
2. Click on the gray arrow next to the *Process Subgrouping* header to view options.
3. Select *Subgroup ID Column*.
4. Press the Shift key, and then click *sample* in the *Select Columns* section and click *chamfer depth* and *threading depth* in the *Cast Selected Columns into Roles* section.
5. Click *Nest Subgroup ID Column* to complete the subgrouping, shown in Figure 3.12.

Figure 3.12: Process Capability Window

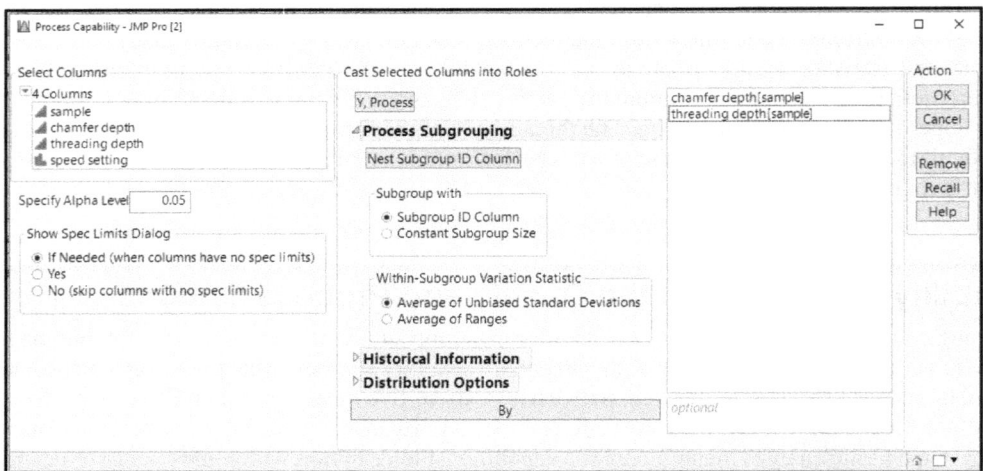

6. Click *OK* to open the *Set Specification Limits* window, shown in Figure 3.13.

Figure 3.13: Specifications Window

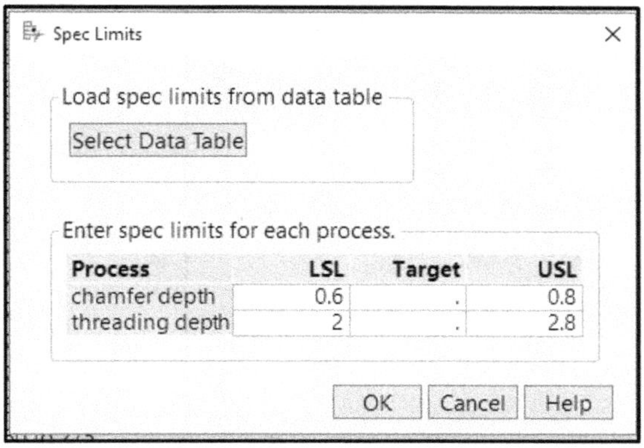

7. Enter *0.6* as *LSL* and *0.8* as *USL* for *chamfer depth* and enter *2.0* as *LSL* and *2.8* as *USL* for *threading depth*. Then, click *OK* for results.

The default capability plots illustrate display the information that results from the capability analysis. The goal plot shown in Figure 3.14 provides information about the capability values respective to the goal of 1 PpK. The default index for this problem differs from the Cpk capability that was used with the fill weights problem. Overall capability involves a variability term calculated from the entire sample of the data, resulting in the Ppk value. Within capability creates a summarized standard deviation representing the amount of variation within the subgroups, resulting in a Cpk value. In general, overall capability is more sensitive to shifts in average results between subgroups and is typically lower than within capability.

In a goal plot, a point that is vertically distant from the red goal triangle has excessive variability. A point that is horizontally distant from the goal triangle has a mean value that is shifted away from specifications. Chamfer depth is high and to the left of the goal triangle, which means that excessive variation and a low mean shift combine to result in a low capability value. The goal can be adjusted with the slider to represent higher or lower targets for capability, depending on organizational requirements.

Figure 3.14: Process Capability Output: Goal Plot

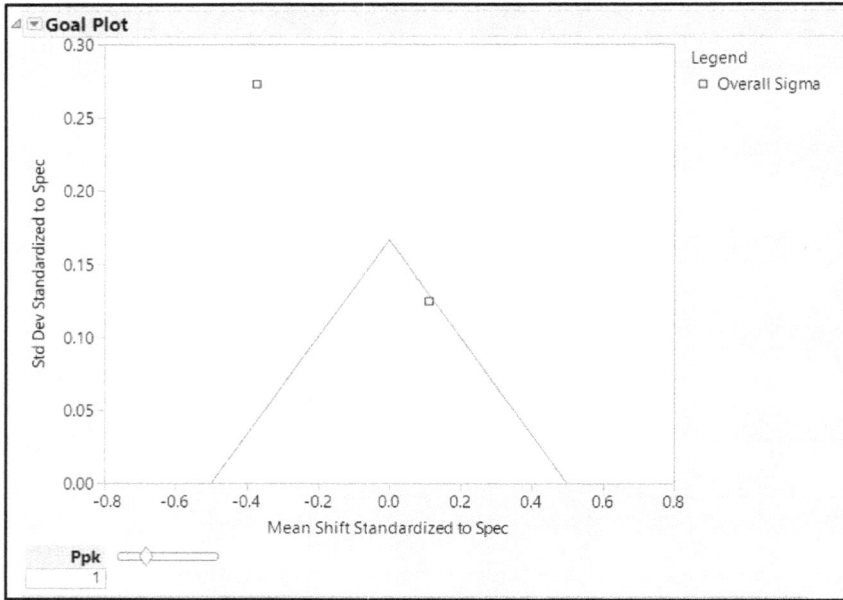

The middle plot of the default window, shown in Figure 3.15, contains standardized box plots used to compare capability. Notice that chamfer depth has a wide box and tails and is shifted toward lower values. The shape of the box plot indicates excessive variability and shift of the mean toward lower values. The same conclusions can be made from the interpretations of the goal plot and the box plots. The differing views of the results provide variety for reporting purposes.

Figure 3.15: Process Capability Output: Box Plots

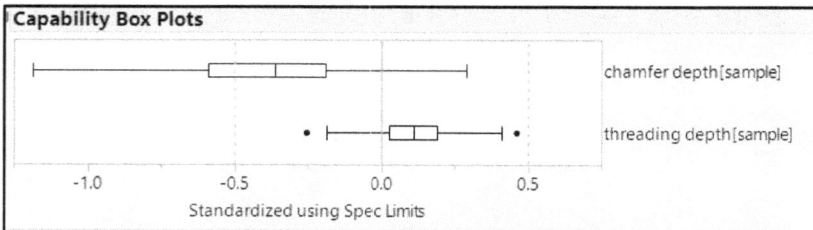

The lower plot shown in Figure 3.16 is limited to illustrating the value of the capability indices (Ppk) on a vertical axis and does not provide much information as to why the results might be lower than what is desired.

Figure 3.16: Process Capability Output: Index Plot

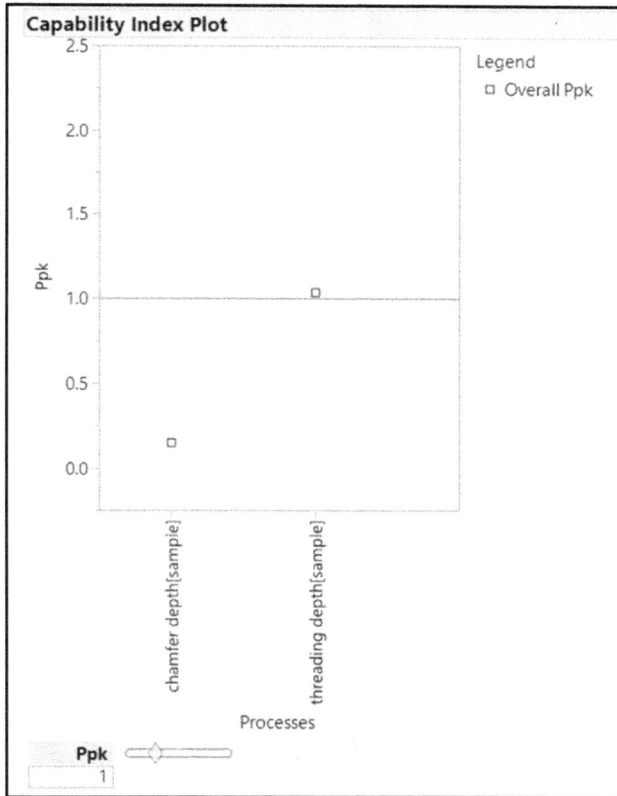

The plots provide a compelling picture of the comparative process capability for each of the measurements. Options are available for adding detail to the analysis, which can complement the value of the analysis.

Select the red triangle menu next to the *Process Capability* header, and select *Summary Reports ▶ Within Sigma Summary Report* and *Summary Reports ▶ Overall Sigma Summary Report*, as shown in Figure 3.17. The summary reports are added to the output.

Figure 3.17: Choosing Summary Report Options

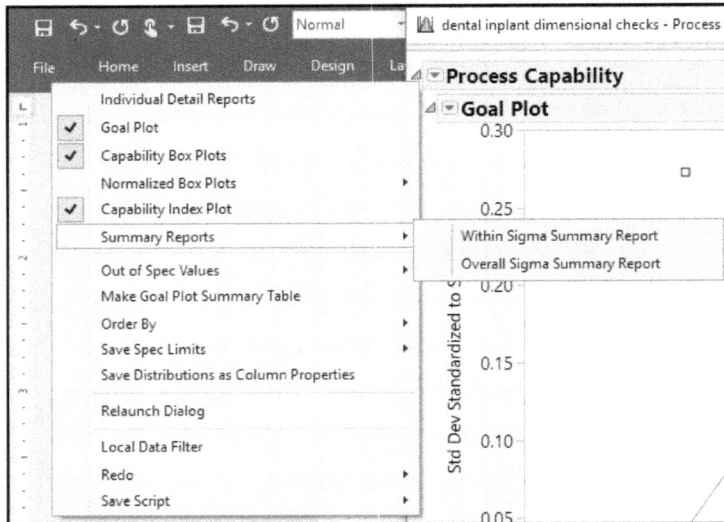

Capability Analysis Summary Reports

The overall sigma capability summary shown in Figure 3.18 was created from a pool of the data that does not take group size into account. As noted previously, overall capability tends to be more conservative than within capability since the variability used in calculations includes both mean shifts between groups as well as the variability within groups. The dental implant data include bilateral specifications, so more information is given regarding the performance of the process than what resulted for fill weight. The targeting of the distribution is assessed for both the lower and upper specifications (Ppl and Ppu) with the minimum chosen as the targeted capability (Ppk) of the process.

Figure 3.18: Summary Report

Overall Sigma Capability Summary Report																	
Process	LSL	Target	USL	Sample Mean	Overall Sigma	Stability Ratio	Ppk	Ppl	Ppu	Pp	Cpm	Expected % Outside	Expected % Below LSL	Expected % Above USL	Observed % Outside	Observed % Below LSL	Observed % Above USL
chamfer depth[sample]	0.6		0.8	0.625165	0.054591	0.987376	0.154	0.154	1.068	0.611		32.3089	32.2409	0.0681	35.0000	35.0000	0.0000
threading depth[sample]	2		2.8	2.488321	0.100192	0.973753	1.037	1.625	1.037	1.331		0.0933	0.0001	0.0933	0.0000	0.0000	0.0000

Overall Capability for Chamfer Depth

$$Ppl = \frac{xbar - low\ spec}{3 \cdot s} \qquad Ppu = \frac{high\ spec - xbar}{3 \cdot s}$$

$$Ppl = \frac{0.625 - 0.6}{3 \cdot 0.0546} \qquad Ppu = \frac{0.8 - 0.625}{3 \cdot 0.0546}$$

$$Ppl = 0.154 \qquad\qquad Ppu = 1.068$$

$$PpK = \min(Ppl, Ppu)$$

$$Ppk = 0.154$$

Overall Capability for Threading Depth

$$Ppl = \frac{2.49 - 2.0}{3 \cdot 0.1002} \qquad Ppl = \frac{2.8 - 2.49}{3 \cdot 0.1002}$$

$$Ppl = 1.625 \qquad Ppu = 1.037$$

$$PpK = \min(Ppl, Ppu)$$

$$Ppk = 1.037$$

The Ppk index identifies that a small portion of the chamfer depth distribution is likely to be within the specifications (Ppk=0.154). Process difficulties in making acceptable implants are likely to be present. Slightly more than a full distribution of threading depth will be within the specifications (Ppk=1.037), which is a decent result. The practical interpretation of the machining process is that 35% of the observed implants have chamfer depth values that are below the 0.60 mm limit. The long-run expectation of the process suggests that approximately 32% of the population values are expected to be less than 0.60 mm. The threading depth of the implant has a Ppk index that indicates minimal risk (<1/10th of a percent) for parts that do not meet the specifications.

An additional index is apparent in the overall capability report: potential capability. Pp is an index that indicates potential capability of a process based solely on variability. The value treats the mean of the data as if it is in the middle of the specification, so it is perfectly targeted. Potential capability is useful because it provides the best possible result that can be achieved without putting forth effort to reduce variation in the process. It is typically much easier to shift the mean of a process than it is to reduce variation. Mean shifts are usually achieved through simple adjustments of the process inputs. Variation reduction typically involves engineering efforts and the allocation of significant resources in an attempt to improve a process.

Overall Potential Capability for Chamfer Depth   *Overall Potential Capability for Threading Depth*

$$Pp = \frac{upper\ spec\ -\ lower\ spec}{6 \cdot s}$$

$$Pp = \frac{0.8 - 0.6}{6 \cdot 0.0546} \qquad\qquad Pp = \frac{2.8 - 2.0}{6 \cdot 0.1002}$$

$$Pp = 0.611 \qquad\qquad Pp = 1.331$$

The potential capability summaries (Pp) are roughly the same values as the actual capability (Ppk). This result is good in that the process seems to be stable regarding mean shifts over time. There is no point in the team attempting to achieve robust process capability of chamfer depth through simple adjustments in the process inputs. Too much variation is present in the results to achieve a minimum process capability. The minimum process capability is one distribution of results that are within the specifications, shown in the summary as a Ppk of 1.0. The capability for thread depth is marginally acceptable and will not be the focus of process improvements. It is entirely possible that improvements to the process to gain acceptable chamfer depth capability will also have beneficial effects for threading depth. The team needs to prepare for work on the process to reduce variation in chamfer depth.

Sampling intervals that involve multiple units allows for additional capability analysis. The within sigma capability report gives information that is focused on the amount of variability present within the five samples. Within capability focuses on repeatability of the process among the subgroups of data collection. Recall that within capability summaries should be applied only to processes that have demonstrated stability over time. Therefore, you typically need to use control charting to establish that a process is stable. The within sigma capability report is shown in Figure 3.19.

Figure 3.19: Summary Report

Process	LSL	Target	USL	Sample Mean	Within Sigma	Stability Ratio	Cpk	Cpl	Cpu	Cp	Cpm	Expected % Outside	Expected % Below LSL	Expected % Above USL	Observed % Outside	Observed % Below LSL	Observed % Above USL
chamfer depth[sample]	0.6	.	0.8	0.625165	0.054938	0.987376	0.153	0.153	1.061	0.607	.	32.4187	32.3456	0.0730	35.0000	35.0000	0.0000
threading depth[sample]	2	.	2.8	2.488321	0.101533	0.973753	1.023	1.603	1.023	1.313	.	0.1072	0.0001	0.1071	0.0000	0.0000	0.0000

Within Sigma Capability Summary Report

In JMP, there are three possible approaches for calculating within sigma. (Detail about the within sigma calculations is available in the *Quality and Process Method* book available in the Help menu.) The report indicates that the overall capability performance is slightly better than the within capability.

The summary capability reports provide so much detail that the stakeholders cannot use it as a report to the stakeholders of the project team. Ngong would like to reduce the reports and include only the desired detail. He does not want to get side-tracked into an explanation of values that do not add to the message that resources need to be allocated for machining process improvement.

Place the pointer on the report, right-click, and select *Columns*, as shown in Figure 3.20. The columns *Target*, *Stability Ratio*, *Cpm*, *Observed % Outside*, *Observed % Below LSL*, and *Observed % Above USL* are deselected to clean up the report. The abridged report is shown in Figure 3.21. The abridged report provides the concise information that will be published to the project stakeholders. The outputs can easily be copied and pasted for reporting purposes.

Figure 3.20: Column View Options

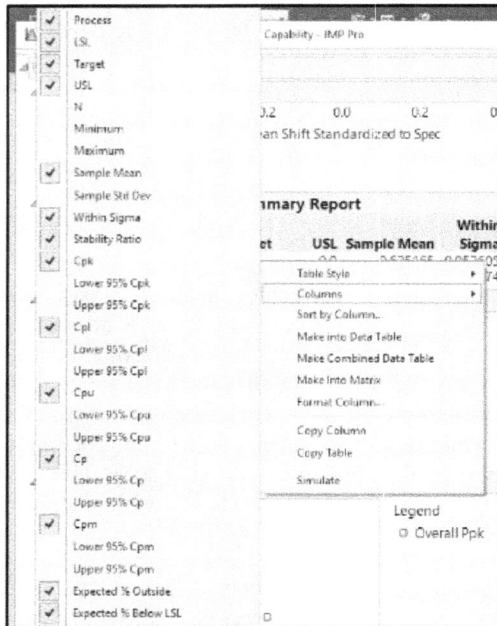

Figure 3.21: Summary Report

Within Sigma Capability Summary Report											
				Within					Expected	Expected %	Expected %
Process	LSL	USL	Sample Mean	Sigma	Cpk	Cpl	Cpu	Cp	% Outside	Below LSL	Above USL
chamfer depth	0.6	0.9	0.625165	0.052605	0.159	0.159	1.742	0.950	31.6192	31.6192	0.0000
threading depth	2	3	2.488321	0.096974	1.679	1.679	1.759	1.719	0.0000	0.0000	0.0000

To obtain a copy of the report that you can paste into a document or slide presentation, position your pointer on the grey arrow to the left of the report header, right-click, and select *Edit ▶ Copy Picture*.

Capability Analysis for Non-normal Distributions

The information about capability studies has involved parametric statistics. Therefore, sample data must meet the assumptions for a normal distribution. There are many situations that either involve normally distributed data or data that can transformed to conform to a normal distribution. Samples that have some non-normal tendencies and that are of a very large size can be studied with parametric statistics. Averages taken from samples of the same size randomly selected from a population form a normal distribution. This sampling distribution takes on a normal shape more readily for large samples, even if the samples are not themselves distributed normally. The central limit theorem is the basis for the use of parametric statistics with non-symmetric distributions that include many samples; you can use the Help menu to explore these concepts. The fill weight data is known to have minor skew toward high values, but the large sample size allows for the use of typical parametric statistics.

Small samples that have non-normal tendencies are more challenging. The application of parametric statistics in such cases is likely to include an unacceptable amount of error. JMP includes a number of continuous distribution options to deal with small samples that are not symmetric in shape. This example uses a small subset of the fill-weight data (n=25) to illustrate the use of capability studies calculated for a continuous distribution that is not normal.

1. Open *Subset of initial fill weight data.jmp*.
2. Select *weight*, and then select *Analyze ▶ Distribution* to visualize the sample.
3. Use the red triangle menu beside the *weight* header to select *Normal Quantile Plot* (Figure 3.22).

Figure 3.22: Distribution Summary of Fill Weight Subset

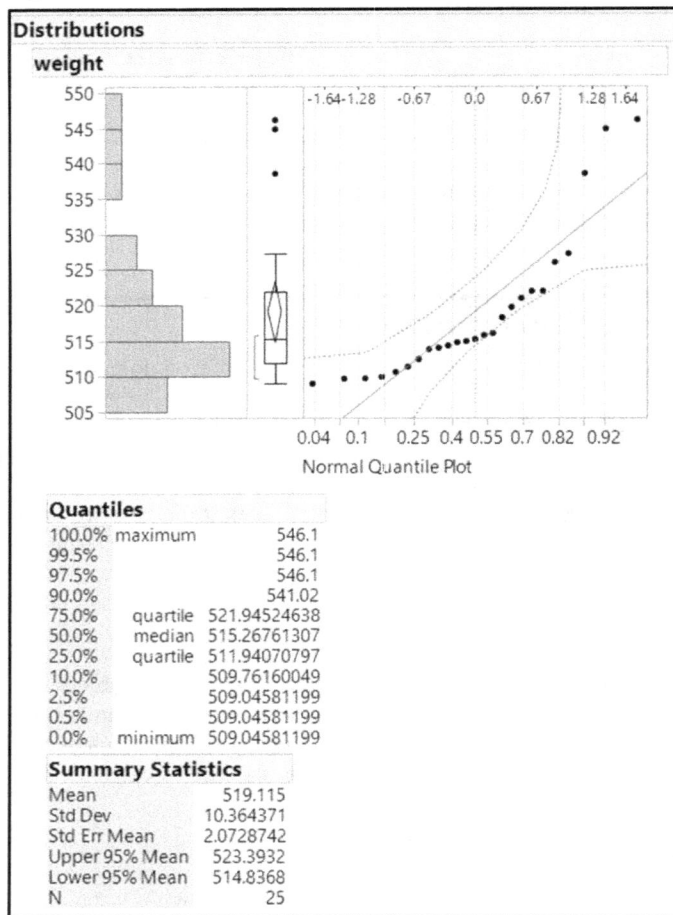

The distribution of the fill-weight subset is clearly skewed to high values, with points that stray significantly away from the normal model line in the Normal Quantile plot. The small sample size might contribute to error in capability based on parametric statistics. JMP includes a useful tool to automatically fit a number of continuous distributions and rank them by fit statistics.

Use the red triangle menu next to the *weight* header to select *Continuous Fit* ▶ *All*. Then add continuous fit detail to the *Distributions* output, as shown in Figure 3.23.

Figure 3.23: Distribution Summary with Continuous Fit Detail

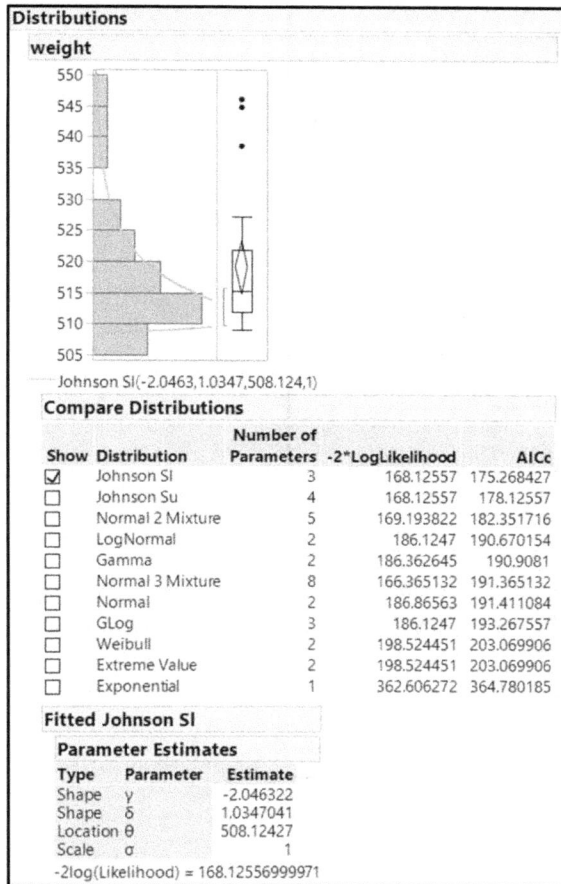

Show	Distribution	Number of Parameters	-2*LogLikelihood	AICc
☑	Johnson Sl	3	168.12557	175.268427
☐	Johnson Su	4	168.12557	178.12557
☐	Normal 2 Mixture	5	169.193822	182.351716
☐	LogNormal	2	186.1247	190.670154
☐	Gamma	2	186.362645	190.9081
☐	Normal 3 Mixture	8	166.365132	191.365132
☐	Normal	2	186.86563	191.411084
☐	GLog	3	186.1247	193.267557
☐	Weibull	2	198.524451	203.069906
☐	Extreme Value	2	198.524451	203.069906
☐	Exponential	1	362.606272	364.780185

Fitted Johnson Sl

Parameter Estimates

Type	Parameter	Estimate
Shape	γ	-2.046322
Shape	δ	1.0347041
Location	θ	508.12427
Scale	σ	1

-2log(Likelihood) = 168.12556999971

JMP tried to fit the data to 11 continuous distributions. The results are sorted by AICc fit statistic, as shown in the *Compare Distributions* table. Lower values for the fit statistic indicate better fits of the data to the function. The best fit distribution is the Johnson Sl transformation with the lowest AICc fit of 175.3. It is worth noting that the normal distribution ranks 7th with an AICc fit of 191.4. JMP automatically provides the parameters of the best fit distribution and illustrates the function on the histogram plot. You can select different distributions, multiple distributions, or both in order to compare results. In general, small differences in fit statistics are not practically relevant. Therefore, a less complex distribution with fewer parameters might be a better choice for similar fit statistics. The normal distribution seems to be quite different, and the capability study will include the best fit Johnson Sl distribution to mitigate error by completing the following steps:

1. Use the red triangle menu next to *weight* and select *Capability Analysis*. The *Capability Analysis, Setting Specification Limits 'weight'* window appears, as shown in Figure 3.24.

Figure 3.24: Capability Specification Setting

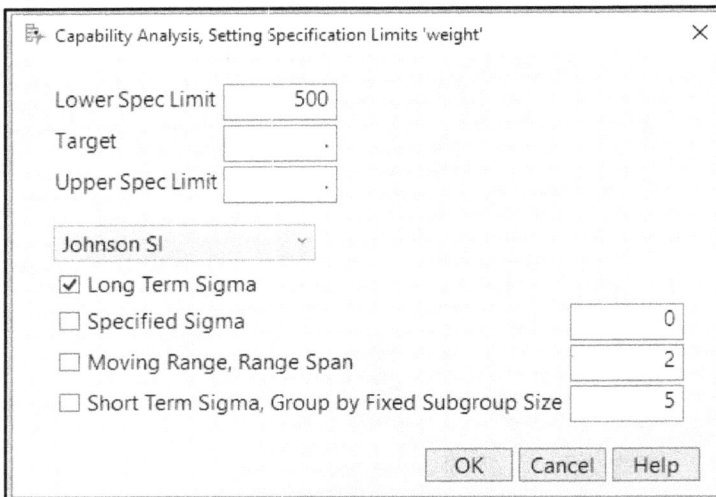

2. Enter 500 for the *Lower Spec Limit*.
3. Change option in the drop-down field from *Normal* to *Johnson Sl*.
4. Since subgroups are no longer relevant in the subset, the deep the *Long Term Sigma* check box selected.
5. Click *OK* to get the output shown in Figure 3.25.

Figure 3.25: Capability Summary of Johnson Sl Distribution

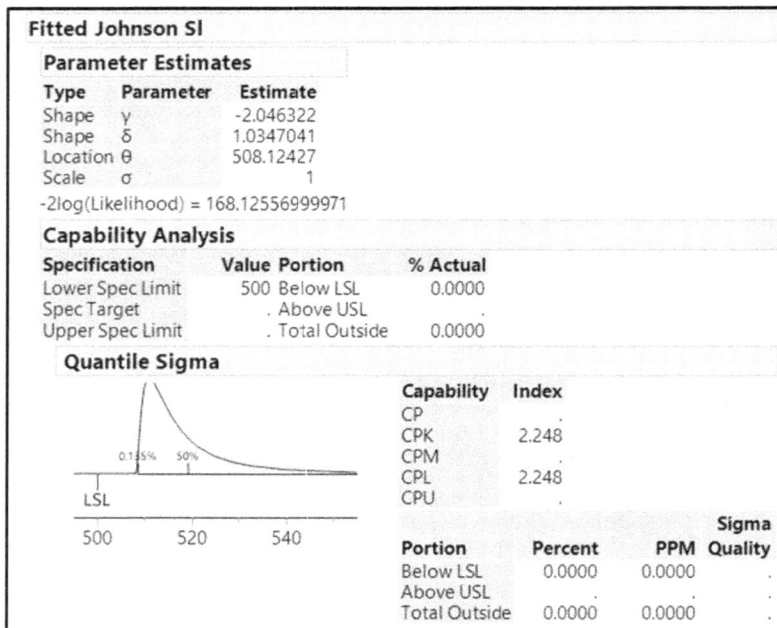

The capability results of the fill weight for the Johnson Sl distribution shown in Figure 3.25 is excellent with a Cpk of 2.2. Results like this make sense for the process since procedures and controls are in place to ensure that the risk of underfilling to the label claim is minimized. Quality requirements have been met with little regard to the business costs of allowing overfilling of units. The Johnson Sl distribution is skewed toward higher weights and truncated in variation for lower weights. Therefore, the risk of not meeting the label claim is minimal. To compare these results to normal capability estimates that can highlight the amount of error possible due to a poor fit of the distribution to the sample, complete the following steps:

1. Use the red triangle menu next to *weight* and select *Capability Analysis*.
2. Leave the default *Long Term Sigma* check box selected.
3. Enter 500 for the *Lower Spec Limit*.
4. Keep the *Normal* distribution selected in the drop-down options.
5. Click *OK* to get the output, shown in Figure 3.26.

Figure 3.26: Capability Summary of Normal Distribution

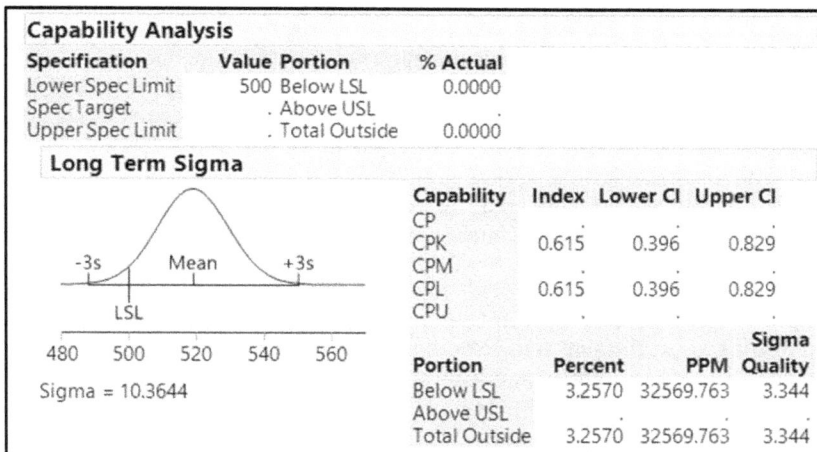

Capability Analysis

Specification	Value	Portion	% Actual
Lower Spec Limit	500	Below LSL	0.0000
Spec Target	.	Above USL	.
Upper Spec Limit	.	Total Outside	0.0000

Long Term Sigma

Capability	Index	Lower CI	Upper CI
CP	.	.	.
CPK	0.615	0.396	0.829
CPM	.	.	.
CPL	0.615	0.396	0.829
CPU	.	.	.

Sigma = 10.3644

Portion	Percent	PPM	Sigma Quality
Below LSL	3.2570	32569.763	3.344
Above USL	.	.	.
Total Outside	3.2570	32569.763	3.344

The capability based on a normal distribution is much lower and indicates a far greater risk of units than might be produced with weights less than the label claim. The results indicate that 3.3% of units will be outside of the label claim minimum. A normal distribution includes symmetric tails, so the distance of the extremely high observations from the mean creates the expectation of a similar distance of extreme low observations from the mean. This is the reason that the result is so much lower than the Johnson Sl capability result. The Johnson Sl distribution is a better representation of the actual process regardless of sample size.

Go back to the full set of fill-weight data and run the capability for the best fit continuous distribution. Determine which capability analysis provides the best representation of the process. Suzanne needs to consider this information and whether it would change the summary that is reported to the stakeholders of the fill-weight improvement project.

Practical Conclusions

Capability studies are very useful for comparing the voice of the process with specifications for the results. Robust quality control plans typically include regular intervals of capability study, along with time-based charting, to ensure that processes produce products that consistently meet and exceed specifications. Interpretations from capability studies can vary in the level of detail provided due to organizational needs and business concerns. JMP provides the necessary tools to optimize the message to be conveyed by the analyst.

The fill-process improvement project utilizes single specification studies to determine the robustness of the process while illustrating the extent of variation in the process. Capability studies provide quantifiable evidence of the overfilling problem to the stakeholders of the improvement project. The project team is interested in reducing the variation in filling to enable them to shift the target fill and mitigate overfilling. The studies are useful to communicate with the quality team and get them on board so that business needs can be better met without adding the risk of not meeting the minimum fill specification. The overall goal is to maintain adequate capability, without adding to the costs of production through overfilling. At a minimum, the team should try to obtain capability values that are over the default goal of Ppk= 1.0 in order to meet both the quality and business goals of the organization.

Bilateral specifications are used for the key physical features of the dental implants. The goal of the engineering improvement project is to reduce customer complaints through improvements in meeting the specifications of the key features. The highest targeted capability values are desired to ensure that the machining processes are producing implants with chamfer depth and threading depth that are centered within specifications with the least possible variation. Increases in targeted capability for bilateral specifications typically improve both quality compliance and business results. Parts that robustly meet specifications create happy customers who buy more product as well as maintain low quality costs.

Observations over time should be analyzed through control charting to establish it to be stable and free of non-random trends prior to running capability analysis. Although not shown in this chapter, control charts that you run from the Analyze ▶ Quality and Process menu include a check box option in the window where you set up the analysis. This is not available through the Control Chart Builder; you must choose a specific chart design to be able to add a capability study. Be sure to explore the option for an efficient combination of studies.

Exercises

E3.1—The annual product review (APR) data for a capsule drug product were analyzed for trends over time in chapter 2. Use this set of data with the specifications below to determine the capability for each of the five critical quality attributes (CQAs).

Product Release Specifications for Product Z54AC	
Uniformity of Beads	90% to 110% of label claim, RSD < 5%
Dissolution in Acid	No more than 10% of label claim
Dissolution in Buffer	No less than 75% of label claim in 60 minutes

Product Release Specifications for Product Z54AC	
Content Uniformity (capsules)	per USP <905> (Acceptance Value <= 15)
Assay	90% to 110% of label claim

1. Open the data file *capsule APR data.jmp*.
2. Create a summary table to check capability for the *uniformity of beads* RSD and *content uniformity (capsules)*.
    a. Select *Tables ▶ Summary* to get the coefficient of variability (CV, aka RSD%) for *uniformity of beads*, and the mean and standard deviation of *content uniformity (capsules)*, grouped by *batch*.
    b. Create a new column named *k* with the constant value *2.4* for all rows.
    c. Create a new column named *M* and use the formula for a set of conditional IF statements with Figure 3.27 shown as the example.
        i. For 98.5< mean(content uniformity(capsule))<=101.5, keep the mean(content uniformity(capsule)).
        ii. For mean(content uniformity(capsule))<98.5, assign the value *98.5.*
        iii. All else, assign the value *101.5* .

Figure 3.27: Formula Editor Window

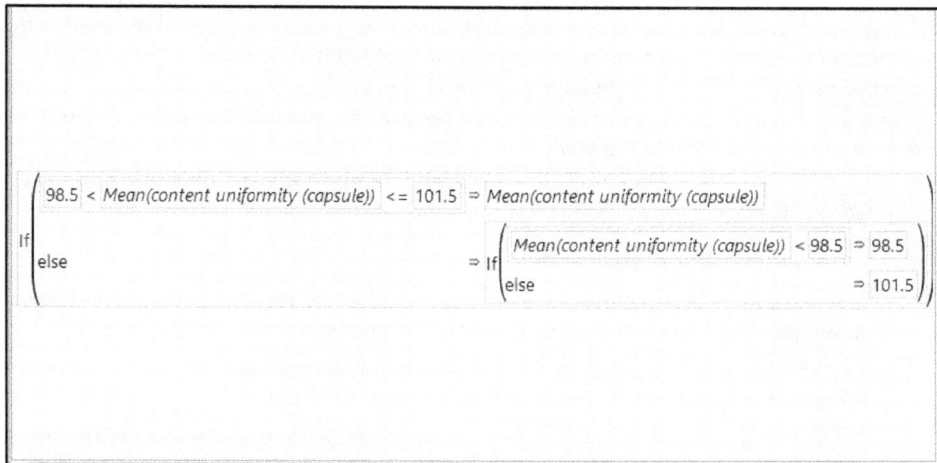

```
    / 98.5 < Mean(content uniformity (capsule))  <= 101.5  ⇒ Mean(content uniformity (capsule))                          \
If |                                                                   / Mean(content uniformity (capsule))  < 98.5  ⇒ 98.5  \ |
   |  else                                                     ⇒ If |                                                          | |
    \                                                                  \ else                                       ⇒ 101.5  / /
```

    d. Create a new column named *AV* and include the following formula:

$$|M - mean\,(content\,\_\,uniformity\,(capsule\,))| + K * Std\,\_\,Dev\,(content\,\_\,uniformity\,(capsule\,))$$

    e. Select *Process ▶ Analyze ▶ Quality and Capability Studies* on the CV(uniformity of beads) and AV variables per the specifications listed in the table at the beginning of this exercise.

3. Select *Analyze ▶ Distributions* for the remaining four CQAs to check for normality and to run capability studies by using the hotbox option for each variable.

   a. The Normal quantile plot illustrates the potential normality.

   b. Use the red triangle menu and select *Continuous Fit ▶ All* to determine the appropriate distribution for the non-normal variables.

   c. Use the red triangle menu option to run *Capability* for each with the appropriate distribution.

4. How would you summarize this information to the group of stakeholders?

E3.2—The surgical tubes data analyzed in the chapter 1 exercises enabled visualization of the units that were outside of internal limits for outer diameter for three sizes of tubes. With capability studies, you can use the distribution of actual values to gain a more precise estimate of the likelihood for tubes to be outside of limits.

1. Open the data file *surgical tubes.jmp*.

2. Select *Analyze ▶ Quality and Process ▶ Capability Studies* to calculate capability for the six tube size variables to the internal specifications noted (3 mm tubes +/- 0.1, 1.5 mm tubes +/-0.05, 5 mm tubes +/-1.5 mm).

3. Do the capability studies provide a different estimate than the OOL proportions in chapter 1?

4. How would you summarize this information to the group of stakeholders?

E3.3—A liquid medication manufacturer is responding to the challenge of ensuring that the dosage cups used are accurate for the medications they produce. Previous cups included a printed scale for the dose level and were found not to be accurate enough for all suggested dosages of the medication. A supplier has developed a new dosage cup with the dosage scale molded into the side of the plastic cup. All calculations were based on the volume of deionized water held at each dose level. with conversions made for the density of the actual product. The supplier just pulled a random sample of 50 dosage cups from an early run of the molding tool and sent them to the analytical laboratory to test the amount of medication delivered for the three dose levels: 2.5 ml, 5 ml, and 10 ml. The standard of accuracy is that the dose must be within +/- 10% of the target dose amounts to be acceptable.

1. Open the *dosage cups.jmp* data file.

2. Use *Analyze>Distribution* to determine whether the results for the three dosage levels meet the assumptions of normality needed for typical, parametric capability studies.

3. Select *Analyze ▶ Quality and Process*, and then run capability studies in either the distributions platform or the capability analysis function.

4. How would you summarize the overall capability to the stakeholders of the new dosage cups project? Is the cup design ready to be approved?

E3.4—The medical device manufacturer has data about the seal strength of surgical kit trays over time. The managers of the product line are interested in how capable the trays are in exceeding the minimum requirement of exposure to 23 inches of mercury. They want to know whether they might be able to change to a thinner foil or plastic laminate seal to save money on costs per unit. The data from the study of random sample testing over time can be used again to estimate the overall capability of the kits.

1. Open the *burst testing.jmp* data file.

2. Use the distribution platform to analyze *burst test results inHg*. Use the red triangle menu options to test capability.

3. Is the product capable of meeting the pressure requirements and exceeding them enough to reduce the thickness of the foil or plastic laminate s?

4. The data over time indicated that there are differences in seal strength possible between the three production shifts. Select *Tables* ▶ *Split* to create a new table of *burst test results inHg* split into three columns by shift.

5. Run capability studies to determine whether a difference exists in overall capability between shifts.

6. How would you summarize this analysis to ensure that the best decisions can be made? Is it appropriate to move forward with the cost savings project? Is there a call to action needed due to the results?

Chapter 4: Using Random Samples to Estimate Results for the Commercial Population of a Process

Overview

JMP is an invaluable tool for building a technical understanding of your work from a set of data. Previous chapters demonstrated the power of data visualization through plots and tables that are easy to create. Even though a great deal of information is available, the interpretations are limited to the trends inherent to the set of sample data used for the analysis.

Information about a sample is interesting, but many projects strive to gain knowledge about all possible results of a process. Inferential techniques extend the interpretation of analyses beyond the sample with statistical tools to provide precise estimates for the population of interest. Knowledge about the population parameters enables decision makers to achieve higher levels of success by taking inferential estimates into account. This chapter is an important link between descriptive statistics and inferential techniques.

The Problems: A Possible Difference between the Current Dissolution Results and the Historical Average

Sudhir is faced with the concern that 45-minute dissolution testing for a tablet formula differs due to a recent change in the source of an active pharmaceutical ingredient (API). The historical average for 45-minute dissolution is 90% for batches made with the previous API source. Results from testing 12 months of product batches, including the newly sourced API, are available. Batches are selected from the order in which they have been made to make the sample. A random number generator is used to ensure that potential bias is mitigated for these batches. Inferential statistics are used to determine whether a significant difference in 45-minute dissolution exists, as compared with the prior historical average.

Steps for a Significance Test for a Single Mean

Random samples of data, representing a population of interest, can be explored with distribution tools in JMP. As noted in previous chapters, summary statistics and data visualization of the distribution of results is a good first step. The summaries that come from a sample are point estimates that offer only a gross approximation of the population parameters. Continuous data can be summarized for both location (median, mean) and spread (range, standard deviation) to describe the distribution. Point estimates of the population parameters are not precise because the estimates vary randomly from sample to sample. Inferential statistics provide for robust estimates of the population since both the sample location and spread are involved. Statistical tools used for inferential estimates are based on a sampling distribution, which is the theoretical distribution of a summary statistic made from many identically sized samples chosen from the same population. There are many statistical models that use a sampling distribution as a model to estimate the parameters of a population. Sampling distribution theory is not covered in this book. However, the book *Practical Data Analysis with JMP* by Robert H. Carver includes an excellent chapter on the subject.

Making estimates with inferential statistics is very easy to do in JMP. Analyses are completed, and results generated in an instant. However, the practical value of the information is directly proportional to the quality of the thought and structure that go into the work. You should think about the appropriate level of precision required for a problem before delving into analyses of data. Statistical techniques that are used to make estimations of a population are based on a specific level of precision, known as the level of significance (α). The context of the problem and the associated consequences of making an incorrect estimate must be considered to ensure that the level of significance is appropriate. You can think of this as the shower valves vs heart valve consideration. The tolerance for making a mistake when dealing with average failure rates of heart valves is extremely small since a mistake can be lethal for a patient. An incorrect estimate made for the average rate of failure of shower valves is not as critical, so a larger level of significance can be used. Keep in mind that as significance levels decrease, more precision is required, and so the sample size must be increased to meaningfully estimate the mean. The default level of significance used by JMP and most other statistical software applications is a tolerance of making an incorrect estimate 5% of the time. A 5% level of significance equates to a 95% level of confidence.

It is important to focus on the statistical principles for making inferences and ensure minimal error. A structured, stepped process helps keep you from making costly errors. The structure preferred by the author to execute the inferential statistics of hypothesis testing is built on five steps (Gabrosek and Stephenson 2016).

First step: Gather the details necessary to define the purpose of the work with regard to the population of interest and the sample selected. As noted previously, the quality management team of a pharmaceutical manufacturing company notices that a number of out-of-specification dissolution results have been occurring in lab testing over the last six months. The team is interested in making estimates for the population of all tablets produced and marketed for the subject product. The team studied tablet product that was made in three separate facilities in order to obtain the 45-minute dissolution values, so the sample collected includes batches from all three. This sampling method ensures that there is an equal chance that any tablet sample comes from one of the three because samples are randomized per facility.

Second step: Put some thought into the guess about the population parameter prior to the analysis. This guess can come from prior historical information or could be a claim from an authority on the subject. The 45-minute dissolution data has a historical average of 90%. If the sample average is close to 90%, you would not expect that a true difference exists beyond random variability. Random variability is always present among the averages of many samples of the same size chosen from the population. No change to your guess about the population parameter is stated as the null hypothesis:

$$H_o: (null\ hypothesis)\ \mu = 0.90$$

A change from the guess about the population can take one of three different scenarios: the population mean is different from the guess; the population mean is less than the guess; or the population mean is greater than the guess. The context of the problem provides you with the information needed to test for the desired outcome. In the case of 45-minute dissolution, the team is interested in the existence of a simple difference. A change to the guess about the population parameter is stated as the alternate hypothesis:

$$H_a: (alternative\ hypothesis)\ \mu \neq 0.90$$

Third step: Check the assumptions that must be met in order to utilize the statistical methods selected for the inferential test and determine the test statistic from the inferential model. Since the team is dealing with dissolution measurements that are continuous random variables, a parametric model is the default option for inferential statistics. Parametric models for making inferences of a single mean involve a normal distribution—a single peak of high frequency with symmetric tailing of values for the upper and lower sides of the peak (a bell-shaped curve). Selecting *Analyze ▶ Distribution* results in a plot and a table of descriptive statistics that provide quick information about the behavior of the sample. A symmetric histogram with a median that is similar to the mean justifies that the distribution is approximately normal. The sample size also provides information regarding the meeting of assumptions; large samples are not as sensitive to the shape of the distribution due to the central limit theorem. Large samples are defined as consisting of approximately more than 30 observations for a sample distribution with a single peak, some skew, and few extreme outliers existing on one side of the curve. Samples with greater than 50 observations allow for meeting assumptions of a parametric test when more skew is present in the distribution.

The test statistic indicates a standardized distance between the sample mean made from the inferential model and the value of the population mean obtained from process history. Sample means chosen from a population vary randomly about the population parameter, but most are located relatively close to the population mean. The bell-shaped sampling distribution model includes the guess of the population parameter at the peak with random variation about the mean, defined as two standard deviations for a 95% confidence level. The frequency of sample means is reduced as the distance increases from the population mean, as shown in Figure 4.1. The test statistic is the standardized distance from the mean expressed in units of standard deviation. The parametric model used for one mean is the t-distribution. A t-distribution can be explained as a standard normal distribution with the spread of the function corrected for sample size. Small sample size inferences ($n \leq 30$) are made from a t-distribution that is spread out more widely than inferences made for large samples ($n > 30$). The t-distribution becomes the standard normal distribution (Z) for large sample sizes.

Figure 4.1: Sample Means

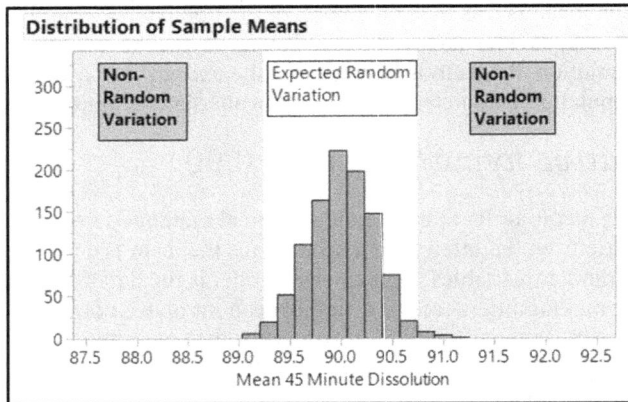

Fourth step: Consider whether the difference between the sample mean and the null mean is statistically significant. A summary value used for the distance is the test statistic. A single mean hypothesis test involves the Student's t distribution as a model. The mean distance of the sample to the null is expressed in t units of standard deviation for the model. As the absolute value for the t-statistic becomes large, the percentage of data that might be contained in the tail or tails becomes small.

Recall that the alternate hypothesis noted in the second step is chosen from the three possible scenarios in order to match the context of the problem. The alternate that involves a simple difference allows for the test mean to be either smaller or larger than the guess for the population parameter; for this reason, the test is two-tailed. If the alternate is testing for a difference of either less than or greater than the guess of the population parameter, the test is considered one-sided. You must choose from the results for all three alternate hypothesis scenarios that are generated by JMP.

The probability value for each test result is the amount of the model distribution that has as much or more difference from the guess for the population mean. The probability is the area of the data that is left in the tail or tails of the population distribution. The practical interpretation of the significance is represesnted by a p-value, which is the chance that the estimated mean could be as distant from the guess for the population parameter due to random variability. As the p-value gets smaller, the chance that the estimated mean comes from the population described by the guess of the population parameter is reduced. When the p-value of the t-test is equal to or less than the level of significance, determined in the first step, the assumption that the guess of the population parameter is true is rejected. Evidence of a significant difference exists between the estimated mean and the guess for the population parameter when the null hypothesis is rejected. JMP color-codes significant results by making the p-value red or orange.

Fifth step: The final step of the inference test is to explain the results in common terms for the stake holders of the project. When the p-value of the t-test is smaller than the significance level, it is said that "significant evidence exists" of a difference between the sample mean and the population parameter. Large probabilities indicate that "no significant evidence exists" of a difference. Differences that are statistically significant might not be of practical relevance. For instance, a sample with minimal variability might show significance with little difference between the sample mean and the null. Significant results should always be interpreted with the help of subject matter experts to determine the practical relevance of the results.

Importing Data and Preparing Tables for Analysis

Sudhir had a random sample collected of 45-minute dissolution results from 10 batches that were made after a change in the source for the active pharmaceutical ingredient (API) was implemented. The source change was implemented to realize significant cost savings in making the product. The procurement team worked to gain quality approval for the change, and testing was completed by analytical research and development to demonstrate chemical similarity. Dissolution data collected from actual batches produced provides the best evidence for practical similarity to or difference from the historical average. Sudhir receives the data as a formatted Excel spreadsheet, and he is concerned about the amount of time it will take to organize the data in a format that can be used by JMP. Sudhir learns of the Excel Import Wizard and decides to give it a try on the data sheet seen in Figure 4.2.

Figure 4.2: Dissolution Data Sheet

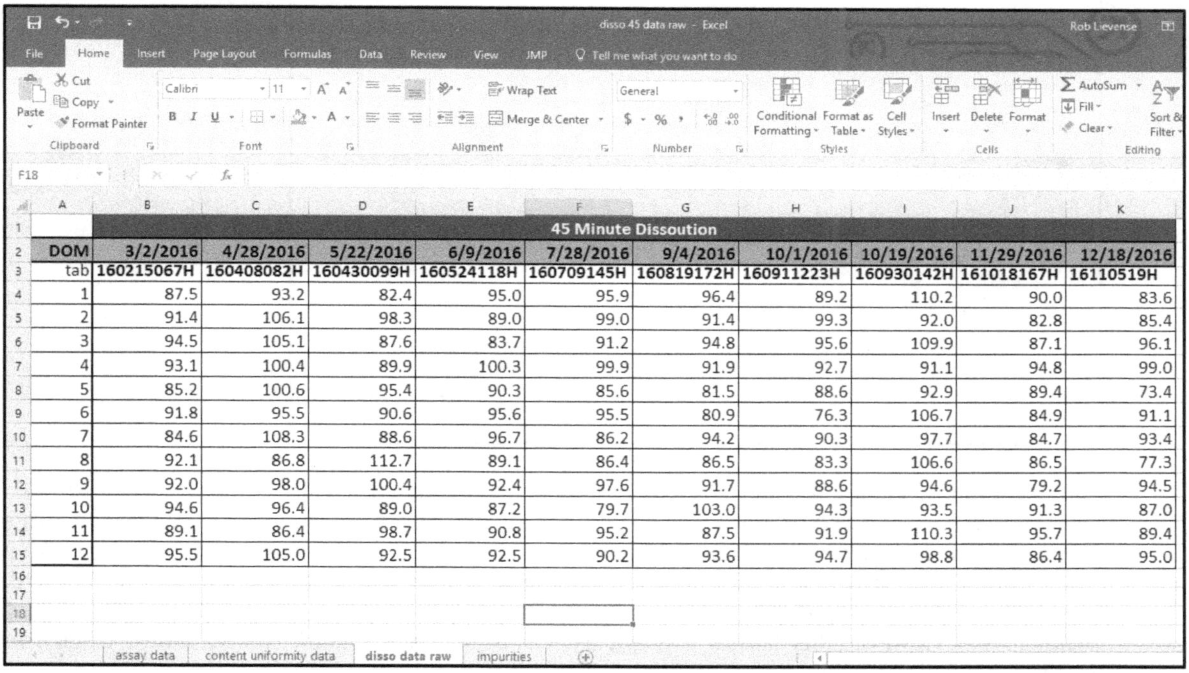

DOM	3/2/2016	4/28/2016	5/22/2016	6/9/2016	7/28/2016	9/4/2016	10/1/2016	10/19/2016	11/29/2016	12/18/2016
tab	160215067H	160408082H	160430099H	160524118H	160709145H	160819172H	160911223H	160930142H	161018167H	16110519H
1	87.5	93.2	82.4	95.0	95.9	96.4	89.2	110.2	90.0	83.6
2	91.4	106.1	98.3	89.0	99.0	91.4	99.3	92.0	82.8	85.4
3	94.5	105.1	87.6	83.7	91.2	94.8	95.6	109.9	87.1	96.1
4	93.1	100.4	89.9	100.3	99.9	91.9	92.7	91.1	94.8	99.0
5	85.2	100.6	95.4	90.3	85.6	81.5	88.6	92.9	89.4	73.4
6	91.8	95.5	90.6	95.6	95.5	80.9	76.3	106.7	84.9	91.1
7	84.6	108.3	88.6	96.7	86.2	94.2	90.3	97.7	84.7	93.4
8	92.1	86.8	112.7	89.1	86.4	86.5	83.3	106.6	86.5	77.3
9	92.0	98.0	100.4	92.4	97.6	91.7	88.6	94.6	79.2	94.5
10	94.6	96.4	89.0	87.2	79.7	103.0	94.3	93.5	91.3	87.0
11	89.1	86.4	98.7	90.8	95.2	87.5	91.9	110.3	95.7	89.4
12	95.5	105.0	92.5	92.5	90.2	93.6	94.7	98.8	86.4	95.0

Select the Excel file type, and click *File ▶ Open* to locate the Excel sheet *disso 45 data raw.xlsx* in the file directory (see Figure 4.3). The default settings are appropriate, so click *Open* to use the *Excel Import Wizard*.

Figure 4.3: Data Open Window

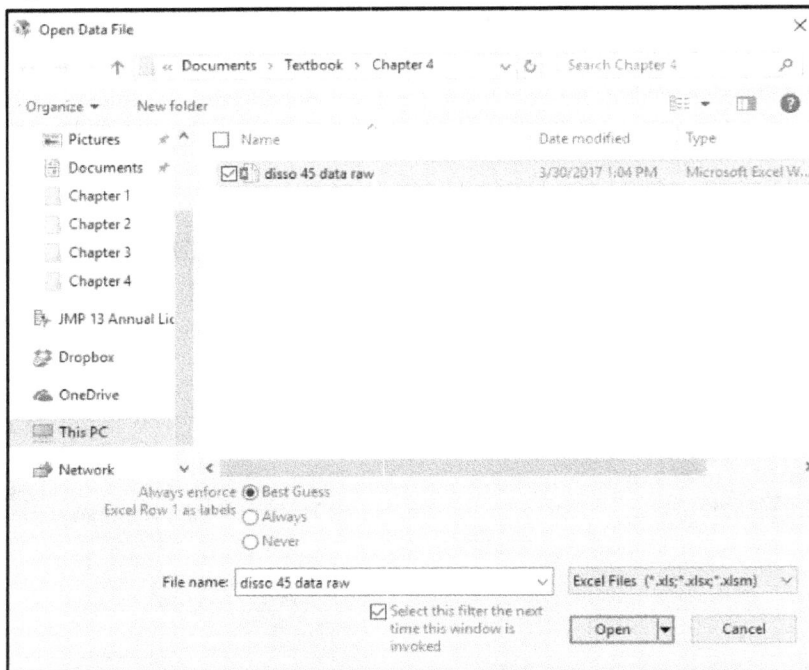

The Excel sheet includes multiple tabs with different types of analytical data for the sample of 10 batches. Select the worksheet *disso data raw* is selected in the *Worksheets* section of the window, as shown in Figure 4.4. The *Data Preview* portion of the window helps you visualize what the data will look like when it is imported into JMP. Modify the options in *Individual Worksheet Settings* until the headers and data rows in the preview are appropriate for the import. More detail is available for additional options by clicking *Next*. However, this data set is simple and can be imported with the choices shown. Click *Import* to create the JMP data table shown in Figure 4.5.

Figure 4.4: Excel Import Wizard Window

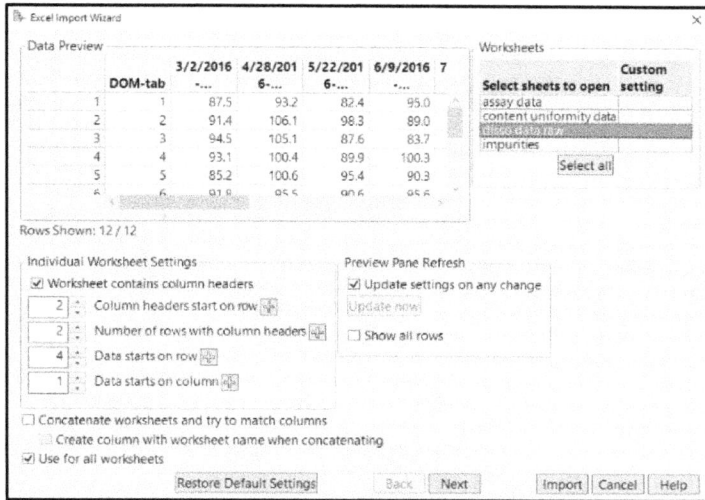

Figure 4.5: JMP Data Table

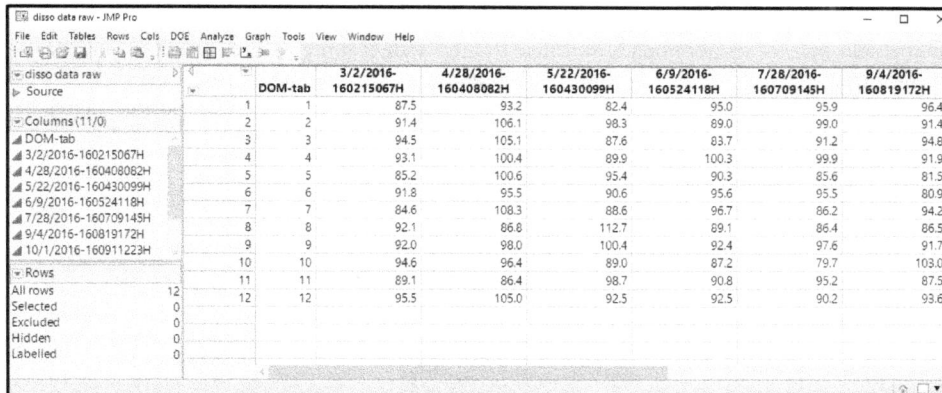

Figure 4.5 that the imported data sheet is in an unstacked format with column names that include the date of manufacturing (DOM) with the batch number separated by a hyphen. The 45-minute dissolution values for each of the 12 individual tablets per batch make up the rows of the raw data table. Dissolution is measured as a batch average summary value. The table can be easily manipulated using the Tables menu to create a table of average 45-minute dissolution for each of the 10 batches. Figure 4.6 illustrates the use of *Tables* ▶ *Summary* to initiate the process.

Figure 4.6: Tables Menu

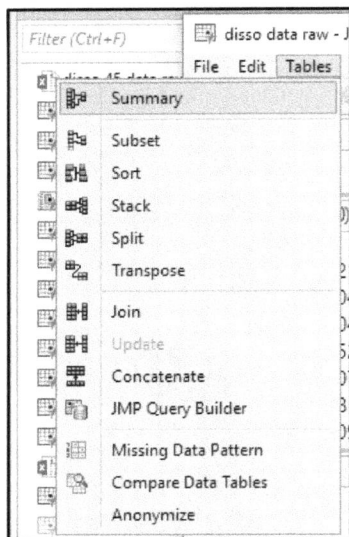

In Figure 4.7, the 10 data columns are highlighted, and *Mean Summary* is selected in the *Statistics* drop-down list. The *DOM-tab* column is not needed because the individual tablet values are not of interest. You can select *Keep dialog open* so that you can redo the summary selections in case the summary table that results from the action is not satisfactory. Click *OK* to obtain the table of average dissolution values by *DOM/batch*, shown in Figure 4.8.

Figure 4.7: Summary Dialog Box

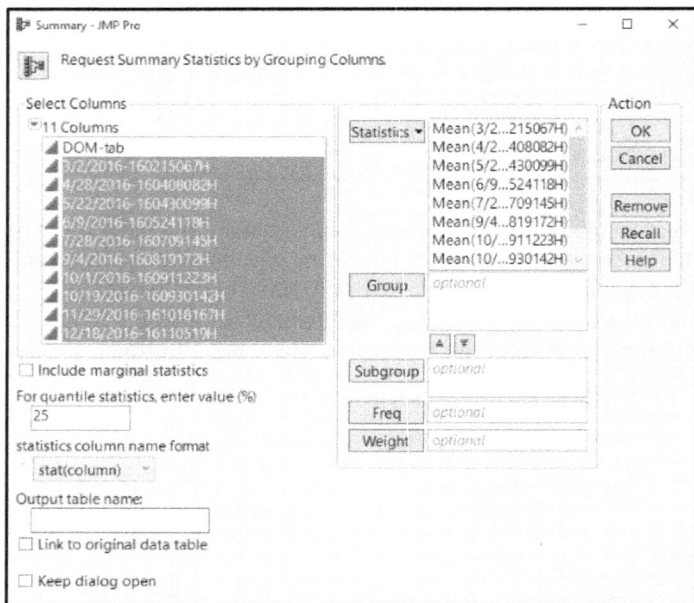

Figure 4.8: Summary Data Table

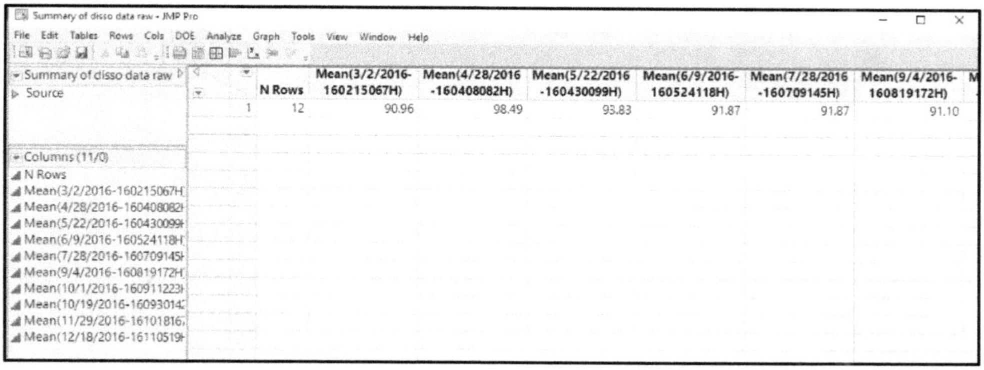

The data table must be rearranged so that average dissolution is a column variable, with batch and date of manufacturing (DOM) as grouping variables for the inferential analysis. Use the transpose function within the tables menu to fine-tune the table format. Figure 4.9 shows that selecting *Tables ▶ Transpose* creates a stacked table with one row of batch dissolution means. All but the *N Rows* variable are chosen in order to make the transpose table shown in Figure 4.10. Click *OK* to get the stacked data table shown in Figure 4.11.

Figure 4.9: Summary Table Menu

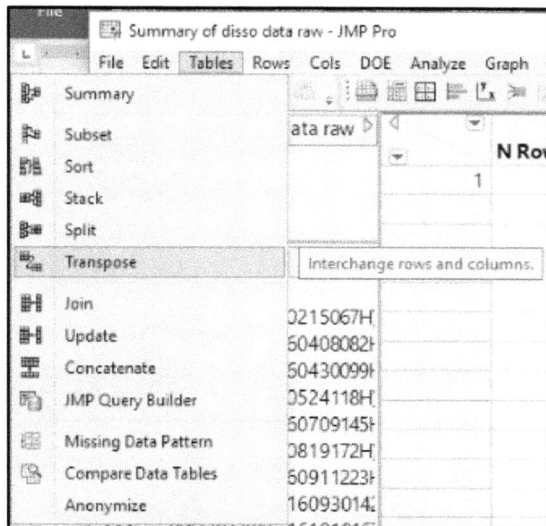

Figure 4.10: Transpose Dialog Box

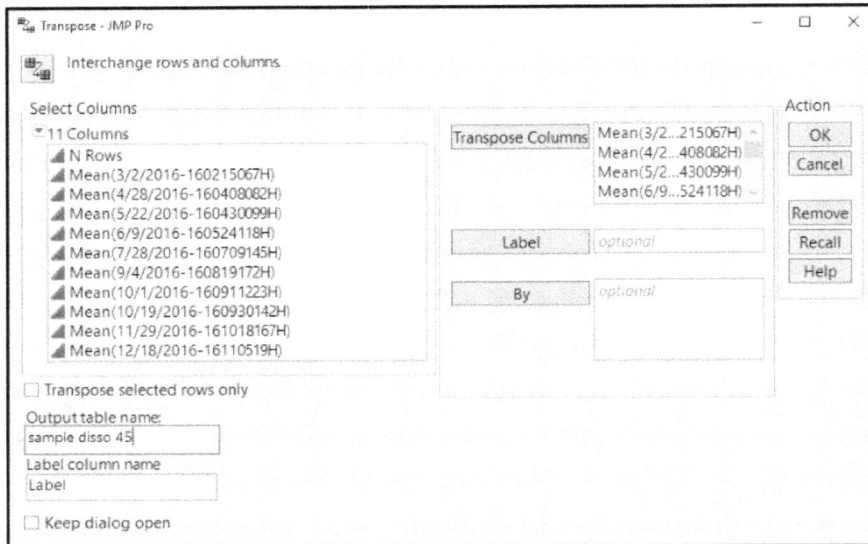

Figure 4.11: Summary Stacked Data Table

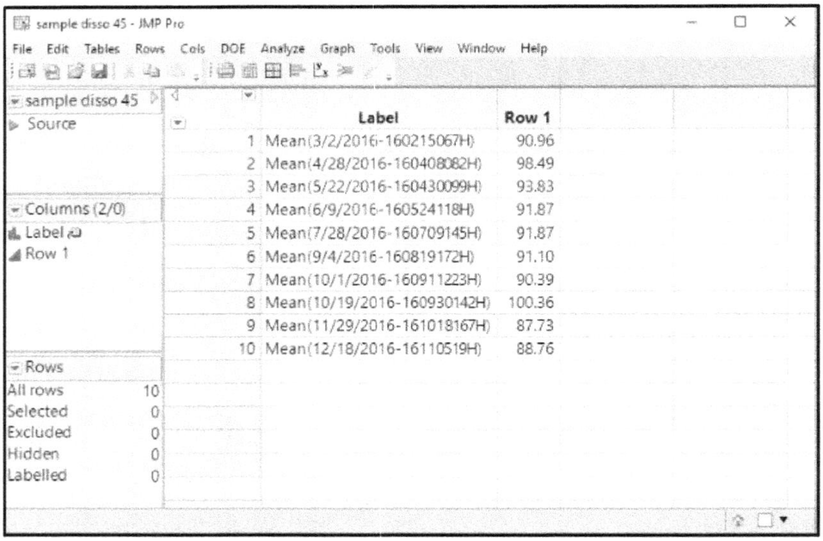

The team could use the summary data table as is. However, the *Label* variable has the previous column names combined with the grouping variable, which is confusing to the team. You use column utilities in JMP to improve a table for clear analysis output. Select *Cols ▶ Utilities ▶ Text to Columns*, as shown in Figure 4.12, to separate the *Label* variable with open and closed parentheses included as delimiters, shown in Figure 4.13.

Figure 4.12: Text to Columns Utility

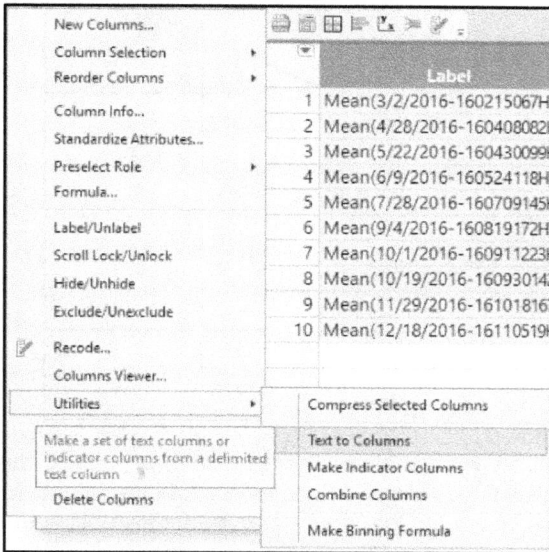

Figure 4.13: Text to Columns Dialog Box

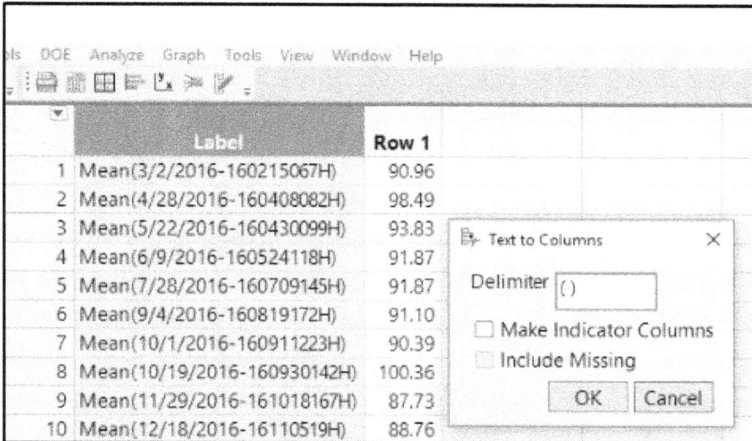

Figure 4.14: Data Table with Separated Columns

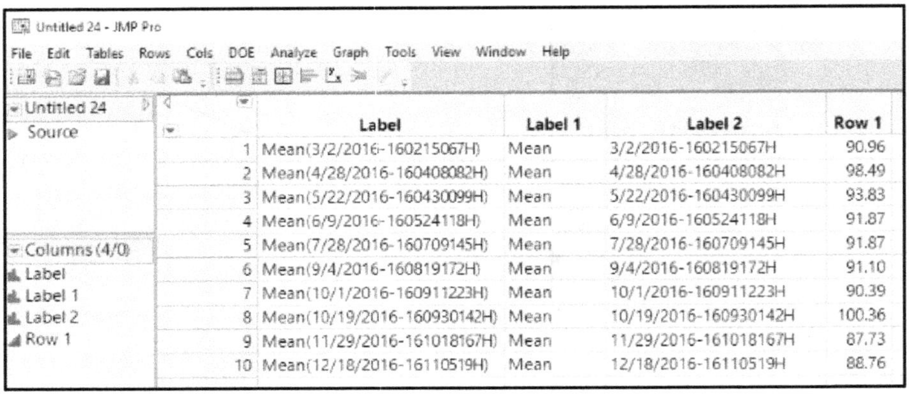

	Label	Label 1	Label 2	Row 1
1	Mean(3/2/2016-160215067H)	Mean	3/2/2016-160215067H	90.96
2	Mean(4/28/2016-160408082H)	Mean	4/28/2016-160408082H	98.49
3	Mean(5/22/2016-160430099H)	Mean	5/22/2016-160430099H	93.83
4	Mean(6/9/2016-160524118H)	Mean	6/9/2016-160524118H	91.87
5	Mean(7/28/2016-160709145H)	Mean	7/28/2016-160709145H	91.87
6	Mean(9/4/2016-160819172H)	Mean	9/4/2016-160819172H	91.10
7	Mean(10/1/2016-160911223H)	Mean	10/1/2016-160911223H	90.39
8	Mean(10/19/2016-160930142H)	Mean	10/19/2016-160930142H	100.36
9	Mean(11/29/2016-161018167H)	Mean	11/29/2016-161018167H	87.73
10	Mean(12/18/2016-16110519H)	Mean	12/18/2016-16110519H	88.76

The summary data table with separated columns is shown in Figure 4.14. However, more work is needed to get the DOM separated from the batch number. Use *Cols ▶ Utilities ▶ Text to Columns* on the *Label 2* variable with a hyphen as the delimiter to obtain the data table shown in Figure 4.15. Add the column names for *DOM*, *Batch*, and *45 minute disso* by specifying values in *Column Properties* for each. Then, delete the surplus variables to get the completed summary table shown in Figure 4.16.

Figure 4.15: Data Table with Second Set of Separated Columns

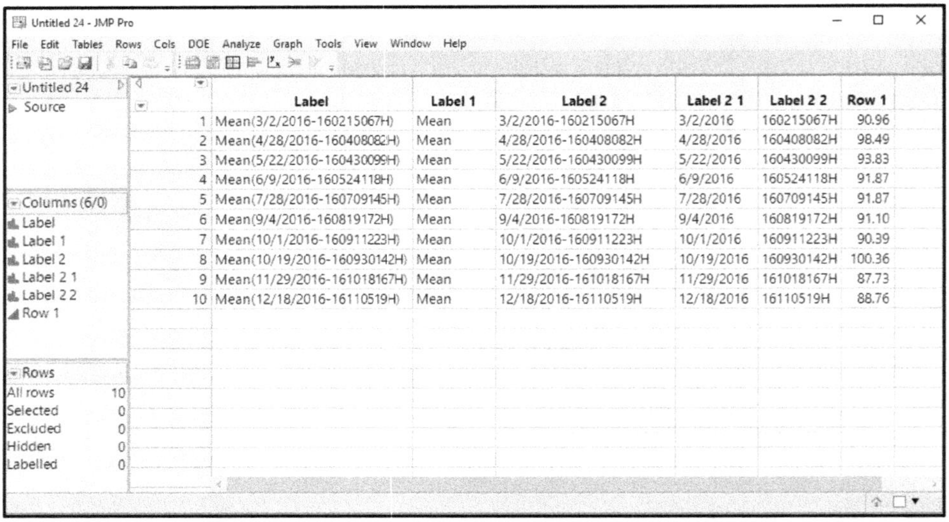

	Label	Label 1	Label 2	Label 2 1	Label 2 2	Row 1
1	Mean(3/2/2016-160215067H)	Mean	3/2/2016-160215067H	3/2/2016	160215067H	90.96
2	Mean(4/28/2016-160408082H)	Mean	4/28/2016-160408082H	4/28/2016	160408082H	98.49
3	Mean(5/22/2016-160430099H)	Mean	5/22/2016-160430099H	5/22/2016	160430099H	93.83
4	Mean(6/9/2016-160524118H)	Mean	6/9/2016-160524118H	6/9/2016	160524118H	91.87
5	Mean(7/28/2016-160709145H)	Mean	7/28/2016-160709145H	7/28/2016	160709145H	91.87
6	Mean(9/4/2016-160819172H)	Mean	9/4/2016-160819172H	9/4/2016	160819172H	91.10
7	Mean(10/1/2016-160911223H)	Mean	10/1/2016-160911223H	10/1/2016	160911223H	90.39
8	Mean(10/19/2016-160930142H)	Mean	10/19/2016-160930142H	10/19/2016	160930142H	100.36
9	Mean(11/29/2016-161018167H)	Mean	11/29/2016-161018167H	11/29/2016	161018167H	87.73
10	Mean(12/18/2016-16110519H)	Mean	12/18/2016-16110519H	12/18/2016	16110519H	88.76

Figure 4.16: Completed Summary Data Table

		DOM	Batch	45 minute disso
	1	3/2/2016	160215067H	90.96
	2	4/28/2016	160408082H	98.49
	3	5/22/2016	160430099H	93.83
	4	6/9/2016	160524118H	91.87
	5	7/28/2016	160709145H	91.87
	6	9/4/2016	160819172H	91.10
	7	10/1/2016	160911223H	90.39
	8	10/19/2016	160930142H	100.36
	9	11/29/2016	161018167H	87.73
	10	12/18/2016	16110519H	88.76

Untitled 24 - JMP Pro

File Edit Tables Rows Cols DOE Analyze Graph Tools View Window He

Untitled 24
Source

Columns (3/0)
DOM
Batch
45 minute disso

Practical Application of a t-test for One Mean

The data is in stacked format for the random sample of 10 batches of dissolution results. The sample mean will be compared to the historical average of 0.90, which is the best guess for the population mean. The test results are used to determine whether there is significant evidence of an average difference. Sudhir will use the five-step structure to maintain focus and mitigate the potential for error.

First step: The population of interest includes all commercial batches of the subject tablet product. The sample is the random selection of 10 batches that have been tested since the change was made to the source of the API.

Second step: The team needs to define the condition of no change, which is the null hypothesis. They expect that the population of results represented by the random sample of 10 batches is the same as the population of historical results from all batches made with the prior source of the API. Recall from the problem description that Sudhir could determine that the population mean for all 45-minute dissolution testing completed with the prior supplier of API is 90%. A change of average results is the alternate hypothesis; basically, they want to know whether the batches made from the new API source have a different average dissolution test value.

Third step: The sample information must be analyzed to determine whether the assumptions are met for the inferential technique that is being used. Select *Analyze ▶ Distribution* and create a distribution for the *45-minute disso*. Use the red triangle menu next to the *45-minute dissolution* header, and select *Normal Quantile Plot* to create the output shown in Figure 4.17.

Figure 4.17: 45-Minute Disso Summary

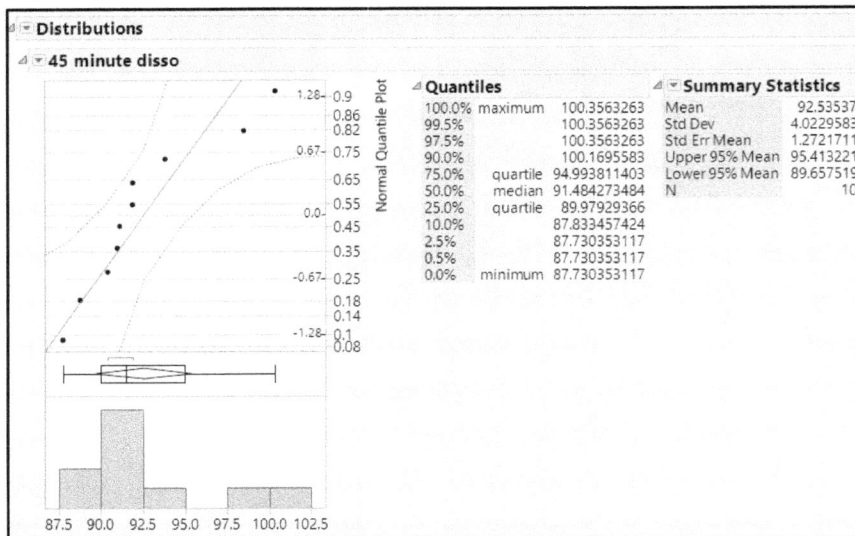

The histogram and outlier box plot show that the 45-minute dissolution averages have a distribution that is slightly skewed toward higher values. The descriptive table includes the median of 91.5, which is close to the mean value of 92.5; recall that a normal distribution has a mean and median that are approximately equal. The normal quantile plot illustrates some deviation of the points from the normal model line, but all points are within the red segmented confidence interval lines. The sample size is small, and the distribution marginally meets the assumption of normality. The team decides to test the mean to get initial results even though the assumptions are marginally met.

The following steps explain how to test the mean.

1. In the *Distributions* analysis, use the red triangle menu options to select *Test Mean*.
2. Enter the guess for the population parameter (90) into the *Specify Hypothesis Mean* field (Figure 4.18).
3. Select the nonparametric test option (*Wilcoxon-Signed Rank Test*) to address the marginal assumptions for the test based on normality of the distribution.
4. Information about the standard deviation of the 45-minute dissolution averages is lacking, so leave that field blank.
5. Click *OK* to get the output.

Figure 4.18: Test Mean Dialog Box

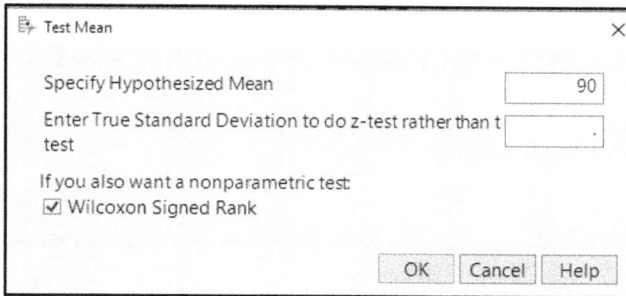

Figure 4.19 lists the parametric test statistic (t=1.99, df=9), which indicates that the estimated mean of 92.54% is two units of standard deviation greater than the 90% population parameter guess. The nonparametric signed-rank test has a test statistic of 16.5. The signed-rank test statistic is more difficult to interpret in simple terms. The test sample is 16.5 units of signed-rank away from where it is expected to be, due to simple random variability.

Figure 4.19: Test Mean Output

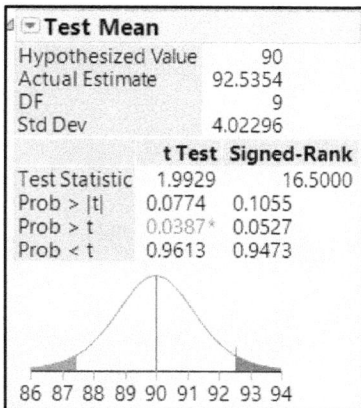

Test Mean		
Hypothesized Value	90	
Actual Estimate	92.5354	
DF	9	
Std Dev	4.02296	
	t Test	Signed-Rank
Test Statistic	1.9929	16.5000
Prob > |t|	0.0774	0.1055
Prob > t	0.0387*	0.0527
Prob < t	0.9613	0.9473

Fourth step: The goal of the subject inferential test is to determine whether the population of test results from batches made with the new API source differs from the population of results collected from batches made with the previous API source. The probability value for a simple difference is Prob > |t|=0.08. Obtaining a sample average result that is as much as 2.5 percentage points distant from the expected population mean (92.5% - 90.0%) can happen 8% of the time when random samples of size 10 are selected. This is not considered significant to the default significance level limit of 5% or less. There is insufficient evidence to reject the null hypothesis of the population mean being 90%. Statistics always involve some amount of error in the conclusions made. You must keep in mind that the possibility that a true difference exists, and that you might not have enough information to conclude the difference to be statistically different.

The nonparametric test results are evaluated due to the slight asymmetry of the distribution small sample size (n=10). The nonparametric signed-rank test provides insufficient evidence (Prob >|t|=0.08) of a difference between the estimated median and the guess of the population parameter. Sudhir could go with either option (parametric or nonparametric) since the conclusions of the test are the same and the assumptions are met. The p-value of 0.08 is not much different from the significance level of 0.05, and the data includes some skew.

Fifth Step: The last major step of hypotheses testing typically involves a practical conclusion made from the significance test results. Before Sudhir draws a conclusion from the inferential testing, he decides to ask the analytical team to randomly choose five more batches of data from their records and repeat the inferential test to hopefully gain a more robust result.

This process continues into the next section.

Using a Script to Easily Repeat an Analysis

The first and second steps for hypothesis testing of 45 minute dissolution data have not changed due to the desire to collect a larger sample. The plan to repeat the analysis on a larger set of data can be quickly carried out by saving the current analysis as a script to the data table. Use the red triangle menu next to the *Distributions* header, and select *Save Script ▶ To Data Table*, as shown in Figure 4.20.

Figure 4.20: Distributions Red Triangle Menu Options

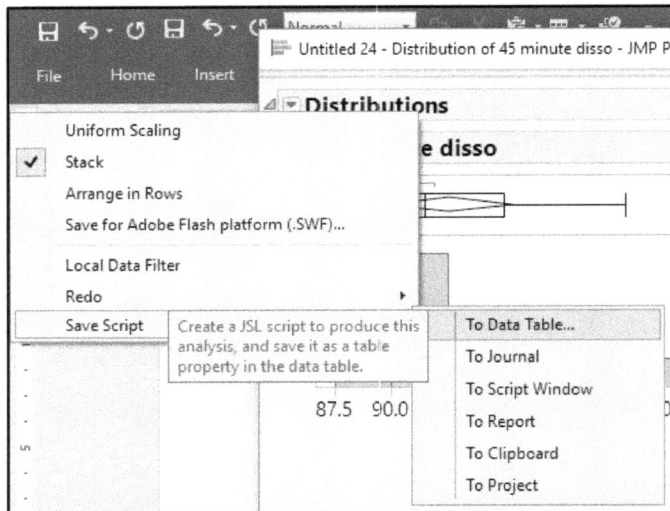

Sudhir adds five new rows of 45-minute dissolution means to the table of data. Open *updated disso summary data.jmp*, shown in Figure 4.21, to view the data set including 15 batches. You can also see the saved script for *Distribution of 45-minute disso* in the data table options list in the upper left of the table view. Click on the green arrow next to *Distribution of 45-minute disso* to execute the script, repeating the mean testing with the updated data table. If you are using a JMP version earlier than JMP 13, you must select the *run* option to execute the script.

Figure 4.21: Updated Data Table

	DOM	Batch	45 minute disso
2	4/28/2016	160408082H	98.5
3	5/22/2016	160430099H	93.8
4	6/9/2016	160524118H	91.9
5	7/28/2016	160709145H	91.9
6	9/4/2016	160819172H	91.1
7	10/1/2016	160911223H	90.4
8	10/19/2016	160930142H	100.4
9	11/29/2016	161018167H	87.7
10	12/18/2016	161105191H	88.8
11	02/25/2016	160205054H	94.6
12	03/19/2016	160227086H	90.6
13	08/16/2016	160728154H	92.7
14	11/04/2016	161001157H	87.3
15	12/04/2016	161030172H	98.3

Figure 4.22: Distributions Output

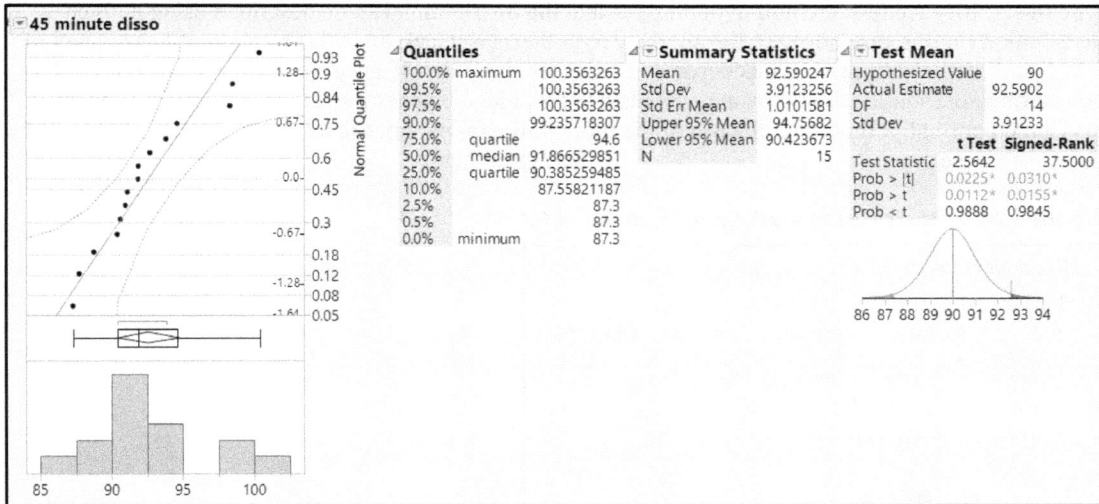

Third step (redone for the larger sample size): JMP runs the inferential test for one mean with the settings that the team specified in the previous steps to automatically generate the output. Notice that the five additional batches helped mitigate the nonsymmetric distribution of the dissolution results. The mean of 92.59% now differs very little from the median of 91.87%, indicating a symmetric distribution. The normal quantile plot also has an improved trend over the 10-batch sample. The parametric test now indicates a test statistic that is greater than the original test (t=2.56, df=14) with more degrees of freedom to add robustness to the test. Nonparametric testing has a signed-rank test statistic of 37.5, which is very different from the initial sample of 10 batches. The added data helps ease the lacking assumptions, but more detail is needed to ensure that the parametric result is appropriate. You can gain more definitive detail by using the red triangle menu next to *45-minute disso* and selecting *Continuous Fit* ▶ *Normal*. The *Fitted Normal* analysis is added to the *Distribution* output. In the red triangle menu to the left of the *Fitted Normal* output, select *Goodness of Fit*, as shown in Figure 4.23.

Figure 4.23: Fitted Normal Analysis

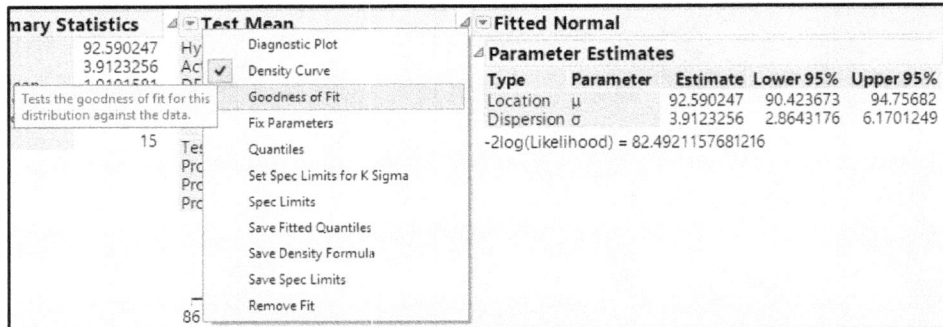

The required detail regarding the assumptions for using the parametric test is shown in Figure 4.24. The significance value of the goodness of fit test (Shapiro-Wilk W=0.924, Prob<W=0.23) leads to the conclusion that insufficient evidence exists that the distribution differs from a normal function. JMP utilizes the Shapiro-Wilk to test normality for samples that have up to 2000 observations. A goodness of fit test is a type of significance test; the null hypothesis is that the distribution can be described using a given distribution (in this case normal), the alternate hypothesis is a lack of fit. Using a significance level of 0.05, it is clear that the data can be considered as distributed normally. If the p-value were to be 0.05 or less, a non-normal distribution would be the conclusion. The added batches meet the assumption of being distributed normally; therefore, the use of the more powerful parametric test is used for the inferential test for a difference.

Figure 4.24: Fitted Normal Analysis/Goodness of Fit

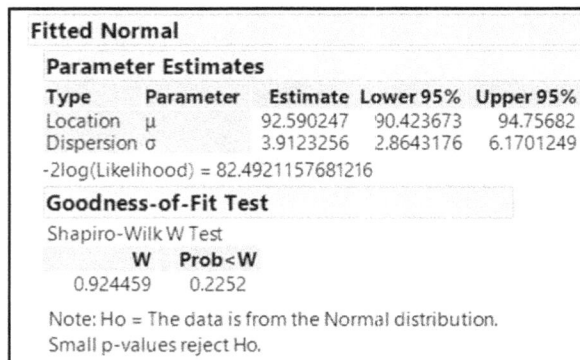

Fitted Normal

Parameter Estimates

Type	Parameter	Estimate	Lower 95%	Upper 95%
Location	μ	92.590247	90.423673	94.75682
Dispersion	σ	3.9123256	2.8643176	6.1701249

-2log(Likelihood) = 82.4921157681216

Goodness-of-Fit Test

Shapiro-Wilk W Test

W	Prob<W
0.924459	0.2252

Note: Ho = The data is from the Normal distribution.
Small p-values reject Ho.

Fourth step (redone for the larger sample size): Results for the sample of 15 randomly selected batches, shown in Figure 4.25, now show a small p-value (Prob > |t|=0.022). It is appropriate to reject the null hypothesis idea that the sample of batches come from the historical distribution of results that have the 90% average.

Figure 4.25: t-Test Output (n=15)

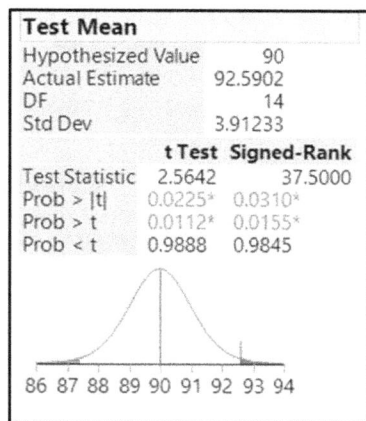

Test Mean

Hypothesized Value	90
Actual Estimate	92.5902
DF	14
Std Dev	3.91233

	t Test	Signed-Rank
Test Statistic	2.5642	37.5000
Prob > |t|	0.0225*	0.0310*
Prob > t	0.0112*	0.0155*
Prob < t	0.9888	0.9845

86 87 88 89 90 91 92 93 94

Fifth step (redone for the larger sample size): The inferential test results have robust, significant evidence that an average difference exists between the population of dissolution results obtained from the batches of the new API source and the previous historical distribution with the mean of 90%. This result is based on a significance level of 5% ($\alpha=0.05$).

Sudhir completed the analysis and summarized the information in a report to the stakeholders of the improvement project. The JMP output, with a p-value much smaller than 0.05, clearly shows significant evidence that the average 45-minute dissolution differs from the previous average of 90%. He knows that this conclusion is robust since the data met the assumptions to be able to use powerful parametric statistics for the inference on the population. The chemical testing indicated equivalence between the API regardless of source. However, differences in analytical results were uncovered through hypothesis testing.

The question "how different is the current average 45-minute dissolution from 90%?" came up during the peer review of the work. A confidence interval provides a practical interpretation of the inference made of the population mean. JMP answered the question before it was asked through the output in the summary statistics table shown in Figure 4.23. The team can be 95% confident that the population mean for 45-minute dissolution is between 90.4% and 94.8%. The lower limit of the interval exceeds the 90% guess for the population parameter and concurs with the significant difference noted previously.

Practical Application of a Hypothesis Test for One Proportion

The management team is impressed by the detailed information provided by Sudhir's analyses. The presentation of the hypothesis test results is clear and understandable because it uses JMP output. Previous concerns that a change in 45-minute dissolution results has occurred since the API source change are now confirmed. The quality manager wants to know whether the increase in dissolution results has had any effect on the level of quality for the tablet product.

The dissolution requirements are historically based on a sample of 24 tablets from a batch, but staged testing is allowed by regulatory authorities (USFDA). Testing the product in stages enables producers to utilize smaller sample sizes while testing to tightened acceptance criteria. Products that have CQAs that perform on target with minimal variability will regularly pass the tightened criteria. Staged testing is popular among drug manufacturers because it saves costs and resources. Sudhir researches the lab information and determines that there were 35 batches that did not meet the stage 1 testing requirements out of 293 batches manufactured with the newly sourced API (12%). The proportion of recent batches in the sample, which did not pass stage 1 testing, is to be compared to prior knowledge of the commercial population. Process records indicate 15% of the batches made from the previously sourced API did not pass stage 1 testing. Inferential statistics for one proportion will enable the team to determine whether the proportional difference is significant.

Create a new table by selecting *File* ▶ *New* ▶ *Data Table* in the JMP menu. The count of *batches* for the out of specifications is entered on row 1. The count of acceptable batches is entered on row 2, noted as *pass*. Add a new column variable named *dissolution test result* to type *OOS* in the cell of row 1. Type *pass* in the cell of row 2. The data sheet with batch counts by dissolution test result categories is shown in Figure 4.26.

Figure 4.26: New Table of Stage 1 Acceptance Data

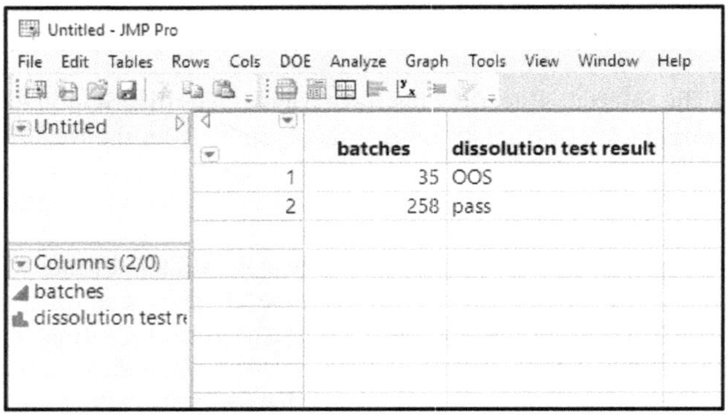

The population the team expects to explain includes all batches of the subject product that were tested for 45-minute dissolution. The sample includes the 293 test results that have been completed for batches made since the source change of the API.

If there is no significant change in the stage one testing, you can expect that the sample came from the population of prior batches that had the OOS rate of 15%. The team is interested in the possible change of a lower stage 1 OOS rate since the average 45-minute dissolution is of a significantly greater mean than the prior historical average. The quality specification for 45-minute dissolution is only a minimum.

Assumptions for the testing of one proportion are simple because they involve checking the sample size with the null proportion as well as with the alternate. The null value of 15% is multiplied by the sample size to determine whether the product is at least 5 ($n*p_o>=5$); the complement of the null value is also multiplied by the sample size to see if the product is at least 5 ($n*[1-p_o]>=5$). The products of 44 and 249 are much greater than 5. Therefore, the assumption of adequate sample size has been met.

In JMP, it is easy to perform inferential testing for proportions. The test is to determine if the proportion of OOS batches tested from product with the new source of API differs from the prior history. Select *Analyze* ▶ *Distribution* to obtain the statistics for the sample of 293 batches that were tested. Specify *distribution test results* in the *Y, Columns* dialog box and *batches* in the *Freq* dialog box shown in Figure 4.27.

Figure 4.27: Sample Statistics for Stage 1 Testing

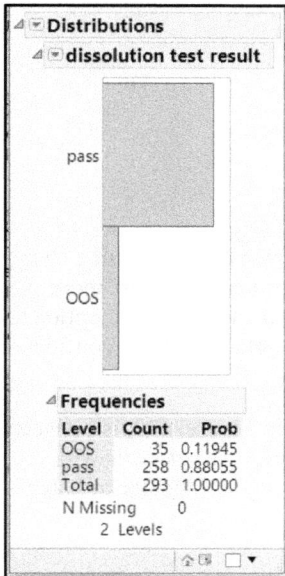

The proportion of nearly 12% stage 1 failures for the sample is less than the 15% failure rate realized in prior history, but is it different enough to be statistically significant?

Use the red triangle menu options next to *dissolution test result*, select *Test Probabilities*. Complete the information in the *Test Probabilities* dialog box by entering *15%* as the hypothesized probability and selecting the option *probability less than value*, as shown in Figure 4.28. Click *Done* to get the results.

Figure 4.28: Test Probabilities Dialog Box

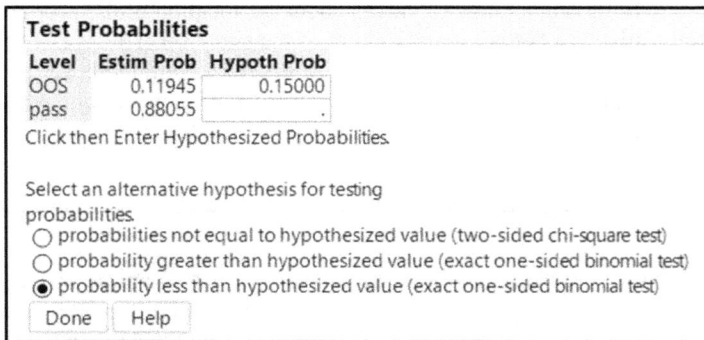

The results shown in Figure 4.29 point out that there is no significant evidence that the stage 1 reject OOS rate has been reduced with a p-value of 0.08. The team cannot reject the hypothesis that the sample comes from the prior history distribution of stage 1 test OOS failures that are 15%.

Figure 4.29: Test Probabilities Results

Test Probabilities		
Level	Estim Prob	Hypoth Prob
OOS	0.11945	0.15000
pass	0.88055	0.85000

	Level	Hypoth	
Binomial Test	Tested	Prob (p1)	p-Value
Ha: Prob(p < p1)	OOS	0.15000	0.0804

There is no significant evidence that the stage 1 OOS rate has been reduced since the API source changed. A conclusion for the potential significant difference of the stage 1 OOS proportion that offers more practical value is a 95% confidence interval for the true proportion. Use the red triangle menu options next to *Stage 1 Dissolution Testing* and select *Confidence Interval* ▶ *0.95* to obtain the 95% confidence intervals about the OOS proportion.

The precise estimate for the true proportion of OOS events at stage 1, shown in Figure 4.30, is between 8.7% and 16.2%, which includes the historical parameter of 15%. Since the prior parameter is within the 95% confidence limits, you cannot say that the change in API source is affecting the number of batches likely to be outside of the stage 1 specifications.

Figure 4.30: 95% Confidence Intervals

Confidence Intervals					
Level	Count	Prob	Lower CI	Upper CI	1-Alpha
OOS	35	0.11945	0.087157	0.1616	0.950
pass	258	0.88055	0.8384	0.912843	0.950
Total	293				
Note: Computed using score confidence intervals.					

It is easy to run hypothesis tests for one proportion in JMP. However, confidence intervals are a more popular option in industry. The practicality of having a range of values about the location of the proportion parameter gives consumers of inferential statistics a better mental picture of the situation. This is due to the restricted domain for proportions between 0 and 1 (0% to 100%), which differs from continuous variables with an unlimited range of values. Hypothesis testing tends to be a more common initial technique when dealing with one mean; confidence intervals tend to be the initial approach for proportions.

Practical Conclusions

The inferential tests have given Sudhir valuable information about the change in the API source and the effects on 45-minute dissolution. The team is able to report to the leadership team the significant evidence that the average 45-minute dissolution differs from the average realized with the previous API source. The evidence indicates that the sample taken after the API change is from a different population than the previous data. Even though the population average differs significantly from the historical parameter, the effect of the change is insignificant with regard to reductions in the proportion of stage 1 test results that are outside of specifications. Sudhir knows that the evidence-based conclusions that he shared regarding the effects of the API source change are robust because the statistics support his claims. JMP has become an invaluable tool for the team to be able to easily determine real changes in the population from random variation that is present among samples. This information provides for better decisions and improved focus of resource allocation, so the team can reduce wasted time chasing trends in random variation and more time getting results by focusing on significant trends.

Exercises

E4.1—A packaging plant has a new piece of equipment on the bottle filling line that is designed to re-torque the twist on caps after each bottle passes under the induction sealer (to weld the foil liner in the cap to the bottle rim). The technical information for the equipment indicates that the settings used during equipment qualification (EQ) resulted in completed bottles with an average removal torque of 8.5 inch pounds. The unit has been running for several weeks, and the packaging team collected data from several days of random samples to determine whether the results have changed since the unit was installed. You have been asked to use the data to compare to the sample to the initial average expected for the population of results identified during EQ.

1. Open the file *retorquer data.jmp*.
2. Use the Distributions platform to visualize removal torque (in-lbs).
3. Create a normal quantile plot to determine whether the data looks normal.
4. Use the red triangle menu options to test the mean against the 8.5 in-lbs hypothesized mean.
5. Is there significant evidence of a change in performance in the equipment over the time period?
6. How would you present your findings to leadership?

E4.2—The annual product review was prepared for the previous year for a capsule drug product. A slight change in processing has taken place and the manufacturing order has been revised. The first 25 batches have been processed under the new revision, and quality leadership wants to confirm that the critical quality attributes have not changed. This is not a random sample, but the team needs to use it to determine whether any significant differences are present.

1. Open the file *capsule APR data.jmp* and run the distributions script to get the summary statistics.
2. Open the file *capsule data new revision.jmp*.
3. Use the Distributions platform to evaluate the normality assumption for *content uniformity (capsule)*.
4. Use the red triangle menu options to test the mean against the hypothesized mean from the summary statistics that are found in the *capsule APR data.jmp*. Is there significant evidence that the content uniformity changed?
5. Run tests on the other CQAs. You need to use the test information for distributions that cannot be considered normal on some of the CQAs.
6. Different is not always bad. Summarize the information for quality in the terms of overall risk for being outside of specifications for the CQAs.

E4.3—Burst testing of sealed surgical trays was analyzed over time in chapter 2. The project stakeholders explain that the design is expected to have a seal strength that will burst on average at 25 inches of mercury. The data is collected from all three shifts of the manufacturing operation because there is interest in testing for a significant difference from the expected population average for each shift's results.

1. Open the file *burst testing.jmp*.
2. Use the Distributions platform and the By box to create a distribution of results for each shift.
3. Use the red triangle menu options to test the mean against the hypothesized mean of 25 inches of mercury.
4. Do any of the shifts produce trays that have significantly less burst strength than what is expected?
5. What will you report to the stakeholders? Be sure to include any suggestions for additional study.

Chapter 5: Working with Two or More Groups of Variables

Overview

Technical professionals regularly deal with problems that involve more than one variable. It is typical for a team to be interested in either a relationship between two measurable entities or in a measurable entity that might come from two groups. The pharmaceutical and medical device industries typically do not lack data. Information about several aspects of the production of products is collected regularly for evidence of quality and compliance. Comparisons of the sample information are useful because they provide a rough estimate of possible relationships. The visualization of the data and related summary statistics offer good sample information, but they do not offer precise estimates for the operational trends of all products produced. The Fit Y by X platform in JMP offers a rich array of options to visualize data and run inferential tests for the robust determination of comparative relationships and trends in the population of all products produced.

The Problems: Comparing Blend Uniformity and Content Uniformity, Average Flow of Medication, and Differences Between No-Drip Medications

Kim is a quality engineer who oversees annual product reviews for a pharmaceutical manufacturer. She is working on a tablet product that has been commercially produced over the last year. The quality controls for the product include analytical tests for both the uniformity of the blend and the content uniformity of the compressed tablets. Regulatory requirements for content uniformity are in place that require evidence that tablets produced throughout the batch are uniform with regard to the amount of active pharmaceutical ingredient (API) contained in each dose. Tablets are collected into a large sample bag from in-process checks occurring every 15 minutes. The tablets from the bag are dumped out onto a large tray in the quality control lab, and 10 random tablet samples are chosen to be individually tested for API content.

Blend uniformity is expected to be a predictive check that represents the content uniformity that can be expected from the batch. Mix samples containing 1 to 3 times the tablet dose by weight are collected with a sample thief from 10 tote locations in three replicates. The 10 blend samples are then tested by the quality control lab. The blend uniformity sampling and testing has been problematic and complicated because it is extremely difficult to obtain good samples from the powder bed of mix while not interfering with later samples.

The team noticed that the results of the content uniformity differ from the blend uniformity results regularly. Kim would like to compare the results to determine just how predictive the blend uniformity results are for the expected content uniformity. Blend uniformity is tested with the assumption that batches lacking a uniform amount of active in the blend will be detected prior to compressing tablets. This ideal is not currently realized as tablets from batches with suspect blend uniformity are regularly compressed at risk. The compression of suspect blends into tablets typically results in content uniformity that is well within specifications.

Hue is working with a development team on a medical device that meters regular dosage events of a thick, liquid medication delivered to the patient through a feeding tube. Multiple doses need to be delivered throughout a 24 hour period, and each is followed by a feeding event. The surgical tubing used is specified by the outside diameter (O.D.), and a couple of different sources of tubes are being considered. Hue is concerned that the flow of the medicine might be altered by the inner diameter of the tubing.

The last problem involves a liquid medication applied by a metered spray device. The product will have a no-drip claim on the label because patients need the sprayed dose to stay on the skin after application to ensure that the full dose is received. Tanya is a Senior Statistician working with the scientific team to analyze data from a new test method that measures the percentage of the dose retained within an inverted test tube after a sprayed dose has been added. The project stakeholders are interested in whether a difference exists between candidate formulas, and between candidate formulas and a regular, existing product that does not have a no-drip claim.

Comparison of Two Quantitative Variables

The quality control team has compiled a set of lab data for several batches, including both the blend uniformity (BU) and content uniformity (CU) results for each batch. Open *B26 API Test Data.jmp* to access the data. Select *Tables* ▶ *Summary* to get the summary statistic of the mean for *B26 API blend uniformity %* and *B26 API tablet content uniformity %* grouped by Lot. Figure 5.1 shows the summary table, which includes 4 columns and 46 rows of data. Save the summary table of results as *B26 API Test Data By (Lot).jmp* before using it for further analysis. Close the *B26 API Test Data.jmp* file to avoid confusion.

Figure 5.1: Summary Table of Test Results

	Lot	N Rows	Mean(B26 API blend ...	Mean(B26 API tablet content ...
1	AF235	10	100.5	99.35
2	AF236	10	101	98.55
3	AF237	10	102.5	100.15
4	AF238	10	103.16666666667	99.72
5	AF275	10	102.16666666667	98.99
6	AF276	10	103	99.55
7	AF277	10	99.833333333333	97.13
8	AF278	10	101.5	99.32
9	AF279	10	104.66666666667	98.20
10	AF280	10	103.83333333333	100.91
11	AF314	10	103.83333333333	99.86
12	AF315	10	105.16666666667	101.54
13	AF316	10	103.5	101.63
14	AF317	10	102.16666666667	100.88
15	AF318	10	101.66666666667	100.12
16	AF347	10	100	100.59
17	AF348	10	101.16666666667	100.82
18	AF349	10	104.16666666667	100.79
19	AF350	10	103.66666666667	99.28
20	AF351	10	101.16666666667	99.18
21	AF352	10	101.33333333333	100.04
22	AF353	10	104.5	99.08
23	AF378	10	104.16666666667	100.62
24	AF379	10	103.5	102.31
25	AF380	10	105.5	100.08
26	AF381	10	101.83333333333	101.56

Panel labels (left side of window): B26 API Test Data By (Lot) — JMP Pro; File Edit Tables Rows Cols DOE Analyze Graph Tools View Window Help; B26 API Test Data By (Lot); Source; Columns (4/0): Lot, N Rows, Mean(B26 API blend uniform, Mean(B26 API tablet content; Rows: All rows 46, Selected 0, Excluded 0, Hidden 0, Labelled 0

Be sure that the summary table *B26 API Test Data By (Lot).jmp* is open, and select *Analyze ▶ Fit Y by X*. Move *Mean(B26 API blend uniformity %)* into the *X, Factor* box and *Mean(B26 API tablet content uniformity%)* into to the *Y, Response* box. Click *OK* to launch the analysis. Figure 5.2 shows a simple scatter plot of the location of each lot for the BU and CU results.

Figure 5.2: Scatter Plot of BU and CU

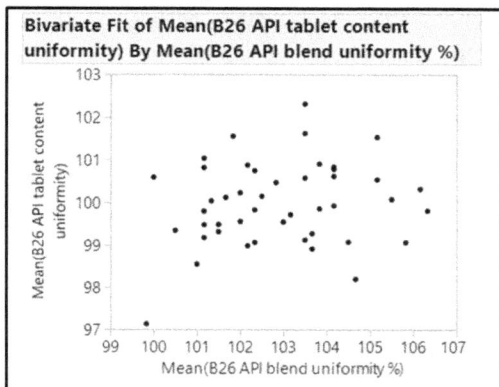

The scatter plot is not showing much of a trend between the two continuous variables. You would expect that increases in BU would be related to increases in CU. An example of this lack of trend are the three points that have CU values just under 102 that are in a horizontal orientation across the plot. The BU results range between 101.5 to 105.5. However, you would expect that the three would be clustered closely about a single value of BU.

One argument for the lack of a relationship between BU and CU is that the test results for each variable have some kind of unnatural skew in the distribution of results that interferes with the trend of a relationship that is expected. You can add histograms to the axis to visualize the distributions of BU and CU and assess this possibility. Use the red triangle menu to the left of *Bivariate Fit of Mean* to create histogram borders on the plot, as shown in Figure 5.3.

Figure 5.3: Scatter Plot of BU and CU with Histogram Borders

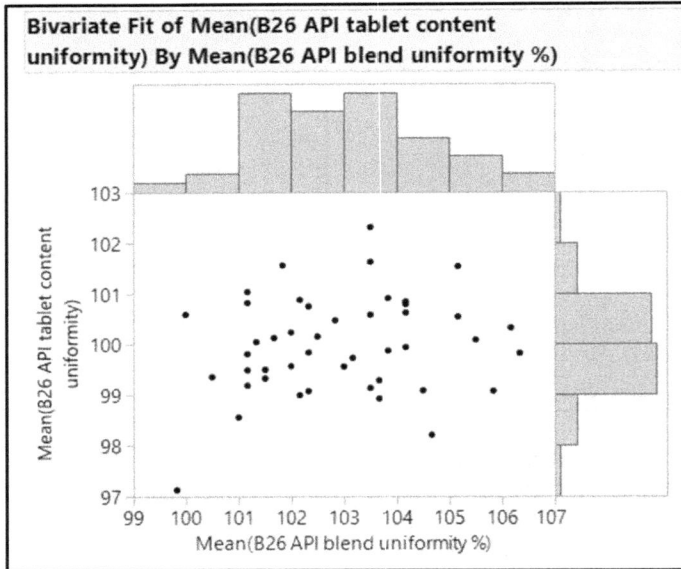

The distributions of the variables have a single, centered peak with symmetrical reductions in frequency as distance from the mean increases on either end of the distribution. The shape of the distributions is not likely influencing the analysis results since they are symmetric. Futher analysis can be completed without concern for error due to distribution shape. Use the red triangle menu again to deselect *Histogram Borders* and select *Fit Line* to obtain the linear regression analysis, shown in Figure 5.4.

Figure 5.4: Linear Regression Analysis

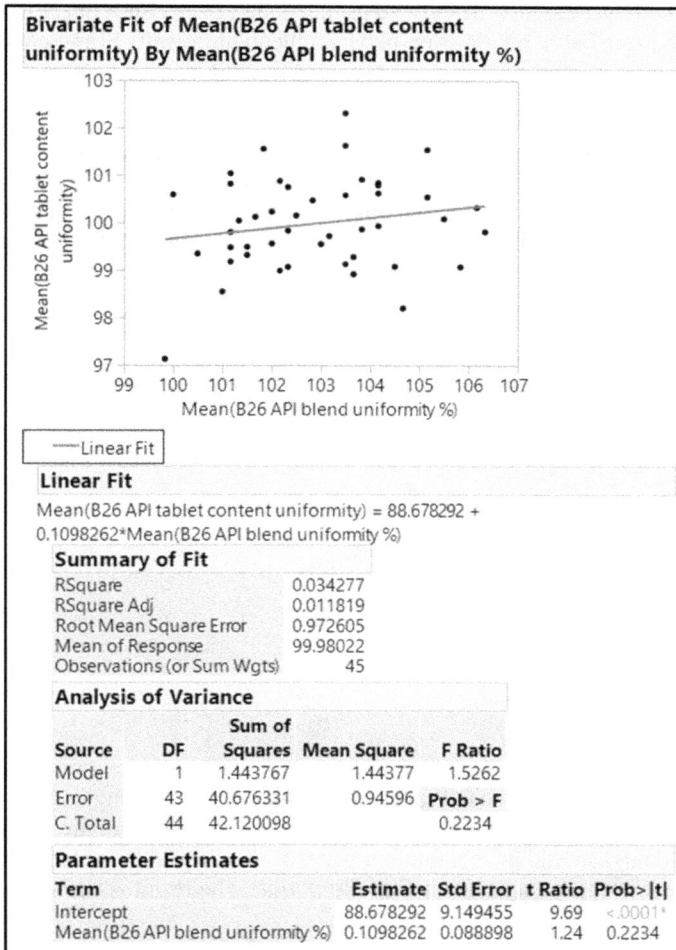

Bivariate Fit of Mean(B26 API tablet content uniformity) By Mean(B26 API blend uniformity %)

— Linear Fit

Linear Fit

Mean(B26 API tablet content uniformity) = 88.678292 + 0.1098262*Mean(B26 API blend uniformity %)

Summary of Fit

RSquare	0.034277
RSquare Adj	0.011819
Root Mean Square Error	0.972605
Mean of Response	99.98022
Observations (or Sum Wgts)	45

Analysis of Variance

Source	DF	Sum of Squares	Mean Square	F Ratio
Model	1	1.443767	1.44377	1.5262
Error	43	40.676331	0.94596	**Prob > F**
C. Total	44	42.120098		0.2234

Parameter Estimates

Term	Estimate	Std Error	t Ratio	Prob>|t|
Intercept	88.678292	9.149455	9.69	<.0001*
Mean(B26 API blend uniformity %)	0.1098262	0.088898	1.24	0.2234

There is no obvious pattern of fit that can be seen in the scatter plot, which now includes the sum of squares linear model as the bold red line. The *Summary of Fit* table shows the Rsquare fit 0.0343, which means that the linear model explains only 3.4% of the variability in content uniformity. Basically, the blend uniformity values cannot be used to predict the content uniformity results and offer little value as an upstream check of the content quality of the tablets produced.

The information indicates that the strength of the linear model for the 45 batches studied is very poor, providing doubts about the predictive value of blend uniformity%. Inferential techniques are included in the output, enabling you to determine how the linear model might work to estimate the population of values for the commercially produced batches of the tablet product. The five-step guideline for inferential testing can help to organize your thoughts, possibly preventing embarrassing mistakes.

First step: The population of interest is all batches of the tablet product that will be produced in commercial production. The sample for the test includes analytical testing data for 45 batches that have been randomly selected from the annual product report that was created for the first year of commercial production.

Second step: The null hypothesis is that there is no relationship between blend uniformity% values and content uniformity% values. The null hypothesis can also be defined as the slope of zero for a linear model. The alternate hypothesis is that there is a quantifiable relationship between blend uniformity% and content uniformity%. A slope that is greater than zero is expected since the relationship will likely be increasing content uniformity% as blend uniformity% increases.

Third step: The assumptions of a linear model have a more complex form than the simple distribution checks that you completed for means. The linear model is created about the cloud of points. Calculations find the line representing the least squares distance between actual observations and a model line. The least squares line minimizes the average distance (in the Y axis) between all points and the line. In the scatter plot, a roughly equal number of points are located above and below the red model line. The vertical distance of a given point from the model line is defined as a residual and is illustrated in Figure 5.5.

residual=actual observed value – linear model estimate (at the same x value)
$$residual = (y - \hat{y})$$

Figure 5.5: Residual Illustration

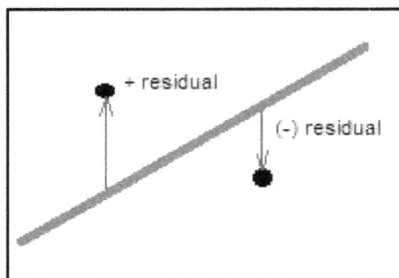

The behavior of the residuals provides the information needed to test the assumptions for using the inferential test. The model line is expected to split the cloud of observation points equally, and the relative distances of the points from the line (residuals) should have a random pattern. Use the red triangle menu located to the left of *Linear Fit* under the scatter plot, and select *Plot Residuals.* The result is shown in Figure 5.6.

Figure 5.6: Linear Regression Diagnostics

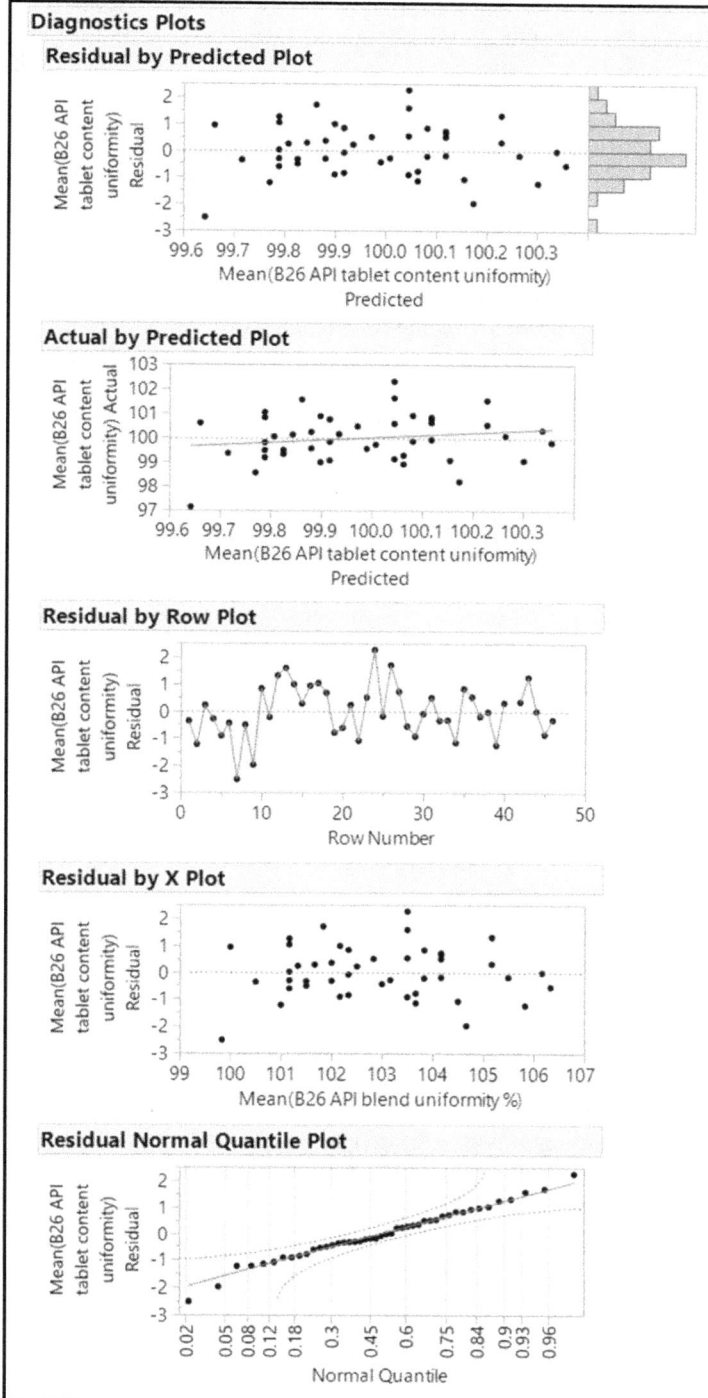

The Residual by Row Plot and the Residual by X Plot both have a mean of zero, indicating an equal sum of positive and negative residuals. The Residual by X Plot illustrates a random pattern of residuals about the span of predicted values with a distribution that is of a normal shape. An important aspect of this plot is that the width of the residuals across the range for the mean is constant; if a cone shape were present, it would indicate changes in variance. Constant variance mitigates error in predictions made from a linear model. The Residual Normal Quantile Plot confirms that the residuals fit the assumption of normality extremely well because points are mostly along the red normal model line and are well within the parabolic confidence interval illustrated by the red segmented lines in an hourglass shape. If there were questions regarding the assumption of normally distributed residuals, the Residual by Row Plot and Residual by X Plot can help identify the observations that are causing the problem.

Since the assumptions for the inferential tests have been met robustly through residual analysis, the test statistic for two techniques can be identified. Figure 5.4 includes an Analysis of Variance table and a Parameter Estimates table to interpret the inferential test results. The ANOVA test statistic of F=1.53 and the slope parameter estimate of 0.110 are the values of interest from the output. The intercept of the linear model has no practical value because it is used to test against the null value of 0 for the y-intercept. Because you cannot ever obtain a blend uniformity% (X axis) value of zero, there is no need to test for the significance of the intercept.

Fourth step: The significance value from the ANOVA table and the Parameter Estimate table slope give the same result of Prob>F=0.2234. There is insufficient evidence to reject the null hypothesis (no relationship between the variables) due to a p-value that exceeds the default significance level of 0.05. The minimal slope of 0.011 seen in the linear fit analysis is close to zero and can occur due to random variation of sampling data from the population more than 22% of the time.

Fifth step: There is no significant evidence of a relationship between blend uniformity% and content uniformity% for the tablet product. The conclusions from the inferential test of the linear model confirm what is seen in the scatterplot. Blend uniformity values offer very poor, unreliable predictions for content uniformity due to the insignificant relationship between the two.

Comparison of Two Independent Means

The example project for this section is the thick liquid medication that must flow through surgical tubing. A set of data has been compiled by the project leader. Hue needs to analyze the results for evidence of a difference in the flow of the liquid medicine due to two unique sizes of tubing.

The tubing manufacturers maintain tight controls on the O.D., but the inside diameter (I.D.) is known to vary. A comparative test involving inferential statistics is very useful in this case in order to determine whether there is evidence of a significant difference in the average inner diameters between the sources. Hue needs to obtain robust information regarding possible differences in I.D. to determine whether the device needs to include the capability to adjust flow based on the tube source used to ensure the consistency in the delivery of the medication. A data file was created from a random sample of 3 mm O.D. tubes grouped by source. JMP makes it very easy to analyze for differences between the means of two groups by using Fit Y by X with a discrete group variable as X and a continuous variable as Y.

Open *surgical tubing.jmp* and select *Analyze ▶ Fit Y by X*. Move *tube ID* to the *Y,Response* box and *source* to the *X, Factor* box. Click *OK* to get the output shown in Figure 5.7.

Figure 5.7: Initial Oneway Analysis

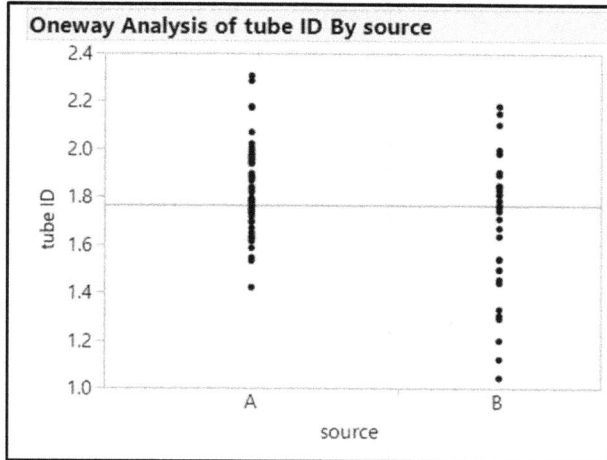

A dot plot of tube ID by source group is the initial output. The horizontal line in the plot represents the overall mean regardless of group. The spread of the dots shows the amount of variability in I.D. measurements within each group. It is difficult to tell by the plot alone if the amount of overlap in points indicates that a difference is likely for the population of tubes produced. An inferential test comparing the two group means can provide the clarity needed to determine whether the difference is significant. The five-step process for hypothesis testing helps keep the analysis organized and on track.

The population of interest includes all 3 mm OD tubes available from both sources. The sample is the random collection of 48 tube sections from source A and 33 tube sections from source B.

The null hypothesis is that no difference exists between the mean ID parameters of the two sources. The alternate hypothesis is that the mean ID parameters that differ.

The information needed for the hypothesis test is obtained by using the hotbox options. The detail is conveniently added to the basic output of the Fit Y by X platform. Use the red triangle menu next to the *Oneway Analysis* header and select *Quantiles, Means and Std Dev,* and *t-Test* for the inferential testing.

Figure 5.8: Expanded Oneway Analysis

Oneway Analysis of tube ID By source

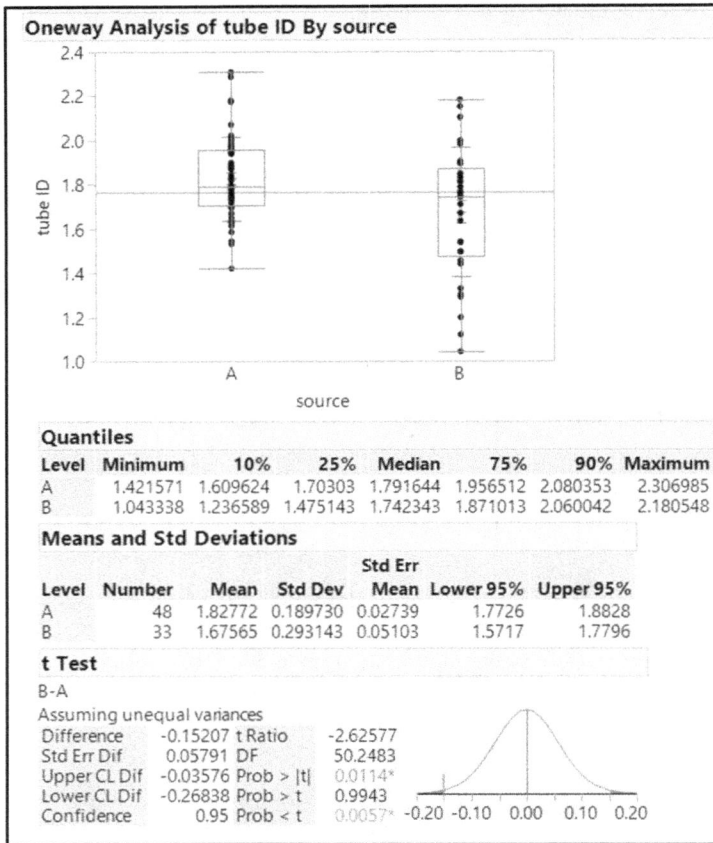

Quantiles

Level	Minimum	10%	25%	Median	75%	90%	Maximum
A	1.421571	1.609624	1.70303	1.791644	1.956512	2.080353	2.306985
B	1.043338	1.236589	1.475143	1.742343	1.871013	2.060042	2.180548

Means and Std Deviations

Level	Number	Mean	Std Dev	Std Err Mean	Lower 95%	Upper 95%
A	48	1.82772	0.189730	0.02739	1.7726	1.8828
B	33	1.67565	0.293143	0.05103	1.5717	1.7796

t Test

B-A

Assuming unequal variances

Difference	-0.15207	t Ratio	-2.62577
Std Err Dif	0.05791	DF	50.2483
Upper CL Dif	-0.03576	Prob > |t|	0.0114*
Lower CL Dif	-0.26838	Prob > t	0.9943
Confidence	0.95	Prob < t	0.0057*

The assumptions for a test of two independent means includes independent groups as well as sample data that follows a normal distribution, adequate sample size, or both. Hue's team can clearly define the independence between the sources because the tube sections differ in color. The red box plots in Figure 5.8, which provide added detail about the shape of distributions, were added to the Oneway Analysis Plot by the *Quantiles* option. The output in Figure 5.8 shows the symmetry of the box plots as well as medians that are equivalent to the means for each group. The sample sizes of more than 30 are more than adequate for the distributions in I.D. that are generally normal.

The t-test table of the output includes the test statistics for the difference in means of $t = (-2.63)$. The t-test provides results that are not dependent on the assumptions that the variances in I.D. results are equal for the two groups. The red triangle menu option *Means/ANOVA/Pooled t* is selected to obtain the results. However, the need to provide evidence of equal variance creates the potential for error. It is always best practice to start with the t-test that allows for unequal variance in order to mitigate the potential for error.

The t-test table indicates that an average difference of (-0.15) exists between the I.D.s of the samples, with a standard error for the difference of 0.058. The amount of difference is statistically significant (Prob > |t|=0.0114). In other words, Hue can expect the amount of difference only 1.1% of the time due to random sampling alone when the null hypothesis is true. The null hypothesis of no difference between the means should be rejected with such a low probability. If you place your pointer over the subject significance value and move it back and forth slightly, additional detail for the interpretation of the significance value appears, as shown in Figure 5.9.

Figure 5.9: Pop-Up Help

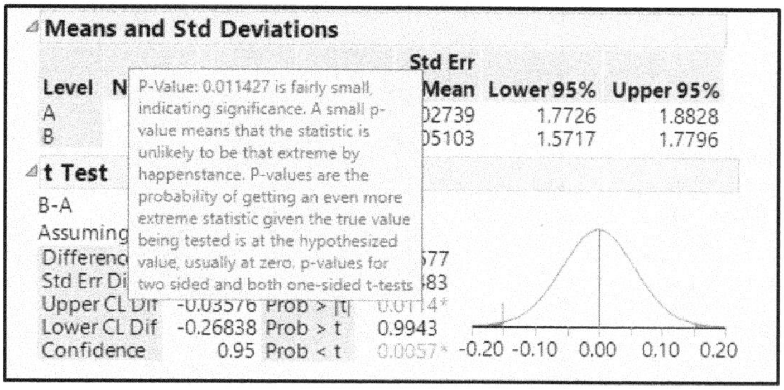

There is significant evidence of an average difference between the I.D.s for each of the tube sources. The information from the comparative testing of I.D.s indicates a result that is of statistical significance, but the team is unsure that the average difference will create a practical difference in the flow of medicine that is delivered by the device. The practical difference is the range of values included in the 95% confidence intervals for the difference in means included in the results.

The output for the inferential test comparing the two means, shown in Figure 5.8, includes a 95% confidence interval for the difference. The team can expect that the tubes from source B will be between 0.04 mm and 0.27 mm smaller in I.D. than the source A tubes. Hue shares the range of expected differences for the population with the engineering team. They use the expected difference interval values to calculate the resulting flow rate differences. The flow rate difference for tubes that differ by as much as 0.27 mm in I.D. is not enough to change the dose delivered by enough to be practically meaningful. In short, the consistency of tube I.D. within the sources is much better that expected, and there is no need to make flow adjustments based on the source of the tubes.

Unequal Variance Test

The t-test compares average results of two independent groups to determine whether significant evidence of a difference exists. Another aspect of difference that could be relevant to project stakeholders involves the variation within each group. If the variation within each group is significantly different, a conclusion on differences in the means might be prone to error. You can add a test for unequal variances to the output by using the red triangle menu next to *Oneway Analysis of tube ID By source* and selecting *Unequal Variances*. The output is shown in Figure 5.10.

Figure 5.10: Unequal Variances Analysis

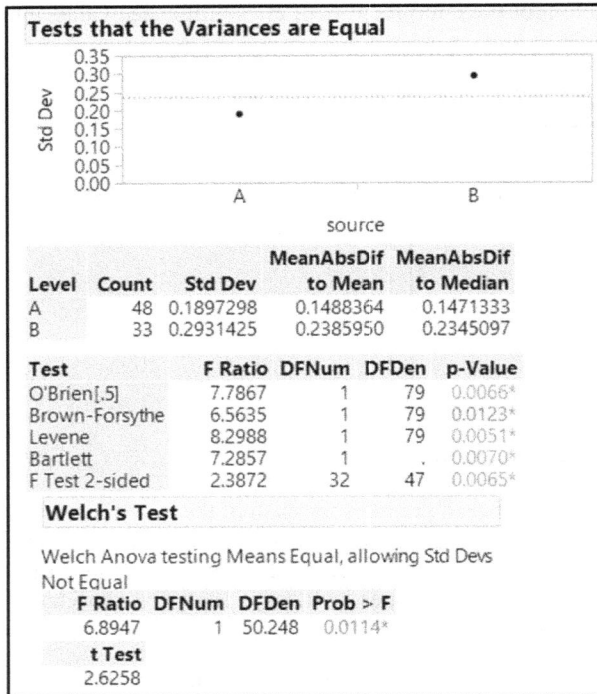

Tests that the Variances are Equal

Level	Count	Std Dev	MeanAbsDif to Mean	MeanAbsDif to Median
A	48	0.1897298	0.1488364	0.1471333
B	33	0.2931425	0.2385950	0.2345097

Test	F Ratio	DFNum	DFDen	p-Value
O'Brien[.5]	7.7867	1	79	0.0066*
Brown-Forsythe	6.5635	1	79	0.0123*
Levene	8.2988	1	79	0.0051*
Bartlett	7.2857	1	.	0.0070*
F Test 2-sided	2.3872	32	47	0.0065*

Welch's Test

Welch Anova testing Means Equal, allowing Std Devs Not Equal

F Ratio	DFNum	DFDen	Prob > F
6.8947	1	50.248	0.0114*

t Test

2.6258

The population and sample have been defined previously for the t-test and do not be repeated here.

The null hypothesis for an unequal variance test is that the standard deviations for the groups are equal. The alternate hypothesis is that the standard deviations differ.

The unequal variance test includes the same assumptions as the t-test for means. It has been previously established that the data are distributed normally, are of adequate sample size, and the groups are known to be independent. The Tests that Variances are Equal plot provides a comparative view of the standard deviation of each group compared with the pooled standard deviation for all of the data. The results of several tests are provided in the summary table below the plot. To see detailed information about the output, press the Shift and ? keys on your keyboard to change the pointer to a question mark. Then position the pointer over one of the tests, and left-click to see detailed information from the JMP documentation. This quick reference is available in JMP for the output of all analyses.

The Levene test works well for the example, which indicates a test statistic of F=8.2988 with 1,79 degrees of freedom for the respective numerator and denominator. The numerator is the number of group comparisons; the denominator is the number of comparisons of individual observations after group comparisons are subtracted. The Levene test has a small significance value (p=0.0051); therefore, the null hypothesis is rejected. There is evidence of a significant difference in variances in the inner diameters of tubes for source A and B.

The Welch's test adjusts for the unequal variance and indicates evidence (p=0.0114) of a significant difference between the inner diameters of tubes for source A and B. The conclusion of the significance test for a difference between means does not change since the unequal variance did not change the amount of evidence that resulted from the test. This is not always the case and running a test for unequal variance is best practice for analysts to mitigate the potential for statistical error.

Matched Pairs Tests

The report leads to additional work by the engineering team to identify whether the differences in tube sources result in relevant flow differences. The team used a target medicine formula for the flow calculations, but there are formulas with varying physical properties that are used with the device. The differences in physical properties of the medicine will likely change the flow characteristics of the liquid moving through the tubes. A very effective way to test for an average difference in flow is to test a sample from each tube source with each formulation and determine the differences in flow. The team obtained 35 samples that represent the population of liquid medicines dosed by the device. A tube from source A that has an I.D. of 1.83 mm and a tube from source B with an I.D. of 1.68 mm are selected. The tubes are cleaned with a flush of water before the trials to remove all residue from the flow trials. The order in which the tubes are used is random to mitigate the potential for error due to order of treatment.

The following steps use data from flow testing to create a distribution of differences in flow between the tube sources for various liquid medications.

1. Open the data set *flow testing of medicines.jmp* to visualize the average difference in flow that exists between the tubes for the various medicines.
2. Select *Cols ▶ New Columns* to add a column to the table. Type the variable name *difference* in the *Column Name* field.
3. Click on the *Column Properties* box to access the Formula Editor.
4. Set up the formula to get the difference between *tube A flow* and *tube B flow*, as shown in Figure 5.11.

Figure 5.11: Formula Editor

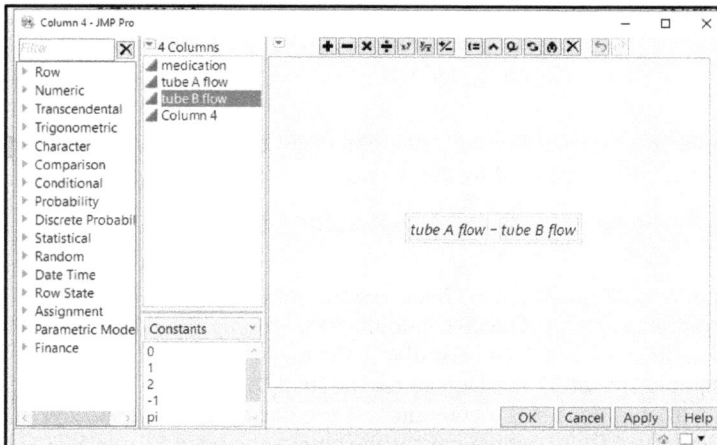

The *difference* column provides information about the difference in flow between tube sources for each of the 35 formulations. Select *Analyze* ▶ *Distributions* and the *Stack* red triangle menu option to obtain the distribution of the differences, as shown in Figure 5.12.

Figure 5.12: Formula Editor

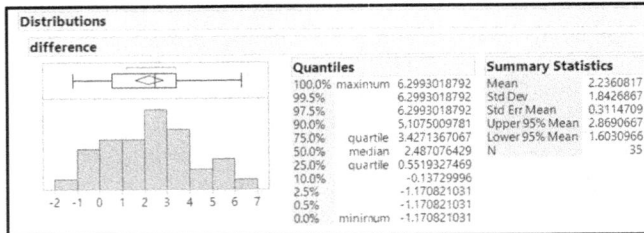

The visualization of sample differences shows that the distribution is basically normal since the median of 2.5 is very similar to the mean of 2.2 and the histogram is symmetrical. The team can be 95% confident that the expected average difference for the population is between 1.6 and 2.9. The extremes include a flow for tube A that is 6.3 units greater than tube B, as well as a tube A flow rate that is 1.2 units less than tube B. Using the distributions platform to visualize the sample before inferential testing is started is always good practice since you can use it to test assumptions.

The flow in each tube is likely to be dependent on the medicinal formula due to changes in physical properties. Inferential testing for an average difference of two dependent groups is known as a Matched Pairs test. The groups are the two sources of tubes. Working through the five steps of a hypothesis test keeps the team organized to ensure that error is minimized.

The population is all medications that will be metered through the device using both tube sources. The sample includes 35 random samples of the different medicinal formulas.

The null hypothesis is that zero average difference in flow exists between tube sources for the population of medicines that will be metered by the device.

$$H_o : mean \_ diff \ (M_d) = 0$$

The alternate hypothesis is that an average difference other than zero exists (either negative or positive) for the population of medicines that will be metered by the device.

$$H_a : mean \_ diff \ (M_d) \neq 0$$

The team can use the information in Figure 5.12 to check assumptions, which are the same as they are for a test for a single mean, covered in chapter 4. The distribution of the sample should be normal or have a large sample size. The median difference of 2.49 is similar to the mean of 2.24, the histogram and box plot display symmetry, and the sample size of 35 is adequate for the shape of the distribution that is close to normal with no outliers. With assumptions met, obtain the test statistic by selecting *Analyze* ▶ *Specialized Modeling* ▶ *Matched Pairs*, which creates the output shown in Figure 5.13. The test statistic of t = (-7.18) is obtained for the sample average difference of (-2.2361) and standard error for the difference of 0.3115 for the flow of tube A subtracted from tube B.

Figure 5.13: Formula Editor

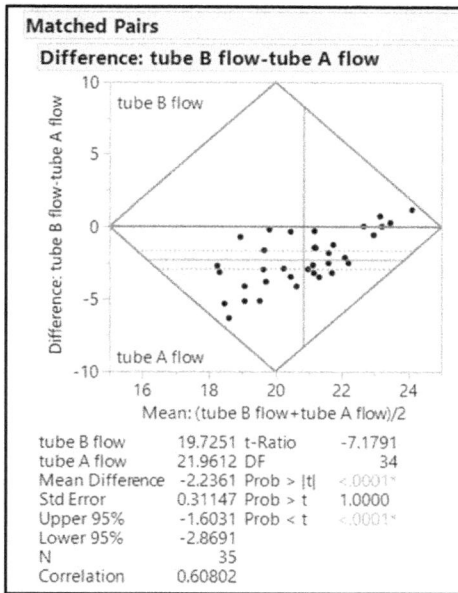

```
Matched Pairs
Difference: tube B flow-tube A flow

tube B flow        19.7251   t-Ratio      -7.1791
tube A flow        21.9612   DF                34
Mean Difference   -2.2361    Prob > |t|   <.0001*
Std Error          0.31147   Prob > t     1.0000
Upper 95%         -1.6031    Prob < t     <.0001*
Lower 95%         -2.8691
N                      35
Correlation       0.60802
```

The value of (Prob > |t| <0.0001) for an average difference existing between the tubes for the population of all medicinal formulas is highly significant. Therefore, the null value of no difference can be rejected.

There is significant evidence of an average difference in flow for all medicines that will be used by the metering device. The practical interpretation of the difference can be completed using the 95% confidence interval for the difference. The entire range of differences is negative indicating that source A tubes have a larger inner diameter than source B tubes. It is reasonable to expect that tubes from source B have an inner diameter that is between 1.6 mm and 2.9 mm smaller than source A tubes.

More Than Two Groups

Situations arise regularly involving measurable data that contains more than two groups. An example is the manufacture of a pharmaceutical product that involves outcome measurements that come from any of four units of processing equipment. Tanya is working with the scientific team responsible for the no-drip drug formulation project involving multiple groups. Stakeholders of such projects typically want to determine whether a real difference exists in outcomes between the groups. If differences exist, the amount of difference needs to be known. The example explored in this section includes data from a liquid product that is applied to a patient's skin with a metered spray. The product label includes a no-drip claim since the dose is designed to cling to the skin surface. The team has data from multiple candidate formulas and is interested in running a statistical test to determine whether a difference exists.

The first inclination might be to run several comparisons between two groups and compile the information. The problem is that the level of significance used to detect evidence of a difference applies to one comparison. The default of 0.05 was used in the previous examples as the level of significance. When more than two groups are involved, each group involves more than one comparison. The 0.05 level of significance can suggest evidence of significance that is in error for multiple groups because it is too high for multiple comparisons. The problem of detecting significance in error for several comparisons is called multiplicity. A more appropriate technique for comparing more than two groups is analysis of variance (ANOVA).

ANOVA is a popular technique. This example uses the simplest type of the technique known as oneway ANOVA. Oneway refers to one type of grouping variable, which is product candidate in the example. If the team were interested in the product candidate and the shift of production, the technique would be a twoway ANOVA. ANOVA compares the average amount of variability present within each group as a result of differing individual observations to the variability between the group averages. Evidence of significance builds as the variability between groups exceeds the average variability within the groups. The Help menu in JMP includes information about the topic; just search for oneway analysis.

The analysis for evidence of a significant difference in percentage retained between the candidate drug groups starts by opening *liquid no drip testing.jmp* shown in Figure 5.14.

Figure 5.14: Data Table for No Drip Testing

	test product	% retained
1	ND1	84.3
2	ND1	74.5
3	ND1	86.9
4	ND1	86.4
5	ND1	89.1
6	ND1	96.1
7	ND1	91.3
8	ND1	78.7
9	ND1	82.0
10	ND1	81.4
11	ND2	98.0
12	ND2	101.6
13	ND2	89.7
14	ND2	99.0
15	ND2	98.3
16	ND2	101.0
17	ND2	103.6
18	ND2	87.9
19	ND2	90.7
20	ND2	95.1
21	ND3	88.2
22	ND3	95.8
23	ND3	93.1
24	ND3	95.0

liquid no drip testing - JMP Pro

File Edit Tables Rows Cols DOE Analyze Graph Tools View

liquid no drip te...
Source

Columns (2/0)
test product
% retained

Rows
All rows	40
Selected	0
Excluded	10
Hidden	10
Labelled	0

Notice that the data table excludes the observations for the group D1. The Rows panel at the lower left of the data table indicates that 10 of 40 rows have been excluded and hidden from the analysis. Make sure that the data table has the exclusion present before continuing.

It is advisable to perform high-level exploration of the data by selecting *Analyze ▶ Distribution*, including the *test product* grouping variable and the *% retained* data. However, this discussion does not include that exploration for brevity's sake. The data is in a stacked table format, which is suitable for running the oneway ANOVA. Select *Analyze ▶ Fit Y by X* to open the dialog box shown in Figure 5.15. Another quick way to launch *Fit Y by X* is to click on the icon under the main menu that includes a Y- and X-labeled axis.

Figure 5.15: Fit Y by X Dialog Box

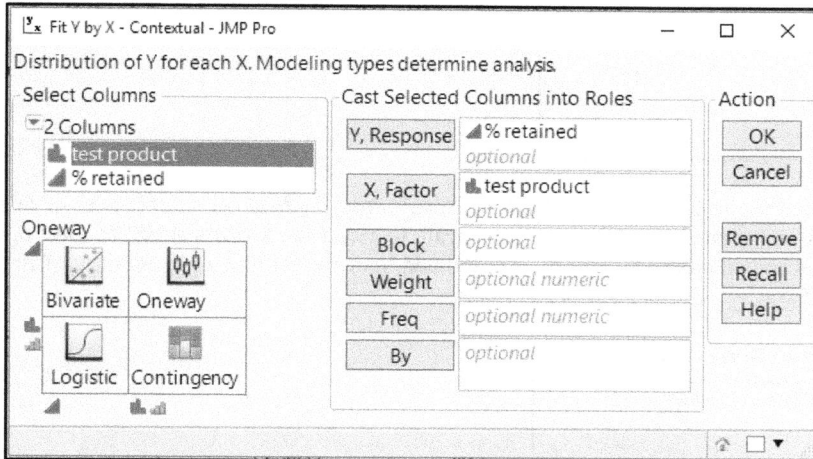

A quick guide is located in the lower left of the dialog box to indicate the type of analysis that will be launched based on the column types of the *X, Factor* and the *Y,Response*. Note the blue ramp indicator of a continuous modeling type shown as a response in the upper half of the vertical axis of the guide. The red bars indicator of a nominal modeling type is shown in the right half of the horizontal axis as the factor. The upper right cell of the guide indicates that a oneway analysis is used. Move *% retained* to the *Y,Response* box and *test product* to the *X, Factor* box, and click *OK* to get the oneway dot plot shown in Figure 5.16.

Figure 5.16: Oneway Analysis

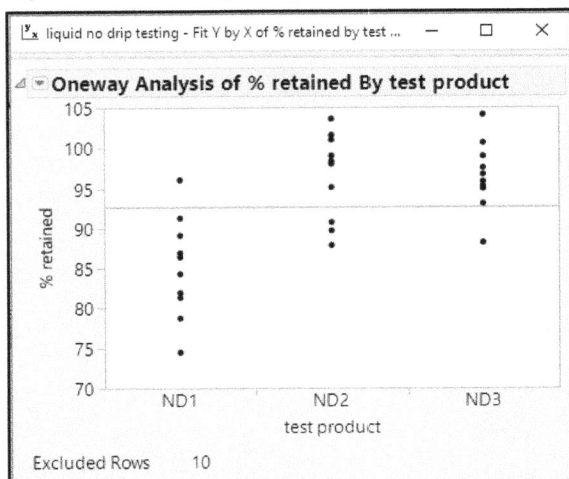

Use the red triangle menu next to the *Oneway Analysis of % retained By test product* header to select the *Means/Anova* option. Use the same red triangle menu again to select the *Means and Std Dev* option to get a table of group means and standard deviations. An updated oneway dot plot and summary of fit results are shown in Figure 5.17.

Figure 5.17: Oneway Plot and Summary of Fit Table

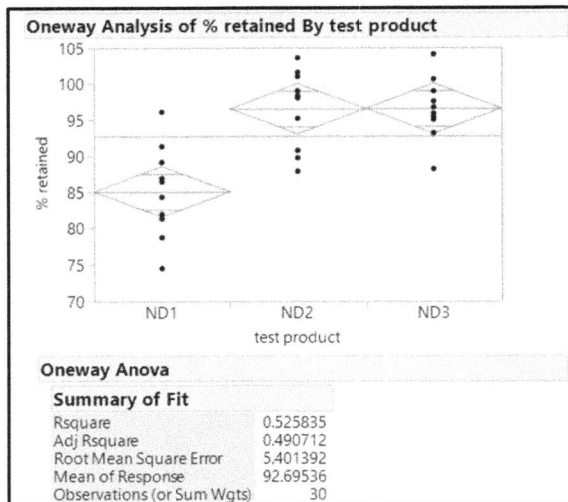

A set of diamonds is added to the dot plot to represent the group mean as the middle horizontal line, with 95% confidence interval horizontal lines at the vertical edges of the diamond. Overlap of the confidence intervals for the groups is a quick visual indicator of a lack of difference between the group averages. The *Summary of Fit* results indicate how well the groups explain the variation in the *% retained* output; an Rsquare of 0.53 is a 53% explanation of variation. The adjusted Rsquare is used to compare models of differing numbers of groups on the *% retained* output, which is not needed in the example. The average of

% retained regardless of groups is 92.7, the amount of variability is the root mean square error of 5.4, and 30 observations are included in the analysis. The table of group means and standard deviations is shown in Figure 5.18.

Figure 5.18: Oneway Plot and Summary of Fit Table

Means and Std Deviations

Level	Number	Mean	Std Dev	Std Err Mean	Lower 95%	Upper 95%
ND1	10	85.0640	6.30475	1.9937	80.554	89.57
ND2	10	96.4943	5.41193	1.7114	92.623	100.37
ND3	10	96.5277	4.29957	1.3596	93.452	99.60

The five-step method for hypothesis testing applies for oneway ANOVA. The population of interest for the example is all commercially produced batches for the three candidate formulas. The sample includes ten product units that have been tested from each of three candidate batches, which totals 30 observations. The null hypothesis is that the averages of % retained for the three candidate formulas are equal. The alternate hypothesis is that at least one of the candidate batches has an average % retained that differs from the others. The most important assumption for ANOVA is that the group variances do not differ. A quick check of the equal variance assumption reveals that no standard deviation for any group is more than two times more than the others. The means and standard deviations table includes the smallest standard deviation for ND3 of 4.3. The largest standard deviation of 6.3 for the ND1 group is less than two times the standard deviation for ND3, so the assumption is not violated. The test statistic and evidence of significance are shown in Figure 5.19.

Figure 5.19: ANOVA Table and Means Table

Analysis of Variance

Source	DF	Sum of Squares	Mean Square	F Ratio	Prob > F
test product	2	873.5656	436.783	14.9711	<.0001*
Error	27	787.7260	29.175		
C. Total	29	1661.2916			

Means for Oneway Anova

Level	Number	Mean	Std Error	Lower 95%	Upper 95%
ND1	10	85.0640	1.7081	81.559	88.57
ND2	10	96.4943	1.7081	92.990	100.00
ND3	10	96.5277	1.7081	93.023	100.03

Std Error uses a pooled estimate of error variance

The first line of the Analysis of Variance table includes details on the variance between the groups. The degrees of freedom value is calculated by the number of comparisons minus 1, which is 2 for the three groups compared. The sum of squared differences of the two comparisons between group means is 873.6, and the mean square is 436.8. The second line of the table includes detail for the within groups variability, which is also known as the model error. The total degrees of freedom is the sample size minus 1, which is 29. The within term of the analysis is the difference between the total degrees of freedom and the degrees of freedom for between group comparisons, which is 27. The within sum of squares is 787.73 and the mean square for within variability is 29.2. The test statistic for the hypothesis test is the F ratio, which is calculated from the between mean square divided by the within mean square, which is 14.97. The evidence of significance increases as the F ratio increases. However, the F distribution is dependent on the between and within degrees of freedom. The Prob>F value is the probability that the difference between sample means of the groups is extreme as it is, and you can expect the population to have equal means, which is <0.0001. The result indicates that there is highly significant evidence that at least one of the group means differs from the other group means. The means and standard deviations table indicates that ND1 has less average % retained than the other two groups. Keep the oneway analysis open for the next analysis.

The data table includes a candidate formula that is not expected to meet a no-drip label claim. Tanya expects that the average percent retained of the three no-drip formulas differ from the regular formula. The next analyses involve running ANOVA on four groups to investigate differences. Position the pointer over the lower row options triangle shown in Figure 5.20, and right-click to get the options.

Figure 5.20: Quick Access to Row Options

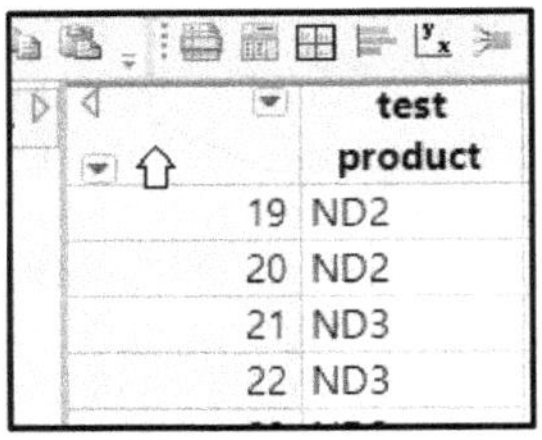

Select the *Clear Row States* option to clear away the excluded rows of the D1 group, as shown in Figure 5.21.

Figure 5.21: Available Row Options

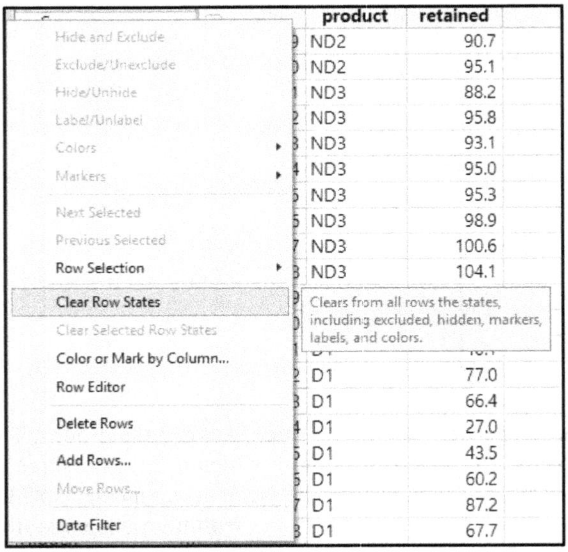

The full data table is used to repeat ANOVA to determine whether evidence of a significant difference exists between the group means. Open the oneway analysis output and use the red triangle menu next to *Oneway Analysis of % retained By test product* to open the analysis options shown in Figure 5.22.

Figure 5.22: Analysis Options

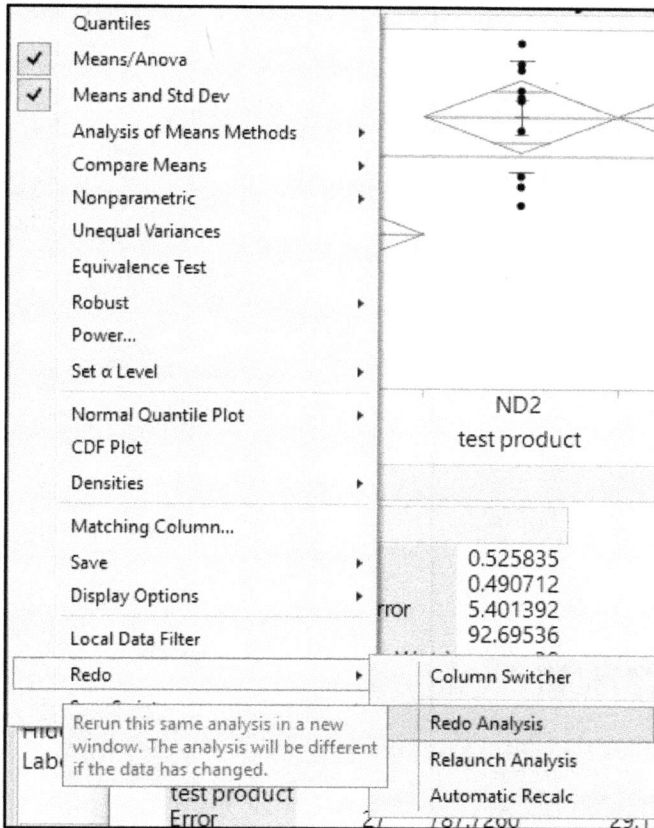

Select *Redo* ▶ *Redo Analysis* to get a new oneway analysis for the full data table with four groups, shown in Figure 5.23.

Figure 5.23: Oneway Analysis of Four Groups

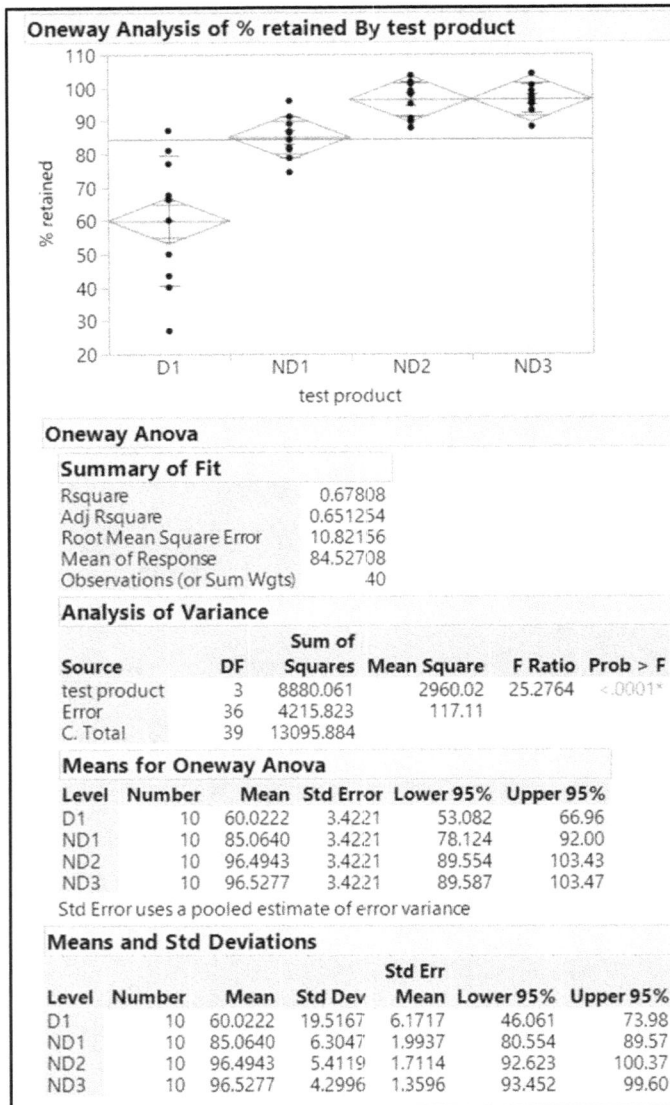

Oneway Analysis of % retained By test product

Oneway Anova

Summary of Fit

Rsquare	0.67808
Adj Rsquare	0.651254
Root Mean Square Error	10.82156
Mean of Response	84.52708
Observations (or Sum Wgts)	40

Analysis of Variance

Source	DF	Sum of Squares	Mean Square	F Ratio	Prob > F
test product	3	8880.061	2960.02	25.2764	<.0001*
Error	36	4215.823	117.11		
C. Total	39	13095.884			

Means for Oneway Anova

Level	Number	Mean	Std Error	Lower 95%	Upper 95%
D1	10	60.0222	3.4221	53.082	66.96
ND1	10	85.0640	3.4221	78.124	92.00
ND2	10	96.4943	3.4221	89.554	103.43
ND3	10	96.5277	3.4221	89.587	103.47

Std Error uses a pooled estimate of error variance

Means and Std Deviations

Level	Number	Mean	Std Dev	Std Err Mean	Lower 95%	Upper 95%
D1	10	60.0222	19.5167	6.1717	46.061	73.98
ND1	10	85.0640	6.3047	1.9937	80.554	89.57
ND2	10	96.4943	5.4119	1.7114	92.623	100.37
ND3	10	96.5277	4.2996	1.3596	93.452	99.60

The first two steps of the hypothesis test are the same as the analysis of three groups. The main assumption for ANOVA of equal variance is not met. The largest standard deviation of 19.5 for the D1 group is nearly five times larger than the standard deviation of 4.3 for group ND3. Since the assumptions are not met, the remaining analysis output might lead to incorrect conclusions. More detail is available to continue the analysis. Use the red triangle menu next to *Oneway Analysis of % retained By test product* and select *Unequal Variances*, as shown in Figure 5.24.

Figure 5.24: Unequal Variances Analysis Option

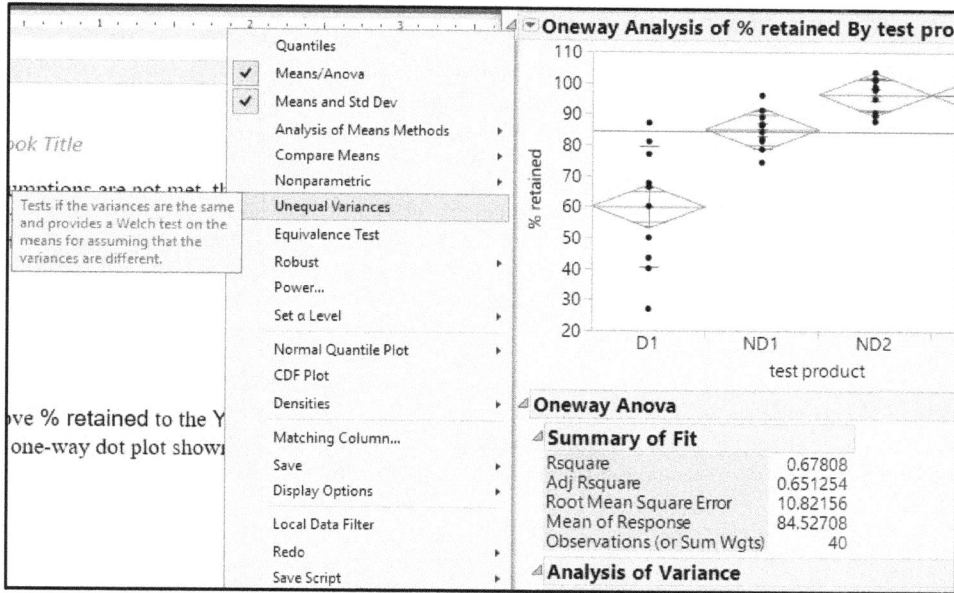

The unequal variances detail is added to the analysis output, as shown in Figure 5.25.

Figure 5.25: Unequal Variances Analysis Detail

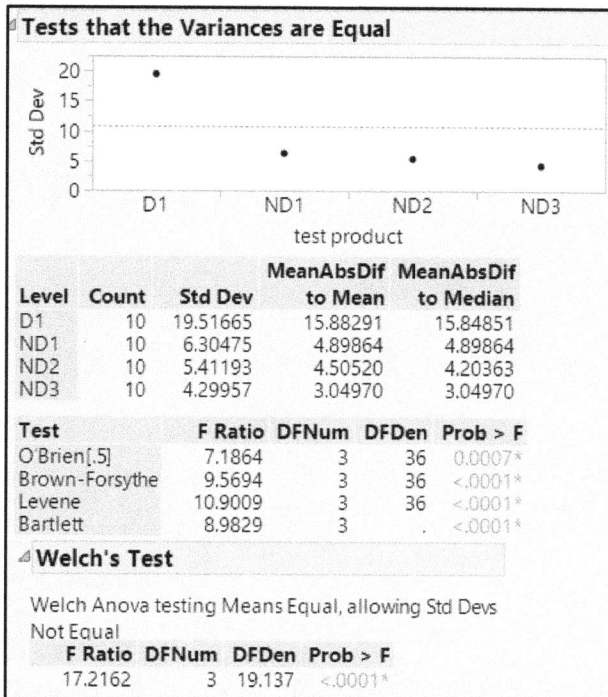

Tests that the Variances are Equal

Level	Count	Std Dev	MeanAbsDif to Mean	MeanAbsDif to Median
D1	10	19.51665	15.88291	15.84851
ND1	10	6.30475	4.89864	4.89864
ND2	10	5.41193	4.50520	4.20363
ND3	10	4.29957	3.04970	3.04970

Test	F Ratio	DFNum	DFDen	Prob > F
O'Brien[.5]	7.1864	3	36	0.0007*
Brown-Forsythe	9.5694	3	36	<.0001*
Levene	10.9009	3	36	<.0001*
Bartlett	8.9829	3	.	<.0001*

Welch's Test

Welch Anova testing Means Equal, allowing Std Devs Not Equal

F Ratio	DFNum	DFDen	Prob > F
17.2162	3	19.137	<.0001*

The output for unequal variances includes a plot of the standard deviations with a horizontal segmented decision limit line. Notice that the marker for D1 exceeds the decision limit by a large amount. The table underneath the plot lists the summary information used for the four tests of significance in the table below the summary information. There are specific scenarios intended for each of the significance tests; you can access these through the Help menu. All four tests indicate significant evidence that at least one of the variance values for at least one group differs from the others. The Welch's test is included below the unequal variance output.

The Welch's test includes an adjustment for unequal variances for the F test for differences between the group means. The adjusted F Ratio of 17.2 for a test with 3 degrees of freedom used for between comparisons and 19.14 degrees of freedom for within comparisons yields a highly significant Prob>F of less than 0.0001. Regardless of unequal variances, there is highly significant evidence of a difference between the group means.

Project stakeholders are most interested in a difference between the regular product and the no-drip formula candidates. Analysis to compare means can provide the needed detail. Use the red triangle menu next to *Oneway Analysis of % retained By test product*, and select *Compare Means ▶ Each Pair, Student's t* to get the detail added to the oneway analysis output, shown in Figure 5.26.

Figure 5.26: Compare Means Detail

Oneway Analysis of % retained By test product

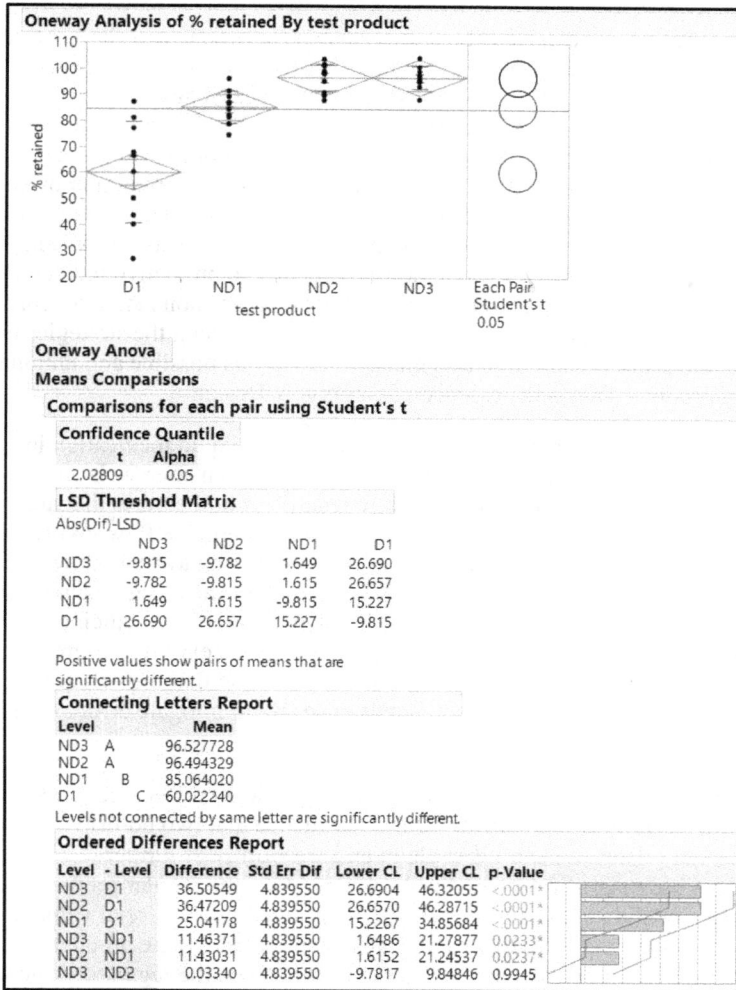

Oneway Anova

Means Comparisons

Comparisons for each pair using Student's t

Confidence Quantile

t	Alpha
2.02809	0.05

LSD Threshold Matrix

Abs(Dif)-LSD

	ND3	ND2	ND1	D1
ND3	-9.815	-9.782	1.649	26.690
ND2	-9.782	-9.815	1.615	26.657
ND1	1.649	1.615	-9.815	15.227
D1	26.690	26.657	15.227	-9.815

Positive values show pairs of means that are significantly different.

Connecting Letters Report

Level		Mean
ND3	A	96.527728
ND2	A	96.494329
ND1	B	85.064020
D1	C	60.022240

Levels not connected by same letter are significantly different.

Ordered Differences Report

Level	- Level	Difference	Std Err Dif	Lower CL	Upper CL	p-Value	
ND3	D1	36.50549	4.839550	26.6904	46.32055	<.0001*	
ND2	D1	36.47209	4.839550	26.6570	46.28715	<.0001*	
ND1	D1	25.04178	4.839550	15.2267	34.85684	<.0001*	
ND3	ND1	11.46371	4.839550	1.6486	21.27877	0.0233*	
ND2	ND1	11.43031	4.839550	1.6152	21.24537	0.0237*	
ND3	ND2	0.03340	4.839550	-9.7817	9.84846	0.9945	

The team uses this wealth of information to fully explore the differences between the means. The team includes two of the more basic analyses to communicate the results to stakeholders. The oneway plot with the *Each Pair Student's t 0.05* is an easily interpreted graphic used to show that the regular product has significantly lower % retained than the no-drip candidate formulas. The *Connecting Letters Report* includes additional detail indicating that the regular product has a lower average % retained than the no-drip candidates because it is the sole member of C. The groups ND2 and ND3 are identified as having the highest average % retained noted by A.

Practical Conclusions

Inferential tools in JMP easily provide robust conclusions about relationships between variables. The color highlighting of the significance values and the Help pop-up information provide details to aid in the interpretation of the results. Kim has been able to obtain robust results for the lack of any relationship between the blend uniformity % and the content uniformity% for the tablet formula. The evidence refutes the regulatory expectation that a blend can be sampled and tested to predict the uniformity of active ingredient in the tablets. Kim's team used the robust statistics from JMP to justify the elimination of the ineffective blend uniformity testing for the product. JMP provided the tools needed for the team to realize a significant reduction in quality costs; the request to eliminate blend uniformity testing was approved by the regulatory agency. The down side to this finding is that the operations team has no predictive indicator from the blend intermediate material, which could be used to save on tablet compression resources for product batches that do not meet specifications. The team will be better served by using the strategies and tactics outlined in this book to study the process and remove as much variability as possible and eliminate the potential to make bad batches of product. This is the essence of Quality by Design.

Hue provided a report to the design team of the medical device used to meter the liquid medication in feeding tubes. The significant evidence of an average difference in tube I.D. between sources is an important consideration to ensure that the device operates consistently regardless of the tubes that are available to the patient care teams who use the device. The follow-up study for medicine flow involving a matched pairs design provides robust evidence that the differences in I.D. result in an average difference in flow for the population of medicine formulas that the device is intended to meter. The design team must ensure that they deal with the differences so that customers realize consistent delivery of medicines for patients. The engineering team challenged their initial conclusions by including the expected source of variability due to physical properties differences (that is, viscosity). Hue is learning that the best solutions come from sound statistical analyses complemented by subject matter expertise to mitigate error as findings are applied in the real world.

Tanya was able to quickly analyze the % retained data to conclude that a significant difference exists between the four drug candidate groups. The stakeholders now have confidence that two no-drip drug candidates have higher performance that relates to the no-drip claim and are worthy of further development. All no-drip formulations have significantly higher % retained than the regular product, which confirms the robustness of the formula candidates. The information contains great commercial value to quantifiably justify the no-drip claim. The output from the oneway analysis is added to the product submission to mitigate any questions about the no-drip claim that might come from the regulatory agencies. Mitigation of risk through sound statistical analysis is the definition of Quality by Design.

Exercises

E5.1—The analytical data used for an exercise in chapter 1 includes two different test methods that are explained to be equivalent. Big mistakes can be made when assumptions are accepted without using analysis to confirm. You want to use the techniques learned in this chapter to test for a difference in the response between the two techniques.

1. Open the file *analytical data.jmp*.
2. Use the *Fit Y by X* platform to run a pooled t-test (within the hotbox options) and test for differences between the two methods.
3. The pooled platform assumes that the variances of responses within each method are equal. Use the unequal variance test (within the hotbox options) to confirm. Do the results of this test change your conclusions?

E5.2—You have been involved with a surgical tray manufacturer regarding the seal strength of the units quantified by the amount of pressure that each withstood before bursting. There is evidence of a difference in the results based on manufacturing shift. The techniques of this chapter can provide the evidence needed to determine of the differences in pressure at seal failure are significant.

1. Open *burst testing.jmp*.
2. Use the *Fit Y by X* platform to test for *burst test results in Hg* due to *shift*.
3. Compare the means with the *Each pair, Student's t* option to explain which shifts differ and by how much.
4. There has been talk that the day of the week might influence results. Create a new variable named "day of week" and use the formula menu for *Date and Time* to find the *Day of the Week* function and calculate for *date* and *time*. The user assigns 1 for Sunday and number the days consecutively ending with 7 for Saturday in the formula. Change the column properties to *Numeric, Ordinal* for analysis. Run *Fit Y by X* to test for a difference in *burst test results in Hg* by *day of week*.
5. How would you summarize the results for both statistically significant differences and differences of practical relevance?

E5.3—The API test data comparison between blend uniformity and content uniformity was completed in this chapter. Another relationship is assumed because the average content uniformity is expected to be related to the composite assay values. The method for composite assay involves a random collection of 10 tablets from in-process samples collected throughout the run of the tablet manufacturing process. The samples are ground up and dissolved into a solution, and an aliquot is removed for chromatography testing. It is reasonable to expect that the average of the 10 individual tablet assays will be strongly related to the composite assay.

1. Open the file *B26 API Test Data.jmp*.
2. Create a summary set of data by using the *Tables* ▶ *Summary* menu. Choose *B26 API tablet content uniformity* and *B26 API assay %* with *Mean* as the statistic and grouped by *Lot*. The assay will be a mean of 1 entry and will be the same value as the parent table; you can use column properties to eliminate Mean( ) from the column name to keep things clear.

3. Analyze the two variables for a relationship between the mean of content uniformity and composite assay. Is there significant evidence of a relationship? How strong might the relationship be between the two variables?

4. How would you report the findings to the leadership teams of the quality and analytical groups?

E5.4 — Implant data were studied in chapter 3. The conclusion was that chamfer depth is not capable to the minimum depth specification. The operations team researched their records and was able to find machine data on the RPM of the machining tool for each of the samples. A great first step in unlocking the clues as to why capability is suboptimum is to test for a relationship between the speeds and the depth results.

1. Open the file *dental implant dimensional checks with speeds.jmp*.

2. Analyze the data to test the mean *chamfer depth* for differences due to the *speed setting* groups.

3. Would any differences in variation within groups interfere with your ability to test for a significant difference?

4. How would you summarize the results of your analysis to project stakeholders?

Chapter 6: Justifying Multivariate Experimental Designs to Leadership

Overview

Development of new products and processes is the life blood of a healthy organization. The pharmaceutical industry has gone through a revival since guidelines promoting Quality by Design (QbD) were published by the International Council on Harmonisation (ICH), a consortium of regulatory bodies from around the world. ICH E8 and ICH E9 guidelines were published to promote risk-based development practices with evidence-based justifications for the identification of the design space to be studied for products and processes. Product development teams must demonstrate how to produce a product with robust processes.

The data-driven tools needed to achieve QbD goals have been a mainstay for many other industries. The techniques have been used and greatly developed over the last 40 years for industries including automotive, aerospace, and semiconductor manufacturing. It is more important than ever for industries to adopt data-driven practices to understand the causational relationships between process inputs and outputs. The modernization of the development process for pharmaceutical products requires teams to utilize data visualization techniques as well as planned, structured, multivariate experiments to understand the design space well enough to create effective quality controls. JMP provides a rich set of data visualization and analysis tools in an easy-to-use package that is uniquely qualified to meet the demands of product development through QbD practices.

This chapter emphasizes the value and importance of structured, multivariate experimentation. The term structured, multivariate experimentation is favored over design of experiments (DOE) because it is specific; DOE is a term that has been inappropriately applied to cover all kinds of experiments. You should be able to answer the question of why leadership should support structured, multivariate experimentation to efficiently learn about processes at the earliest possible stage of development.

The Problems: Developmental Experiments Lack Structure

Jorge is a senior scientist and the lead for a development team involved in a creating a new product. He obtained a JMP license and has taken advantage of training on both QbD and the use of structured, multivariate experimentation. Jorge knows that the management team has been directing teams to include elements of QbD into drug applications. Unfortunately, leadership continues to embrace experiments that are based solely on principal science and experience; teams find settings that get results for one output at a time. Current practices are outcome-based; inputs are manipulated randomly until the outputs meet established goals. The trial and error involved can and does extend development times, making it very difficult to estimate when development will be completed.

Jorge knows that he can dramatically reduce the development cycle while optimizing outputs by incorporating the structure of multivariate designs. This chapter deals with the challenges of working against a developmental culture that is based on unstructured, outcome-based experimentation by justifying the hidden value of structured, multivariate experimentation to leadership.

Why Not One Factor at a Time?

Scientific experimentation, throughout the academic career of a technical professional, is taught by manipulating one input (factor) at a time (OFAT) while all others are held constant. Science educators stress this concept with the hope that the student will stay on task troughout the "scientific journey"and not lose track of the relationships between the manipulations of each input and the changes in outputs. While this methodology sounds reasonable for empirical science, it is lacking and inefficient with regard to the analysis of the data. The comparison of a small, three-input hierarchical experiment to a multivariate design quickly illustrates the shortcomings of OFAT.

Figure 6.1 includes the table of values from a design that is intended to explore a pharmaceutical process involving the spray rate of a liquid solution on a bed of powder, the amount of airflow used to fluidize the powder, and the temperature that is set and controlled by the equipment. Each of the inputs is studied at low (-1), medium (0), and high (1) settings. The output is a critical quality attribute (CQA), which is expected to be maximized in value for best results. The team decides to start experimenting in a sequential set of steps for each factor. The best result will be retained and passed on for the study of the other two factors. For instance, run 1 includes the low level of spray rate with the air volume and temperature at medium values. Run 2 includes the medium level of spray rate with the other two held at the medium level. Run 3 includes the spray rate at the high level with the other two remaining at medium. The best result from the first three runs came from the high spray rate, so the high spray rate is fixed for the remainder of the runs. This sequence is repeated for the air volume and temperature variables.The last run includes settings that are estimated to have produced the highest CQA output of 13.66.

Figure 6.1: 3-Factor, 10-Run OFAT Data

run		spray rate	air volume	temperature	CQA
1	1	-1	0	0	8.5
2	2	0	0	0	9.1
3	3	1	0	0	11.61
4	4	1	-1	0	12.99
5	5	1	0	0	11.14
6	6	1	1	0	10.18
7	7	1	-1	-1	11.55
8	8	1	-1	0	11.69
9	9	1	-1	1	11.03
10	10	1	-1	0	13.66

An alternate design including a multivariate structure is shown in Figure 6.2. This design also includes 10 runs, but JMP software is utilized to find a randomized set of factor levels that best explore the 10-run space. Notice that the combinations of factor levels vary randomly throughout the 10 runs. The best result from the set of experiments is 13.67, which admittedly does not seem much different from the OFAT design. The subtle details of how the inputs work to change the CQA will come from the statistical analysis of the data. This is where the issues of the OFAT design are brought into the light of day.

Figure 6.2: 3-Factor, 10-Run Multivariate Custom Design Data

run		spray rate	air volume	temperature	CQA
1	1	1	0	0	11.62
2	2	-1	1	1	13.35
3	3	1	-1	-1	12.27
4	4	0	0	-1	9.13
5	5	0	-1	1	4.46
6	6	1	1	-1	10.14
7	7	1	1	1	8.57
8	8	-1	-1	0	0.09
9	9	0	1	0	12.47
10	10	-1	1	-1	13.67

Before the team works to detect the differences in design, it will be helpful to list the basic concepts of what is being studied. All possible inputs at all possible levels make up the universe of possible studies that can be designed for a subject process. Development teams must narrow down factors in this space to a small number through principal science, professional experience, and industry knowledge in order to assess quality risks and mitigate them. There is a balance between the amount of information desired from the experiments and the amount of resources available for experimentation. This balance is reflected in how many factors and runs can be incorporated.

The experiments are intended to study the space created by the number of factors and levels included in the design plan. This space is a portion of the infinite universe of possible factors and levels and is known as the design space. A good experimental design should provide robust information about how changes in the factors affect changes in the output for a large proportion of the design space. The ultimate goal of experimentation for a commercial product is to find the optimum combinations of factors and ranges of settings within the factors that are likely to produce robust results that meet and exceed all requirements. The optimized combination of process factor settings typically forms the process recipe documented for manufacturing. An experimental design that limits the amount of information that can be gained adds risk and uncertainty to the process and will likely produce process controls that are not optimum. A suboptimized process is likely to add cost to production and increase the potential for products that do not meet requirements. Added risk and uncertainty in the highly regulated pharmaceutical and medical device industries is always problematic.

One simple way for a team to assess how well an experimental design covers the design space is a scatterplot matrix. The following steps create the plot:

1. Open the data set *OFAT 3F 10R.jmp*.
2. Select *Graph ▶ Scatterplot Matrix*. Move the variables *spray rate*, *air volume*, and *temperature* into the *Y,response* box, and click *OK* to activate.
3. Double-click the title *Scatterplot Matrix* and change it to "OFAT 3F 10R – Scatterplot Matrix."
4. Open the data set *multivariate custom design 3F 10R.jmp* and create a scatter plot as in step 2.
5. Double-click the title *Scatterplot Matrix* and change it to "multivariate custom design 3F 10R – Scatterplot Matrix."
6. There is a checkbox with a black downward arrow located in the lower right corner of plots that is used to arrange plots. Click on the checkbox of the *multivariate custom design 3F 10R – Scatterplot Matrix* window to select it.
7. Open *OFAT 3F 10R – Scatterplot Matrix* to add a check in the checkbox and to click on the downward arrow at the lower right of the plot to view the arrange options.
8. Select the *Combine Windows…* option to get the Combine Windows dialog box window used to create a dashboard of plots.
9. Use the default options of the Combine Windows dialog box window, and click *OK* to get the combined scatterplot, shown in Figure 6.3.

Figure 6.3: Comparative Scatterplot Matrices

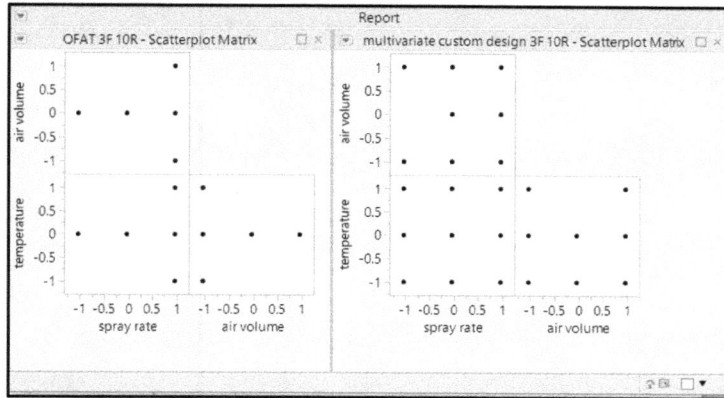

Figure 6.3 includes a comparative set of scatter plots for the OFAT and multivariate custom design. The team does not need a graduate-level statistician for them to quickly identify the shortcomings of the OFAT design. Large areas of the design space have no factor combination points in the OFAT design, and the multivariate custom design space covers all but a relative few factor combinations. The design space is used to make predictions on how the inputs studied have an effect on outputs. Prediction error is greatly reduced as more of the design space is covered by the set of experiments. Moving forward with a marginal design space is likely to add significant error to the conclusions made by the scientific team and should be avoided.

The analysis of statistical models is covered in detail in later chapters. However, the simple model plot and effect summary for each design are relatively easy to interpret and provide detail regarding the model comparison. The following steps create a model with the OFAT data:

1. Make sure that *OFAT 3F 10R.jmp* is the open and active window.
2. Select *Analyze ▶ Fit Model* to open the *Model Specification* window.
3. Select *CQA* and drag it to the *Y* box.
4. Select *spray rate*, *air volume*, and *temperature*; move all three to the *Construct Model Effects* box, as shown in Figure 6.4.
5. Leave the default options in place, and click *Run* to get the output.

Figure 6.4: Model Specifications

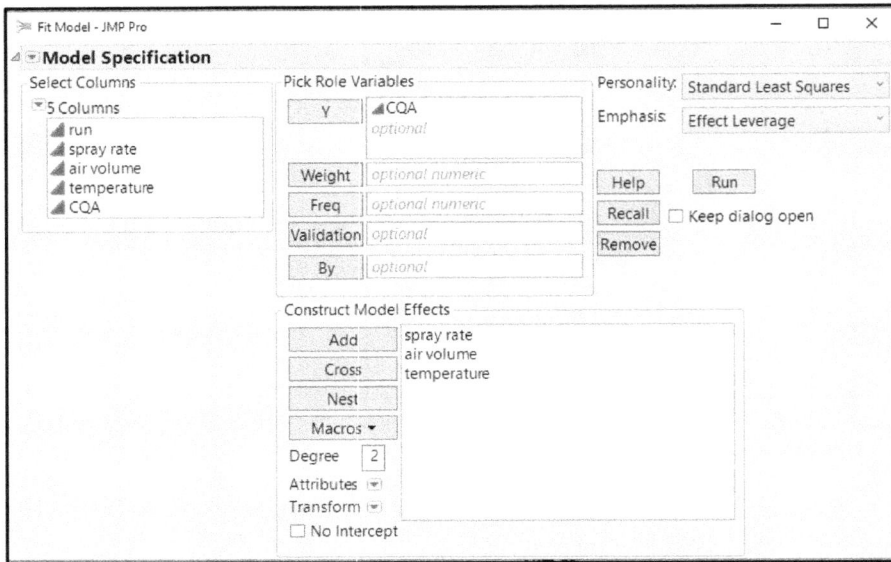

The Actual by Predicted Plot illustrates the strength and direction of the influence of the model on the CQA output. The model output is summarized in Figure 6.5 to illustrate the fit (r-square = 0.77), significance (p=0.026), and the effects of the three process inputs and of the CQA output.

Figure 6.5: OFAT Model Results

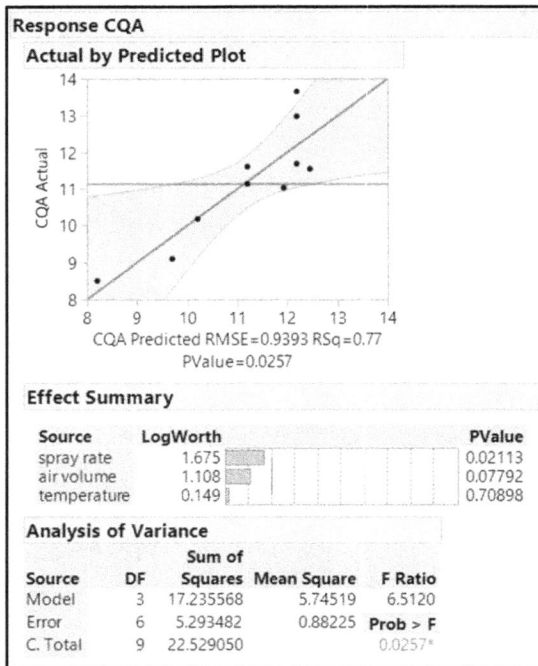

Source	DF	Sum of Squares	Mean Square	F Ratio
Model	3	17.235568	5.74519	6.5120
Error	6	5.293482	0.88225	**Prob > F**
C. Total	9	22.529050		0.0257*

Next, the team creates a model for the multivariate custom design. The OFAT data includes only enough information to study individual inputs. The design includes enough information to study the individual inputs as well as the combined effects for two variable combinations (two-way interactions).

1. Make sure that *multivariate custom design 3F 10R.jmp* is the open and active window.
2. Select *Analyze ► Fit Model* to open the *Model Specification* window.
3. Select *CQA* and drag it to the *Y* box.
4. Select *spray rate*, *air volume*, and *temperature* so that they are shaded in blue. Select *Factorial to degree* in the *Macros* drop-down box so that the individual inputs and interactions show in the *Construct Model Effects* box, as shown in Figure 6.7.
5. Leave the default options in place, and click *Run* to get the output.

Figure 6.6: Model Specification Window with Macros Listed

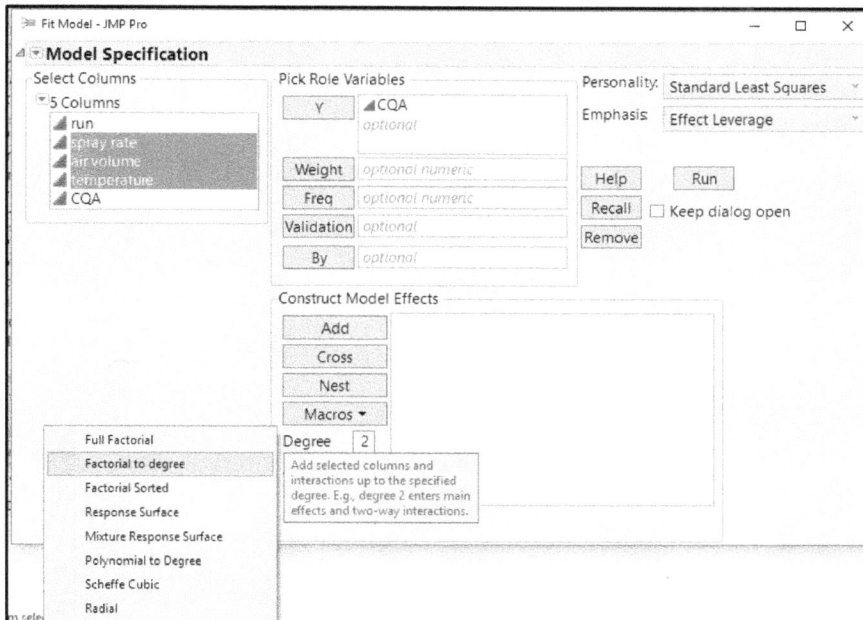

Figure 6.7: Model Specification Window with Individual Inputs and Interactions

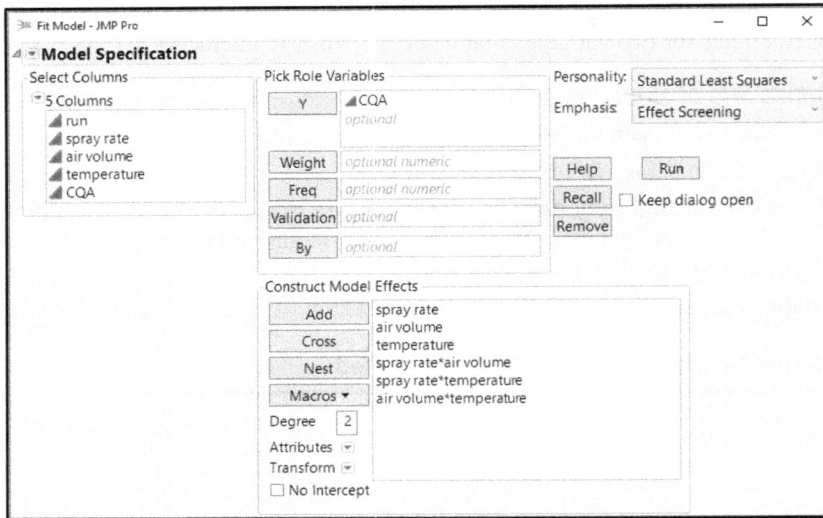

Figure 6.8: Multivariate Model Results

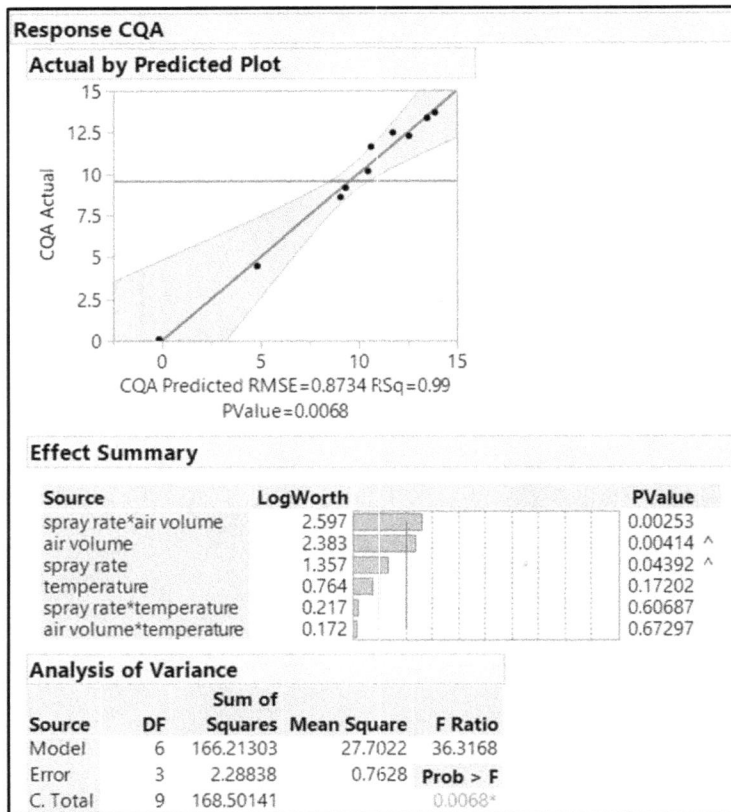

Response CQA

Actual by Predicted Plot

CQA Predicted RMSE=0.8734 RSq=0.99
PValue=0.0068

Effect Summary

Source	LogWorth		PValue
spray rate*air volume	2.597		0.00253
air volume	2.383		0.00414 ^
spray rate	1.357		0.04392 ^
temperature	0.764		0.17202
spray rate*temperature	0.217		0.60687
air volume*temperature	0.172		0.67297

Analysis of Variance

Source	DF	Sum of Squares	Mean Square	F Ratio
Model	6	166.21303	27.7022	36.3168
Error	3	2.28838	0.7628	**Prob > F**
C. Total	9	168.50141		0.0068*

The multivariate model output is summarized in Figure 6.8 to show the fit (r-square = 0.99), significance (p=0.0068), and the effects of the three process inputs and of the CQA output.

The comparison of the two models illustrates the value of the multivariate approach. The fit of the models provides the first advantage of using the multivariate custom design: OFAT modeling explains 77% of the variability in the CQA, and the multivariate custom design explains 99%. With the same number of runs, the multivariate custom design explains nearly 30% more of the variation in CQA, which is much more efficient. The OFAT design allows for the team to study only the three individual factors; the multivariate custom design studies the three individual factors as well as the interactions between factors. The spray rate in the OFAT design is shown as providing the most influence (PValue=0.021, significant) on CQA, followed by air volume (PValue=0.078, marginally significant) and temperature (PValue=0.709, no significance). The results of the multivariate custom design analysis indicate that the interaction between spray rate and air volume is the most influential factor affecting the CQA. The interaction (PValue=0.0025) is 10 times more significant than the spray rate factor identified by OFAT. Incorrectly focusing on changing the spray rate to optimize the CQA will not be reliable without understanding the influence of air volume. Costly errors are much more likely to occur from reliance on OFAT as opposed to the multivariate custom design. More detailed comparisons are available and are discussed in future chapters. The full diagnosis of experimental models is discussed in chapter 9 and analysis of experimental data in chapter 11.

Data Visualization to Justify Multivariate Experiments

Justification for using multivariate experiments should include data visualization to illustrate the value that will be added to the product development process. One way to accomplish this task is to collect information on a sample from the historical records of development projects and compare that to the what can be expected once multivariate methods are utilized. The data for projects utilizing multivariate experiments can be actual results from studies that are part of a phase-in strategy. Another option is to choose data from a small number of projects and simulate the use of multivariate studies through the experimental design plans. The retrospective approach can quantify a reduction in the number of experiments possible to gather the same amount of or more information than was found with the previous OFAT development strategy. JMP includes a seemingly infinite number of graphical options to illustrate the trends; the following example below is only one set of techniques.

The data set development overview.jmp includes a random sample of 26 projects with a product type variable, the number of batches made by developmental stage, a variable to note which projects used multivariate experimentation, and the total weeks that elapsed during the product development process. Open the development overview.jmp data set in JMP and select *Cols ▶ New Column* to create a new column. Type Batches Made as the *Column Name*, and select *Column Properties ▶ Formula* to open the Formula Editor. Enter the formula to add the following columns together: *Lab Batches + Scale Up Batches + Confirmation/Validation Batches.*

It is useful to group the data into the type of product for the visualization. The data does not include product type, but the projects variable includes the name of project with the product type. You can recode the variable to create a new column that groups the three products by type. Click on the *project* column to highlight it. Then, select *Cols ▶ Recode* to open the recode dialog box. Leave the default for *New Column*, and change the *Name* by entering *product type*. Use the red triangle menu next to the *projects* header to select *Split On*, as shown in Figure 6.9, to get the *Split On* options window. Select the *Last Word* radio button and *Recode* to execute the choice, as shown in Figure 6.10. The new grouping is shown in Figure 6.11.

Figure 6.9: Column Recode Options

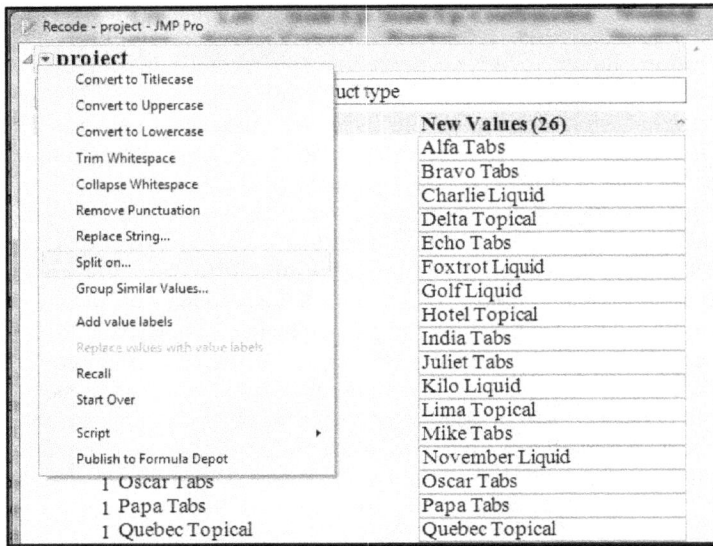

Figure 6.10: Split On Options

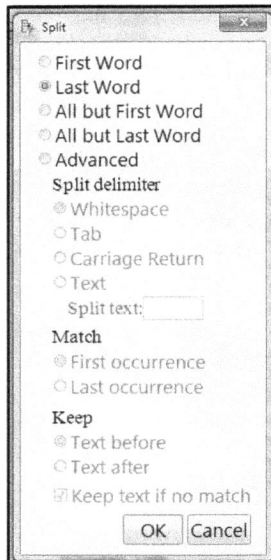

Figure 6.11: Recode Window Ready to Execute

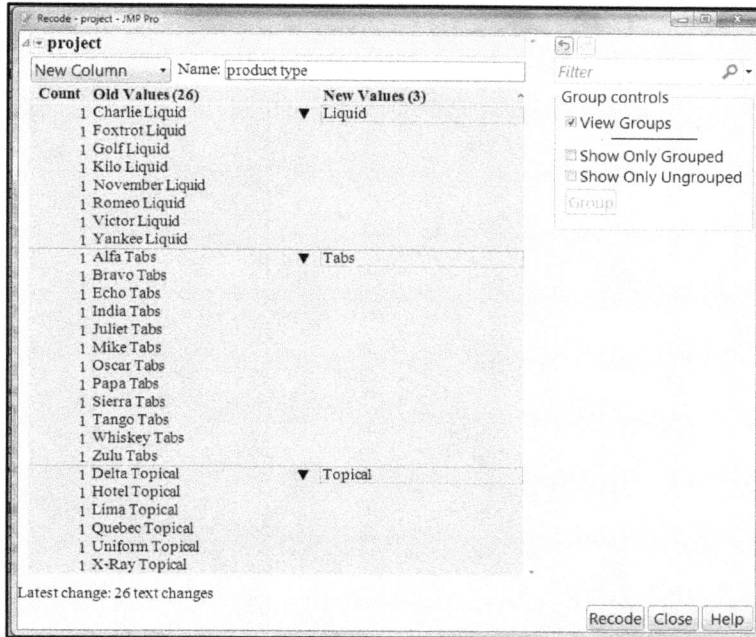

The data now includes a grouping variable that identifies the three product types developed by the R&D team. A simple graphic of the differences between the number of experimental batches as well as develop time (in weeks) by the product type helps to illustrate the value of structured, multivariate experimentation. Use the Graph Builder to create a graph matrix by completing the following steps:

1. Select *Graph ▶ Graph Builder*.
2. Click *Dialog* to open the dialog window shown in Figure 6.12.
3. Select the *Graph matrix* check box in the *Options* area.
4. Move *Weeks of Development* and *Batches Made* into the *Y* box, move *Product Type* to the *Group X* box, and move *Multivariate Experiments* into the *color* box.
5. Click *OK* to produce the plot shown in Figure 6.13.

Figure 6.12: Graph Builder Dialog Box

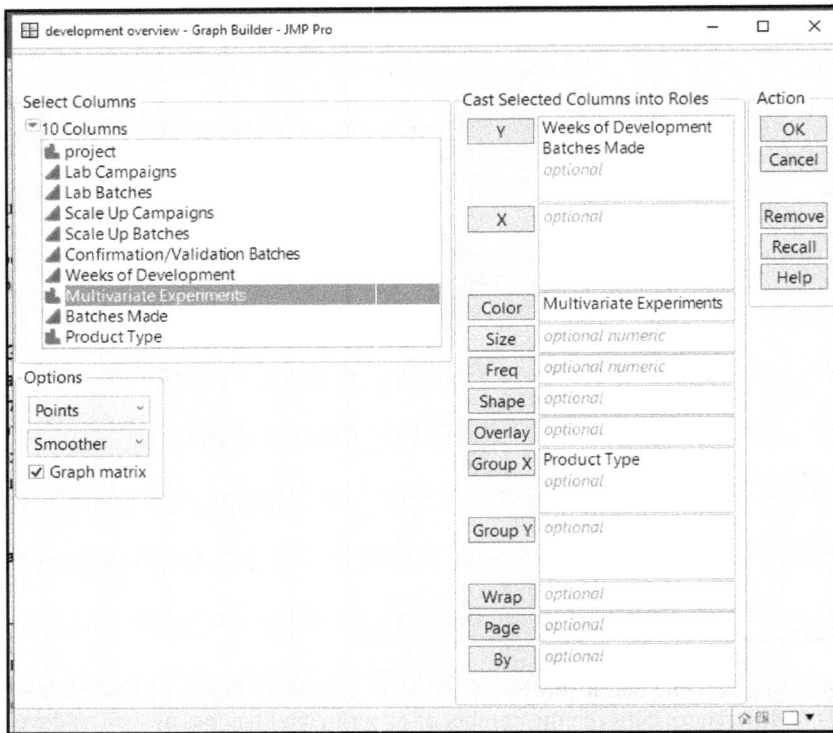

Figure 6.13: Graph Builder Plot

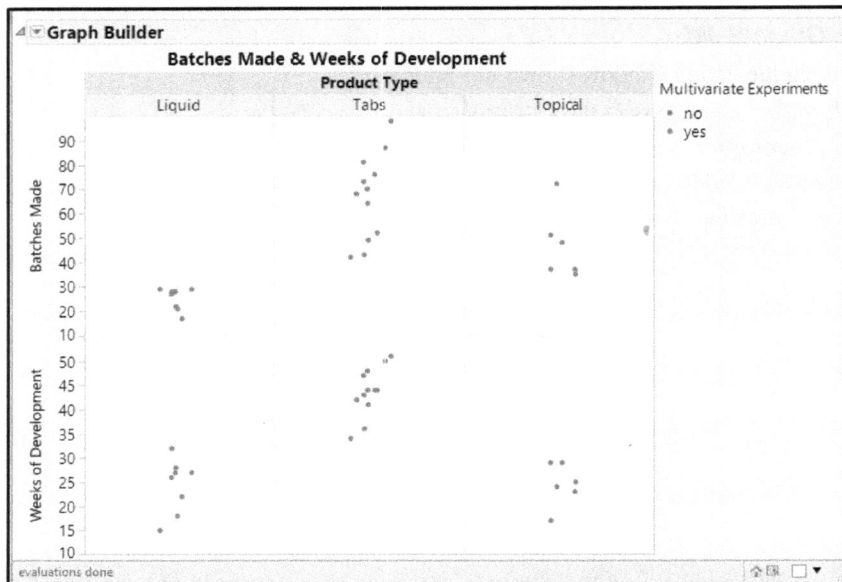

The color is more vivid in the JMP application. But there is value in identifying batches made with multivariate experimentation. You can customize the points in the plot with unique shapes to clarify the interpretation of the plot regardless of where it is published. Place your pointer on *no* in the *Multivariate Experiments* legend and right-click to get options. Select *Marker*, and then select the open circle shape shown in Figure 6.14.

Figure 6.14: Changing Markers

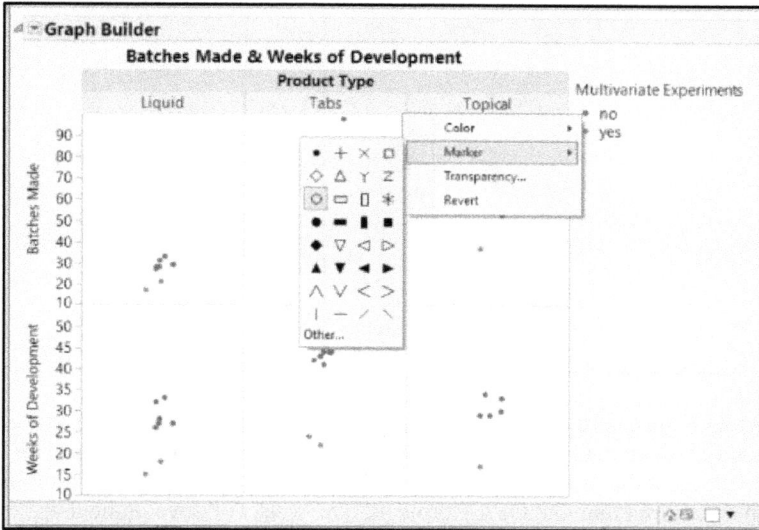

Place your pointer in the white space of the plot, press the Ctrl key, and right-click to access additional options for the entire graph matrix. Select *Graph ▶ Marker Size ▶ XXXL* to enlarge the marker (observations) of the plot, shown in Figure 6.15. This change will make the differences as clear as possible for leadership regardless of how the plot is published.

Figure 6.15: Changing Marker Size

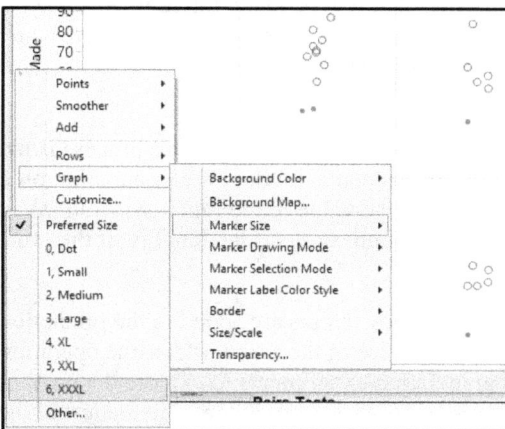

Figure 6.16: Customized Matrix Plot

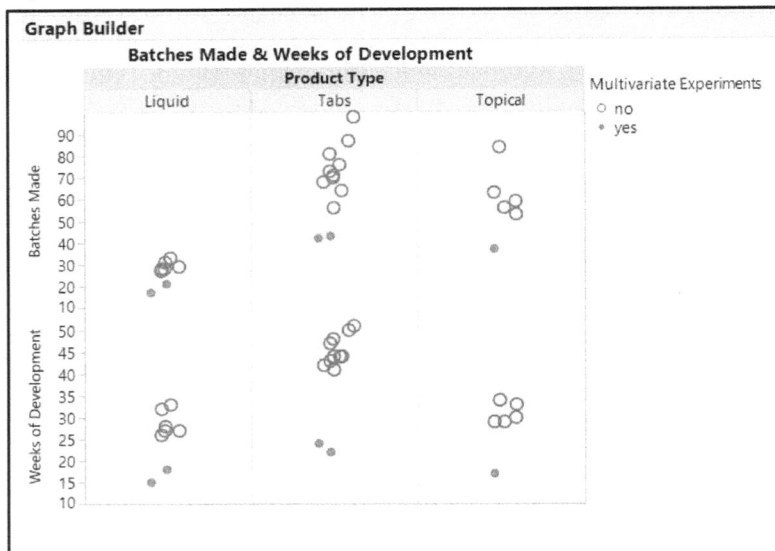

The customized matrix plot shown in Figure 6.16 provides to leadership a picture illustrating the value of using multivariate experimentation to reduce both development time as well as the resources needed to create batches. It is clear from the plot that the projects involving multivariate experimentation result in between 50% to 70% fewer batches required than the OFAT projects. The weeks of development are reduced by at least 43%.

Using the Dynamic Model Profiler to Estimate Process Performance

The value of multivariate experimentation becomes very clear when you use the profiler options available in JMP. Once a model has been created for each of the outputs of interest, you can use the profiler to gain insights on what can be expected from the population of batches that will be made for commercial scale. You can explore a seemingly infinite number of input combinations without using the extensive resources required to make actual batches of product. Utilization of an experimental model in this way is the most efficient way to optimize products and processes.

A desirability function in the profiler offers a practical interpretation of the space where a process is likely to provide the best results. This section returns to the simple, experimental model for a granulation process introduced in the first section of this chapter. This example shows the value of the functions from the perspective of leadership. More detail on how to incorporate the demonstrated functionality of the profiler is included in later chapters.

The goals of the development efforts and random variation of process inputs are added to the prediction profiler of the model to fine tune the visualization to be representative of the real world of the operational process. This example uses results of the executed set of granulation experiments to explore the profiler. Open *multivariate custom design 3F 10R outputs.jmp* and click on the green arrow to the left of the *Fit Least Squares* script to run the model analysis, which are shown in Figure 6.17.

Figure 6.17: Model Profiler with Simulation Options

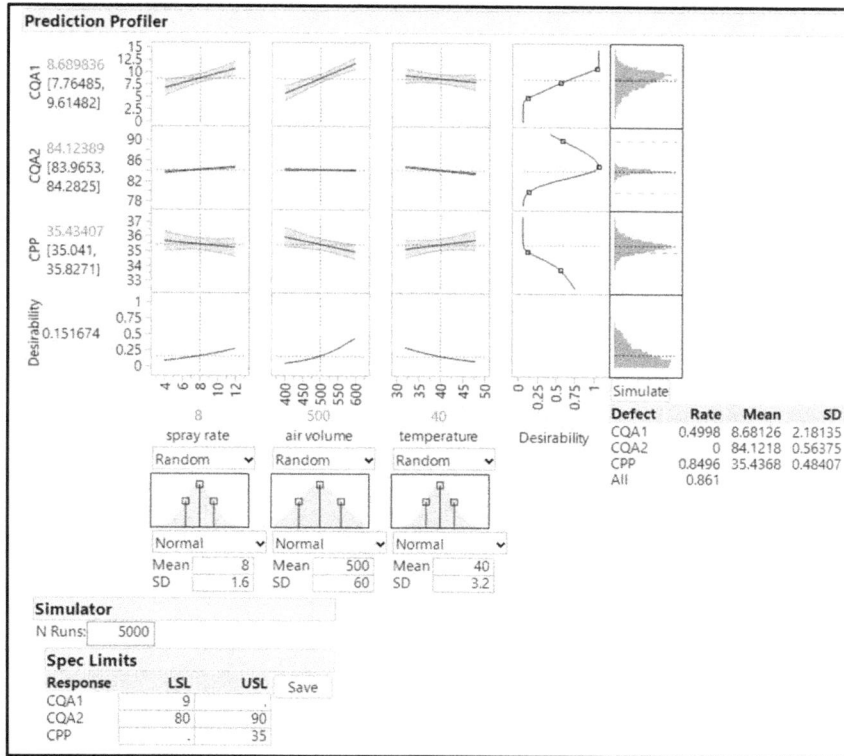

The plots above illustrate the model of the process set to the input targets in the middle of the design space. The desirability indicates how well all of the output goals are met at the set combination of inputs. Note that the process set to target values leaves much to be desired; the desirability value 0.1517 indicates that only 15% of the goals can be met. In addition, the profiler indicates that with random variation added to the process, the target settings are likely to create batches out of specifications 86% of the time.

The profiler output is dynamic. You can change the red input settings in one of two ways: select the numeric value and enter a new value; or right-click on the vertical red segmented line and drag it to higher or lower input amounts. Project leadership is interested in the optimum results as defined by the highest amount of desirability and lowest predicted rate of defects.

JMP offers options for the profiler to maximize the model and optimize the desirability for all outputs. Click on the red triangle menu to the left of the *Prediction Profiler* header, and select *Optimization and Desirability ▶ Maximize Desirability*. The input settings automatically change to illustrate the combination that will result in the highest level of desirability, as shown in Figure 6.18.

Figure 6.18: Modeling Profiler with Simulation (Maximized Desirability)

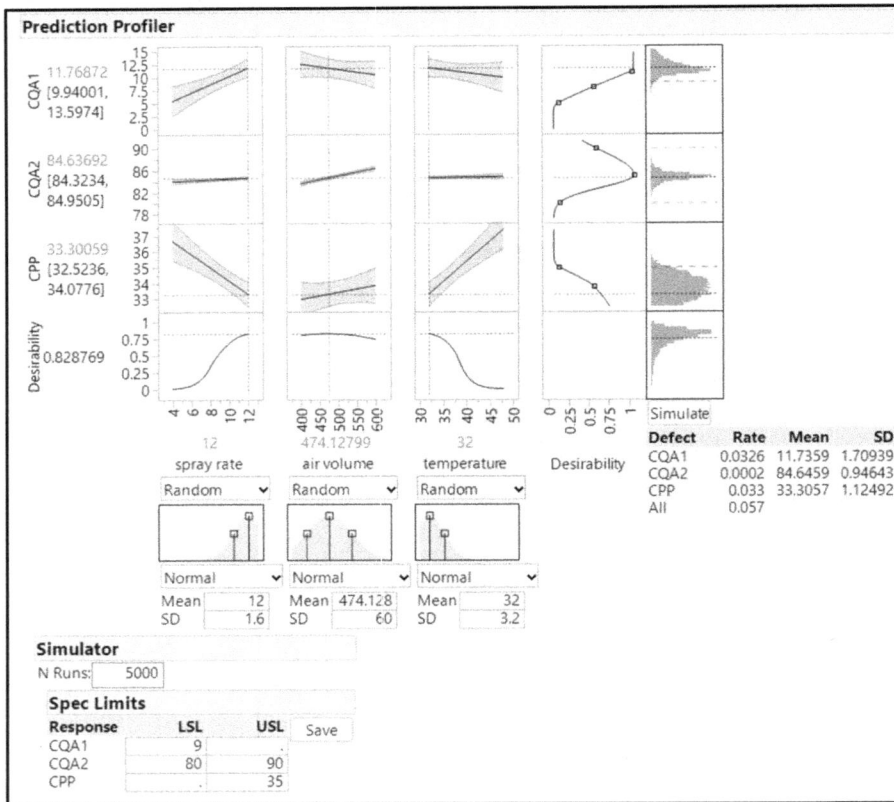

The profiler now displays the result of the maximized models with a desirability value of 0.8288, which indicates that 83% of the goals can be met. The optimized process is predicted to yield less than 6% of future batches produced that are out of specifications. The profiler, augmented with the desirability function and simulated view of the process, can save time by enabling you to analyze an infinite number of factor combinations. The model simulations guide leadership with information from early in the process so that they can make decisions that are critical to the development process.

The results of the modeling reduce the potential that unforseen challenges occur when the operations team is ramping up commercial production to meet customer demand. Risks to production at the early commercial stage are expensive, extremely disruptive, and can risk the loss of credibility with customers caused by product shortages. Modeling with simulation adds value by supporting better decisions made from the information gathered during developmental experimentation. The upcoming chapters provide the details needed to implement structured, multivariate experiments, as well as the dynamic profiling of the process.

Practical Conclusions

There is a wealth of evidence in support of structured, multivariate experimentation (design of experiments [DOE]) from several industries. The pharmaceutical and medical device industries have begun to realize the value of incorporating DOE into the product development process through QbD initiatives. The International Council on Harmonisation (ICH) promotes the use of sound DOE practices to study a design space and the use multivariate models of materials and process inputs to determine robust controls to reduce quality risks. The QbD elements promoted through ICH E8 and ICH E9 can be easily developed from tools and techniques inherent to JMP.

The first step to creating value through structured, multivariate experimentation is to justify the concept to leadership. This chapter provides a few examples that can be utilized to gain acceptance and support of the practice. Resources are always in limited supply, and incorporation of the methodologies noted are proven to efficiently develop robust processes. The excellent graphics provided by JMP are easily produced and should be used to guide management to embrace the value of structured, multivariate experimentation. Regular use of the tools shown will greatly improve the processes and support a robust developmental culture in an organization.

Exercises

E6.1—There is a constant theme of leadership strategy for new product development within the pharmaceutical and medical device industries: the desire to get products to market as fast as possible with the fewest resources expended. A proposal involving structured, multivariate experimentation is often met with resistance due to the multiple batches involved early in the development cycle. Reliance upon principal science and experimentation falsely appears to be a more expedient approach since leadership lacks a comprehensive view of all the batches that need to be made to get to a robust product. Researching the developmental history generally provides a large amount of useful data. Visualization of the data is used to gain support for structured, multivariate experimentation. Be sure to consider the following:

1. Review lab notebooks and operational records for a feasible number of recent new products.
2. Record the start date of development and the date when operational batches were able to be run in a routine manner with regularly acceptable results for all critical quality attributes CQAs.
3. Create groups for experimental batches made through structured, multivariate experiments (if any) and unstructured experiments.
4. Use the tools noted in the chapter to create visualizations of the developmental trends for the organization to illustrate where improvements can be made.
5. Pay special attention to deviations and other problems encountered in scale up and early commercial production. Which problems could have been mitigated through improved knowledge of the design space?
6. Share this information with colleagues and work to use plots for communicating to the leadership team.

Chapter 7: Evaluating the Robustness of a Measurement System

Overview

This book is about helping JMP users to visualize data and use statistical techniques to gather insight about problems and solve them. People often assume that data collected from an instrument is the actual measurement of interest. However, the value obtained is just a *representation* of the actual value. This representation can vary due to limitations of the instrument, the methods utilized for measurement practices, and the environmental conditions present during the time of the measurement trials. Teams should always assume that the values in a data table include some level of uncertainty due to the measurement system. Uncertainty can originate with variance in the instrument, methods, and environment. The calibration of an instrument and the manufacturer's certified amount of instrument accuracy does not ensure that the data collected is "error-free". It is important to procure evidence from a study performed for onsite use of the instrument to represent the population of actual results. Such a study quantifies the quality of the measurements obtained from a measurement system for real-world use. This chapter deals with a few basic ideas of how technical professionals can study measurement systems with JMP to determine the quality of data obtained from an instrument.

The Problems: Determining Precision and Accuracy for Measurements of Dental Implant Physical Features

Chapter 1 includes the problem of reported difficulties in the threading of dental implants and the collection of dimensional data to determine where improvement is needed. Ngong is the processing engineer directing efforts that have indicated that an insufficient chamfer feature is the source of the problem. Ngong realizes that no evidence is on record regarding the quality of the measurement system used to gather data on chamfer and threading depth of the tiny implants. Time and resources are in short supply (as always) and the team decides on the simple plan of including replicate measurements on some of the implants to get a basic idea of measurement quality.

Sudhir is working on dissolution analytical data for tablet products. Analytical results rely upon accurate and precise measurements of the weight of a tablet to ensure that the test values appropriately reflect reality. Digital scales are utilized in the quality laboratory to obtain weights. A detailed analysis of measurement systems is planned and executed by Sudhir's team to ensure the highest possible level of quality for tablet weights. The study focuses on the repeatability of the measurement system as well as on reproducibility. Repeatability is obtained by measuring the same tablet multiple times. Reproducibility is the ability of multiple analysts to get measurements of the same sample that are equivalent. There are other factors that could be studied such as various instruments or various days. However, the available resources result in factors limited to replicates and analysts because they are of the highest priority to the team.

Qualification of Measurement Systems through Simple Replication

A quick and easy way to estimate the uncertainty in measurement is to include replicate measures. Replicates added randomly to the data collection plan allow for the comparison of multiple measurements of the same observational unit to illustrate the variability in the measurement process. In chapter 3, a set of measurement data from dental implants was analyzed to determine how capable two key dimensions were to meeting specifications. Recall that the chamfer feature of the implant was found to have poor capability. The capability analysis assumes that the measurement values in the data are an accurate representation of the specific physical feature of the implant.

For the purposes of this chapter, the clock is rolled back to the early stages of the proposed data collection plan. The team discuss the challenges in measuring the features of such a small object and the need to study the measurement system. A detailed measurement systems analysis cannot be done at this time, so they decide to randomly include replicate measurements of the same object within the plan for comparisons. The variability among the replicate measurements is analyzed to give the team objective evidence of how well the measurement system is working. Open *Subset of dental implant dimensional check.jmp* to initiate the analyses.

The data set is shown in Figure 7.1 and includes 85 measurements that have been broken up into 17 sampling events. The technicians who measure the parts have listed the implant ID along with the two measurements for each unit. The replicate column is included to identify the units that have replicate measurements.

Figure 7.1: Implant Data Table

	sample	implant ID	chamfer depth	threading depth	replicates
1	1	Z47	0.69	2.48	no
2	1	Z91	0.60	2.53	yes
3	1	Z146	0.66	2.52	no
4	1	Z17	0.68	2.46	no
5	1	Z114	0.64	2.41	no
6	2	Z84	0.66	2.37	no
7	2	Z153	0.64	2.62	yes
8	2	Z80	0.63	2.40	yes
9	2	Z94	0.65	2.41	no
10	2	Z91	0.63	2.54	yes
11	3	Z197	0.60	2.47	no
12	3	Z109	0.58	2.69	no
13	3	Z80	0.80	2.40	yes
14	3	Z188	0.58	2.57	no
15	3	Z166	0.65	2.55	no
16	4	Z104	0.61	2.46	no
17	4	Z153	0.61	2.61	yes
18	4	Z124	0.67	2.34	no
19	4	Z77	0.58	2.43	no
20	4	Z97	0.64	2.38	no
21	5	Z31	0.53	2.26	no
22	5	Z123	0.72	2.53	yes
23	5	Z55	0.65	2.25	no
24	5	Z2	0.67	2.51	no
25	5	Z79	0.65	2.61	no
26	6	Z153	0.62	2.61	yes
27	6	Z91	0.65	2.54	yes

Window panel details:

- Subset of dental inplant dimensional checks - JMP Pro
- File Edit Tables Rows Cols DOE Analyze Graph Tools View Window Help
- Subset of dental i...
  - Source
- Columns (5/0)
  - sample
  - implant ID
  - chamfer depth
  - threading depth
  - replicates
- Rows
  - All rows 85
  - Selected 0
  - Excluded 0
  - Hidden 0
  - Labelled 0

A plot and summary statistics for measurements without replicates illustrates the location (mean), spread (standard deviation), and shape of the measurements from independent samples. The following steps provide the required plot and summary statistics.

1. Select *Analyze* ▶ *Distributions* to move *chamfer depth* and *threading depth* to the *Y, Columns* box.

2. Move *replicates* to the *By* box to get separate results for measurements with and without replicates.

3. Use the red triangle menu next to the *Summary Statistics* header under *Distributions replicates=no/ chamfer depth*. Press the Ctrl key, select Custom *Summary Statistics*, and then select the *Variance* check box.

4. Click *OK* to add *Variance* results to all of the *Summary Statistics* tables for later use.

5. Use the red triangle menu next to *Distributions replicates=no* header, right-click and select *Stack* to stack the output. Figure 7.2 shows the distribution of measurement results without replicates.

Figure 7.2: Distributions of Implant Measurements

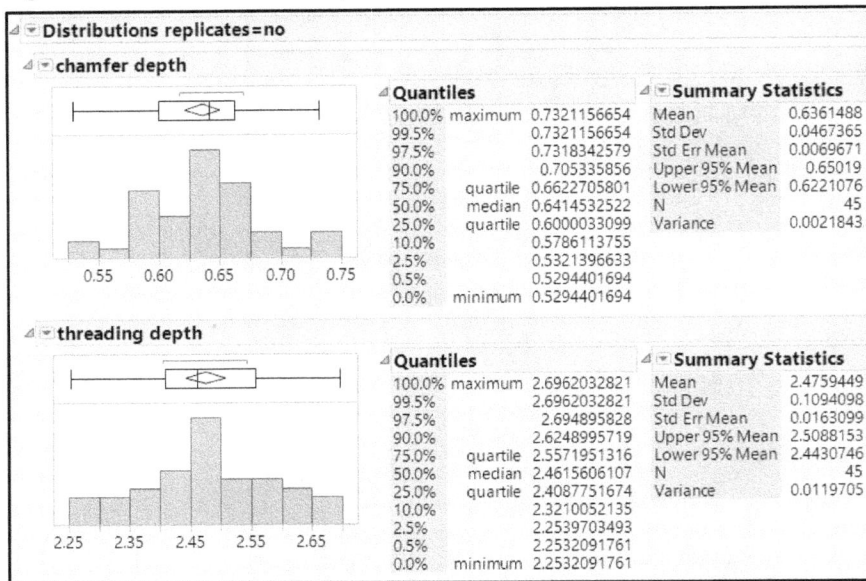

The non-replicated results provide a useful summary of the location and spread of chamfer and threading depth measurements. The chamfer depth average is 0.636 with a standard deviation of 0.0467, and the threading depth average is 2.476 with a standard deviation of 0.109. The shape of both distributions is symmetric with no outliers, meeting the normality assumption.

It is reasonable to expect that the replicate measurements come from the same population and have a location and spread that is very similar to the non-replicate distribution. The measurements with replicates provide information about the variability present when the same implant is measured multiple times. It is likely that some variation will be present between replicate measurements. However, the amount of variation among replicates should be much less than the variation between implant units. When the variation among replicate measurements is large with respect to variation between different units, the robustness of the measurement system is questionable. The next set of steps filters the table on replicate measurements for a comparison between implant ID units.

1. Click the *Data Filter* icon, or select *Row ▶ Data Filter* and select *replicates* as a filter column and click *Add*.
2. Click *yes* for the presence of *replicates* in the *data filter* window.
3. Select the *Show* and *Include* check boxes to ensure that only the replicate measurements are used for the analysis.
4. Select *Analyze ▶ Fit Y by X*. Move *chamfer depth and threading depth* to the *Y,Response* box, and move *implant ID* to the *X, Factor* box.
5. Click *OK* to get the output shown in Figure 7.3.

Figure 7.3: Oneway Plots of Implant Data

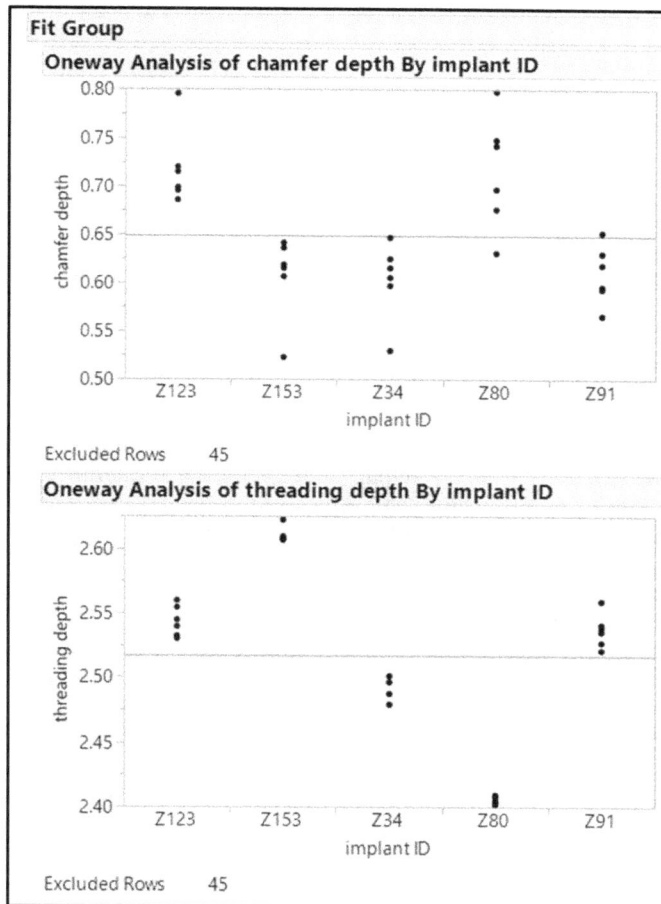

Plots of the five units with replicate measurements of chamfer depth and thread depth are shown in Figure 7.3. The view of the measurements illustrates the replicate variability of chamfer depth and threaded depth by the vertical spread of black dots for each implant ID. The team is interested in evaluating measurement uncertainty and in additional analysis of measurement variability.

Analysis of Means (ANOM) for Variances of Measured Replicates

The dot plots from the Fit Y by X platform indicate the presence of measurement variability. More detailed analysis is beneficial to interpret the variability and the potential effect on error included in the implant data. Analysis of means allows for the comparison of the variance in measures for each unit back to the summary variance. Initiate the analysis by pressing the Ctrl key and clicking the red triangle menu option next to *Oneway Analysis of chamfer depth By implant ID* header. Select *Analysis of Means Methods* ▶ *ANOM for variances* to add to the output shown in Figure 7.4.

Figure 7.4: ANOM for Variances Plot for Chamfer Depth

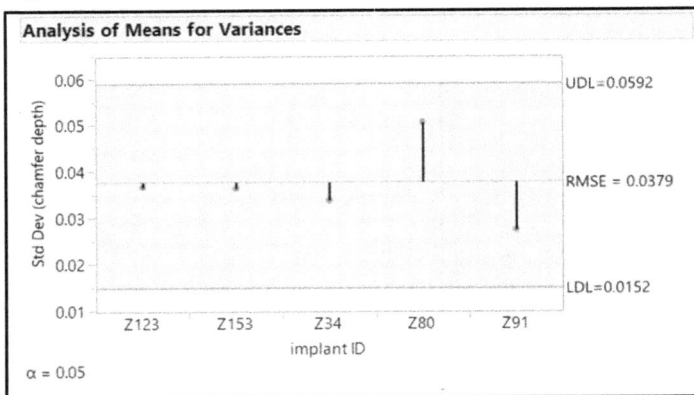

The Analysis of Means for Variances plot expresses the information about replicate variability in units of standard deviation. The root mean standard error (RMSE) is a summary of the replicate variability for all groups. The green dots with vertical lines originating from the RSME represent the standard error in measurement within each group and the distance each is from the RMSE. The light blue shaded region extending equally above and below the RMSE represents the amount of random variability that can be expected for +/- 2 standard deviations, which is a significance level of 0.05. There is no evidence that the variance for any of the five groups is significantly more than the RMSE.

The comparison of the within variance to the between variance converted to a percentage provides a useable summary to use as a simple estimate of measurement robustness. The scale of the analysis of means plot must be changed to variance to allow for the comparison. Use the red triangle menu beside the *Analysis of Means for Variances* header and select *Graph in Variance Scale*. The plot changes to that shown in Figure 7.5.

Figure 7.5: ANOM for Variances Plot for Chamfer Depth (in Variance Scale)

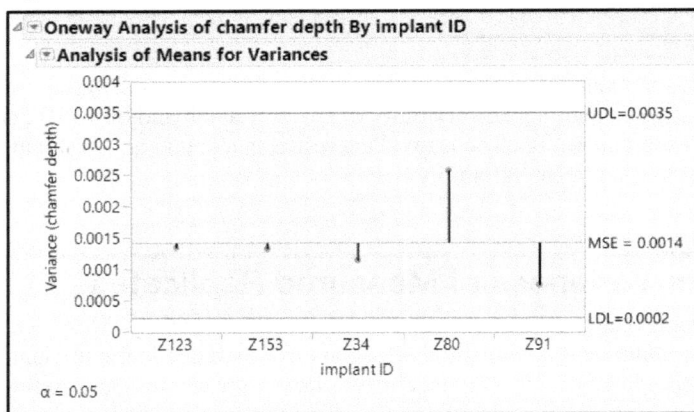

As noted previously, it is reasonable to expect that a robust measurement system will have variance within repeated measurements that is less than the variance between implants measured without replication. It is highly desirable for the variance within replicate measurements to be as small as possible compared to the non-replicate variance between implant IDs so that the system can precisely detect real differences between individual units. Recall from Figure 7.2 that the non-replicated chamfer depth measurements have a variance of 0.0021843. The mean standard error (MSE) from the analysis of means plot is compared to the variance of the non-replicate subgroup in order to create an estimate for the percentage of uncertainty that is present in the measurement process. The following calculations below are used estimate measurement uncertainty.

$$\frac{MSE_{replicates}}{variance_{nonreplicates}} = uncertainty$$

$$\frac{0.0014}{0.0022} = 0.636$$

$$\%uncertainty = 0.636 * 100$$

$$\%uncertainty = 63.6\%$$

The replicate variance for chamfer depth is approximately 64% of the variance between non-replicate measurements. With such a large amount of uncertainty present in the measurement system, it is very difficult to obtain a measurement value that accurately represents the physical feature of chamfer depth.

The concern over the ability to measure chamfer depth accurately has been noted. The team is interested in using the technique to evaluate the threaded depth measurements. The team repeats the analysis of means for threaded depth by using the same process they used for chamfer depth. The results are shown in Figure 7.6.

Figure 7.6: ANOM for Variances Plot for Threaded Depth

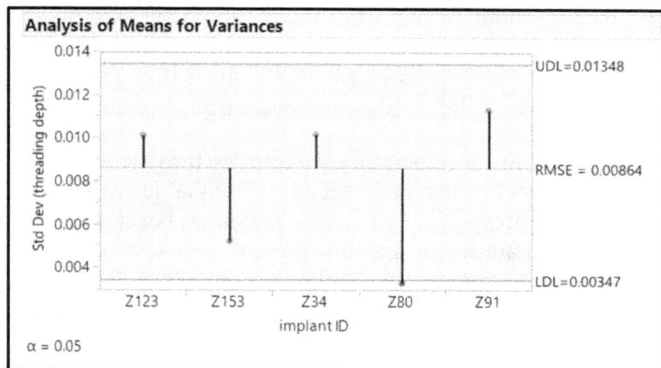

The Analysis of Means for Variances plot provides evidence that the variance in threaded depth for the Z80 implant ID is significantly less than the RMSE. The result is worth noting, but it is not likely to interfere with the quick estimate of %uncertainty. Use the red triangle menu beside the *Analysis of Means for Variances* header, and select *Graph in Variance Scale*. The plot changes to that shown in Figure 7.7.

Figure 7.7: ANOM for Variances Plot for Threaded Depth (in Variance Scale)

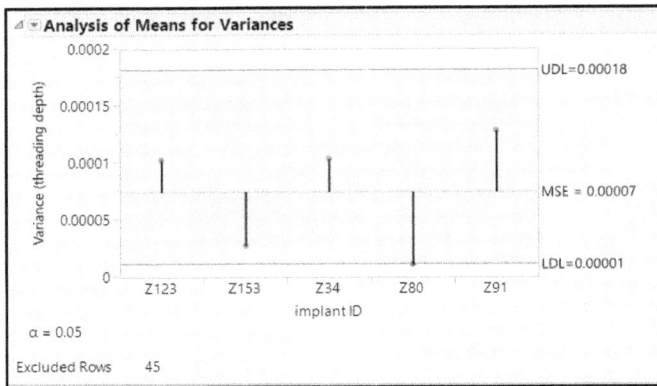

Recall from Figure 7.2 that the non-replicated threaded depth measurements has a variance of 0.0119705. The %uncertainty estimate indicates that the replicate variance for chamfer depth is approximately 0.5% of the variance between non-replicate measurements. The measurement system used for threaded depth has minimal uncertainty and can be considered robust.

Measurement Systems Analysis (MSA)

The simple check of the measurement system used for the dental implants indicates issues present with the chamfer depth. However, the analysis did not include enough data to gain details about the excessive measurement uncertainty. JMP offers multiple analysis options for measurement systems analyses in the *Quality and Process* submenu in the *Analyze* menu. The team uses systems analysis (MSA) techniques on data that is from a digital scale used to measure tablet weight in the quality control laboratory.

Accurate and precise tablet weight is required in order for the laboratory to calculate the dissolution critical quality attribute CQA with minimal measurement error. The team created a set of ten tablets with weights that are known to vary beyond the manufacturing specifications. This is very important because the measurement system must be able to detect tablets that are within specifications as well as outside of specifications. If specifications are not known, the objects measured should have outputs with the widest span of values possible to ensure that the system evaluation represents the entire population of subjects. Each tablet sample is given an identity value and is measured with three replications by three different technicians. Results from the study will illustrate the repeatability of measurements (replicate to replicate) as well as the reproducibility (technician to technician). The team was careful to include technicians who perform typical lab testing, and they designed a randomization plan to ensure that the order of the tablets is randomized for each trial. This example uses the data set *Tablet Scale MSA.jmp*.

Data has been provided and is in the proper structure for immediate analysis, as shown in Figure 7.8. Notice that a stacked table format is used with variables for the trial, standard order, randomized run order, tablet ID, and operators. Results for each combination of variables are noted in the *weight* column. The team uses JMP to create a design for an MSA by using the Design of Experiments platform. Chapter 9 provides detail for creating experimental designs, which is not discussed in this chapter.

Figure 7.8: Tablet Scale MSA data

	trial	std order	run order	tablet ID	operators	weight
1	1	3	1	3	1	0.0897
2	1	5	2	5	1	0.1095
3	1	9	3	9	1	0.0985
4	1	7	4	7	1	0.0996
5	1	6	5	6	1	0.0968
6	1	10	6	10	1	0.0998
7	1	2	7	2	1	0.0987
8	1	1	8	1	1	0.1213
9	1	8	9	8	1	0.0832
10	1	4	10	4	1	0.1021
11	1	17	11	7	2	0.0987
12	1	13	12	3	2	0.0896
13	1	14	13	4	2	0.1037
14	1	18	14	8	2	0.0738
15	1	12	15	2	2	0.0999
16	1	16	16	6	2	0.0967
17	1	15	17	5	2	0.1081
18	1	11	18	1	2	0.1225
19	1	20	19	10	2	0.0987
20	1	19	20	9	2	0.0981
21	1	24	21	4	3	0.1039
22	1	26	22	6	3	0.0962
23	1	23	23	3	3	0.0996
24	1	28	24	8	3	0.0748
25	1	22	25	2	3	0.0982
26	1	27	26	7	3	0.0999

Start by making sure that *Tablet Scale MSA.jmp* is open. Select *Analyze* ▶ *Quality and Process* ▶ *Measurement Systems Analysis* to open the *EMP Measurement Systems Analysis* window, shown in Figure 7.9. Move *weight* to *Y, Response*; *tablet ID* to *Part; Sample ID*; and *operators* to *X, Grouping*; and click *OK*. There are many options available in the *Measurement Systems Analysis* window, but the default settings are used for the remaining options.

Figure 7.9: EMP Measurement Systems Analysis Window

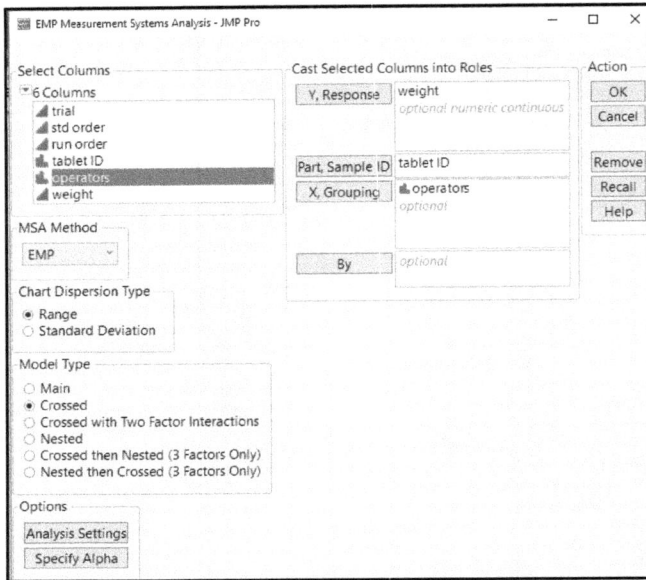

Figure 7.10: Measurement Systems Analysis for Weight Plots

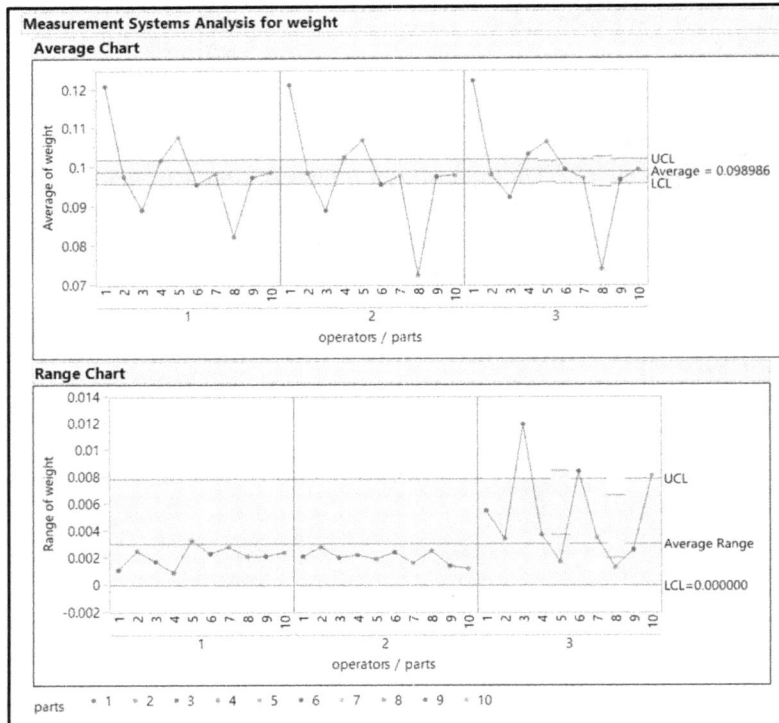

The initial plot of the measurement systems analysis output is of the averages of the measurement (weight) by part (tablet ID), grouped by operator. The figure shows the initial output for the MSA, but much more detail is available via red triangle menu options.

The expectation for the average chart is that most of points will be located outside of the zone of random variation, noted by the blue shaded area. The differences due to tablet ID should be significantly greater than random variation. If this is not the case, the measurement system includes excessive variation within replicates.

The range chart provides a large amount of information: the amount of variation within replicates; comparison of variation in results across the tablet IDs; and comparison of variation trends between operators. It is expected that the range of replicate values is minimal, so the values obtained represent real results. The average range in values is approximately 0.003, and the majority of measurements for operators one and two have less range than the summary average. It is clear that 4 of the 10 measurements taken by operator 3 include a range that is twice as much as the average range.

Interpretation of the results concludes that the trend in average weights for each tablet can be reproduced reliably, with an exception being tablet 8. The ranges in weights for the third operator indicate an inability to measure with precision similar to that of operators 1 and 2. Utilization of analysis options provides greater detail for diagnosing the measurement system. Use the red triangle menu option on the *Average Chart* and select *Show Data* to add individual measurements to the plot, as shown in Figure 7.11.

Figure 7.11: MSA Average Chart

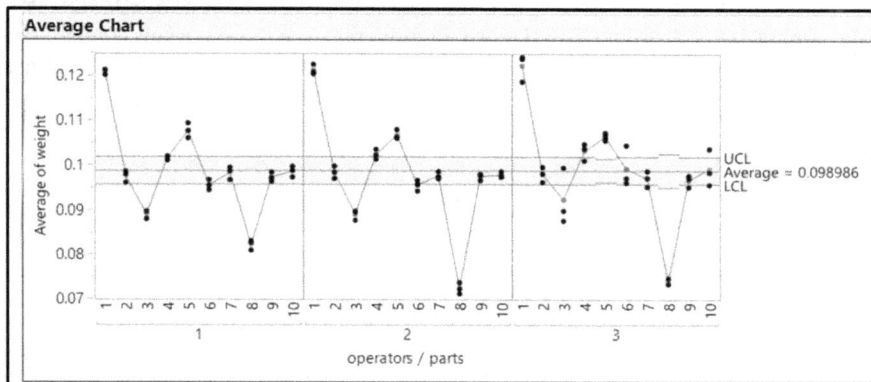

The plot with individual values enables the team to visualize data for both average values and the spread among values for each tablet ID. It is clear that differences are likely among the operators. Another option provides additional detail for the analysis. Use the red triangle menu option on the *Measurement Systems Analysis for weight* header and select *Parallelism Plots*. The resulting plot is shown in Figure 7.12.

Figure 7.12: Parallelism Plots

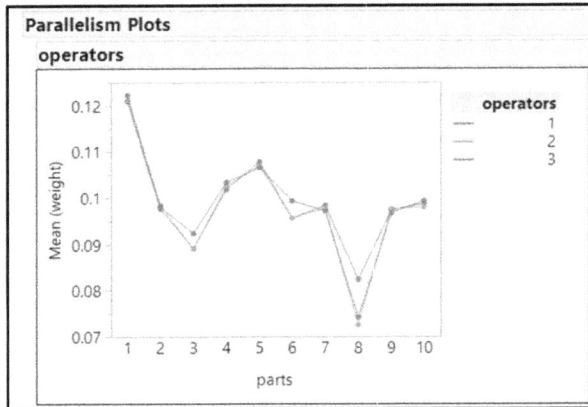

The parallelism plot is an overlaid comparison of the average values of the tablet IDs by operator. The plot in Figure 7.12 illustrates that tablet 8 has averages that differ between operators more than any other tablet. The average obtained by operator 1 is approximately 0.01 greater than the average from operators 2 and 3.

Detailed Diagnostics of Measurement Systems through MSA Options

One of the most valuable summaries is the plot of Variance Components. Use the red triangle menu option on the *Measurement Systems Analysis for weight*, and select *Variance Components* to obtain the plot shown in Figure 7.13.

Figure 7.13: Variance Components

Component	Variance Component	% of Total	20406080	Std Dev
operators	0.00000000	0.0		0.00000
parts	0.00012940	95.1		0.01138
operators*parts	0.00000242	1.8		0.00156
Within	0.00000418	3.1		0.00204
Total	0.00013600	100.0		0.01166

The variance components explain the source of the spread in measurement values. The variance component for *parts* (Tablet IDs) of 0.00012940 is 95% of the total of 0.00013600. The *Within* component represents the overall precision of tablet-to-tablet variation among replicate measurements and is 3.1% of the total. The *operators* and interaction of *operators*parts* make up 0% and 1.8% respectively. A high-performance measurement system can be identified by having the clear majority of variance among the parts and minimal variation among the measurement system components. The measurement system used to obtain tablet weights looks to be robust.

Evaluation of the Measurement Process (EMP) is the default technique of measurement systems analysis utilized by JMP. The technique is largely based on the methods presented in Donald J. Wheeler's book *EMP III Using Imperfect Data* (2006). The EMP method provides the information needed to get optimal performance from a measurement system. Another popular technique is Gage Repeatability and

Reproducibility, which is available by using the red triangle menu options on the *Measurement Systems Analysis for weight* header. You are encouraged to research the information about the techniques that is available in the JMP documentation and elsewhere to better understand the merits of each option. The following example uses EMP for the analysis. Use the red triangle menu next to the *Measurement Systems Analysis for weight* header, and select *EMP Results* to gain additional detail.

Figure 7.14: EMP Results

```
EMP Results

  EMP Test                                              Results  Description
  Test-Retest Error                                      0.002  Within Error
  Degrees of Freedom                                        60  Amount of information used to estimate within error
  Probable Error                                        0.0014  Median error for a single measurement
  Intraclass Correlation (no bias)                      0.9687  Proportion of variation attributed to part variation without including bias factors
  Intraclass Correlation (with bias)                    0.9687  Proportion of variation attributed to part variation with bias factors
  Intraclass Correlation (with bias and interactions)   0.9515  Proportion of variation attributed to part variation with bias factors and interactions
  Bias Impact                                                0  Amount by which the bias factors reduce the intraclass correlation
  Bias and Interaction Impact                           0.0173  Amount by which the bias factors and interactions reduce the intraclass correlation


  System                         Classification
  Current (with bias)            First Class
  Current (with bias and interactions)  First Class
  Potential (no bias)            First Class

  Monitor Classification Legend

                 Intraclass   Attenuation of  Probability of         Probability of
  Classification Correlation  Process Signal  Warning, Test 1 Only*  Warning, Tests 1-4*
  First Class    0.80 - 1.00  Less than 11%   0.99 - 1.00            1.00
  Second Class   0.50 - 0.80  11% - 29%       0.88 - 0.99            1.00
  Third Class    0.20 - 0.50  29% - 55%       0.40 - 0.88            0.92 - 1.00
  Fourth Class   0.00 - 0.20  More than 55%   0.03 - 0.40            0.08 - 0.92

   * Probability of warning for a 3 standard error shift within 10 subgroups
     using Wheeler's tests, which correspond to Nelson's tests 1, 2, 5, and 6.
```

The EMP results provide diagnostic information about the measurement system. Intraclass correlation is the proportion of variation in results that can be attributed to the part, which is the true measurement value. The intraclass correlation (with bias and interactions) of 95.2% indicates that the measurement system for tablet weight is very precise and accurate. The probable error explains that a measured weight value is likely to be approximately 0.0014 off from the true physical weight, which is a very small amount of error. The classification of the system (based on Wheeler's detection tests) is first class since less than 5% of the variation in measurement values can be attributed to the measurement system. The team now has detailed statistical proof that the tablet weights used for analytical calculations are contributing very little to overall error even though differences are known to exist among operators (lab analysts).

Variability and Attribute Charts for Measurement Systems

The tablet weight measurement can be used for the intended purpose without concern for measurement error, but this does not mean that the system is perfect. Opportunities exist for improvement even in high-performing systems. The tablet weight data has been augmented with the standard weights of tablets that have been obtained by a digital scale known to have accuracy of two more significant digits than the bench scale unit used in the MSA. Standard measurements are not always available due to the expense of extremely accurate instruments, but should always be considered to add the important dimension of bias to the analysis. Open *Tablet Scale MSA Data with Standard.jmp* for the next set of steps.

1. Select *Analyze ▶ Quality and Process ▶ Variability / Attribute Gauge Charts*.
2. Move *weight* to *Y, Response*; *operator* to *X, Grouping*; *Tablet ID* to *Part, Sample ID*; and *standard weight* to *Standard* in the *Variability / Attribute Gage*.
3. Use the defaults for all other options, and click *OK* to get the output shown in Figure 7.15.

Figure 7.15: Variability Plots

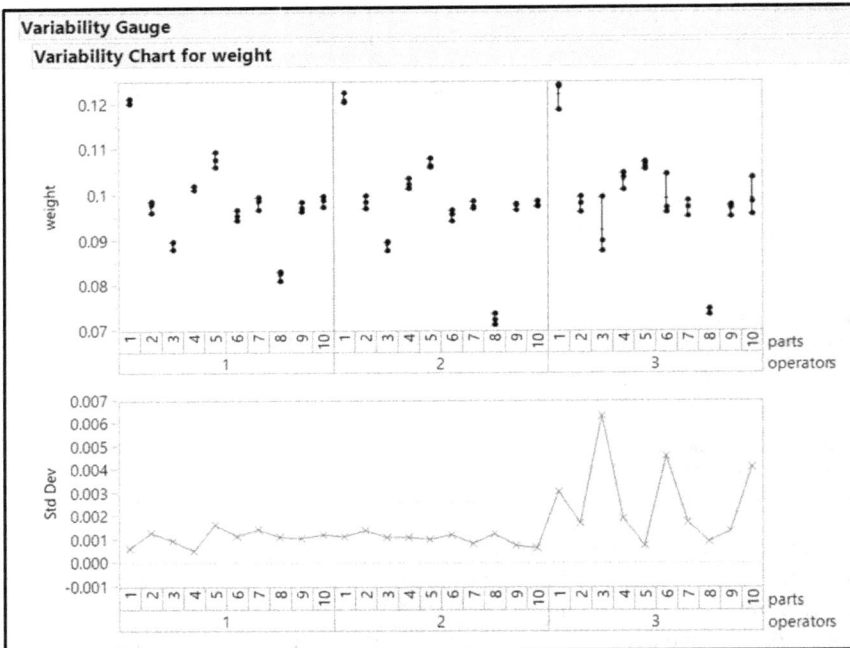

The Variability Gauge plots for weight and standard deviation provide similar information to the plots seen in the previous examples. One thing is very clear: improvement to the measurement system might be possible if the team can investigate and determine why the difference in variability exists between operator 3 and operators 1 and 2.

The trend in bias among measurements is obtained with a bias report. Use the red triangle menu option to select *Gauge Studies ▶ Bias Report* to obtain the output seen in Figure 7.16

Figure 7.16: Measurement Bias Report

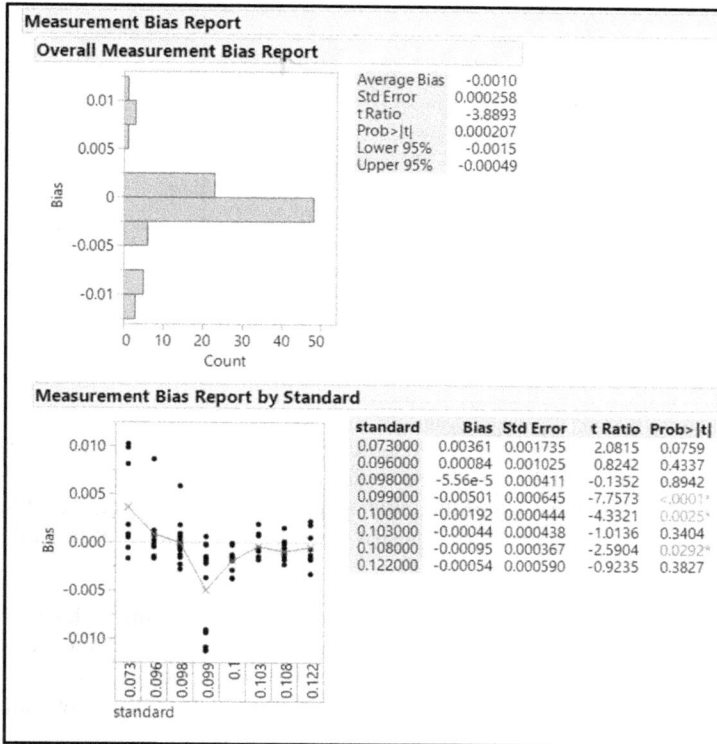

Measurement Bias Report

Overall Measurement Bias Report

	Average Bias	-0.0010
	Std Error	0.000258
	t Ratio	-3.8893
	Prob>|t|	0.000207
	Lower 95%	-0.0015
	Upper 95%	-0.00049

Measurement Bias Report by Standard

standard	Bias	Std Error	t Ratio	Prob>|t|
0.073000	0.00361	0.001735	2.0815	0.0759
0.096000	0.00084	0.001025	0.8242	0.4337
0.098000	-5.56e-5	0.000411	-0.1352	0.8942
0.099000	-0.00501	0.000645	-7.7573	<.0001*
0.100000	-0.00192	0.000444	-4.3321	0.0025*
0.103000	-0.00044	0.000438	-1.0136	0.3404
0.108000	-0.00095	0.000367	-2.5904	0.0292*
0.122000	-0.00054	0.000590	-0.9235	0.3827

The minimal amount of bias is desired for the measurements with results as close to 0 bias as possible. The average bias indicates that the system tends to measure weights that are slightly below the standard (-0.001). The *Measurement Bias Report by Standard* plot illustrates that tablets with a standard weight of 0.099 have the most negative bias and might be influencing the average. Tablets with standards lower than 0.099 tend to measure a bit higher than the standard weight.

Further investigation might provide opportunities for improved accuracy through reductions in bias. It would be very helpful to know the Tablet ID of the measurements with the most bias since some tablets share the same standard weight. The following steps differentiate the observations by using the row legend trough graphic options.

1. Right-click in the white space of the *Bias by Standard* plot and select *Row Legend*.
2. Select the *tablet ID* column and notice that the points are colored with the JMP default color scheme. Use the drop-down menu to choose an optimum color scheme, and explore the markers to best define the points. The example uses the default options.
3. Click *Make Window with Legend* to define the colors by *Tablet ID*. The example uses options that allow for visual differentiation in this book.

Figure 7.17: Measurement Bias Report with Tablet ID Colors

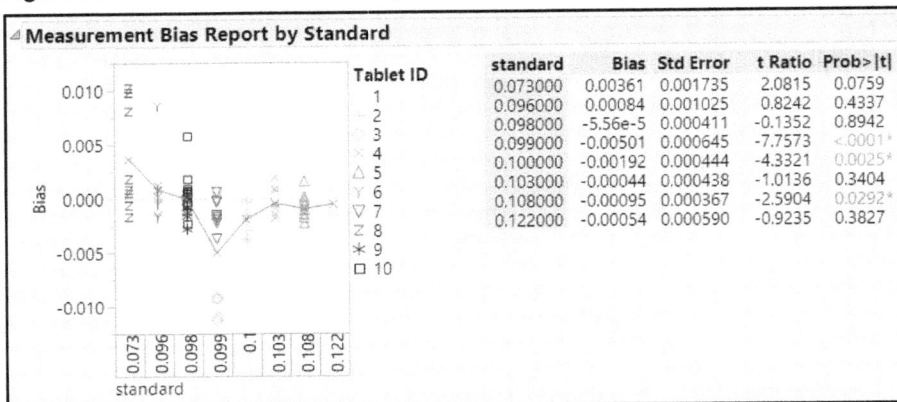

standard	Bias	Std Error	t Ratio	Prob>|t|
0.073000	0.00361	0.001735	2.0815	0.0759
0.096000	0.00084	0.001025	0.8242	0.4337
0.098000	-5.56e-5	0.000411	-0.1352	0.8942
0.099000	-0.00501	0.000645	-7.7573	<.0001*
0.100000	-0.00192	0.000444	-4.3321	0.0025*
0.103000	-0.00044	0.000438	-1.0136	0.3404
0.108000	-0.00095	0.000367	-2.5904	0.0292*
0.122000	-0.00054	0.000590	-0.9235	0.3827

The low bias for tablet 3 was identified by discussions with the technicians who executed the studies. The technicians explained that the tablet was dropped to the table surface during initial handling of the specimen, and it is expected that a small chip on the edge of the tablet might have resulted from handling damage after the standard weight was taken. The reason for the positive bias for the lighter weight tablets could not be determined.

Through further investigation, the team found out that operator 3 uses a sonicating mixer that is located on the bench top near the scale to enable efficient multitasking. Operators 1 and 2 did not use as they weighed the tablets. The team moved the mixer to another location in the lab as a preventive action to improve precision in weights. The team had operator 3 repeat the weight trials and noted a pattern that now matched operators 1 and 2, confirming the improvement.

Practical Conclusions

It is tempting for teams to assume that the measurement values obtained from instrumentation accurately and precisely represent the true physical value. Without knowledge of MSA, technicians explain that the instruments are calibrated and assumed to be free of measurement error. Measurement systems analysis involves both the instrument or device as well as the environment and methods utilized to obtain values and is a more reliable gage of measurement performance. It is always a best practice to ensure that measurement systems are at least verified through replication or are run through a full MSA to make sure that conclusions made from the measurement data do not include excessive measurement uncertainty and error.

The tablet weight example illustrates that even a good measurement system can be improved. The EMP results illustrate that the system is acceptable even though error was identified for tablet 8 and for the replicates of operator 3. The team was able to make significant improvements to the measurement process through the details offered by MSA.

JMP offers many tools for a team to easily quantify the amount of error and variability that are inherent to a measurements system and that qualify the system for use to make the best decisions possible. All data collection events should include planning for verification of a measurement system.

Exercises

E7.1—A project is coming up in later chapters involving a thin, plastic molded cover that provides a sterile barrier to surgical handle covers. A technique was developed for technicians to follow so that they can obtain the minimum wall thickness of covers. The timing is very tight for the improvement project that oversees the manufacturing process for the handle covers. A world-class measurement system includes 10% measurement uncertainty; an acceptable measurement system might have up to 25% measurement uncertainty. A random sample of 42 handle covers was collected from the process and labeled as test units. Twelve of the test units were measured four times each to obtain some replicate values.

1. Open *surgical handle thickness measurements.jmp*.
2. Use the analysis that was completed for the dental implant measurements to determine the % gross uncertainty.
3. How would you present this information to the project stakeholders who are very motivated to move forward with improvements?

E7.2—A new drug product is in the late stages of development, and production readiness activity for the line that fills bottles with capsules is ongoing. The process includes a checkweigher device to ensure that no bottles get through that have fewer capsules than the count noted on the label. Ten bottles were marked and added to the flow of the process just ahead of the checkweigher. Bottles are retrieved and run through at random until each bottle has gone through the weigher at least three times. The average weight of a capsule is 900 mg, and each bottle includes a count of 50; the target weight of a filled bottle with 50 capsules is 66 grams. Four of the bottles were intentionally manipulated to either contain one or two missing capsules or one or two extra capsules to ensure that the full range of possible measurement is explored. Each bottle was weighed by a high-precision digital scale in the analytical laboratory to record a standard weight.

Use the file *bottle checkweigher data.jmp* for this problem.

1. Use the Distribution platform to get an overall view of *test weight*. What is the location and spread of the weights?
2. Select *Analyze ▸ Quality and Process ▸ Variability Charts* to study the measurement process. Be sure to include *std weight* as the standard. Is there excessive variability for any of the bottle numbers? Is the bias excessive? Which of the bottle numbers were the ones with extra or missing capsules?
3. Select *Analyze ▸ Quality and Process ▸ Measurement Systems Analysis* to run an EMP study. Keep in mind that because the feed of the checkweigher is automated, there is no reproducibility involved, so the reproducibility value is 0.
4. Divide the probable error and by the capsule weight. The goal is to ensure that the system can detect a bottle with a missing capsule. How would you use the probable error as a percent of capsule weight as evidence of a robust measurement system?

E7.3—We have been studying differences in the outer diameter (O.D.) of surgical tubes in previous chapter exercises. There is concern that variation in the measurement process could be creating bias in statistical analysis. Technicians report that the toughest tube to measure is the 3 mm O.D. tube. Running a measurement systems analysis on the 3 mm O.D. tube will provide the most conservative results, so it is used as the basis for the study. Process technicians worked with the engineers to produce two tubes at the extremes of the range of O.D. values. The extreme tube samples have an O.D. that is a span that is 125% of the specification limits. Two other tubes are mad e with an O.D. that is at the minimum O.D. specification and maximum. The six remaining tubes are selected at random from production. The engineered and randomly selected tubes represent the range of measurement studied includes O.D.s that are within and outside of the 3.0 mm +/- 0.1 mm O.D. specifications.

1. Open *surgical tube OD measurements.jmp*.
2. Run measurement systems analysis by using the EMP techniques and grouping on technician.
3. How would you present the results to the project stakeholders?
4. Are there any additional evaluations that you can suggest that will result in the most precise and accurate measurement system possible?

Chapter 8: Using Predictive Models to Reduce the Number of Process Inputs for Further Study

Overview

Structured, multivariate experimentation allows for the study of many inputs and outputs simultaneously to efficiently gain insight into a process. The Design of Experiments (DOE) platform offers many powerful options for statistical methods used to execute randomized experiments. The major drawback of structured, multivariate experimentation is the amount of planning and resources needed to design and run the experiments. Teams are often faced with many more inputs of potential interest than can be studied with a reasonable amount of resources. This chapter covers the use of predictive modeling of observational process data to narrow down the number of input variables for further study. There are many techniques for predictive modeling available in JMP. A few of the more common approaches are examined for the purpose of variable reduction.

The Problem: Thin Surgical Handle Covers

Michelyne works within the medical device industry and is managing a project involving surgical kits used by hospitals and surgical clinics. The kits include thin, disposable plastic sterile covers that go over the handles of surgical lamps. Her team has been alerted to a growing number of complaints received from customers who report that lamp handle covers are not staying in place as designed. A cover that is easily detached from the handle creates unacceptable risk to patients because it can drop onto them during a procedure, and at minimum it will break the sterility barrier.

Examination of customer returns identifies that the bad handle covers have areas of wall stock that are thinner than the minimum specification of 0.50 mm. Many of the handle covers split in the thinned areas. Therefore, the quality team has identified the root cause for the loose covers as thin material. Michelyne has contacted the thermoforming facility that manufactured the covers to request a set of random process data. The team plans to use the data from several process inputs to create predictive models for material thickness. The hope is that the modeling results will narrow the list of inputs to a reasonable number so that structured experimentation can be designed and run. The manufacturer has a growing interest in process improvements because they must pay for containment measures to protect against more covers with thin material from being received by customers.

Data Visualization with Dynamic Distribution Plots

The surgical handle cover manufacturer submitted a table of more than 110 random in-process samples with information about 14 process inputs. Open the file *Surgical handle cover data.jmp* to see the table shown in Figure 8.1.

Figure 8.1: Surgical Handle Cover Data Table

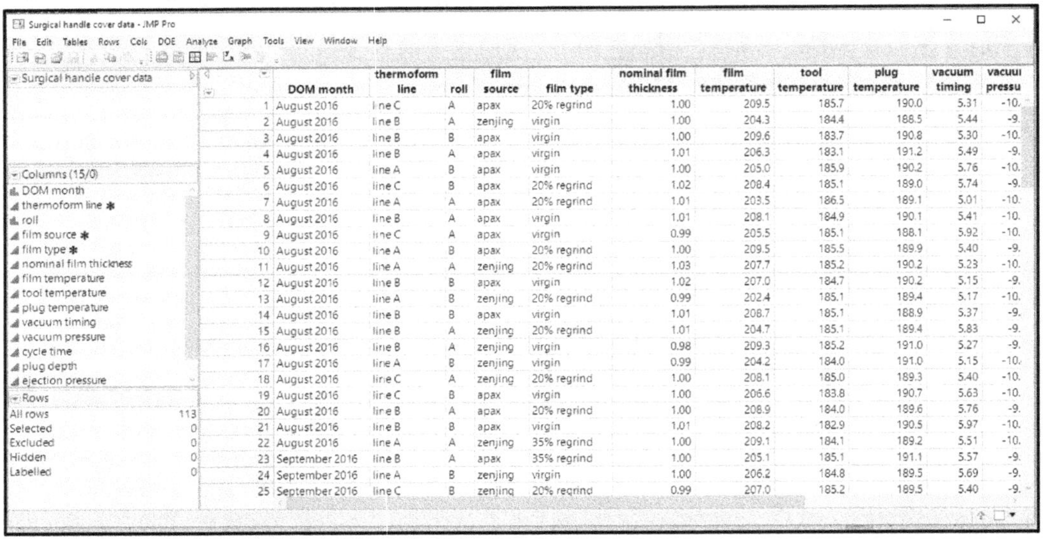

	DOM month	thermoform line	roll	film source	film type	nominal film thickness	film temperature	tool temperature	plug temperature	vacuum timing	vacuui pressu
1	August 2016	line C	A	apax	20% regrind	1.00	209.5	185.7	190.0	5.31	-10.
2	August 2016	line B	A	zenjing	virgin	1.00	204.3	184.4	188.5	5.44	-9.
3	August 2016	line B	B	apax	virgin	1.00	209.6	183.7	190.8	5.30	-10.
4	August 2016	line B	A	apax	virgin	1.01	206.3	183.1	191.2	5.49	-9.
5	August 2016	line A	B	apax	virgin	1.00	205.0	185.9	190.2	5.76	-10.
6	August 2016	line C	B	apax	20% regrind	1.02	208.4	185.1	189.0	5.74	-9.
7	August 2016	line A	A	apax	20% regrind	1.01	203.5	186.5	189.1	5.01	-10.
8	August 2016	line B	A	apax	virgin	1.01	208.1	184.9	190.1	5.41	-10.
9	August 2016	line C	A	apax	virgin	0.99	205.5	185.1	188.1	5.92	-10.
10	August 2016	line A	B	apax	20% regrind	1.00	209.5	185.5	189.9	5.40	-9.
11	August 2016	line A	A	zenjing	20% regrind	1.03	207.7	185.2	190.2	5.23	-10.
12	August 2016	line B	B	apax	virgin	1.02	207.0	184.7	190.2	5.15	-9.
13	August 2016	line A	B	zenjing	20% regrind	0.99	202.4	185.1	189.4	5.17	-10.
14	August 2016	line B	B	apax	virgin	1.01	208.7	185.1	188.9	5.37	-9.
15	August 2016	line B	A	zenjing	20% regrind	1.01	204.7	185.1	189.4	5.83	-9.
16	August 2016	line B	A	zenjing	virgin	0.98	209.3	185.2	191.0	5.27	-9.
17	August 2016	line A	B	zenjing	virgin	0.99	204.2	184.0	191.0	5.15	-10.
18	August 2016	line C	A	zenjing	20% regrind	1.00	208.1	185.0	189.3	5.40	-10.
19	August 2016	line C	B	apax	virgin	1.00	206.6	183.8	190.7	5.63	-10.
20	August 2016	line B	A	apax	20% regrind	1.00	208.9	184.0	189.6	5.76	-9.
21	August 2016	line B	B	apax	virgin	1.01	208.2	182.9	190.5	5.97	-10.
22	August 2016	line A	A	zenjing	35% regrind	1.00	209.1	184.1	189.2	5.51	-10.
23	September 2016	line B	A	apax	35% regrind	1.00	205.1	185.1	191.1	5.57	-9.
24	September 2016	line A	B	zenjing	virgin	1.00	206.2	184.8	189.5	5.69	-9.
25	September 2016	line C	B	zenjing	20% regrind	0.99	207.0	185.2	189.5	5.40	-9.

The column properties include the option to set specifications for a variable, which is a time-saving feature. Once specifications have been set, they show up on plots that are made for the variable. Minimum thickness has only a lower specification, which is set by selecting *min thickness*. The variable column will be shaded, and the column header will be bright blue in color when selected. To access column options, right-click in the column. Select *Column Properties* ▶ *Spec Limits* and type 0.48 in the *Lower Spec Limit* box located in the lower right of the *min thickness* window. Be sure to select the *Show as graph reference lines* check box, and then click *OK* to confirm. Alternatively, you can select *Cols* ▶ *Column Info*, and then select *Column Properties* ▶ *Spec Limits* to enter the limit.

A great first step for analyses is to visualize the data with the Distributions platform. Use the dynamic linking in JMP to identify non-random patterns in the plots by completing the following steps.

1. Select *Analyze* ▶ *Distributions* to open the *Distributions* window.
2. Move the output variable *min thickness* to the top of the *Y, Columns* box so that it is the first plot.
3. Move all of the remaining variables to the *Y, Columns* box so that they are below *min thickness*.
4. Select the *Histograms Only* check box for clarity of the information shown, and then click *OK* to get the output.
5. With the *Distributions* output in view, use the red triangle menu options next to the *Distributions* header to select *Arrange in Rows*.
6. Enter 5 to get all histograms in view on the plot.

7. Press the Ctrl key while sizing the frames of the histograms to size all 15 plots at the same time.
8. Select the bars that are at and below 0.50 mm from the *min thickness* histogram by holding the Shift key while clicking the bars.

Figure 8.2: Distributions

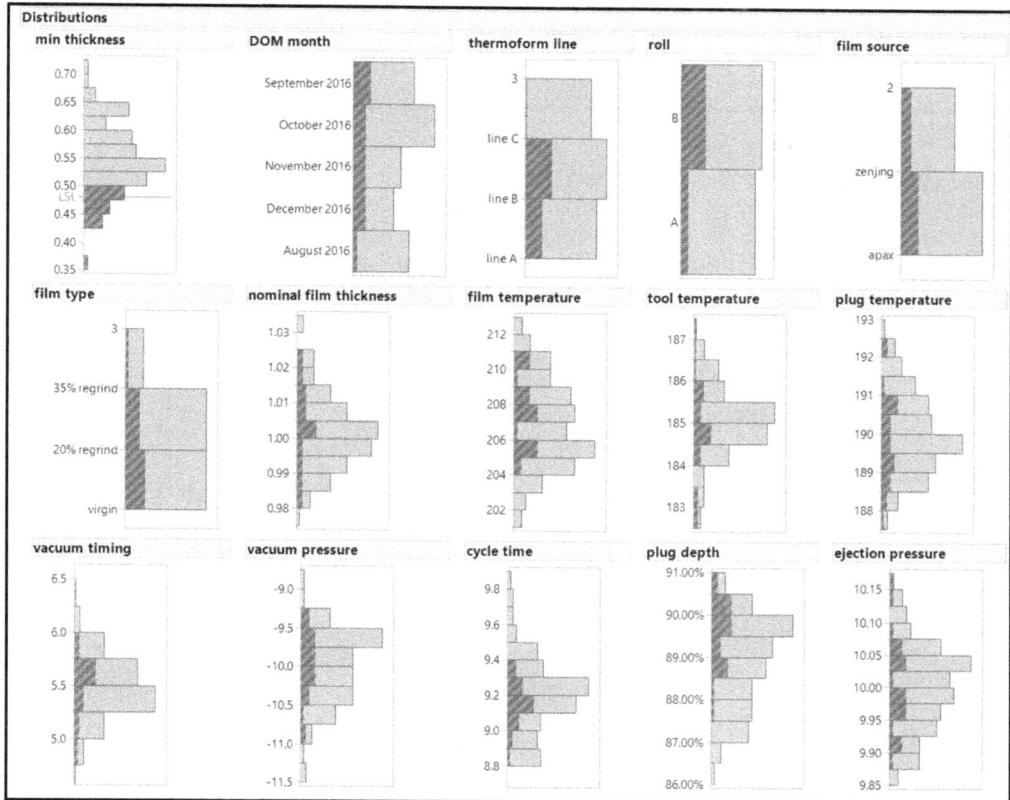

Figure 8.2 illustrates the result of dynamic linking. All 14 process inputs have the portions of the histograms darkened with diagonal lines that relate to covers made with a min thickness that is 0.50 mm or less. The goal of this high-level analysis is to determine whether there are any non-random patterns in the process input histograms. The first one you notice is the thermoform line histogram; line C produced almost no covers with substandard minimum thickness. Another possible pattern is in the plug depth histogram; higher percentages seem to capture most of the thin covers. An example of a histogram with a random pattern of darkened bars and of no interest is DOM month.

The results of visualizing the data using dynamic linking and the Distributions platform offers information yet is subjective. A non-random pattern interesting to one person might not be of interest to another person due to the perceived magnitude of the trend. An advantage of using Distributions to visualize data is the ability to see the general pattern of each variable. Outlier results, skewed distributions, and unequal proportions are quickly evident to the analyst. The data collected for the thermoforming process inputs that were monitored is relatively free of these patterns.

The data visualization exercise illustrates that thin covers might be related to the thermoform line, cycle time, and plug depth. Many options in JMP can provide more detailed statistical evidence of potential relationships through predictive modeling. The following sections are limited to a few approaches. However, you are encouraged to explore the rich resources offered through JMP and beyond to gain confidence in using other techniques. The field of predictive modeling is advancing rapidly, and new releases of JMP and JMP Pro are likely to provide additions to the brief coverage of topics in this book.

Basic Partitioning

Data visualization provided some clues to the process inputs that might be related to changes in thickness of the surgical handle covers. The partition technique in JMP offers a powerful, flexible set of tools that allow for exploration of wide data sets. Wide data sets include multiple variable columns and a relatively limited number of rows. One of the best features of partitioning is the fact that it works for both discrete and continuous variables in one model. The file *Surgical handle cover data.jmp* includes a few rows that are missing values, such as row 76 of the plug temperature variable. Partitioning allows for the use of the full set of data values by including an algorithm that estimates the value for the missing observation. The feature is particularly useful for large sets of process data. Select *Analyze* ▶ *Predictive Modeling* ▶ *Partition* to obtain the *Partition* window shown in Figure 8.3.

Figure 8.3: Partition Window

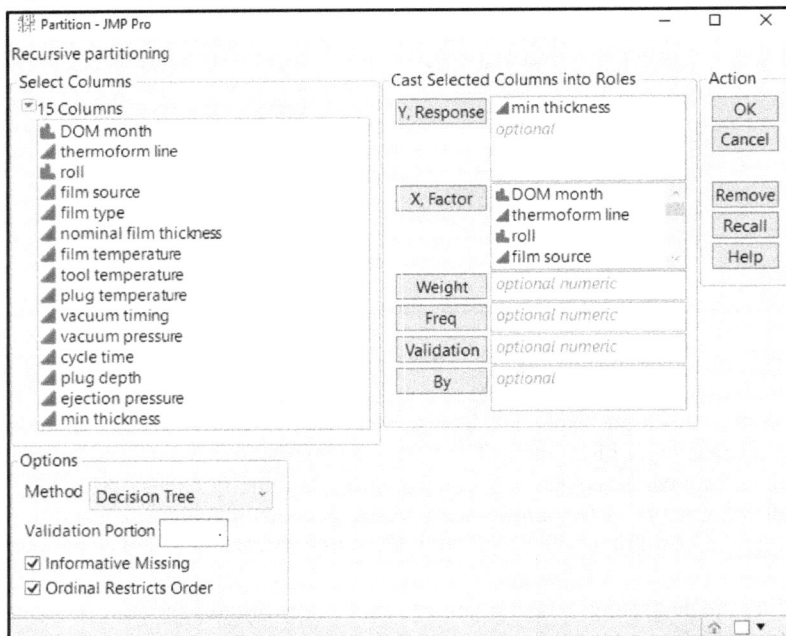

The following steps set up a partition model.

1. Move *min thickness* to the *Y, Response* box.
2. Move all other variables into the *X, Factor* box.
3. Be sure that the *Informative Missing* check box in the *Options* area is selected.
4. Figure 8.3 illustrates the JMP Pro functionality that lets you choose the method. The default method is decision tree.
5. Click *OK* to get the output shown in Figure 8.4.

Figure 8.4: Partition Output

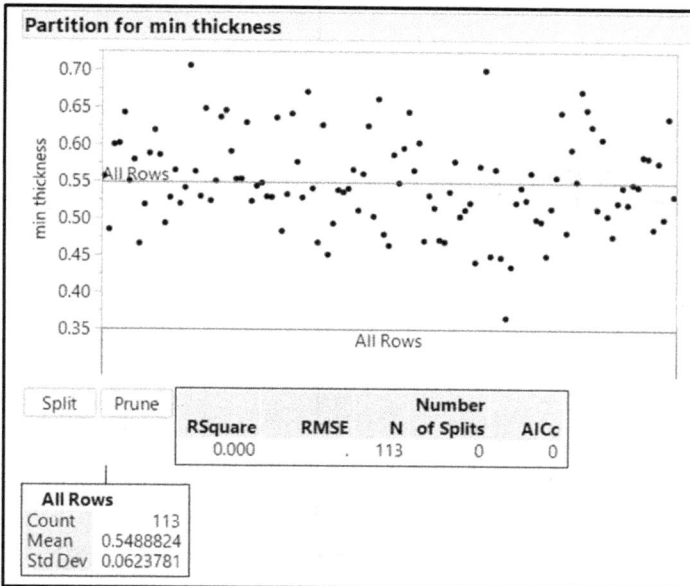

The initial output includes the average of 0.55 mm for the min thickness, with each observation noted as a black dot. The minimum observed value is around 0.37 mm and the maximum value is just over 0.70 mm. Place the pointer over each observation to see the row number label. The table below the plot provides model fit values that initiate as r-square = 0. At the bottom of the output, the decision tree is initiated as an *All Rows* box. As you click the *Split* button, the partition algorithm utilizes all the data to detect a variable that is related to the biggest average difference in min thickness. Click *Split* once to get the output in Figure 8.5, which detects the input of the model related to the largest difference in min thickness.

Figure 8.5: Partition with a Single Split

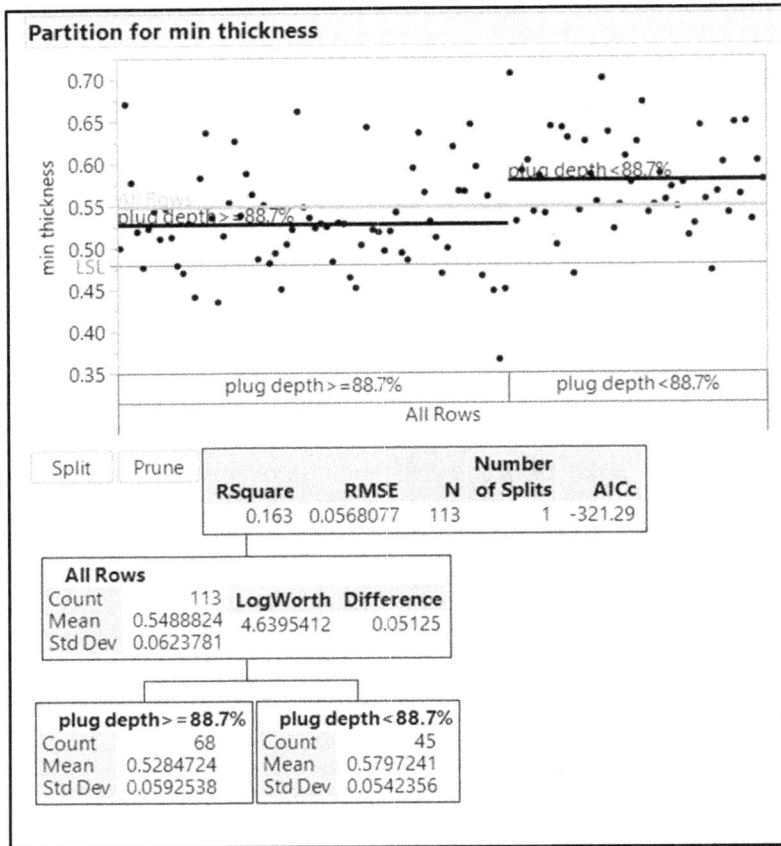

The single split in the output is illustrated by two horizontal lines added to the dot plot. The observations are grouped by the plug depth input: greater than or equal to 88.7%, and less than 88.7%. The r-square = 0.163 model fit indicates that the split explains 16.3% of the variation in min thickness. The split includes a root mean standard error (RMSE) of 0.057. Hence, the random noise in the data is minimal. The decision tree node for higher percentages of plug depth includes 68 observations and a mean min thickness of 0.528 mm. The node for lesser values of plug depth includes 45 observations and a mean min thickness of 0.580 mm. The difference in average min thickness between the nodes is 0.052, large enough to be considered practically relevant by the subject matter experts.

Additional splits provide detail for the potential for other inputs to have an effect on min thickness. Continue to click *Split* until no further splits occur. The modeling of this data set results in a maximum of 17 splits. Use the red triangle menu options by the *Partition* header to select the *Display Options ▶ Show Tree* (*Show Tree* is enabled by default, so this action disables *Show Tree*) to condense the output. Use the red triangle menu options to select *Split History*. The result is shown in Figure 8.6. Figure 8.7 provides the model fit information for the 17-split partition.

Figure 8.6: Split History of Partitioning

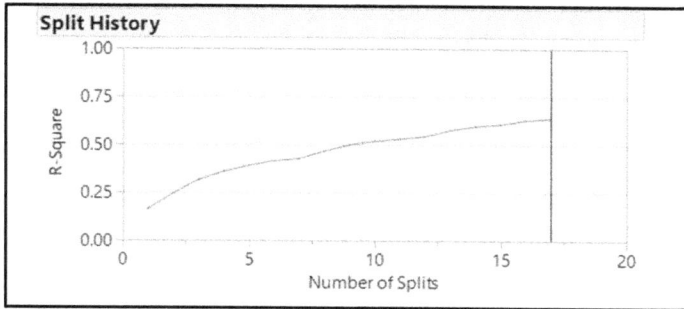

Figure 8.7: Partition Model with 17 Splits

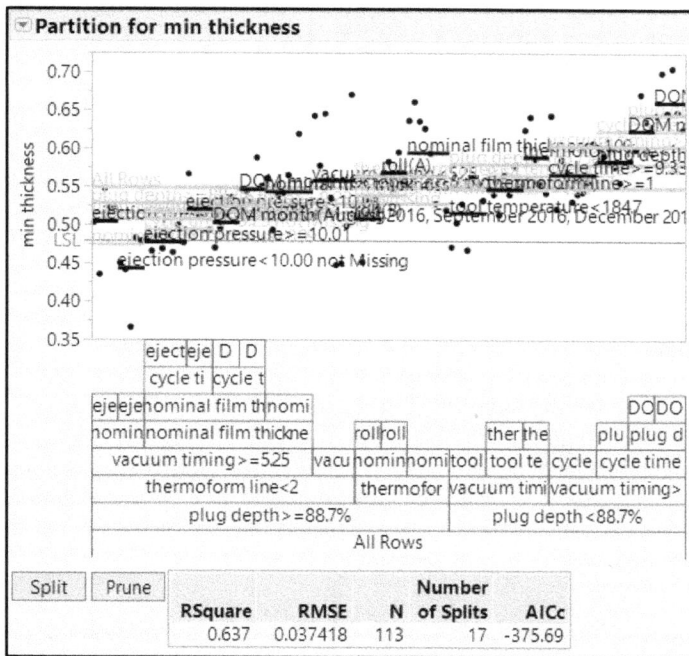

The split history plot shows how the r-square fit of the model increases as more splits are added to the decision tree. Figure 8.7 indicates that the 17-split model explains approximately 64% of the variability in min thickness (r-square = 0.637), and the fit improves steadily from 1 to 17 splits. The root mean square error has decreased to 0.037 for the 17-split model. Use the red triangle menu options one more time to select the *Column Contributions* plot in Figure 8.8.

Figure 8.8: Basic Partition Column Contributions

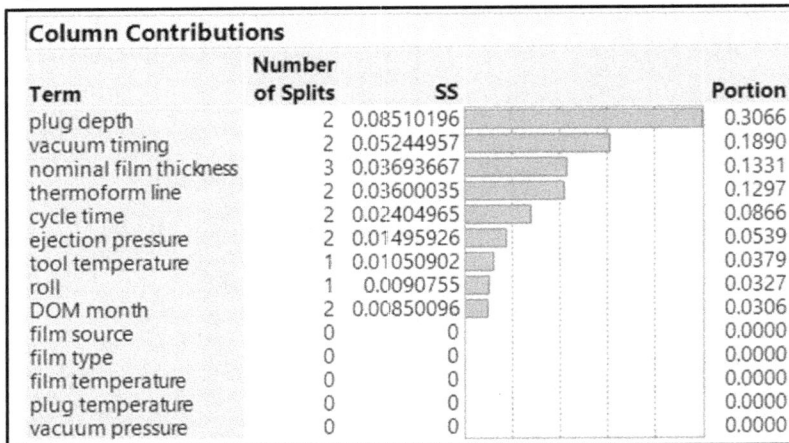

Column Contributions

Term	Number of Splits	SS		Portion
plug depth	2	0.08510196		0.3066
vacuum timing	2	0.05244957		0.1890
nominal film thickness	3	0.03693667		0.1331
thermoform line	2	0.03600035		0.1297
cycle time	2	0.02404965		0.0866
ejection pressure	2	0.01495926		0.0539
tool temperature	1	0.01050902		0.0379
roll	1	0.0090755		0.0327
DOM month	2	0.00850096		0.0306
film source	0	0		0.0000
film type	0	0		0.0000
film temperature	0	0		0.0000
plug temperature	0	0		0.0000
vacuum pressure	0	0		0.0000

The column contributions plot provides a summary of the 14 process inputs used within the model in order of their contribution to explaining changes in min thickness. Plug depth contributes 30.7% of the portion of variation in min thickness that can be explained by the model and is the input of greatest interest. Vacuum timing contributes a 19% portion of the variance that can be explained with the model. Five of the 14 processing inputs have no detectable influence on min thickness and are not of much interest for further study.

Partitioning has quickly and easily reduced the number of process inputs that might have influence on min thickness. The column contributions tell us that the top four process inputs add to a proportional contribution that is more than 75% of the amount of change in min thickness explained by the model. The proportion contributed relates to the fit of the overall model. Therefore, the 75% portion contributed is multiplied by the 64% fit, indicating that four inputs explain up to 48% of the changes in min thickness. Michelyne and her team have a much better chance of convincing the manufacturer to provide resources to study four process variables rather than a large study of 14. Given the cost and severity of the problem at hand, the team uses simple tools to double-check the model and ensure that the estimates are not overfit to random changes in inputs that really have little to do with changes in min thickness.

Partitioning with Cross Validation

One way to adjust a predictive model for potential overfitting is to use the cross validation technique. Cross validation splits the observations randomly into a given number of subgroups (K). A model is created for using each subgroup as a validation set and the remaining data as a training set. The model with the best validation statistic is used as the final model. The process is used to mitigate the potential of overfitting of the model to random variation. Use the red triangle menu option by the *Partitioning* header to select *K Fold Cross validation*. Keep the default value of 5, and click *OK* to add it to the output.

Figure 8.9: Partition with K-fold Cross validation

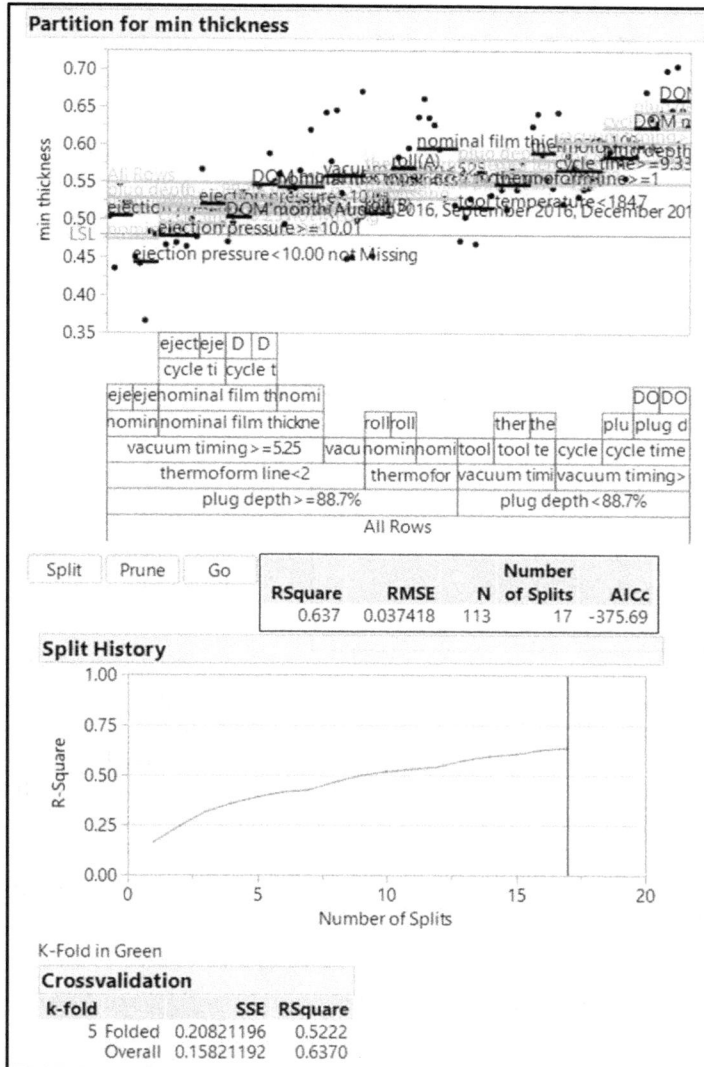

Cross validation indicates that the overall model with 17 splits is overfit; the folded fit of r-square = 0.52 is ten percentage points lower. The split history plot does not change when cross validation is added to an existing partition model, but JMP includes a shortcut to do this. Complete the following steps to update the split history plot with cross validation.

1. Click the red triangle menu for the *Partitioning* output.
2. Select *Redo ▶ Relaunch Analysis*.
3. Do not change the columns in the *Y, Response* or *X, Factor* boxes, and click *OK*.
4. In the *Partition for min thickness* window, use the red triangle menu to select *Split History* and *K-fold Cross validation*. You can deselect all the display options for clarity.

5. Click *Go* to proceed and obtain the output in Figure 8.10.

Figure 8.10: Partition with Automated K-fold Cross Validation

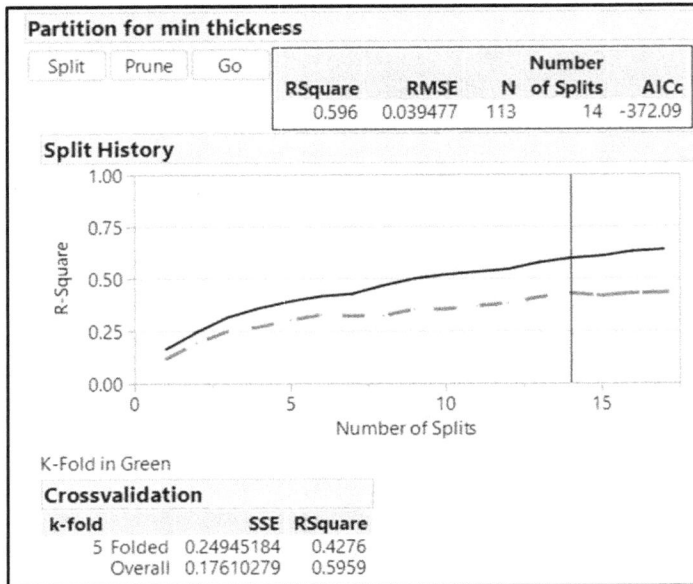

Partition for min thickness

	RSquare	RMSE	N	Number of Splits	AICc
	0.596	0.039477	113	14	-372.09

Split History

K-Fold in Green

Crossvalidation

k-fold	SSE	RSquare
5 Folded	0.24945184	0.4276
Overall	0.17610279	0.5959

Your output will vary each time you use the technique because each cross validation trial is a randomized subgrouping of the data. The automated cross validation you initiate by clicking the *Go* button stops splits when the separation in trends between the two models is significant. The min thickness model with 14 splits is optimum in this example. There is evidence of overfitting since the overall model r-square = 0.60 and the folded model r-square=0.43. The *Split History* figure illustrates that a model with no more than six splits can have the least amount of overfitting. Mitigation of overfitting comes at a cost since a six-split model has a reduced r-square value of less than 0.50. The next point of interest is the amount of change contributed by the inputs for min thickness. Use the red triangle menu options to select the *Column Contributions* plot and add it to the output of the 14-split model.

Figure 8.11: Partition (K-fold Cross validated) Column Contributions

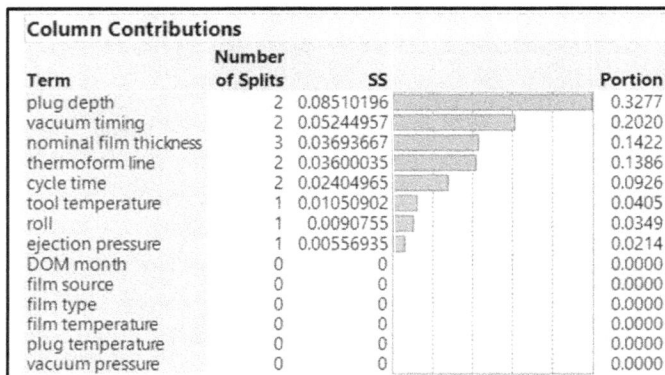

Column Contributions

Term	Number of Splits	SS	Portion
plug depth	2	0.08510196	0.3277
vacuum timing	2	0.05244957	0.2020
nominal film thickness	3	0.03693667	0.1422
thermoform line	2	0.03600035	0.1386
cycle time	2	0.02404965	0.0926
tool temperature	1	0.01050902	0.0405
roll	1	0.0090755	0.0349
ejection pressure	1	0.00556935	0.0214
DOM month	0	0	0.0000
film source	0	0	0.0000
film type	0	0	0.0000
film temperature	0	0	0.0000
plug temperature	0	0	0.0000
vacuum pressure	0	0	0.0000

The plot of column contributions for a cross validated model includes the same top four inputs as did previous modeling. The portion fit differs for the inputs with the cross validated technique. The portions of contributions of the top four inputs total 81%, and the folded model fit is r-square = 0.43. The largest amount of change in min thickness that can be explained by the four top inputs is 35% for a 14-split cross validated partition model.

Partitioning with Validation (JMP Pro Only)

Another method that is used to mitigate the potential for overfitting is holding back a portion of the data set for validation. The subset is run as a separate partition model that is compared to the remaining data set (referred to as the training set; the validation subset is also known as the test set).

To begin, click the red triangle menu for the *Partitioning* output and select *Redo* ▶ *Relaunch Analysis*. Notice that there is a *Validation Portion* field in the *Options* section. You could use an infinite number of potential proportions for validation. This example uses the proportion 0.15 (15%), as shown in Figure 8.12. Click *OK* to get the partition model output.

Figure 8.12: Partition Dialog Box with Validation Portion

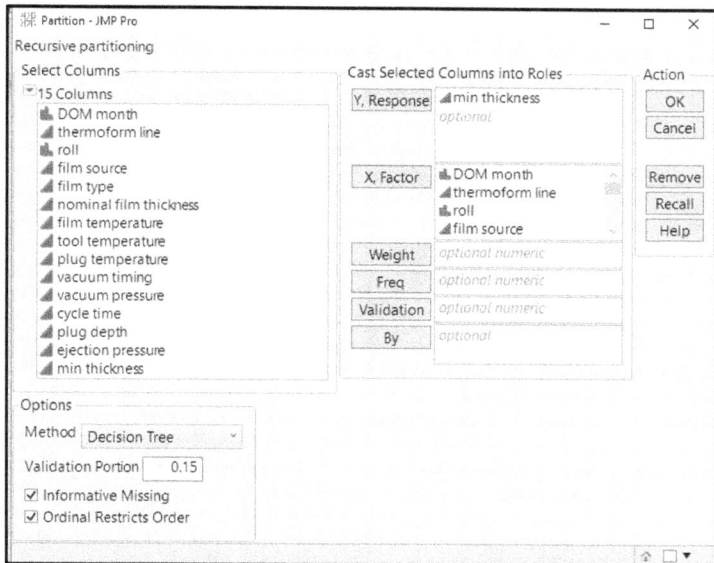

JMP Pro also includes the Validation Proportion field for column selection. In JMP Pro, there is the possibility of creating a validation column by using the random functions of column properties. Once a validation variable is set up within the set of data, move it over to the Validation column selection for a validated model. The validation process automatically creates an increasing number of splits for the data until the model fit statistics for the training and test sets start to differ significantly. The split history plot illustrates why the number of splits has been attained. More than six splits reduces the model fit statistic dramatically. Use the red triangle menu beside the *Partition* header, select *Split History* and deselect *Display Options* ▶ *Show Tree* to condense the output. Click *Go* to automatically create the optimum, validated partition model shown in Figure 8.13.

Figure 8.13: Partition (with Validation) Split History

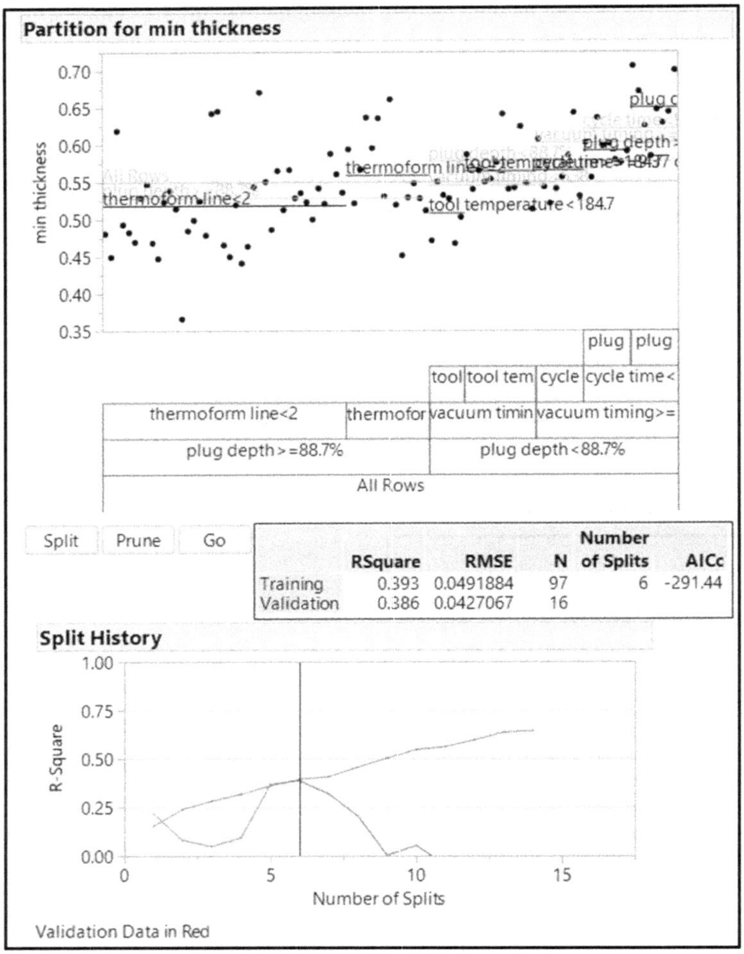

The validation algorithm creates an optimum model by stopping the splits when the validation set stops adding to the r-square fit statistic. Your results will differ from the example because validation is a randomized subset that differs every time it is selected. The pattern of how the r-square fit changes as splits increase shown in figure 8.13 is very different than the crossfit model because the fit is based on a relatively small number of observations. This pattern will change as various proportions are used for the validation set. Use the *Redo* shortcut to experiment with different sizes of validation sets. The validated partition model with six splits has a model fit r-square = 0.39. The model explains 39% of the variation in min thickness. Notice that the fit of the training set is approximately the same for six splits. Therefore, overfitting is mitigated.

The contributions of the inputs listed in order of influence is evaluated next. Use the red triangle menu to select the *Column Contributions* plot, shown in Figure 8.14.

Figure 8.14: Partition (with Validation) Column Contributions

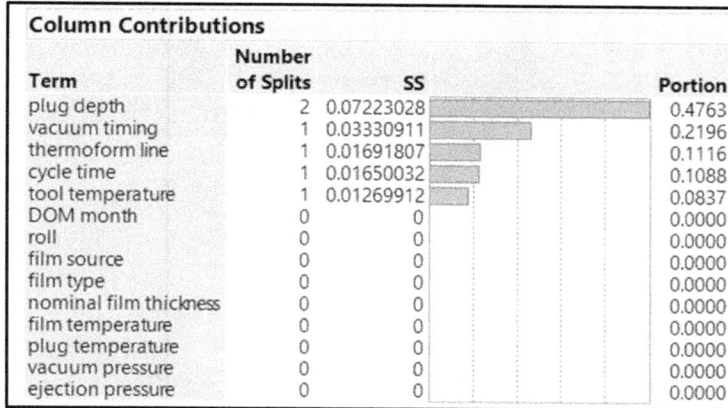

Column Contributions			
Term	Number of Splits	SS	Portion
plug depth	2	0.07223028	0.4763
vacuum timing	1	0.03330911	0.2196
thermoform line	1	0.01691807	0.1116
cycle time	1	0.01650032	0.1088
tool temperature	1	0.01269912	0.0837
DOM month	0	0	0.0000
roll	0	0	0.0000
film source	0	0	0.0000
film type	0	0	0.0000
nominal film thickness	0	0	0.0000
film temperature	0	0	0.0000
plug temperature	0	0	0.0000
vacuum pressure	0	0	0.0000
ejection pressure	0	0	0.0000

There are nine inputs that have no detected influence on min thickness, which is more than the cross validated model. There are four process inputs with portions greater than 10%. However, the inputs are different than previous models since cycle time has made it to the top four. The proportion contributed by the four leading inputs differs from previous models; the total has increased to 92%. The portion contributed multiplied by the r-square fit lets us know that the six-split model can explain 36% of the influence on min thickness. The validation seems to do a good job of choosing important inputs while mitigating over fitting of the model.

Stepwise Model Selection

Partition modeling is an excellent technique for reducing to inputs that are worthy of further study. One big limitation of partitioning is the lack of ability to identify whether relationships might exist in combinations of process inputs and the output of interest. A way to analyze a wide set of data with the ability to detect interactions among inputs is a stepwise selection model. The stepwise model can handle individual inputs, interactions, and even squared terms. Squared relationships are common in chemical processes because the rate of change is not constant across the range of input levels. The team limits interest to two-way interactions and individual inputs because they think it unlikely that squared relationships exist in the thermoforming process. Start by selecting *Analyze ▶ Fit Model* to open the *Fit Model* window. Move min thickness to the *Y (output) box*, as shown in Figure 8.15.

Figure 8.15: Fit Model Dialog Box

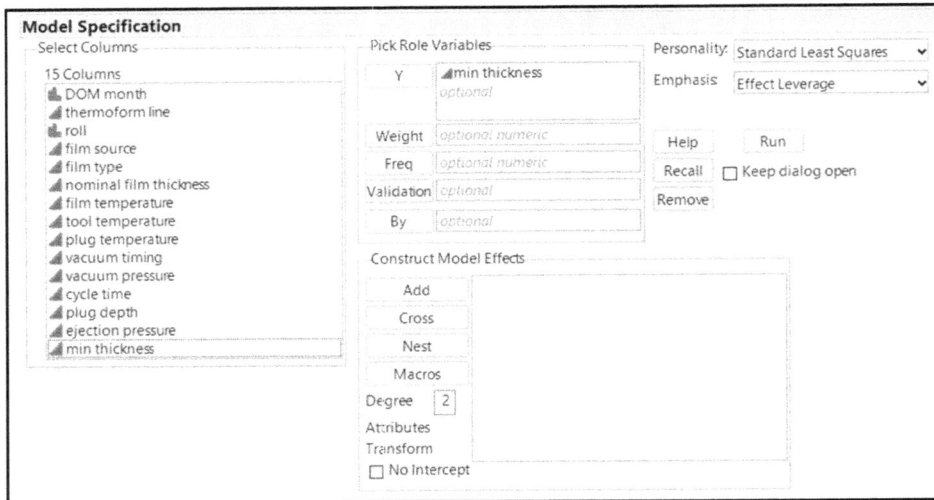

The output has been selected and the model needs to be identified for the analysis. The 14 processing variables are added with all the two-way interactions. Press the Shift key and select the 14 input variables until they are shaded in blue. Leave the default value 2 for *Degree* ; there is no interest in higher-order terms involving interactions of three or more inputs. JMP provides a shortcut to include all possible interactions of degree 2.Click *Macros* and select *Factorial to degree*, as shown in Figure 8.16.

Figure 8.16: Fit Model Dialog Box (Inputs with Two-Way Interactions)

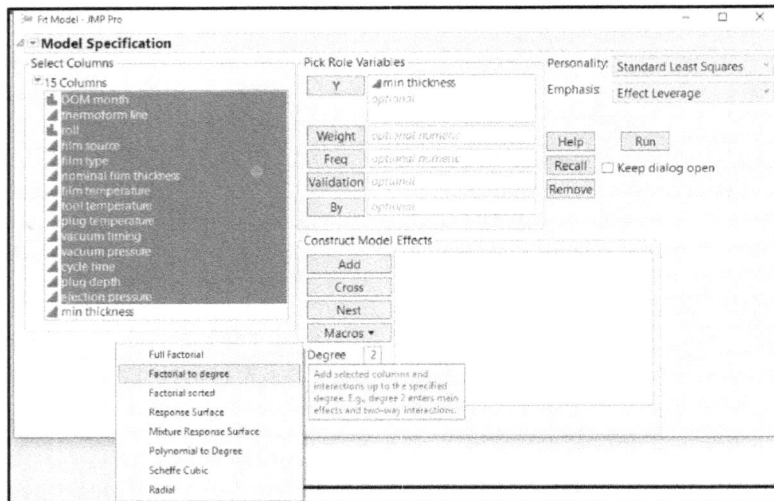

The individual process inputs and two-way interactions automatically populate the *Construct Model Effects* box. The default personality is a standard least squares model for the data to determine the leverage of the effects. The 113 samples collected is not enough data to analyze a wide model with 14 individual inputs and all two-way interactions. JMP would run the model and immediately alert you to singularity issues, which indicates the lack of data to estimate leverage and model predictions for the large number of inputs and interactions. You could manually run a large number of models, alternating the inputs and interactions of each to find the optimum. The stepwise personality is a powerful tool that automatically reduces the model by using an algorithm that provides results quickly and easily. Change the *Personality* in the upper right of the dialog box to *Stepwise*, and then click *Run* to get the initial output shown in Figure 8.17.

Figure 8.17: Stepwise Personality Window

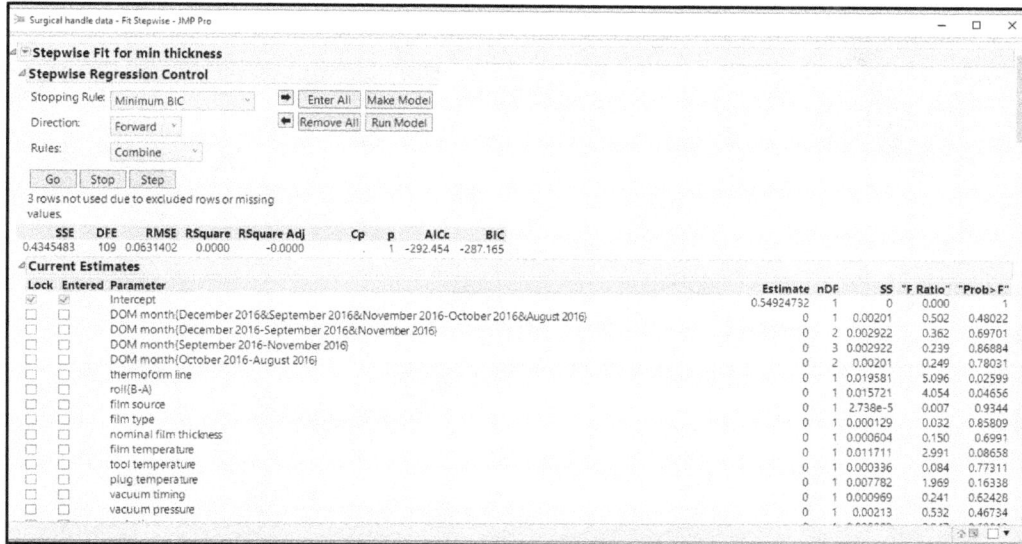

The stepwise fit output requires you to select a technique to execute the stepwise selection. You can explore the many options available in the JMP documentation to determine the most appropriate model for the goal of the analysis. For simplicity's sake, this example uses the default options *Minimum BIC* (Bayesian Information Criterion), *Forward*, and *Combine* rules. These default settings provide results for this analysis that are reasonably robust. Click the *Step* to iteratively evaluate the different models and detect the inputs with the highest likelihood of having influence on min thickness. Click *GO* to efficiently execute the process by letting JMP automate the model selection. A portion of the long list of current estimates output is shown in Figure 8.18, with the step history shown in Figure 8.19.

Figure 8.18: Stepwise Model Selection

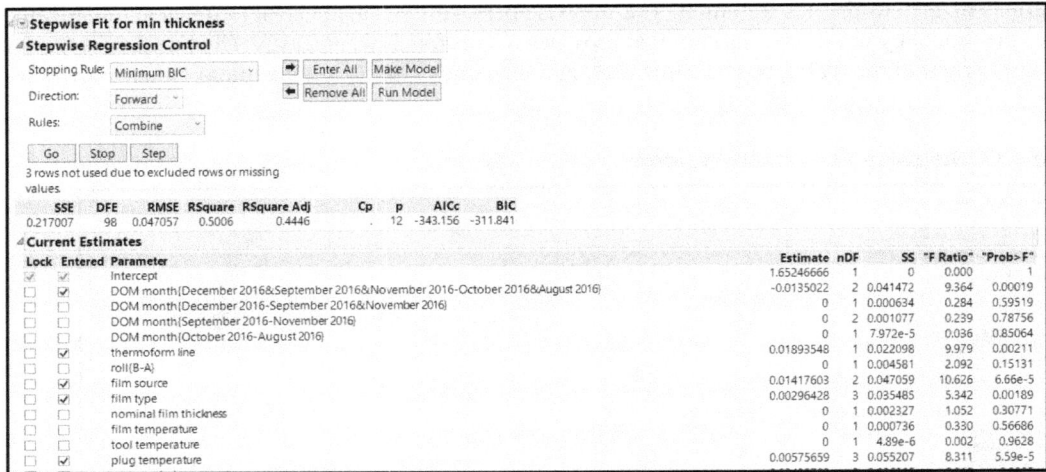

Figure 8.19: Step History

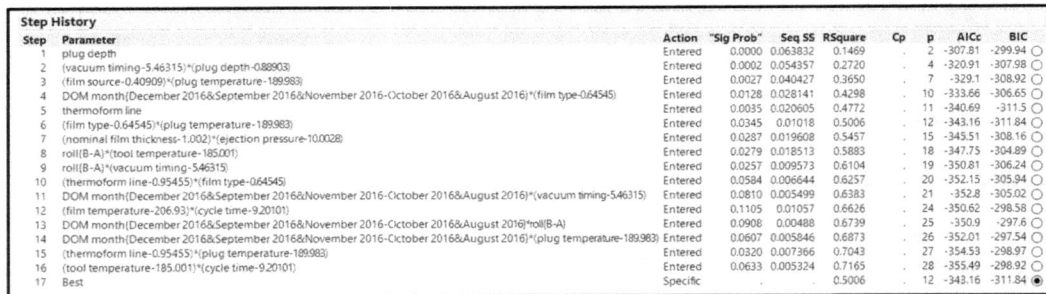

The fit of the final model from the stepwise process is the adjusted r-square =0.44. The adjusted r-square value allows for the comparison of models that include different numbers of inputs. The amount of random error is the root mean square error = 0.047, which compares similarly to the previous partition models. The current estimates illustrate the input predictors that are included in the model with inputs noted by the selected checkboxes located in the left columns of the table shown in figure 8.18. The step history illustrates the forward selection algorithm; up to 28 predictors were iteratively added to explore model fit statistics. The model with 12 predictors is chosen as optimum.

The output includes a *Make Model* and a *Run Model* button in the upper right of the output. These buttons enable you to analyze the model directly from the stepwise selection results. Click *Make Model* to open the *Stepped Model* dialog box, shown in Figure 8.20.

Figure 8.20: Stepped Model Dialog Box

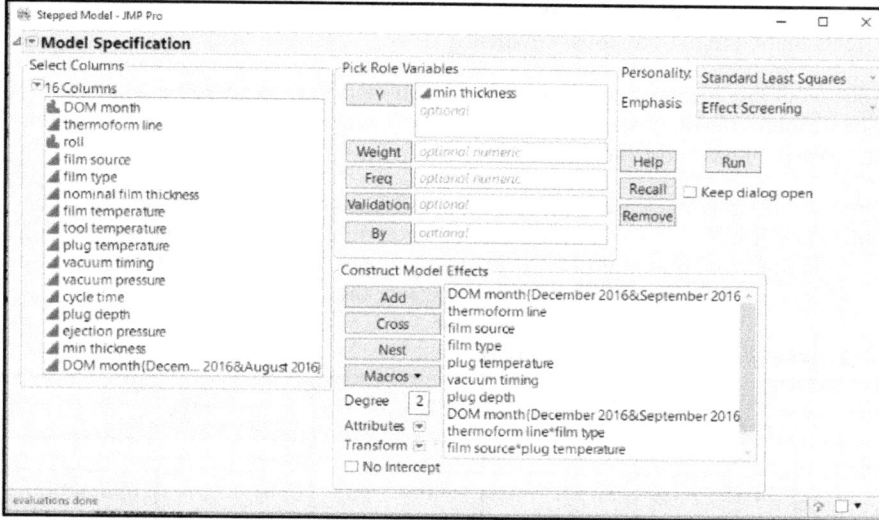

The vertical slider function of the *Construct Model Effects* box can be manipulated to see all of the many listed individual inputs and interactions that were determined to be important. You can add or delete model effects, but keep in mind that both of the individual variables for two-way interactions must be included even if they are not significant. Click *Run* to get the model output shown in Figure 8.21. (You can skip the review of the model and click *Run Model* in the stepwise window.)

Figure 8.21: Stepwise Reduced Model Results

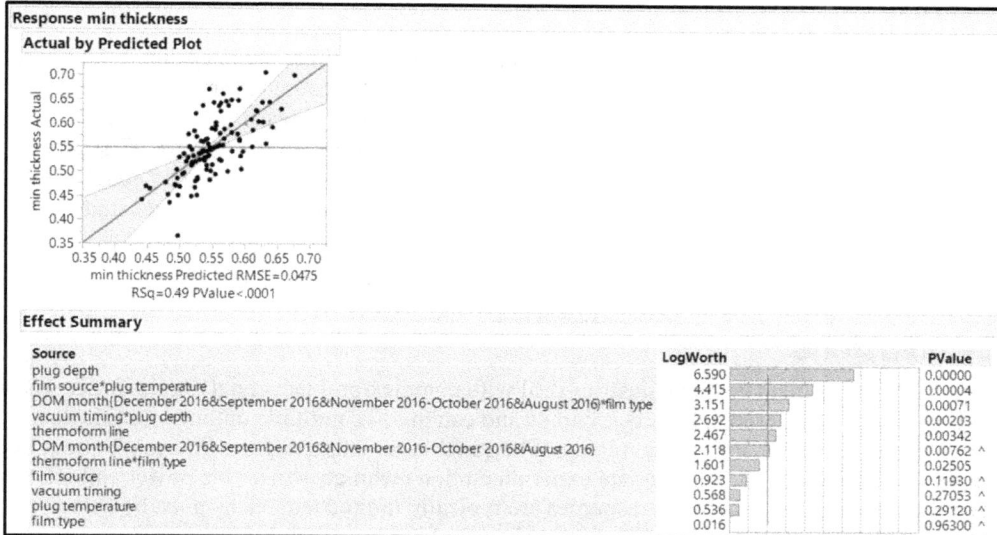

The results of the stepped model indicate that the model explains just under 50% of the variation in min thickness (r-square=0.49). The effect summary in the model output provides information that is similar to the column contributions of partitioning. The difference is that a LogWorth value is used to weight the effects instead of proportion of influence contributed. The more complex model has uncovered the potential for some interactive effects that were not previously evident.

The utilization of various modeling techniques to determine which inputs are likely to be related to changes in min thickness provide different results. The question that must be answered from predictive modeling is: Which inputs are of the greatest interest for further study?

The confusion of different results can be cleared up somewhat by creating a summary table from the various outputs. The information in Table 8.1 indicates that the modeling consistently narrows to plug depth, vacuum timing, cycle time, and nominal film thickness as the few variables that are worth further investigation. JMP Pro offers a model comparison feature to explore the models efficiently.

Table 8.1: The Four Process Inputs of Greatest Interest

		Basic Partition	Partition (K-fold)	Partition (validation)	Stepwise Model	
Input	r- square	0.64	0.45	0.39	0.49	
Plug Depth		30.7%	31.1%	47.6%	6.6*	Interaction 2.7*
Vacuum Timing		18.9%	19.2%	22.0%	2.7*	
Cycle Time		8.7%	8.2%	10.9%	-	
Nominal Film Thickness		13.3%	13.5%	-	-	

*In LogWorth values

You might recall that the thermoform line was a potentially strong contributor because line C had very few issues with thin covers. A significant benefit attained from the organized study of process monitoring information is that the teams have heightened focus on the process. The processing identification mark for the lines was found to be misleading and a cause for mixed labeling of in-process samples. The processing mark was made more definitive, and further study of a random sample of covers found an equal number of thin results among the three lines. The input was removed from consideration once the side study conclusions were reported.

Practical Conclusions

The pharmaceutical and medical device industries deal with complex problems on a regular basis. These problems involve multiple material and process inputs and can involve multiple outputs. Simple data visualization tools and basic statistical tests might not be suitable for extracting the information required to address such problems. Structured, multivariate experimentation techniques offer the power and efficiency needed to gain this information. However, resources are typically limited and teams must be able to narrow the possibilities down to inputs that offer the greatest chance of having relationships with the outputs of interest. JMP offers several predictive modeling techniques, which can be effectively used to model historical data and limit the scope of the inputs utilized for experimental designs. JMP Pro adds to the modeling options with a larger variety of predictive modeling tools to mitigate the chance of overfitting and provide the predictor inputs of the highest potential value.

This chapter touches on a few techniques that are easy-to-use so that teams can justify how they were able to choose the inputs for further study. You should invest some time to research the topic of predictive modeling and take full advantage of JMP documentation, including several excellent books, the Help documentation provided with a license, on-demand webcasts, and by visiting the JMP user community available at www.jmp.com.

Exercises

E8.1—You have been working with technicians and engineers involved in improving the sealed film that is applied to plastic trays that contain surgical kits. The seal is critical because it maintains the sterility of the instruments and materials packaged in the tray. Previous data visualization has identified differences by shift, and the team has been able to obtain a sample of data for 50 individual kits that were removed from the line for burst testing. Process data was recorded for nine factors. The team needs to optimize the process and is interested in limiting focus to the smallest number of potential influential factors.

1. Open *burst testing with process factors.jmp*.
2. Start the analysis with distributions of the output and nine inputs. Use the dynamic features of JMP to look for possible trends and relationships.
3. Run predictive modeling to determine which potential factor should be included in future studies.
4. How would you summarize the analysis to the project stakeholders and suggest factors to be included in structured experimentation?

E8.2—A new tablet formulation is being developed, and the team needs to determine how to target tablet hardness and minimize variation in hardness. A small number of batches have been made during scale-up of the process. The presses used are of a two-sided design; each side has independent controls (other than turret speed) and is treated as a unique observational unit. Predictive modeling is to be completed to determine whether the inputs can be narrowed down to allow for a set of structured experiments that conserve resources.

1. Open *mix and compression process data.jmp*.
2. Run partitioning of the data to detect important inputs to tablet hardness and tablet hardness range; each is a separate model. Hint: There are many inputs that are either duplicate information or information of little predictive value. Mesh screen measurements of particle size are not as accurate as methods that estimate the d (0.1), d (0.5), and d (0.9) particle size values. Dates are also of little importance for modeling.
3. Which inputs would you suggest for further study? How would you summarize this information to the project stakeholders?

Chapter 9: Designing a Set of Structured, Multivariate Experiments for Materials

Overview

Experimental design is a vast subject that should not be approached without a great deal of thought and reliance on process expertise to ensure that the results are of value. JMP provides an excellent platform to guide a novice and eliminate a great deal of stress and frustration for people who are learning the techniques. As with any new endeavor, it is good practice to start with simple problems and basic designs to gain knowledge and understanding of experimental design and execution. The information in this chapter is written with the assumption that the reader is not well versed in design of experiments (DOE). The techniques are basic and not intended to be the most precise for minimizing experimental error. Technical professionals who learn and use multivariate, structured experimentation realize vast improvement in process knowledge over one factor at a time (OFAT) experimentation. Evan low-powered, minimal designs can be augmented with additional runs to mitigate random error and gain stronger signals from the inputs. Augmentation of designs is especially useful in the highly regulated pharmaceutical and medical device industries because the likelihood of equivalent processing conditions for the augmented runs is generally good due to the standard operating procedures and work instructions endemic to the industries.

The Problem: Designing a Formulation Materials Set of Experiments

Sudhir was able to convince management in the value of using structured experiments to optimize an extended release formula. The risk assessment for the formula identified that there are 3 materials considered as critical materials attributes (CMA). The CMAs exert likely influence on the critical quality attributes (CQA), which define the performance of the formulation. The team is most concerned about meeting the goals for tablet dissolution at 4 hours and plans to experiment with 3 materials: a disintegrant, a diluent, and a glidant. Finding the right balance for the amounts of each material is crucial to the success of the formula. The total amount of the three materials is known to be fixed as the weight of the tablet has been established and cannot be changed. The amount of diluent in the target formula is higher by volume than the release controlling agent or glidant. Other materials in the mix include a fixed amount of the active pharmaceutical ingredient (API) and lubricant. This chapter explores two major types of experimental designs for materials.

The Plan

The materials of a formulation need to be determined early in the development process for a new tablet product. A great deal of work goes into the principal science to determine the types of materials needed to robustly meet the goals for CQAs. Sudhir's team has been able to produce the desired results with a target formula shown in Table 9.1 but obtaining acceptable results one time does not ensure that the formula is robust.

Table 9.1: Formulation Plan

Material	Target (mg/tab)	Percent of Total	Low Value (mg/tab)	High Value (mg/tab)	Factor Type
API 1	100.0	19.6%			Fixed
API 2	225.0	44.1%			Fixed
Diluent	126.0	24.7%	Changes are random, complimentary amounts to make up for other materials		Continuous Slack
Disintegrant	32.0	6.3%	20	44	Continuous Independent
Glidant	18.5	3.6%	13	24	Continuous Independent
Lubricant	8.5	1.7%			Fixed
Total Tab	510.0	100.0%			

JMP includes an entire suite of tools in the *DOE* menu that are used to quickly create robust experimental designs. The goal for the experimental activity is to determine which materials have effects on outputs, based on a limited list of candidates. The tablet weight is established as 510 mg, dictated by the required size and shape of the tablet noted by the marketing team. Only three materials are allowed to vary within the powder mix. The team has been through extensive planning to ensure that the processing attributes are controlled to fixed levels as much as possible. Members of the team will be working with manufacturing to ensure that all have an acute awareness of the need for minimal variation in processing. The control of the process is indented to ensure that the variability from materials can be detected as clearly as possible.

One way to create an experimental plan is to treat the materials inputs as independent factors. The plan will allow for only two of the three changing materials to be modeled; the third variable will be used to make up the slack created by independent combinations of the two factors. The principal science and experience of the formulation scientists was utilized to select the diluent as the slack variable. Slack variables randomly make up the difference to ensure that a fixed tablet weight of 510 mg is maintained. A drawback to the independent factors design with a slack variable is the inability to detect a signal from slack material. It is possible that differences noted due to changes in the factors are due to changes in the slack material that create errors in the model. The large percentage of the material in the formulation (24.7%) is believed to be enough to not result in changes to outputs due to the small random changes used to make up the slack.

A great deal of discussion went into the levels of the factors that are to be studied. Best practices include utilizing the largest increments as possible for the experimental factors. Big changes increase the potential that the signal from changes in the factors will overcome the noise of random variation, given the relatively small number of runs that are included in the model. A change of just over 37% for the disintegrant and a change of roughly 30% for the glidant are believed to be appropriate.

The design plan is deceivingly quick and easy to create. The diligence expended in the planning of the set of experiments is directly proportional to the robustness that might result from the analysis results. The custom designer in JMP is an excellent first step to rely on regardless of whether you are at a novice or expert level. The designer utilizes extremely powerful algorithms that optimize designs based on the details you enter to specify the model. Technical professionals no longer need to pour through experimental design textbooks to find a model that creates a palatable compromise between real world needs and limitations and the analysis model needed to produce reasonable results.

Using the Custom Designer

The initial goal of formulation development is to detect the material inputs that are related to changes in outputs. There is also interest in quantifying the effect of the predictor material inputs. Optimal experimental designs are based on criteria that relate to the experimental goal. The D-optimality criterion focuses points at the outer edges of a design space to emphasize the detection of the inputs that are related to changes in outputs. The I-optimality criterion locates points throughout a design space to emphasize the quantification of the amount of effect that inputs have on outputs so that accurate predictions can result. Other optimality criteria are available for more complex goals. You are encouraged to research the information in the *Design of Experiments Guide* in the JMP user documentation for additional information.

The default of the custom designer is the D-optimality criterion to prioritize the detection of which inputs affect outputs. Predictions can be made to quantify the effects when using D-optimal designs; however, more error is likely than a model produced using an I-optimal design. Start the experimental design for materials by selecting *DOE* ▶ *Custom Design* to get to the *Custom Design* window, shown in Figure 9.1.

Figure 9.1: The Custom Designer

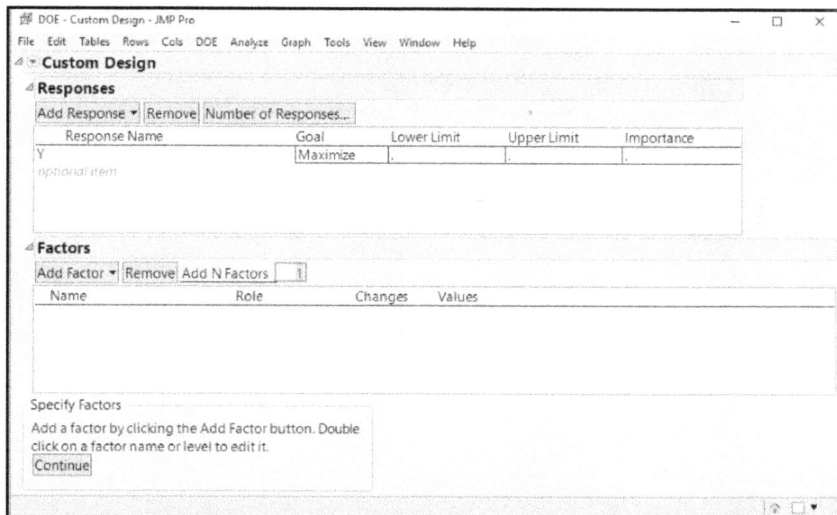

Several outputs will be measured for each of the experimental runs, which you can include in the responses box. It is typical to leave the responses blank during the design phase because they might not be known. For the subject formulation in this example, the responses are added later in the design process, and the limits are defined between the time that the experiments are designed and the results are analyzed.

The inputs of the model are added as factors to build the design. Enter 2 in the *Add N Factors* box for the two inputs being studied, and select *Continuous* in the *Add Factor* field, as shown in Figure 9.2.

Figure 9.2: Adding Inputs (Factors) in the Custom Designer

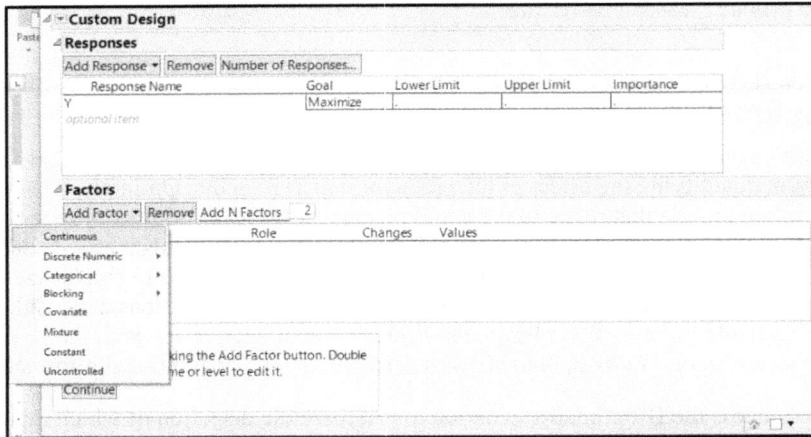

The values of -1 is for the low level and 1 is for the high level of each factor are added as defaults by the platform, as shown in Figure 9.3. The design can be developed with the default factor level values before the actual input values have been settled upon by subject matter experts. Sudhir was able to define the factor levels of each input prior to initiating the design. The default factor level values are changed by selecting each coded value and typing the new value in its place. Having actual factor values included in the design helps the subject matter experts to interpret the experimental plan in practical terms.

Figure 9.3: Factors of Experiment Shown in the Custom Designer

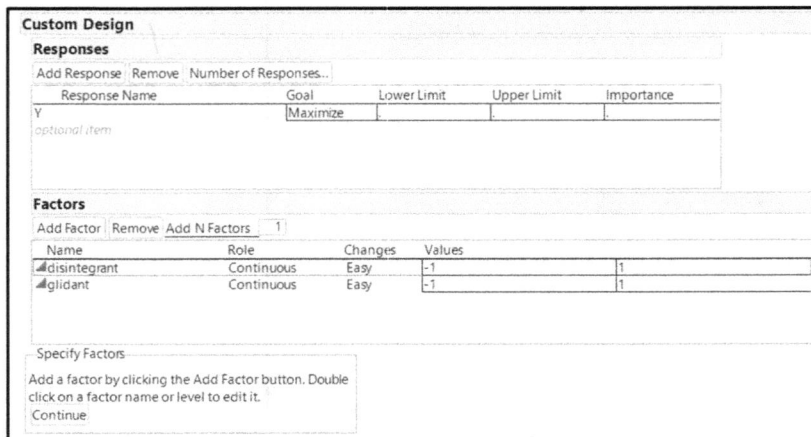

The actual values are known for the experiment (shown in Table 9.1) and are included to add useful information for the team. Modify the coded levels of each input in the *Factors* box with the values shown in Figure 9.4. Click *Continue* to add detail to the design.

Figure 9.4: Uncoded Factor Levels

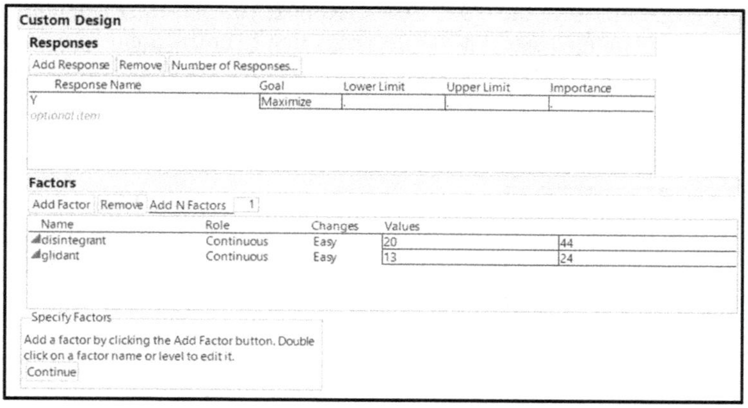

The basic model has been defined but more detail is needed to qualify the materials inputs (factors). At the top of Figure 9.5, the *Define Factor Constraints* options enable you to fine-tune for combinations of factor levels that are known to be a high risk for a functional failure. For instance, the highest level of one material input combined with the highest level of another might deplete the diluent (slack) variable so much as to interfere with lubrication and cause tablets to stick onto tooling. The planning for levels already took the risk of sticking into account, so further restrictions are not necessary. Adding a factor constraint might allow for wider levels of factors to be studied. However, a design with limits on all combinations adds complexity, increasing prediction variance. The default level of no factor constraints is utilized for the project.

There is a good chance that a change in the output is due to a combined effect among two or more variables—an interaction. Interactions should be included in the factor details to ensure detection of the phenomenon. Click *Interactions* and select *2nd*, shown in Figure 9.5. This option includes in the model the interaction between the two inputs. Interactions are possible for more than two inputs. However, the example can include only a *2nd* order interaction since only two inputs are considered in the design.

Figure 9.5: Defining the Model

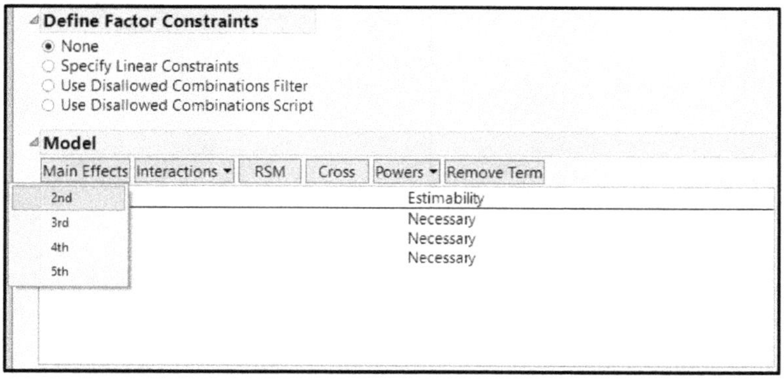

The *Design Generation* section of the window utilizes the model options to guide you to the number of runs that should be included in the plan. The team is interested in a run to represent the target formula, which means that a center point is needed in the model. Including the center point also provides an ability to determine whether the model results are not linear. Enter 1 in the *Number of Center Points* box. *Number of Runs* includes radio buttons for the *Minimum* and *Default* number of runs that are calculated from the algorithm that the Custom Designer uses. In the *User Specified* box, you can enter any number of runs greater than the minimum. For this example, select *Minimum* runs first to create a model with the least amount of resources required, as shown in Figure 9.6. Click *Make Design* to create the design shown in Figure 9.7.

Figure 9.6: Options to Generate the Design

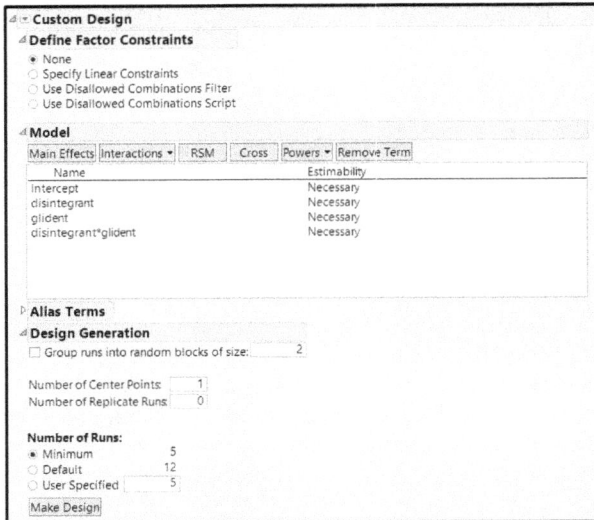

Figure 9.7: Minimum Size Custom Design of Five Runs

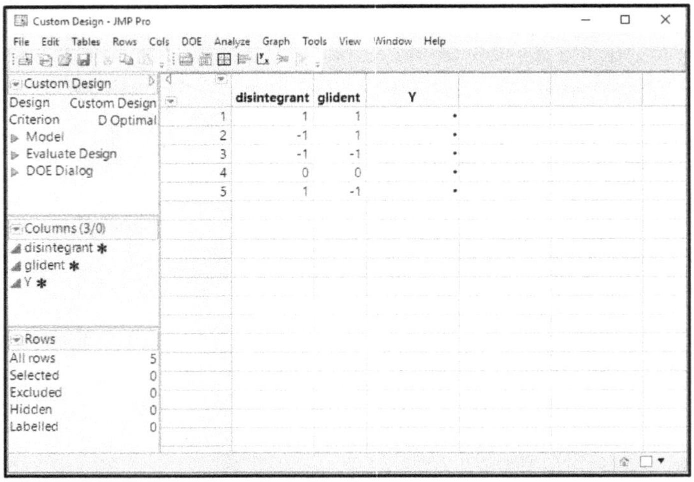

Using Model Diagnostics to Evaluate Designs

The minimum design with five runs is shown in Figure 9.7. This section describes the diagnostic tools that are used for evaluating the quality of the model. Significance might be incorrectly detected from a model that was made with insufficient runs. An important diagnostic is a measurement of the ability to mitigate the mistake of detecting a significant relationship due to a lack of runs. The term used for this measurement is statistical power, which can be between 0% and 100%.

Power is influenced by the balance struck between the influence from inputs that explain changes in an output and the random variation of a model. The factor influence can be thought of as a signal from a model and the random variation as noise. Figure 9.8 assumes that a good level of power is to be maintained, which is typically 80%. The balance on the left represents a model with a large amount of noise and a subtle signal; many runs are needed to maintain good statistical power. The balance on the right represents a model with minimal noise and a strong signal; fewer runs are needed to maintain good power.

Figure 9.8: Visualization of Statistical Power

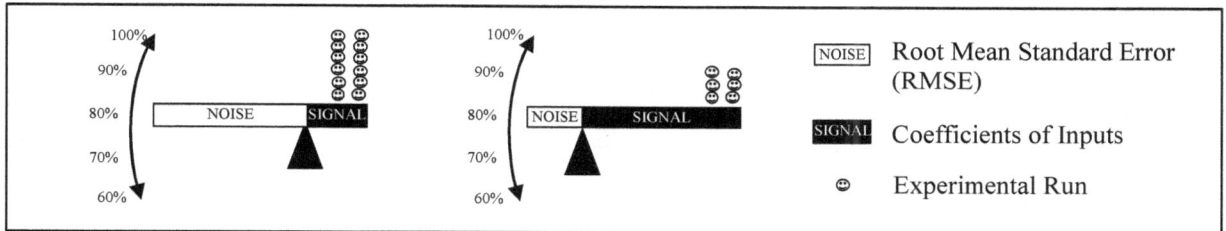

The calculation of statistical power is 1 minus the probability that the model does not conclude significance when a significant relationship exists in real life. Basically, power can be defined as the likelihood that a significant relationship exists in real life for the population of batches that is produced.

The diagnostic tools for the model give an estimated value for power, which is based on the following factors:

- The desired significance level; 0.05 is the default used by JMP.
- The amount of random variation that can be expected; JMP uses one unit of random mean square error (RMSE) as the default.
- The RMSE is the variation that can be expected in the outputs regardless of changes that occur with the inputs.
- If information is available from previous studies, the expected RMSE can be included to make the model evaluation more precise. However, updating the RMSE should be done in concert with expected values for the input coefficients.
- The amount of change in the output that can be expected from the inputs; JMP uses one unit of change in the output as the default.
- Prior studies can be utilized to extract anticipated coefficients that differ from the one-unit default.
- If the anticipated coefficients are changed, it is good practice to also include the related random variation value (RMSE).
- The sample size for the power calculations is derived from how you defined the model in the Custom Designer.

Figure 9.9: Design Evaluation (Statistical Power)

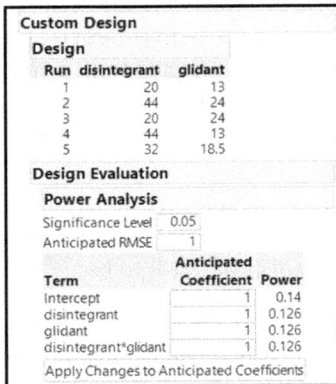

The Custom Designer creates models through the use of random selection. The diagnostics of each trial of creating a model will differ. The examples in this section are not likely to match your results due to the randomness of the technique.

The power calculations for the 5-run model do not look very promising, given the default values used for the calculations. The estimated power is between 12.6% and 14% out of a possible 100%. The actual power will likely change because the actual random variation and the actual coefficients for the inputs are likely to differ from the one-unit default. There is no way to predict whether the power values will be higher than the estimates or lower until the analysis of results is completed.

The next aspect of diagnosing the model is estimating the amount of variance that can be expected for the design space used, shown with Figure 9.10.

Figure 9.10: Design Evaluation (Prediction Variance)

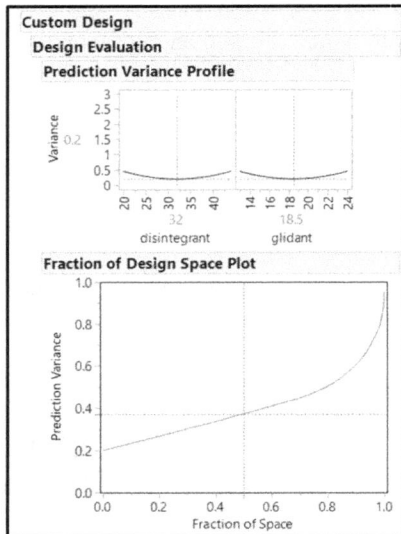

The structure of the model comes from the allocation of runs among the factor combinations. Ample runs allow for a balanced number of runs among input combinations, reducing the amount of prediction variance expected from the model. Flat, bowl-like curves are desired for the individual inputs of the variance profile. You can manipulate the red segmented slider to see how the variance changes for different levels of the materials inputs. The variance is always highest at the extremes (high and low) and minimal in the middle of the design space. The shape of the variance profiles is the reason the widest possible levels are desirable for model planning.

The Fraction of Design Space plot provides a detailed illustration of the design space of all inputs from the center out to the extreme levels. The flattest rate of growth for the variance is desired in the plot, with the 50% value typically used as a standard of comparison among multiple potential models. The 5-run design is expected to have roughly 0.38 units of variance at 50% of the design space. You can get more detail by evaluating the Prediction Variance Surface plot, shown in Figure 9.11.

Figure 9.11: Design Evaluation (Variance Structure)

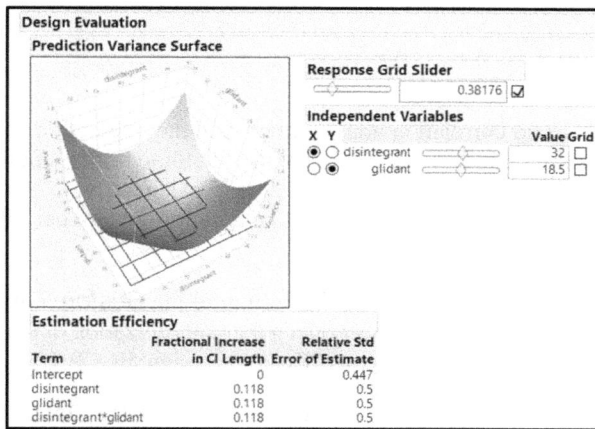

The plot offers a dynamic 3-dimensional view of the model variance. A symmetric bowl shape is desired for the variance structure, with a circular grid showing up as the response grid slider. The Estimation Efficiency table provides estimates for the amount that the prediction variance of the model structure is likely to inflate confidence level estimates made from the analysis of the completed model. The smallest values possible are desired.

One of the most important aspects of model design is the potential for inputs and interactions that "blind" each other due to aliasing and correlations, as shown in Figure 9.12. Each diagnostic is set up as a matrix, which is obvious because the disintegrant is 100% aliased and is correlated with itself. The comparison of disintegrant with the other input and the interaction between the two have no aliasing or correlation. The alias table indicates 0 aliasing; the correlation matrix illustrates no correlation between inputs since the comparison cells have no color. The correlation plot in Figure 9.12 uses the white-to-black color scheme for presentation in this book; the default of the platform is a red-to-blue color scheme.

Figure 9.12: Design Evaluation (Correlations)

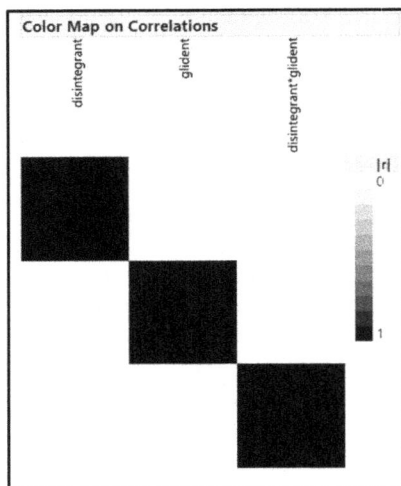

The 5-run model requires minimal resources and has no correlation among inputs and the interaction, which are positive aspects of the diagnostics. The limitations of the model are the likelihood of low power, relatively high prediction of variance, and a variance structure that is not completely balanced and symmetric. The great thing about JMP is that it is very easy to evaluate many options for the number of runs within a very short period of time.

Before you create a new model, click the *Make Table* button located at the bottom of the *Custom Design* window, below the *Diagnostic Evaluation,* to create a 5-run data sheet. Keep the randomization default values; you want a completely randomized design. Save the data sheet as "Custom Design 5R CP."

Go back to the *Custom Design* dialog box, and click *Back* to create a new model. Choose the *Default* choice to create a model with 12 runs, and click *Make Design* to create the design. The design evaluation for the 12-run design is shown in Figure 9.13.

Figure 9.13: Design Evaluation for 12-Run Model

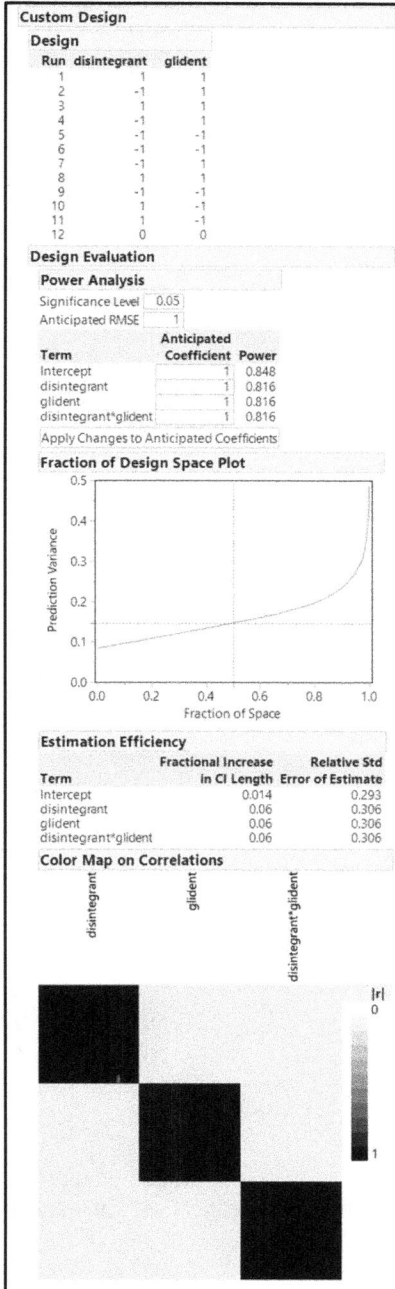

Custom Design

Design

Run	disintegrant	glident
1	1	1
2	-1	1
3	1	1
4	-1	1
5	-1	-1
6	-1	-1
7	-1	1
8	1	1
9	-1	-1
10	1	-1
11	1	-1
12	0	0

Design Evaluation

Power Analysis

Significance Level 0.05
Anticipated RMSE 1

Term	Anticipated Coefficient	Power
Intercept	1	0.848
disintegrant	1	0.816
glident	1	0.816
disintegrant*glident	1	0.816

Apply Changes to Anticipated Coefficients

Fraction of Design Space Plot

Estimation Efficiency

Term	Fractional Increase in CI Length	Relative Std Error of Estimate
Intercept	0.014	0.293
disintegrant	0.06	0.306
glident	0.06	0.306
disintegrant*glident	0.06	0.306

Color Map on Correlations

The estimated power has increased dramatically to more than 80%. The prediction variance is half of what is produced by the minimal design. A small amount of correlation is present among the inputs, and the interaction as combinations of input levels cannot be completely balanced with a multiple of 12 runs.

A third option is to create and evaluate a design with a number of runs that is between the minimal and default designs. Click *Make Table* below the *Diagnostic Evaluation* to create a 12-run data sheet. Save the data sheet as "Custom Design 12R CP."

Go back to the *Custom Design* dialog box, and click *Back* to create a new model. Select the *User Specified* option to create models with various numbers of runs. This example uses a 9-run design as a go between for the 5- and 12-run options. Click *Make Design* to create the design. Figure 9.14 shows that the 9-run design has a greater issue with correlation than the 5- or 12-run designs.

Figure 9.14: Design Evaluation of 9-Run Model with Correlation

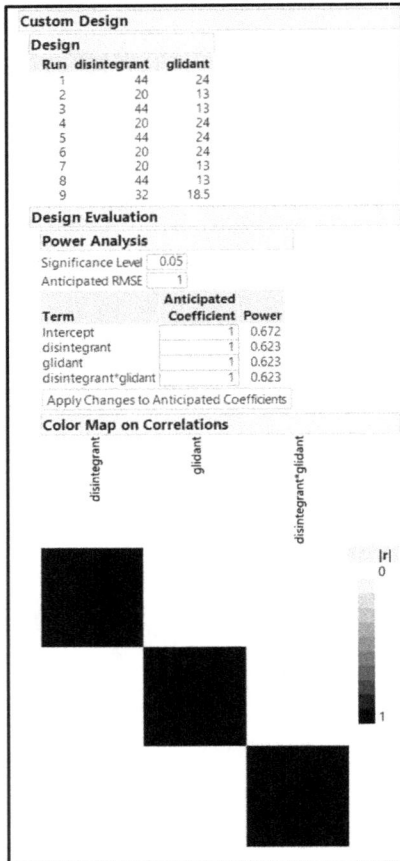

You could fully explore the diagnostics of the 9-run model and compare them to the other models. However, JMP includes an easy way to compare models. Create and save the data sheet as "Custom Design 9R CP".

Compare Designs – An Easy Way to Compare Up to Three Designs (JMP Pro Only)

The Compare Designs platform efficiently compares model diagnostics for up to four designs. The three design data sheets that you have saved (5-run, 9-run, and 12-run) must be open in order to run the design comparison. This example uses the minimal 5-run design as the standard of comparison. Be sure that the 5-run design is on top. Select *DOE* ▸ *Design Diagnostics* ▸ *Compare Designs*, as shown in Figure 9.15.

Figure 9.15: Comparing Designs

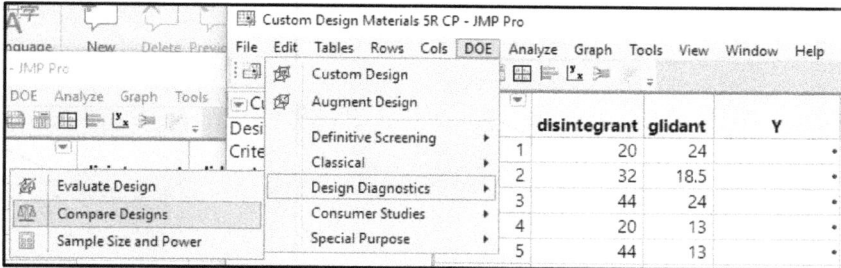

A dialog box opens where you can set up the comparison of the three models that you created. Press the Ctrl key and select the other two models by clicking on the data sheets in the *Compare 'Custom Design Materials 5R CP' With* box. Those sheets appear in the *Source Columns* boxes, as shown in Figure 9.16. Select *disintegrant* in each of the three models, and click *Match*. Do the same for *glidant*. Then, click *OK* to obtain the output.

Figure 9.16: Compare Designs Setup

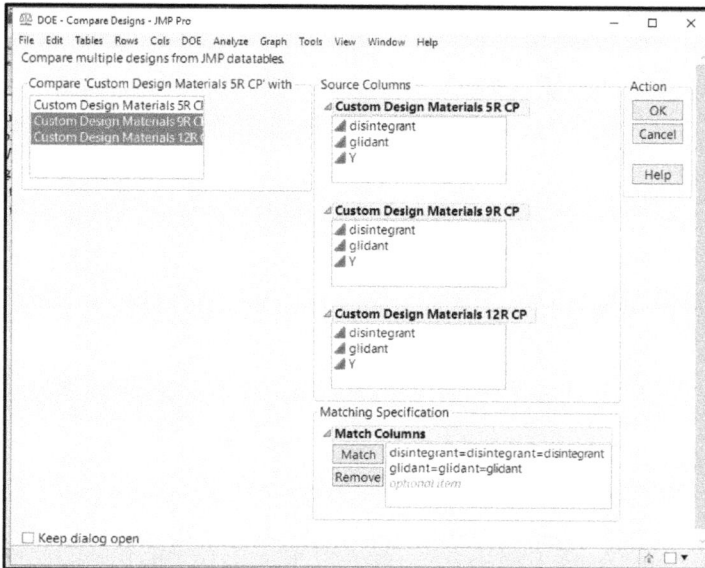

The *Factors* indicate the inputs of the designs, which are shown with uncoded levels regardless of the actual levels that you entered in the Custom Designer. The defaults shown in the model box include the factors and interactions that were identified in the original designs. You can select different options by using the model buttons. In this example, the default model is of interest to the stakeholders of the project, which is shown in Figure 9.17.

Figure 9.17: Compare Designs Model Description

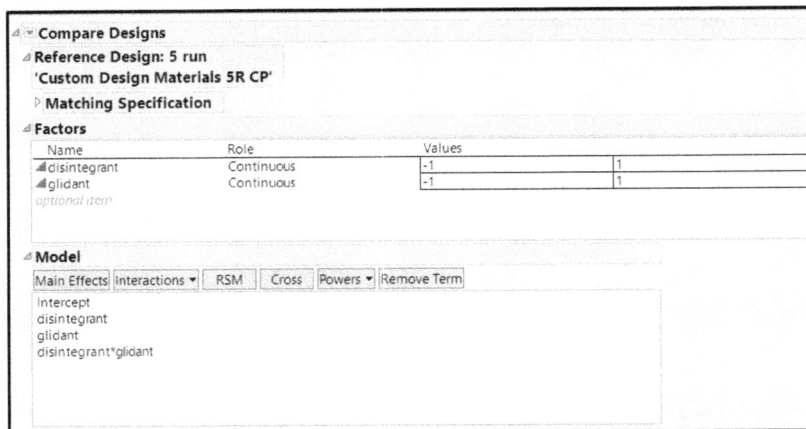

Comparisons of the statistical power for the three candidate models is the first set of diagnostics, shown in Figure 9.18. You can alter the options for RMSE and the amount of the anticipated coefficients with values from previous studies if they are known. This example uses the 1-unit default for RMSE and anticipated coefficients. The estimated power values, a comparative clustered bar chart, and a line plot are included for the models that differ only by the number of runs. The line plot is not included if other differences in modeling are present, such as comparing a custom design to a fractional factorial design. The diagnostics clearly illustrate the superior power of the 12-run design with a center point.

Figure 9.18: Compare Designs (Statistical Power)

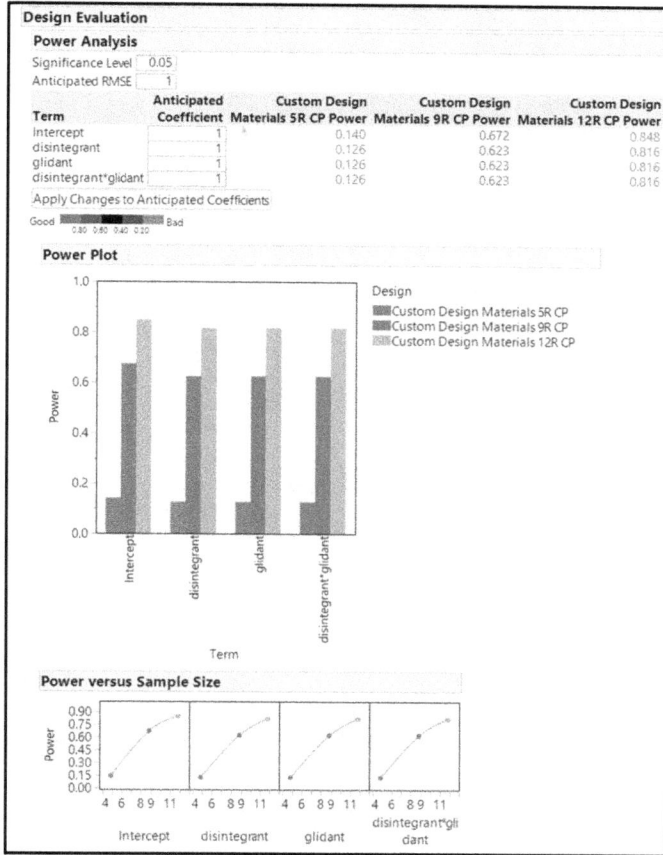

Design Evaluation

Power Analysis

Significance Level 0.05
Anticipated RMSE 1

Term	Anticipated Coefficient	Custom Design Materials 5R CP Power	Custom Design Materials 9R CP Power	Custom Design Materials 12R CP Power
intercept	1	0.140	0.672	0.848
disintegrant	1	0.126	0.623	0.816
glidant	1	0.126	0.623	0.816
disintegrant*glidant	1	0.126	0.623	0.816

Apply Changes to Anticipated Coefficients

Good ▮▮▮▮▮▮▮▮ Bad
0.80 0.60 0.40 0.20

Power Plot

Design
Custom Design Materials 5R CP
Custom Design Materials 9R CP
Custom Design Materials 12R CP

Power versus Sample Size

Figure 9.19 provides the output for comparisons of prediction variance and estimation efficiency among the models. You have seen the Fraction of Design Space plot before. Now, however, you see an overlaid plot of the three models, which clearly indicates the minimum prediction variance of the 12-run design with a center point. The values shown on the relative estimation efficiency reflect a comparative ratio. Values less than 1.0 indicate lower values compared to the reference design, values greater than 1.0 illustrate greater amounts than the reference design. JMP enhances interpretation by color-coding the numbers according to the Good-Bad legend. Since the 5-run design is the reference, the estimation efficiency ratios not as high as what is offered by the larger designs. The bright red numbers illustrate that the 5-run design is 61% worse than the 12-run design.

Figure 9.19: Compare Designs (Prediction Variance and Estimation Efficiency)

Compare Designs

Reference Design: 5 run
'Custom Design Materials 5R CP'

Design Evaluation

Fraction of Design Space Plot

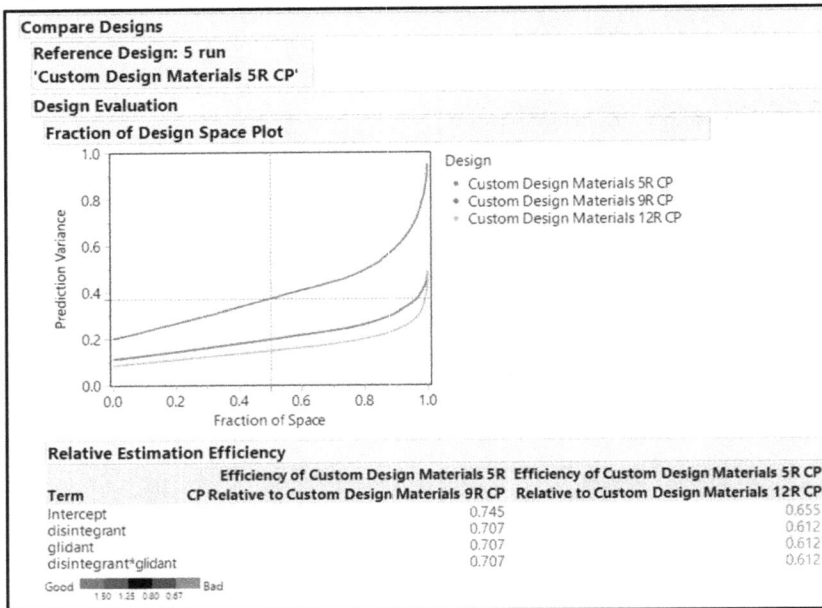

Relative Estimation Efficiency

Term	Efficiency of Custom Design Materials 5R CP Relative to Custom Design Materials 9R CP	Efficiency of Custom Design Materials 5R CP Relative to Custom Design Materials 12R CP
Intercept	0.745	0.655
disintegrant	0.707	0.612
glidant	0.707	0.612
disintegrant*glidant	0.707	0.612

Good ▨▨▨ Bad
1.50 1.25 0.80 0.67

Figure 9.20 provides information about correlations and design diagnostics. The 5-run and 9-run models have no correlations among individual inputs and the interaction. The 12-run design with a center point cannot have runs allocated equally among the input combinations and has less than 10% correlation present. Adding a 13th run or dropping the center point eliminates correlation for a new design. The color maps for correlation present the same information in a plot that is easy to interpret.

Figure 9.20: Compare Designs (Correlations and Efficiency)

Design Evaluation

Absolute Correlations

Model x Model	Average Correlation	Number of Confoundings	Number of Pairs
Custom Design Materials 5R CP	0.000	0	3
Custom Design Materials 9R CP	0.000	0	3
Custom Design Materials 12R CP	0.099	0	3

Good ▨▨▨ Bad
0.10 0.30 0.50 0.70

Color Map on Correlations

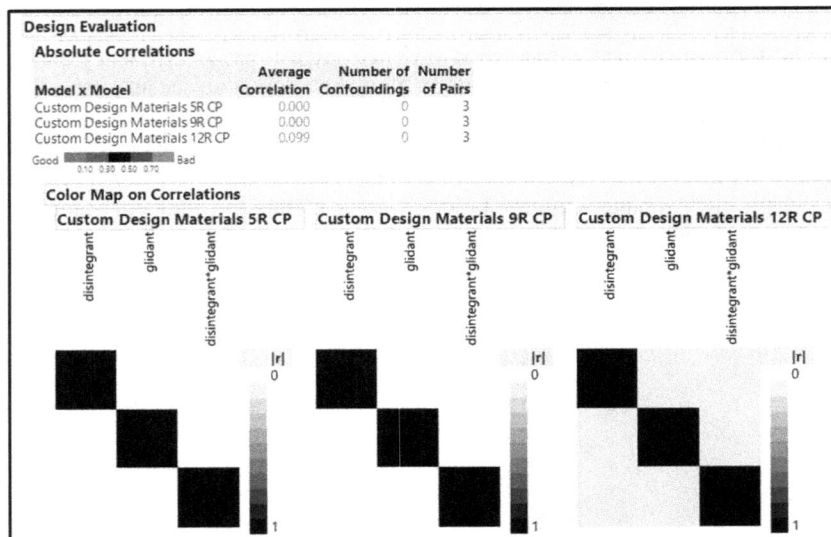

Comparing design through the diagnostic tools of JMP makes the job of balancing the need for information with resources that are easy to interpret, which is especially helpful for decision makers who might not be familiar with statistics. The design comparison helps people to visualize the balance between the information required from the experimental model and resource requirements.

The Data Collection Plan

The team used the comparative information to decide on a 9-run design with a center point. Low power and high prediction variance inherent in the 5-run design is not acceptable, and the need to conserve resources pushed the team to the 9-run design. The great benefit of utilizing structured, multivariate experimentation is the ability to augment models as needed. If the results of the 9-run design do not offer clarity of information, a JMP user can quickly augment a design by adding runs to further mitigate random variability. Augmentation is discussed later in this chapter.

You can easily update the fully randomized experimental design provided in the 9-run data sheet to create a design plan that can be used to execute experiments and collect data. Figure 9.21 presents the final design plan, which includes the following changes:

- new columns are created for fixed materials (mg/dose)
- new columns are created with formulas to show material proportions (%)
- new columns are created to determine the amounts needed from the slack variable (diluent) to balance out the total materials for each run
- new columns are created to convert the material proportions to the amounts needed (kg/mix) for each run

Figure 9.21: Design Plan

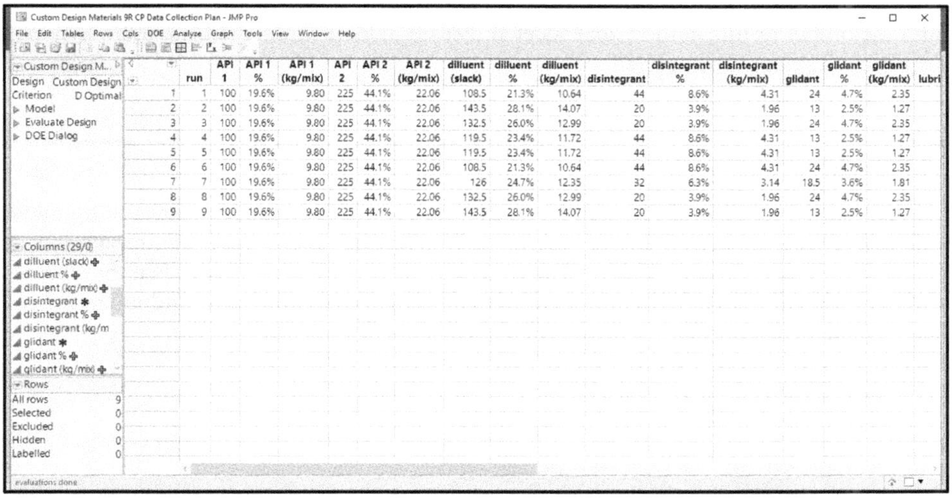

	run	API 1	API 1 %	API 1 (kg/mix)	API 2	API 2 %	API 2 (kg/mix)	diluent (slack)	diluent %	diluent (kg/mix)	disintegrant	disintegrant %	disintegrant (kg/mix)	glidant	glidant %	glidant (kg/mix)	lubri
1	1	100	19.6%	9.80	225	44.1%	22.06	108.5	21.3%	10.64	44	8.6%	4.31	24	4.7%	2.35	
2	2	100	19.6%	9.80	225	44.1%	22.06	143.5	28.1%	14.07	20	3.9%	1.96	13	2.5%	1.27	
3	3	100	19.6%	9.80	225	44.1%	22.06	132.5	26.0%	12.99	20	3.9%	1.96	24	4.7%	2.35	
4	4	100	19.6%	9.80	225	44.1%	22.06	119.5	23.4%	11.72	44	8.6%	4.31	13	2.5%	1.27	
5	5	100	19.6%	9.80	225	44.1%	22.06	119.5	23.4%	11.72	44	8.6%	4.31	13	2.5%	1.27	
6	6	100	19.6%	9.80	225	44.1%	22.06	108.5	21.3%	10.64	44	8.6%	4.31	24	4.7%	2.35	
7	7	100	19.6%	9.80	225	44.1%	22.06	126	24.7%	12.35	32	6.3%	3.14	18.5	3.6%	1.81	
8	8	100	19.6%	9.80	225	44.1%	22.06	132.5	26.0%	12.99	20	3.9%	1.96	24	4.7%	2.35	
9	9	100	19.6%	9.80	225	44.1%	22.06	143.5	28.1%	14.07	20	3.9%	1.96	13	2.5%	1.27	

You can save a JMP file that is the design plan in various formats as a data collection plan. The optimum is keeping the information as a JMP data sheet so analysis can be carried out easily with the default scripts.

Some organizations that Sudhir depends on do not yet have JMP available for all functional groups. The pilot plant that Sudhir will use for experimentation prefers to collect data in an Excel sheet. He quickly creates the Excel version of the plan by using the file save options. Select *File ▶ Save As* and change *Save As Type* to *Excel Workbook (*.xlxs, *.xls)*. Name the file "Custom Design Materials 9R CP Data Collection Plan", choose the location where you want to save the file, and click *Save*. The Excel version of the data collection plan opens, shown in Figure 9.22. The columns other than the amounts needed for the operations team are hidden for simplicity. Sudhir saves the final version and sends it to the team who will be executing the work.

Figure 9.22: Design Plan (Excel Version)

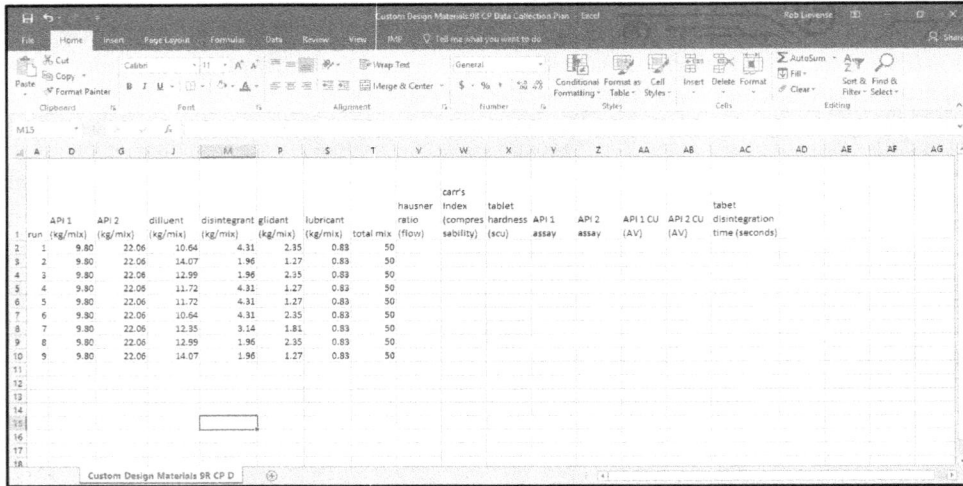

The team can enter the output results in the Excel sheet and easily update those changes to the JMP data sheet for analysis when all runs are completed.

Augmenting a Design

The team collects data for the Hausner Ratio in order to execute the 5-run design, as shown in Figure 9.23. The Hausner Ratio is a measurement of the potential flow of the powder and is important for a robust tablet compression process. Analysis indicates that the disintegrant input is marginally significant, but the power calculation indicates only 67% power. The analysis team is concerned that too much random variation is present in the data to robustly detect the significant input from the minimal 5-run design.

Figure 9.23: Initial Results for 5-run Design

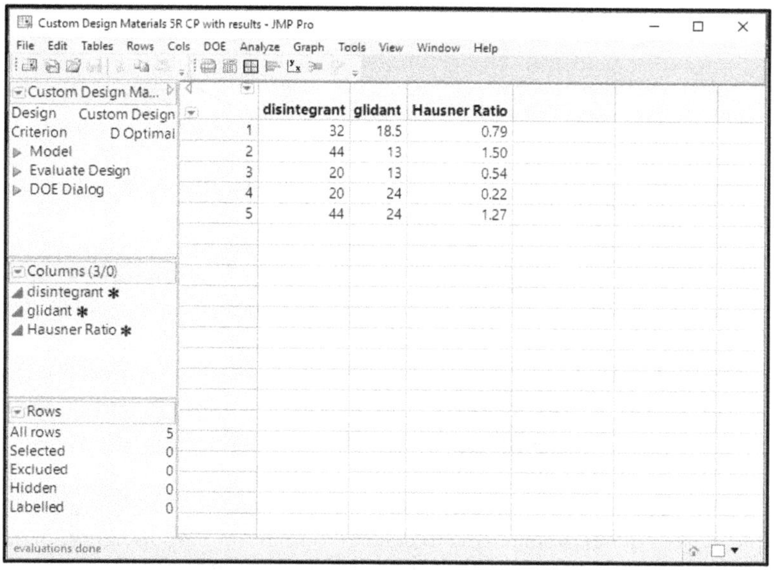

Adding runs is an option since the procedural control of the process mitigates the risk of a lack of reproducibility. There is no guarantee that the added runs will improve the power of a significant result. However, larger designs represent the population more precisely. The end result could be that materials combinations include a large amount of variability for the Hausner Ratio output and do not add to the evidence of a significant relationship. It is decided to augment the design to strengthen the reliability of conclusions made from the analysis of the model.

Open the file *Custom Design Materials 5R CP with results.jmp* to get started. Select *DOE* ▶ *Augment Design*. In the *Augment Design* window, set up the task. Move *Hausner Ratio* to the *Y,Response* box, and move *disintegrant* and *glidant* to the *X, Factor* box, as shown in Figure 9.24. Click *OK* to get to the next window (shown in Figure 9.25) where you select the best augmentation option.

Figure 9.24: Augment Design Setup

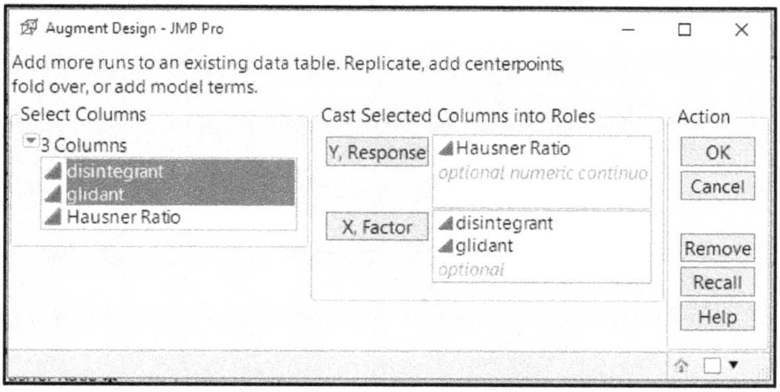

Figure 9.25: Augment Design Window for Choosing Augmentation Choices

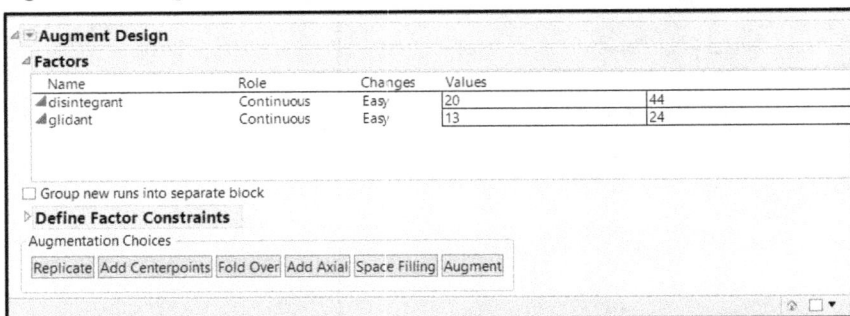

You can choose from several techniques to augment a design. The following augmentation options are available in the Augment *Design* window shown in Figure 9.25:

- *Replicate* makes a copy of the runs of the initial design. If enter 2 for replicates, in the result is a pooled design of 10 runs.
- Use *Add Centerpoints* to focus the model on the midpoint between extremes, which can reduce variability. Recall that the goal of the experiments is to detect the important inputs. Therefore, a concentration of points in the middle of the design space is not a good strategy.
- Use *Fold Over* to create a fold-over design, which mitigates the correlation between the individual inputs and interactions. Because the initial design has no correlation issues, this design does not meet the needs of the team.
- *Add Axial* and *Space Filling* are options that create points to fill within and outside of the design space of the initial experiments. Each is a good option for adding runs to a D-optimal design so that analysts can improve the predictions made from significant factors. The options do not necessarily meet the goal of detecting important inputs and are not used for the example.
- *Augment* is a simple option to utilize the plan for the initial experiments and randomly add more runs to hopefully improve power by reducing random error. This is the option chosen by the team. Click *Augment* to get to the window used to add detail and run the augmentation, shown in Figure 9.26.

Figure 9.26: Augment Design Window for Adding Options and Making the Design

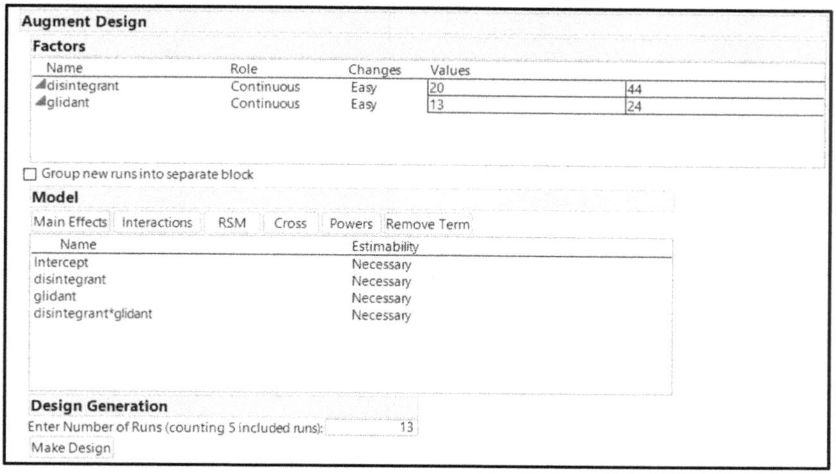

Augment Design				
Factors				
Name	Role	Changes	Values	
◢disintegrant	Continuous	Easy	20	44
◢glidant	Continuous	Easy	13	24

☐ Group new runs into separate block

Model

Main Effects	Interactions	RSM	Cross	Powers	Remove Term

Name	Estimability
Intercept	Necessary
disintegrant	Necessary
glidant	Necessary
disintegrant*glidant	Necessary

Design Generation

Enter Number of Runs (counting 5 included runs):   13

Make Design

The default of 13 runs is provided by JMP at the bottom of the window. You can change the number, but the team decides to use the default. If the team wants to detect a difference that could be due to the two experimental campaigns, they could select the *Group new runs into separate block* check box. It is not selected in this example keep the analyses simple. Click *Make Design* to create a new model with the five existing runs and eight added runs. Next, evaluate the model diagnostics by opening the options underneath the *Design Evaluation* outline header in the output journal, shown in Figure 9.27. Click *Make Table* to create the data sheet shown in Figure 9.28.

Figure 9.27: Augmented Design Model Details

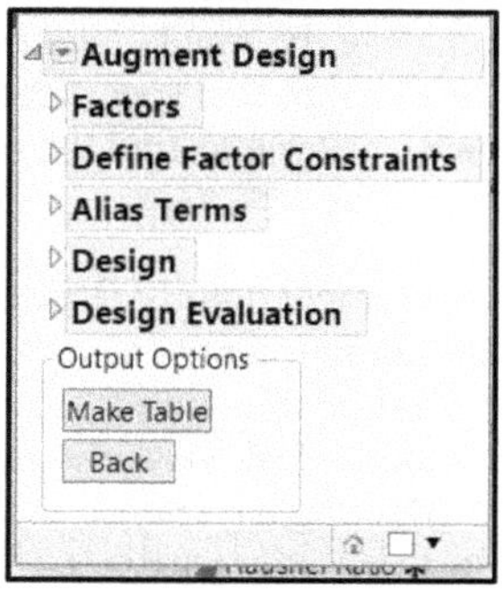

Figure 9.28: Augmented Design Model Data Collection Plan

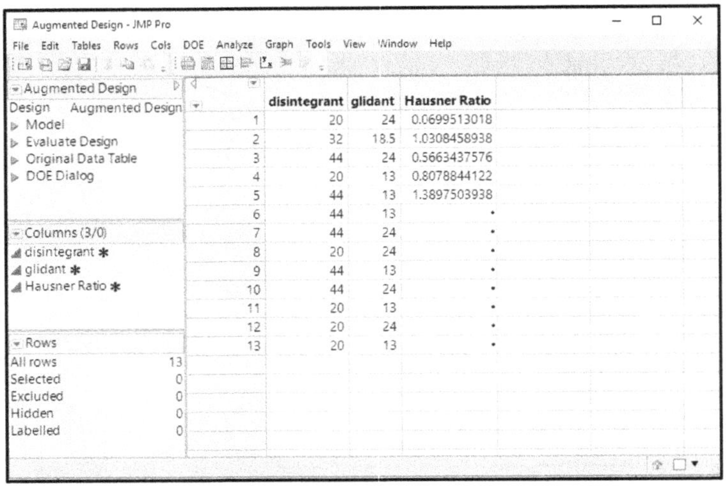

The team sends the new data collection plan to the pilot plant for eight new experiments to pool into a 13-run set of experiments.

Practical Conclusions

The DOE platform tools in JMP can help a novice obtain expert results. Experimental planning used to require a DOE textbook to select a specific design with the subject matter manipulated to fit the model structure. You can run such classic designs can be run in JMP if you want to, but JMP can make things so much simpler. The Custom Designer enables an analyst to simplify work by including the inputs intended for the study and the amount of resources available for the experiments to create a design that specifically meets the experimentational goal. Algorithms within the designer utilize optimality criteria to find the best possible structure for the number of runs that can be afforded.

This chapter involves determining how changes in two key materials might affect critical quality attributes and important process attributes. A slack variable used with the two independent random inputs easily obtains knowledge of material influence on responses of interest. The designer must be tolerant of experimental error and the potential for confounding that is present due to the random slack variable used to make up the total of materials in the mix. If changes in the slack variable are affecting a response, it will appear to be due to leverage exerted by one of the experimental variables. Subject matter expertise is extremely important; evidence of principle science and experience is needed to justify the assumption that the slack variable is not likely to create influence on outputs. A different approach involving a mixture design is described in chapter 14.

The design diagnostics provide required information for teams to find the right balance between information gained from a model and the resources required to run experiments. The goal of this example is limited to the detection of the important factors. Therefore, the resource needs are less than what is required for quantifying effects and making predictions. Even minimal structured, multivariate experiments can be augmented to expand on the original goal or to pursue a more complex goal. The strict procedural environment utilized in the manufacturing of pharmaceutical and medical device products makes augmentation a great option. Care must be exercised to ensure that each experimental campaign is as similar as possible to mitigate the risk of additional experimental error.

Sudhir led the formulation team to quickly create D-optimal designs with the Custom Designer and compare them with the Compare Designs platform. A 9-run, completely randomized design with two material inputs and one interaction should provide the required information while using a manageable amount of resources. This design is formalized into a design protocol to list all inputs, controls, and noise factors, with the randomized set of runs as the data collection plan. The structure of the experimentation is far superior to a trail-and-error approach, and the team looks forward to the analysis of the results.

Exercises

E9.1—Create a set of structured, multivariate experiments. Use the results of the predictive modeling from exercise problem E8.2 with treatment ranges that are slightly wider than the ranges determined from the sample data. There are enough materials to make 18 more batches of product, so be sure to retain enough material to include an adequate number of confirmation batches.

1. Create two comparative designs with the intent of finding the best balance between the information desired and the available resources. Keep in mind that there is a possibility that curved effects are present.
2. Create a report that includes a comparison of a good design with the best design, and be prepared to explain the balance between information and resource requirements.

E9.2—You are working with a formulation team as they try to find the right materials mix for a new extended-release tablet formula. The materials that have the highest potential for affecting dissolution results over time include a super disintegrant, a fast-acting release controlling agent, a slower acting release controlling agent, and a diluent. The formula also includes a filler material that aids the compression process and makes up a large proportion of the dose.

1. Create a slack variable set of structured experiments for materials for the following levels:

Filler (slack variable)	330mg target
Super disintegrant	20mg to 28mg
Fast-release control agent	42mg to 54mg
Slow-release control agent	26mg to 36mg

2. Create at least two comparative designs to balance out the desired information with resource requirements.
3. Create a report that includes a comparison of a good design with the best design, and be prepared to explain the balance between information and resource requirements.

E9.3—A set of experiments for the study of a formulation has been designed and saved as a JMP table. The table includes all materials in the blend. However, the interest of the study involves the proportional mix of two grades of diluent, the amount per dose of a glidant, the amount per dose of a micronized grade of glidant, and the amount of a disintegrant. The initial goal of the design has been to mitigate correlation between inputs and to include at least 80% estimated statistical power. The leadership team is interested in what will be given up in information if the minimum number of 12 runs for the experiments is chosen over the 16 runs in the current plan.

1. Open *Custom Design Materials 5F 16R CP.jmp*.
2. Use the *Evaluate Design* script to look at the design evaluation.
3. Use the *Scatterplot Matrix* script to view how the design space is filled with observational points.
4. Use the *DOE Dialog* script to open the design journal.
   a. Click the *Back* button at the bottom of the journal.
   b. Redo the Design Generation with 12 runs to create a new design table that includes 12 runs.
   c. Select *DOE* ▶ *Design Diagnostics* ▶ *Compare Designs* for a view of comparative diagnostics for the 16- and 12-run designs.
5. How would you present the information to leadership so that they can make the best decision on how to study this important new product?

Chapter 10: Using Structured Experiments for Learning about a Manufacturing Process

Overview

The requirement to have a robust manufacturing process is growing in importance for all industries who want to remain competitive. It is especially true with pharmaceutical and medical device manufacturing as regulatory agencies pursue Quality by Design (QbD) initiatives. Regulatory guidance typically notes the need for robust processes and suggests that elements of QbD need to be included in product submissions. Structured, multivariate experimentation is the gold standard for providing useful information about the relationships between process inputs and outputs. JMP includes an excellent Design of Experiments (DOE) platform for you to easily design a set of experiments and provide the best possible balance between available resources and information about the process. This section expands on previous chapters and provides more guidance for basic materials experiments to develop a design to screen out the critical process parameters (CPPs), as well as a design that is used to accurately quantify the influence of CPPs on multiple outputs of interest.

The Problems: A Thermoforming Process and a Granulation Process, Each in Need of Improvement

Chapter 8 worked through a set of historical process data from a plastic thermoforming process utilized to make surgical handle covers. Michelyene has been notified of customer complaints regarding covers that split or stretch and are at risk for falling off the handle. This is a big concern because they are a component of the sterile barrier needed in an operating room. Through predictive modeling, her team focused on a narrow set of process inputs that might be related to thin minimum wall thickness. The thin minimum wall thickness is the root cause for the loose handle covers. The team decides that a screening DOE is needed to determine the significant process inputs so that they can focus improvement efforts.

The second example involves a pharmaceutical manufacturing process development team lead by Emily. She has been tasked to figure out the process settings needed from four key process inputs for a high-shear wet granulation process and to ensure that goals will be met for five outputs measured from the completed granulation. The task seems daunting, but the team recently acquired JMP licenses and received training on how to use the DOE platform. They will use the Custom Designer to find the balance between the information needed by the stakeholders of the project and the amount of resources that are required.

Screening Experimental Designs for the Thermoforming Process

Process experiments often involve the goals of determining the most influential process inputs from a list of several. Screening designs are efficient at limiting the scope of further study to include only the process inputs that have statistically significant influence on an output. These examples show three design alternatives that can be used to screen the inputs. The design evaluation results for the three models are used to justify the best choice for the experimental goals.

In chapter 8, you learned how Michelyne utilized predictive modeling techniques on a collection of all measured inputs from the thermoforming process. The modeling was used to determine the inputs with the strongest potential of having relationships to changes in minimum thickness. Predictive modeling is an effective technique for planning structured experimentation by limiting the scope of study to the inputs of greatest potential.

The Custom Designer in the DOE menu is a great tool for designing an efficient experimental model with the high potential inputs. Start by selecting *DOE* ▶ *Custom Design*. In the *Custom Design* dialog box, enter 6 in the *Add N Factors* box, and click *Add Factor* to see the menu list shown in Figure 10.1. Select *Continuous*; the process inputs Michelyne plans to study are of this type.

Figure 10.1: Custom Designer: Adding Factors

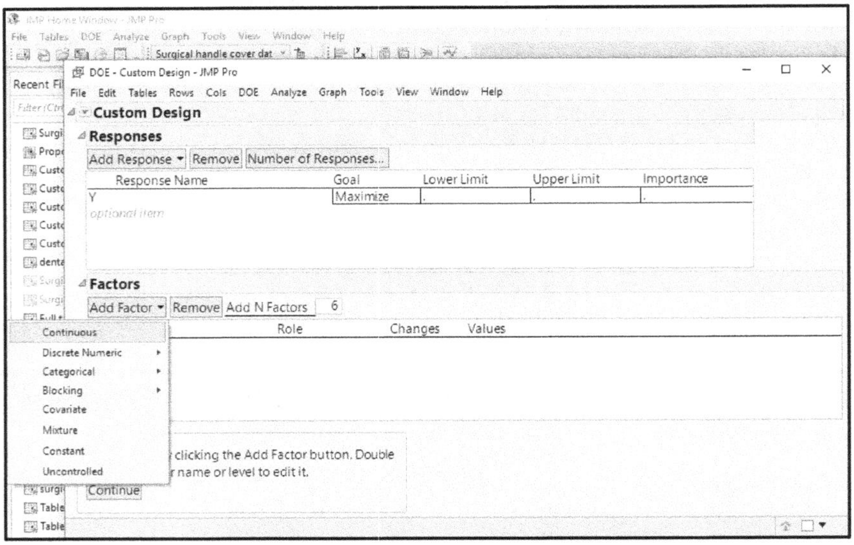

The *Factors* portion of the *Custom Design* window lists variables X1 through X6. To add the names of the process inputs, double-click on each variable to select it and enter the name in the box over the nondescriptive X variable. Enter low and high values so that the table of factors looks the one in Figure 10.2.

Figure 10.2: Custom Designer: Defining Factor Levels

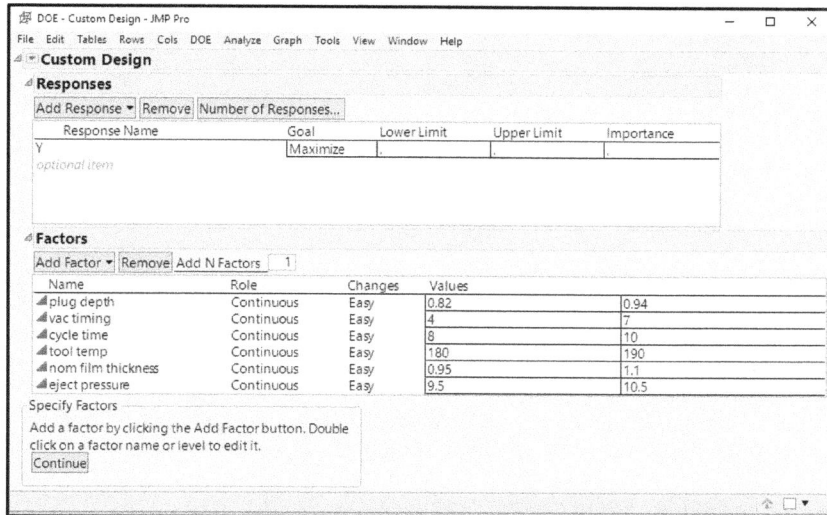

The custom design is not the only model we intend to explore. The Design Evaluation results in the model output is used to evaluate multiple designs. Saving factors as a data file is good practice, and it allows for multiple designs to be created very quickly. Click on the red triangle menu next to *Custom Design* and select *Save Factors*, as shown in Figure 10.3. Select *File* ▶ *Save As* and name the file "Thermoform Process Factors." Now, save the file to the location of your choice.

Figure 10.3: Custom Designer: Save Factors

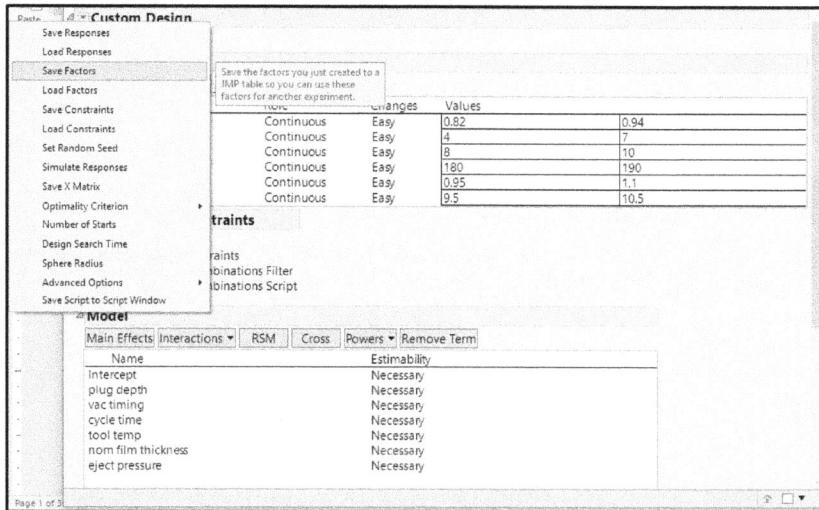

The factors are now named, and the levels are noted and saved. The goals of the experiments can now be reflected in the model portion of the *Custom Design* dialog box. Since the model is to be used for screening purposes, the default values for the main effects are all that will be used.

Figure 10.4: Custom Designer: Model Specification

Custom Design	
Model	
Main Effects	Interactions	RSM	Cross	Powers	Remove Term	

Name	Estimability
Intercept	Necessary
plug depth	Necessary
vac timing	Necessary
cycle time	Necessary
tool temp	Necessary
nom film thickness	Necessary
eject pressure	Necessary

The default process for design generation is an experimental plan that is randomized, which matches the goal of the team. In regulated industries, it is good practice to randomize based on a seed value so that the plan can be replicated. A randomization seed is a number used as the base for the randomization engine within JMP. A randomization seed works very well for the purposes of this book because the data table you obtain from following the instructions matches up with the example shown in the book. To set a randomization seed, click on the red triangle menu next to *Custom Design* and select *Set Random Seed*, as shown in Figure 10.5. Enter 2018 as the randomization seed and click *OK*.

Figure 10.5: Custom Designer: Setting Random Seed

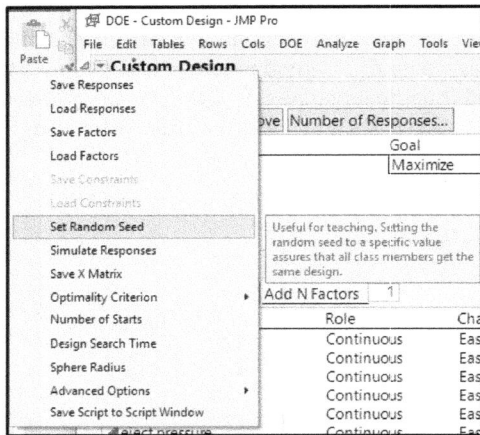

Use the *Default* value of 12 runs in the *Design Generator*, and click *Make Design* (shown in Figure 10.6). The design shown in Figure 10.7 is added to the *Custom Design* window.

Figure 10.6: Design Generation

Figure 10.7: 12-run Randomized Design

Design						
Run	plug depth	vac timing	cycle time	tool temp	nom film thickness	eject pressure
1	0.82	7	10	180	0.95	9.5
2	0.94	4	8	190	1.1	10.5
3	0.82	7	8	190	0.95	10.5
4	0.94	4	10	180	0.95	10.5
5	0.82	4	10	180	1.1	10.5
6	0.94	7	10	190	1.1	9.5
7	0.82	4	8	190	0.95	9.5
8	0.94	7	8	180	1.1	9.5
9	0.82	7	10	190	1.1	10.5
10	0.94	7	8	180	0.95	10.5
11	0.82	4	8	180	1.1	9.5
12	0.94	4	10	190	0.95	9.5

The design table contains the 12 runs with all factor combinations, but the runs are not randomized. The design diagnostics are available for review once you have created the design, but this example compares the evaluation of three competing designs. Comparative evaluations are very helpful to illustrate the advantages and disadvantages of each design and enable a team to achieve an optimum balance between resources and information.

Data table options are available, but this example does not use them for the sake of clarity. Click *Make Table* to create the JMP data sheet that contains a randomized set of runs, shown in Figure 10.8. Save the new JMP file as "Custom Design 12R Main Effects Thermoform Process" and leave it open.

Figure 10.8: Randomized Data Table for 12-Run Custom Design

		plug depth	vac timing	cycle time	tool temp	nom film thickness	eject pressure	Y
	1	0.82	4	8	190	0.95	9.5	
	2	0.94	7	8	180	0.95	10.5	
	3	0.94	4	10	180	0.95	10.5	
	4	0.82	4	8	180	1.1	9.5	
	5	0.94	7	8	180	1.1	9.5	
	6	0.94	4	8	190	1.1	10.5	
	7	0.82	7	10	190	1.1	10.5	
	8	0.94	4	10	190	0.95	9.5	
	9	0.82	4	10	180	1.1	10.5	
	10	0.82	7	8	190	0.95	10.5	
	11	0.82	7	10	180	0.95	9.5	
	12	0.94	7	10	190	1.1	9.5	

Custom Design - JMP Pro

File   Edit   Tables   Rows   Cols   DOE   Analyze   Graph   Tools   View   Window   Help

Custom Design
Design   Custom Design
Criterion   D Optimal
Model
Columns (7/0)
plug depth *
vac timing *
cycle time *
tool temp *
Rows
All rows   12
Selected   0
Excluded   0
Hidden   0
Labelled   0

The next design to consider is a classical screening design, extracted from DOE textbooks. Initiate the design by selecting *DOE ▶ Classical ▶ Screening Design*, as shown in Figure 10.8.

Figure 10.8: Classical Screening Design

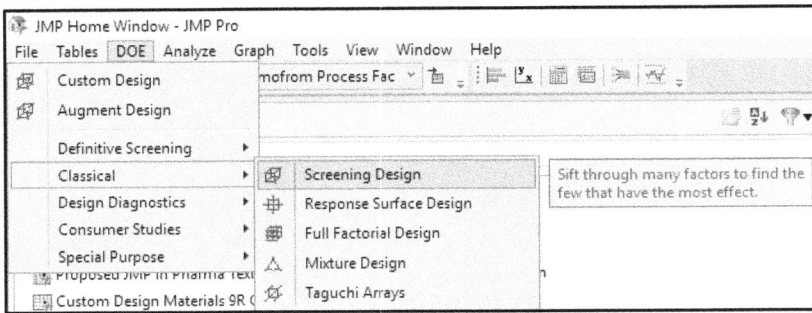

The saved factors file makes the task of creating a new design simple. Use the red triangle menu beside the *Screening Design* header to select *Load Factors*, as shown in Figure 10.9. Open *thermoform process factors.jpg* to add them to the model. When the factors are loaded, the *Screening Design* window appears, as shown in Figure 10.10. Click *Continue* to add the screening type choices to the output.

Figure 10.9: Loading Factors

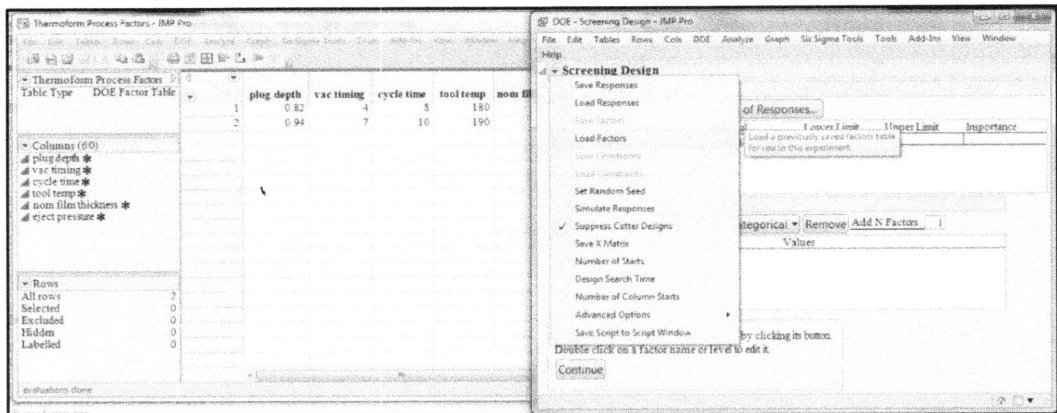

Figure 10.10: Screening Design: List of Factors

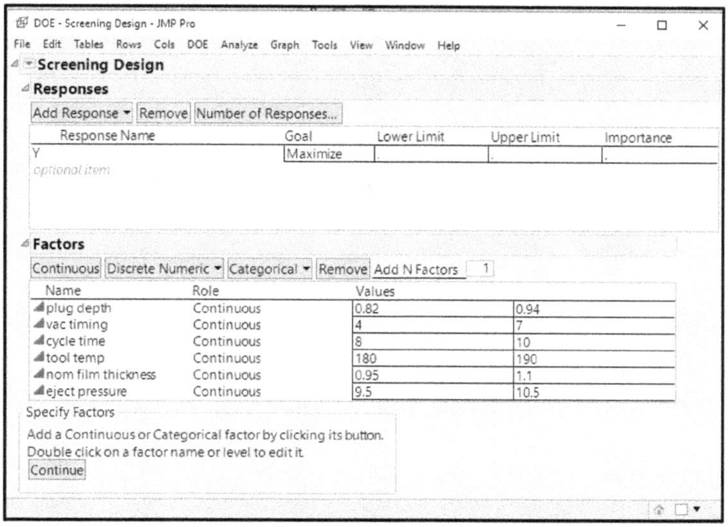

Figure 10.11 shows the *Choose Screening Type* radio buttons in the output window. Select *Construct from a list of fractional designs* to get a list of classic design options, and click *Continue*. This example does not use the main effects screening option because it defaults to the D-optimal design feature used previously.

Figure 10.11: Screening Design: Screening Types

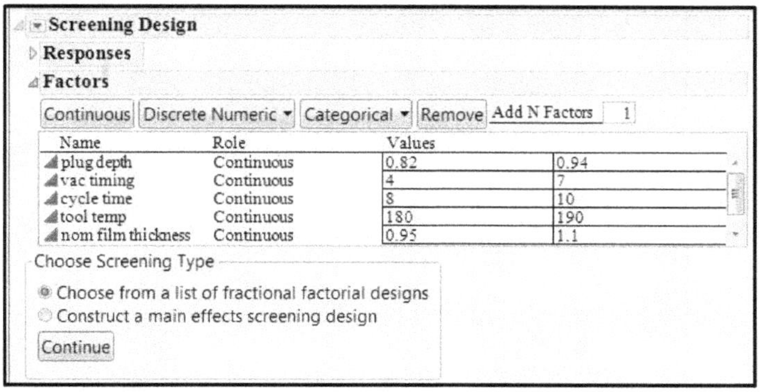

The classic designs differ from the custom designs in that each has a resource requirement for the number of runs. The flexibility is reduced since an optimality algorithm is not utilized. Use the Design List to select the design shown in Figure 10.12. By selecting the Plackett-Burman design with the same number of runs as the custom design, the goal of detecting main effects is maintained.

Figure 10.12: Screening Design: Design List

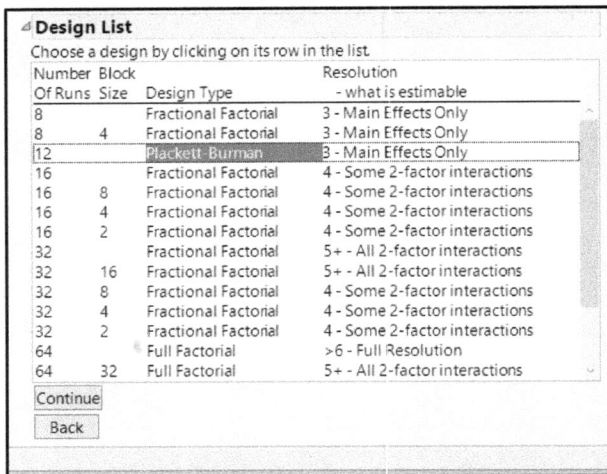

Select the 12-run Plackett-Burman option to study the main effects. The Plackett-Burman main effects design is described as a Resolution 3 design. A Resolution 3 design can be used to detect the main effects of the individual inputs. A note of caution is that two-factor interactions might confound the main effects. Confounding of effects is explained as one effect blinding a portion of another effect. Therefore, the analysis results with a significant main effect could actually be the influence of a combination of two effects. Click Continue to add the design to the output, as shown in Figure 10.13.

Figure 10.13: Screening Design: Design Detail

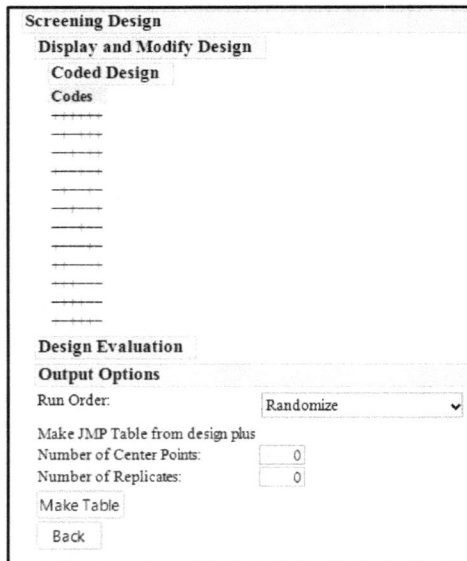

Use the *Design Evaluation* details to review the options such as power, prediction variance, and confounding of factors. Once the team has created the three designs in the example, they will complete a comparison of the designs to help the team determine the best choice. Detailed review of the design evaluation will be held off until the comparison is explained.

The randomization seed needs to be set so that the output matches this example. Click the red triangle menu next to *Custom Design* and select *Set Random Seed*. Enter 2018 as the randomization seed and click *OK*. Click *Make Table* to create a randomized table of runs for the design, shown in Figure 10.14. The pattern of the design is shown, as well as the randomized list of 12 runs in the data table. Select *File* ▶ *Save As* to save the data as "Plackett-Burman 12R Thermoforming."

Figure 10.14: Randomized Plackett-Burnham Design

The next model to create is a Definitive Screening Design (DSD). DSDs are relatively new and offer several advantages over other models. The models work best when you include at least six inputs in the study and the expectation is that only a few of the inputs will have a significant effect on the output. A small number of significant effects out of a large number of candidate factors is known as a parsimonious model.

The thermoforming process experimentation plan works well for a DSD because the team wants to study six continuous variables as inputs. The information from the predictive modeling indicates that three of the inputs are likely to make up between 63% and 81% of the influence on the minimum thickness output. For this reason, the model is likely to be parsimonious. Initiate the design by selecting *DOE* ▶ *Definitive Screening* ▶ *Definitive Screening Design*, as shown in Figure 10.15.

Figure 10.15: Definitive Screening Design

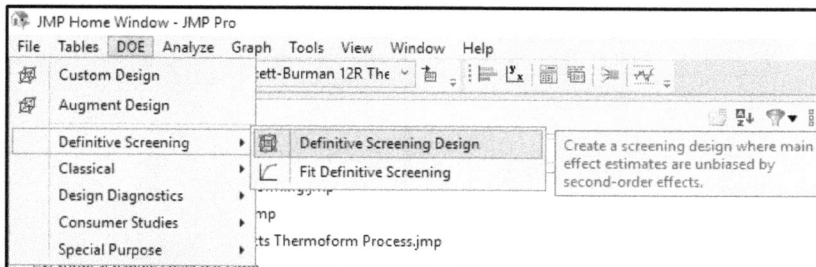

Use the red triangle menu by the *Definitive Screening Design* header to select *Load Factors*. The process is the same as it was for the classic screening design. Select *thermoform process factors.jpg* from the appropriate directory and load the factors as shown in Figure 10.16. Click *Continue* to change the output so that you can specify the design.

Figure 10.16: Definitive Screening Design: Defining Factor Levels

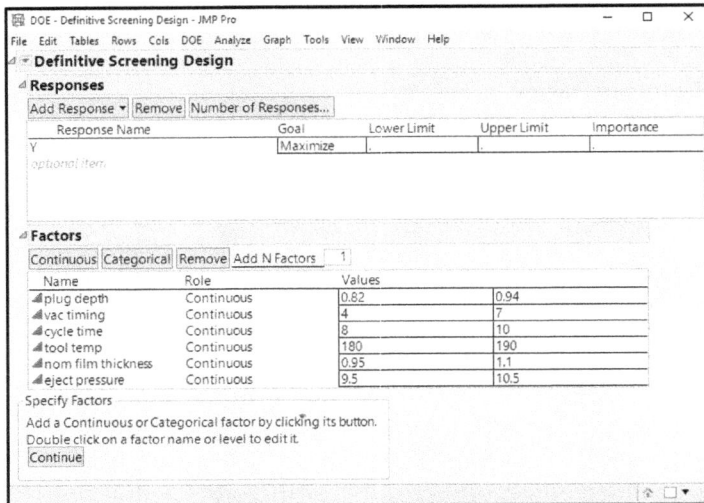

This example requires a basic DSD. For *Design Options*, select No *Blocks Required* and enter 0 in the *Number of Extra Runs* box, shown in Figure 10.17. Click *Make Design* to add the design to the DSD output.

Figure 10.17: Definitive Screening Design: Design Options

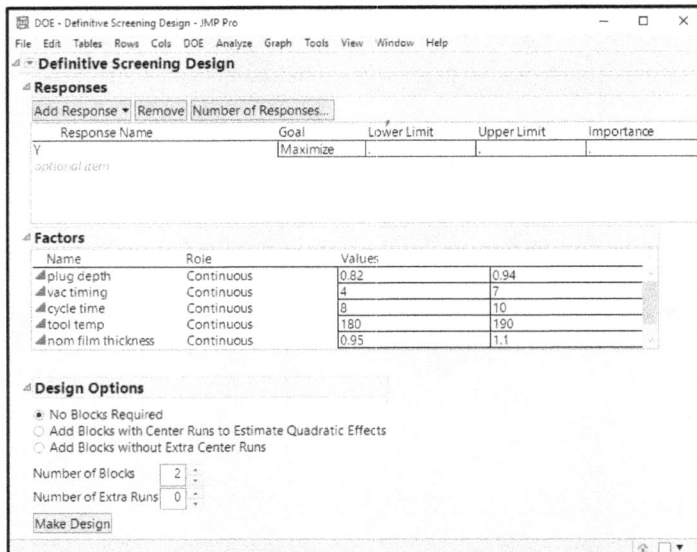

The design shown in Figure 10.18 quickly illustrates one of the most compelling reasons why the DSD should be considered when the number of runs is minimal. A DSD, without added blocks or runs, includes two times the number of factors plus one center point run. The 13-run design is shown and has not been randomized. You should review the design evaluation options. Review the design evaluation results of the three model options to evaluate all three designs.

Figure 10.18: Definitive Screening Design

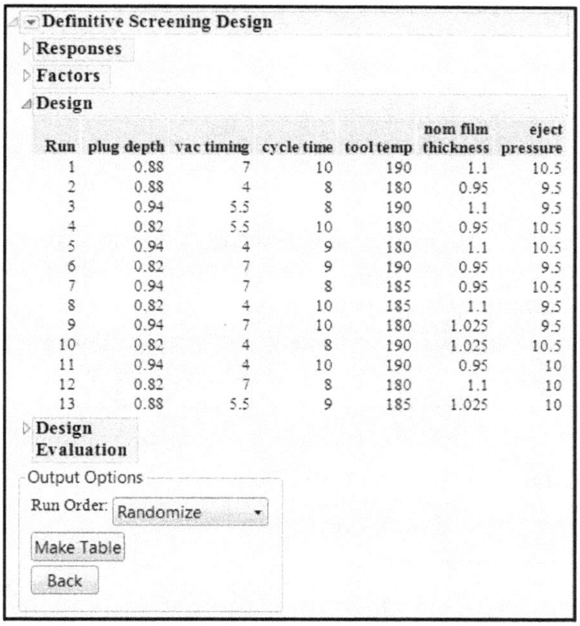

Run	plug depth	vac timing	cycle time	tool temp	nom film thickness	eject pressure
1	0.88	7	10	190	1.1	10.5
2	0.88	4	8	180	0.95	9.5
3	0.94	5.5	8	190	1.1	9.5
4	0.82	5.5	10	180	0.95	10.5
5	0.94	4	9	180	1.1	10.5
6	0.82	7	9	190	0.95	9.5
7	0.94	7	8	185	0.95	10.5
8	0.82	4	10	185	1.1	9.5
9	0.94	7	10	180	1.025	9.5
10	0.82	4	8	190	1.025	10.5
11	0.94	4	10	190	0.95	10
12	0.82	7	8	180	1.1	10
13	0.88	5.5	9	185	1.025	10

Set the randomization seed by clicking the red triangle menu next to *Custom Design* and selecting *Set Random Seed*. Enter 2018 as the randomization seed and click *OK*. Click *Make Table* to create the data table for the 13-run DSD, shown in Figure 10.19. Select *File ▶ Save As* to save the data as "DSD 13R Thermoforming."

Figure 10.19: DSD Data Table

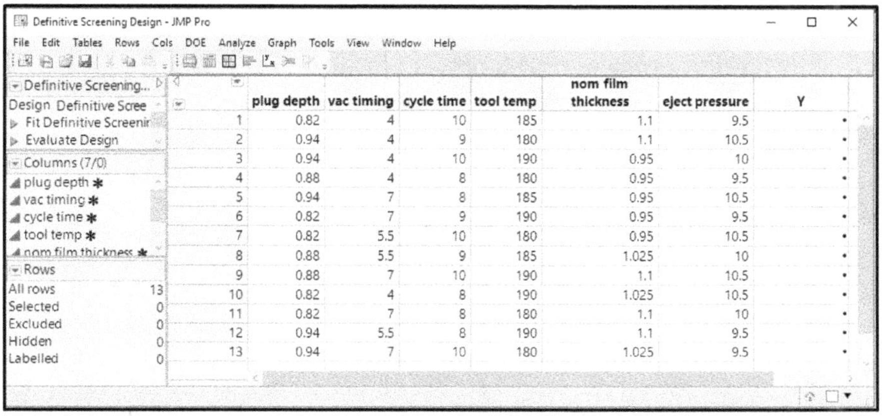

	plug depth	vac timing	cycle time	tool temp	nom film thickness	eject pressure	Y
1	0.82	4	10	185	1.1	9.5	•
2	0.94	4	9	180	1.1	10.5	•
3	0.94	4	10	190	0.95	10	•
4	0.88	4	8	180	0.95	9.5	•
5	0.94	7	8	185	0.95	10.5	•
6	0.82	7	9	190	0.95	9.5	•
7	0.82	5.5	10	180	0.95	10.5	•
8	0.88	5.5	9	185	1.025	10	•
9	0.88	7	10	190	1.1	10.5	•
10	0.82	4	8	190	1.025	10.5	•
11	0.82	7	8	180	1.1	10	•
12	0.94	5.5	8	190	1.1	9.5	•
13	0.94	7	10	180	1.025	9.5	•

Compare Designs for Main Effects with Different Structures (JMP Pro Only)

You can review the design diagnostics of each model by running the Evaluate Design platform in the open experimental model plan output. In later versions of JMP, model creation automatically creates a script that enables you to easily evaluate the design by running it from the saved data sheet. The Compare Designs platform offered starting with JMP Pro 13 is used to evaluate the diagnostics of all three designs. Complete the following steps to efficiently compare the three designs:

1. Be sure that all three design files are open:

    ○   *Custom Design 12R Main Effects Thermoform Process.jmp*

    ○   *Plackett-Burman 12R Thermoforming.jmp*

    ○   *DSD 13R Thermoforming.jmp*

2. Make sure that the design you want to use as the standard of comparison is open as an active window. The *DSD 13R Thermoforming.jmp* is the standard of comparison for the example.

3. Select *DOE* ▶ *Design Diagnostics* ▶ *Compare Designs* to start the comparison.

4. In the *Compare Designs* window, press the Ctrl key while selecting the *Custom Design 12R Main Effects Thermoform Process* and *Plackett-Burman 12R Thermoforming* files.

5. Select the matching inputs and use *Match Columns* to ensure that all six inputs are a highlighted selection for the three designs. You do not need a y output variable to make the comparisons.

6. Click *OK* to get the design comparison output shown in Figure 10.20.

Figure 10.20: Compare Designs: Power

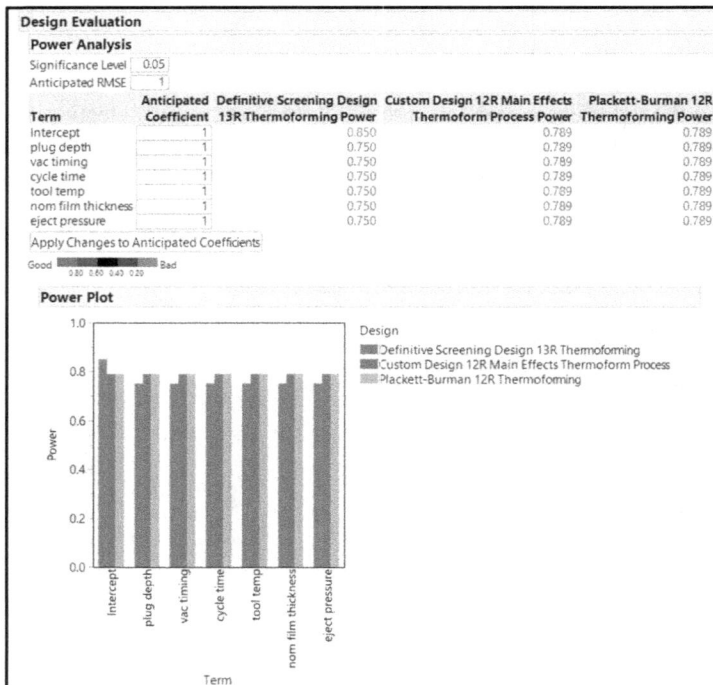

The first question is whether a model can incorrectly indicate an input as significant when it is not significant in the actual population. A relationship that is significant in the actual population is also noted as a real effect. The comparisons of statistical power show no great difference in statistical power among the three models. The custom design and Plackett-Burman screening design have slightly more power than the DSD. However, a difference of less than 4 percentage points of power is of no practical relevance. Small differences in power can result from minimal differences in the effect size or from random variation. The experiments are one sample from a population, and it is known that effect size and random variation differ minimally for each random sample pulled from the population. Analysts and consumers of statistical studies tend to consider power differences to be relevant when they are at least 5%. This is why the minimal difference in estimated power is not relevant to the team.

The next goal is to determine how much error is likely to be included in predictions made from each model. Figure 10.21 provides comparative information about the structure of prediction variance for each model.

Figure 10.21: Compare Designs: Prediction Variance Structure

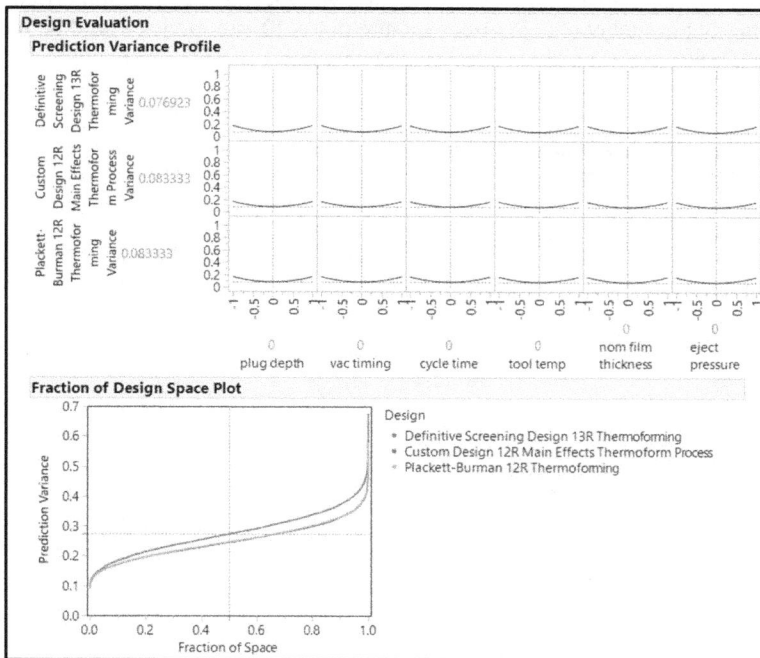

The prediction variance profile of the six inputs (main effects) for the three models illustrates symmetric concave shapes, which are desirable. The Fraction of Design Space plot shows that the DSD has slightly more prediction variance. The difference throughout the design space is no more than 1/10th of a unit of prediction variance. The minimal amount of difference might be due to random sampling error and is not enough to be of practical value.

The comparison of the estimation efficiency of the custom design and the Plackett-Burman screening design to the DSD is shown in Figure 10.22.

Figure 10.22: Compare Designs: Estimation Efficiency

Relative Estimation Efficiency		
Term	Efficiency of Definitive Screening Design 13R Thermoforming Relative to Custom Design 12R Main Effects Thermoform Process	Efficiency of Definitive Screening Design 13R Thermoforming Relative to Plackett-Burman 12R Thermoforming
Intercept	1.041	1.041
plug depth	0.913	0.913
vac timing	0.913	0.913
cycle time	0.913	0.913
tool temp	0.913	0.913
nom film thickness	0.913	0.913
eject pressure	0.913	0.913

Good ▓▓▓▓▓▓ Bad
1.30 1.25 0.90 0.67

The comparison value of 1 indicates a 1:1 ratio of difference between models, which means that the models are equivalent. The comparative estimation efficiency of 0.913 between the custom design and Plackett-Burnam to the DSD indicates a difference of no practical value.

The three models look to have very similar statistical power, prediction variance, and estimation efficiency. You might be thinking at this point that model choice is irrelevant and that a 12-run model would be the best choice to save at least 1 run of resources. The alias matrices and absolute correlations shown in Figure 10.23 prove the great benefit of a DSD.

Figure 10.23: Compare Designs: Alias Matrix and Correlations

Design Evaluation			
Alias Matrix Summary			
Term	Root Mean Squared Values Definitive Screening Design 13R Thermoforming	Root Mean Squared Values Custom Design 12R Main Effects Thermoform Process	Root Mean Squared Values Plackett-Burman 12R Thermoforming
Intercept	0.0000	0.0000	0.0000
plug depth	0.0000	0.3162	0.3162
vac timing	0.0000	0.2789	0.2789
cycle time	0.0000	0.2789	0.2789
tool temp	0.0000	0.2582	0.2582
nom film thickness	0.0000	0.2582	0.2582
eject pressure	0.0000	0.2357	0.2357
Total	0.0000	0.2520	0.2520

Good ▓▓▓▓▓ Bad
0.10 0.30 0.50 0.70

Absolute Correlations			
Model x Model	Average Correlation	Number of Confoundings	Number of Pairs
Definitive Screening Design 13R Thermoforming	0.000	0	15
Custom Design 12R Main Effects Thermoform Process	0.000	0	15
Plackett-Burman 12R Thermoforming	0.000	0	15
Model x Alias	Average Correlation	Number of Confoundings	Number of Pairs
Definitive Screening Design 13R Thermoforming	0.000	0	60
Custom Design 12R Main Effects Thermoform Process	0.222	0	60
Plackett-Burman 12R Thermoforming	0.222	0	60
Alias x Alias	Average Correlation	Number of Confoundings	Number of Pairs
Definitive Screening Design 13R Thermoforming	0.344	0	45
Custom Design 12R Main Effects Thermoform Process	0.126	0	45
Plackett-Burman 12R Thermoforming	0.126	0	45

Good ▓▓▓▓▓ Bad
0.10 0.30 0.50 0.70

As mentioned in the design creation steps, the 12-run designs for the custom design and the Placket-Burman design are focused on detecting the main effects of individual inputs. Aliasing is an amount of estimation bias present for a given effect that is due to another effect. Mitigation of bias is a goal for experimental design. A team might have tolerance for some bias presence for a screening design because they want to minimize resources. The alias matrix indicates that the 12-run designs have portions of the individual inputs that are biased by other terms (from 24% to 32%). The DSD has no aliasing for any of the six inputs. This means that the evidence of significance for an individual input is independent of the evidence of significance for any other input and interactions.

The Absolute Correlations summary provides an average correlation value for each of the three models in three different tables. The terms included in the model correlated to other model terms is shown in the Model x Model summary; this is the only one that is shown if no aliasing exists. None of the three models include an average correlation among the comparisons of 15 pairs made from the six inputs. The Model x Alias summary includes the average correlations between model terms and alias terms. The DSD has zero average correlation between the pairs of the six individual inputs as well as all interactions making up 60 pairs. The 12-run models have an average correlation of 22%. The Alias x Alias summary contains the average correlations between alias terms and other alias terms. The DSD has 34% average correlation between the pairs made from the interactions making up 45 pairs. The 12-run models have an average correlation of 13% among the pairs of interaction alias terms. The DSD does not perform as well among the pairs of interaction terms that alias the individual inputs. However, correlation among interactions is not of great interest for the goal of screening for important process inputs.

The correlation maps shown in Figure 10.24 are great for visualizing the amount of independence present for all effects.

Figure 10.24: Compare Designs: Color Map on Correlation

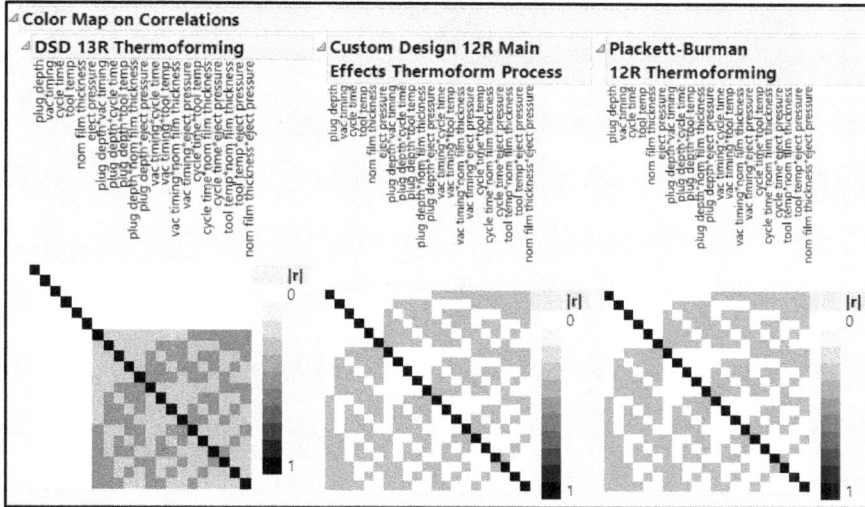

Both of the 2-run designs include up to 50% correlation (medium, gray blocks) between the main effects and the interactions. The DSD illustrates that the main effects have no correlation between main effects and interactions; this is made evident by the white "stripes" of blocks for main effects and interaction combinations running across the top of and down the left side of the matrix plot. The DSD does include moderate correlation among the interactions. Minor higher-order correlations among interaction are not problematic for a goal of screening effects since the team is not primarily focused on minimizing prediction error for making estimates.

Design diagnostics comparing the efficiency of designs is shown in Figure 10.25.

Figure 10.25: Compare Designs: Design Diagnostics

Design Evaluation		
Design Diagnostics		
	Efficiency of Definitive Screening Design 13R Thermoforming Relative to Custom Design 12R Main Effects Thermoform Process	Efficiency of Definitive Screening Design 13R Thermoforming Relative to Plackett-Burman 12R Thermoforming
D-efficiency	0.865	0.865
G-efficiency	0.928	0.928
A-efficiency	0.862	0.862
I-efficiency	0.903	0.903
Additional Run Size	1	1
Good ▓▓▓▓▓▓ Bad 1.50 1.25 0.80 0.67		

The efficiency of the DSD is slightly less than the two 12-run designs, but the amount of difference is minimal and is of no practical value. You might have realized by now that the 12-run default of the custom design created a Plackett-Burman factorial design. The custom design algorithm generally suggests a default model that includes enough runs to be orthogonal and might very well match a classic design model classic design models are based on orthogonality).

The DSD is superior for obtaining clean signals from the process inputs. The trade-off is one more run of resources for the team to be able to obtain such robust information from their set of structured experiments. The DSD has become a very popular option largely due to this fact, but there are even more benefits to consider.

Adding Interactions to Compare Designs (JMP Pro Only)

The goal of the thermoforming experiments is to screen the process inputs to determine which have the most influence on the minimum thickness output. The 12-run, Resolution 3 designs are developed to provide evidence on the main effects only, but you know that interactions between inputs might influence results due to confounding. In this section, you use the Compare Designs platform to see how interactions might be influence the minimum thickness. Complete the following steps to add interactions and compare the three designs:

1. Make sure that the files for the three designs are open and that the DSD design is the active one.
2. Select *DOE* ▶ *Design Diagnostics* ▶ *Compare Designs*.
3. In the *Compare Designs* window, press the Ctrl key and select the screening and Plackett-Burman models. Match all the input columns as you did previously.
4. Under the *Model* heading, click *Interactions* and select 2nd for the interactions between two inputs as shown in Figure 10.26.
5. Click *OK* to get the model comparison output.

Figure 10.26: Compare Designs: Add Interactions

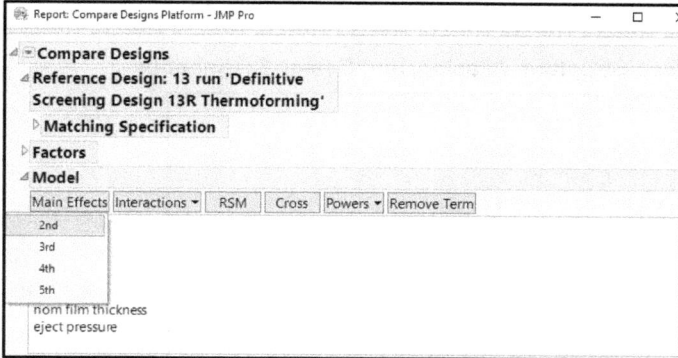

The power values change dramatically when the interactions are included in the model, as shown in Figure 10.27. The relatively small number of runs planned for in the designs are not enough to obtain adequate power to detect the significant factors because the highest estimated power is no more than 20%.

Figure 10.27: Compare Designs: Power with Interactions

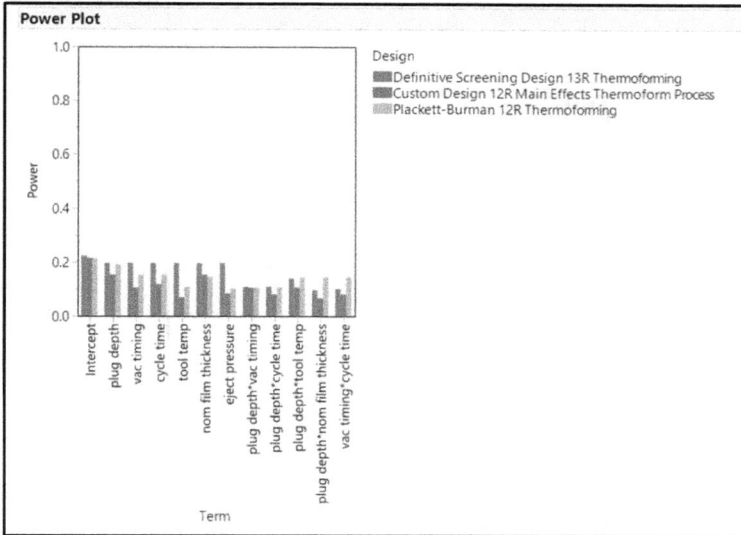

The prediction variance, relative estimation efficiency, alias matrix summary, and design diagnostics are shown in Figure 10.28.

Figure 10.28: Compare Designs: Variance Structure, Aliasing, and Design Diagnostics

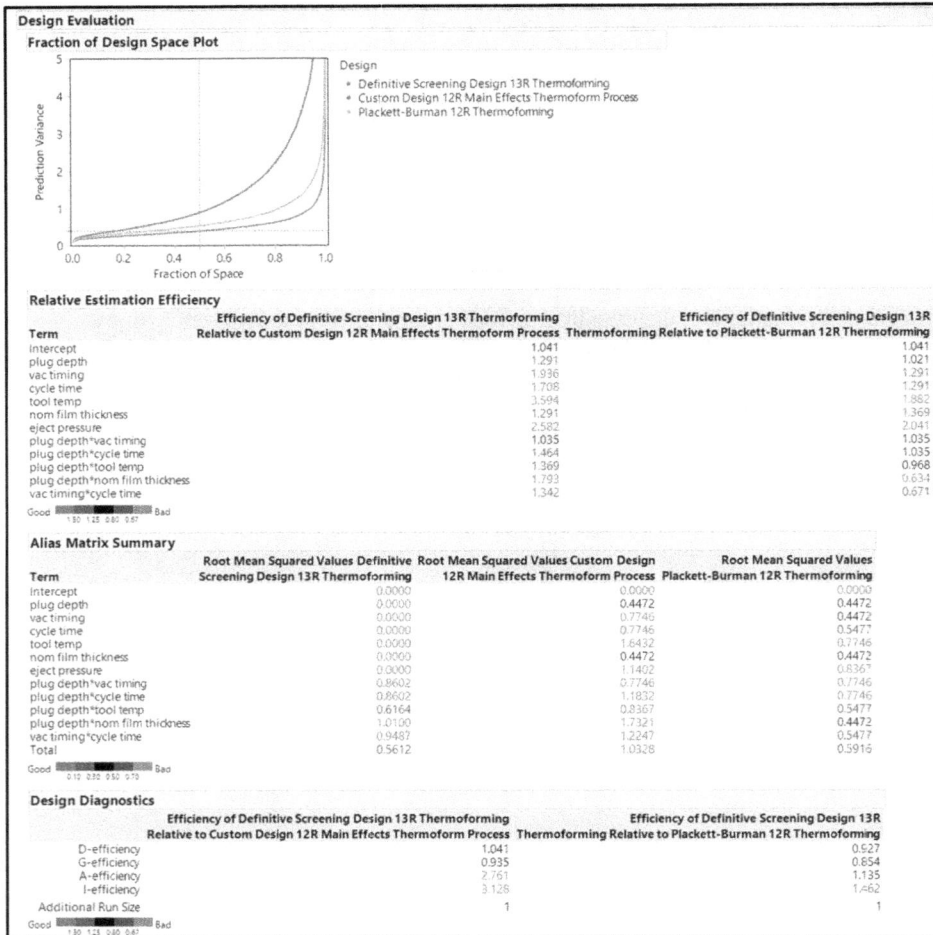

The Fraction of Design Space Plot illustrates that the three models have different patterns of prediction variance across the design space. The overall amount of prediction variance when interactions are included is more than double that of the main effects models, especially at the extremes of the treatment levels of the inputs. Differences among the models are also much more prevalent and it is clear that using the DSD is an advantage; the curve is the lowest of the three and the flattest for most of the space. Estimation efficiency of the DSD is between 2% and 250% better than the 12-run designs when you account for the potential effects of interactions. Notice that many of the comparative values show up in green font, which highlights the advantages of the DSD for making estimations from the model.

This is a great example of the balance between the information gained from the experiments in aliasing amounts and the resources required in the number of runs required. Many of the aliasing values for the 12-run designs are highlighted in red due to the likelihood for error that is included in the analysis results. The alias tables provide more evidence on the superiority of the DSD because no aliasing is possible for the main effects. The efficiency for the algorithm to be able to find optimality for the criteria is not practically different among the models.

Even though interactions between the inputs are not of primary interest for the goal of the experiments, it is well worth the time to use the tools in JMP to evaluate how much they might contribute to error in the analysis of the results. The structure of the DSD and related advantages become clearer as interactions are considered. It is a best practice for the designer of structured experiments to evaluate both the primary goals as well as all possible sources of model error so that stakeholders can best weight the amount of resources to provide with the information desired from the activity.

Visualizing Design Space with Scatterplot Matrices

The concept of design space is important for all stakeholders of an experiment. As noted before, the area within the levels of all inputs makes up the design space for a set of experiments. It is easy to conceive of the space when inputs are limited to two since a simple x,y plot can illustrate the space. You can also view three inputs as a 3-dimensional cube. When a set of experiments involves more than three inputs, the task of conceptualizing the space is much more difficult. You can easily view this space = with a scatterplot matrix available in the *Graphs* menu in JMP. Making a scatterplot matrix with continuous variables results in points that are on top of each other, which hides duplicate runs. When continuous variables are involved, you can copy them and change to a nominal modeling type. This gives you the ability to visualize each run in the design. Start by pressing the Shift key while clicking the column variables for all inputs to select them. While the column variables are shaded as selected, choose *Edit ▸ Copy with Column Names,* as shown in Figure 10.29.

Figure 10.29: Selecting and Copying Column Variables

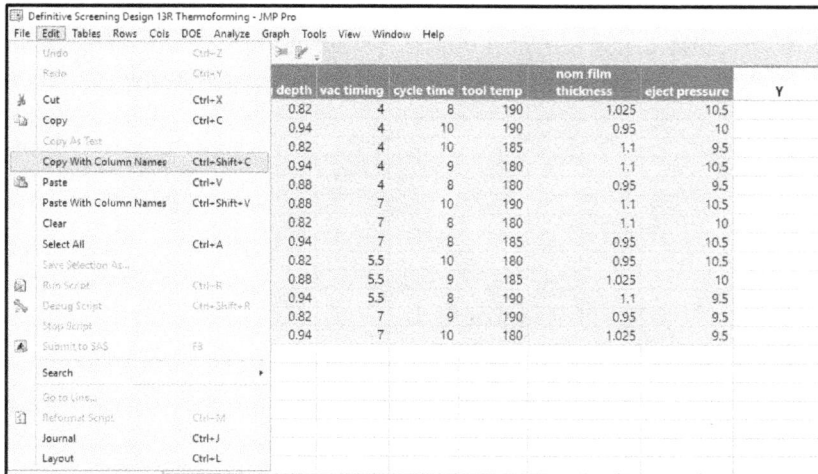

depth	vac timing	cycle time	tool temp	nom film thickness	eject pressure	Y
0.82	4	8	190	1.025	10.5	
0.94	4	10	190	0.95	10	
0.82	4	10	185	1.1	9.5	
0.94	4	9	180	1.1	10.5	
0.88	4	8	180	0.95	9.5	
0.88	7	10	190	1.1	10.5	
0.82	7	8	180	1.1	10	
0.94	7	8	185	0.95	10.5	
0.82	5.5	10	180	0.95	10.5	
0.88	5.5	9	185	1.025	10	
0.94	5.5	8	190	1.1	9.5	
0.82	7	9	190	0.95	9.5	
0.94	7	10	180	1.025	9.5	

Complete the following steps to change the attributes of the variables:

1. Click on the header of the first unused column to the right of the *Y* column to select it so it is highlighted in blue.
2. Select *Edit ▸ Paste with Column Names.* Note that the copied variables are pasted with a "2" to the right of the names as a default. The pasted columns are automatically shown as shaded and selected.
3. With the columns selected, select *Cols ▸ Standardize Attributes*, as shown in Figure 10.30. (Alternatively, right-click in one of the selected columns and select *Standardize Attributes*.

Figure 10.30: Standardize Attributes

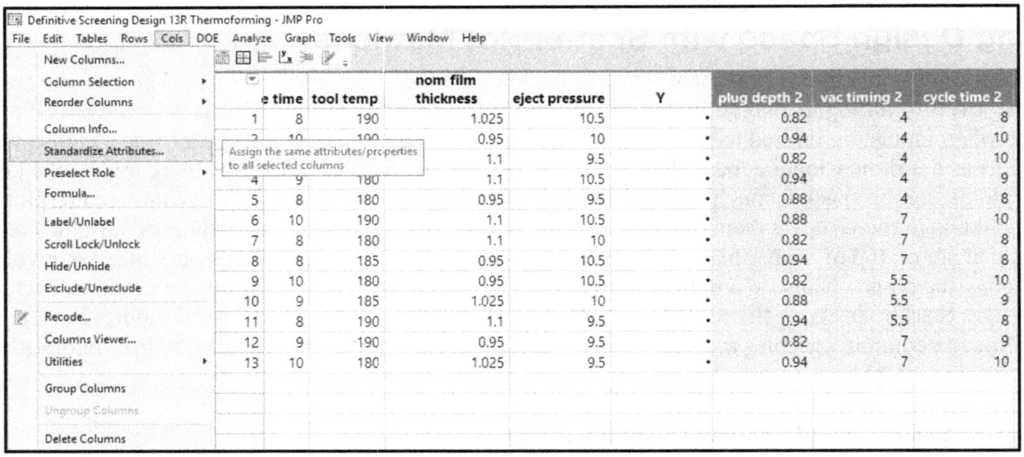

In the *Standardize Columns Attributes* dialog box, click *Attributes* and change the *Modeling Type* from *Continuous* to *Nominal*, as shown in Figure 10.31. Click *Apply* and *OK* to execute the modeling type changes.

Figure 10.31: Set Modeling Type

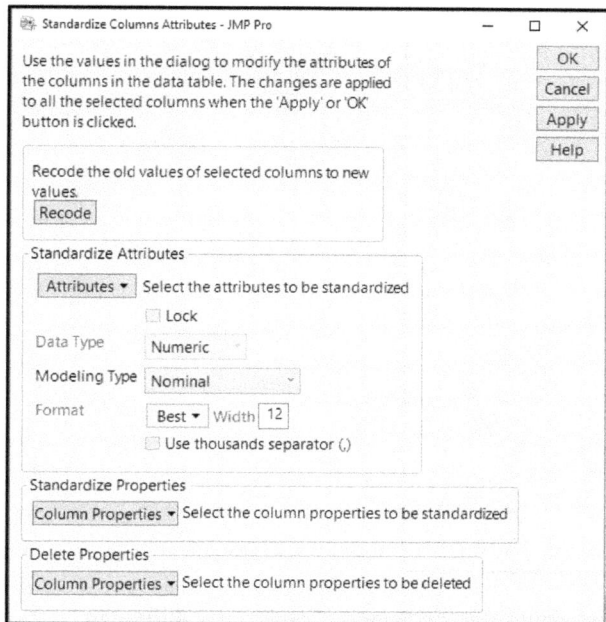

With all six inputs copied and changed to a nominal modeling type, the scatter plot matrix will clearly show individual runs. Select *Graph* ▶ *Scatterplot Matrix* and move the six nominal inputs to the *Y, Columns* box. Click *OK* to get the plot.

The plot shown in Figure 10.32 is the matrix of all 2-variable combinations with each run jittered so that it can be viewed easily. The matrix makes it clear to project stakeholders that in spite of limited resources, the design space is well covered by the design. All of the extreme lows, extreme highs, and centers of the combinations have multiple observations. Figure 10.33 is included to illustrate the space coverage when the original variables (continuous modeling type) are used. The theoretical coverage of the design is evident, but multiple observations are represented by the black dots.

Figure 10.32: Scatterplot Matrix: Definitive Screening Design Space

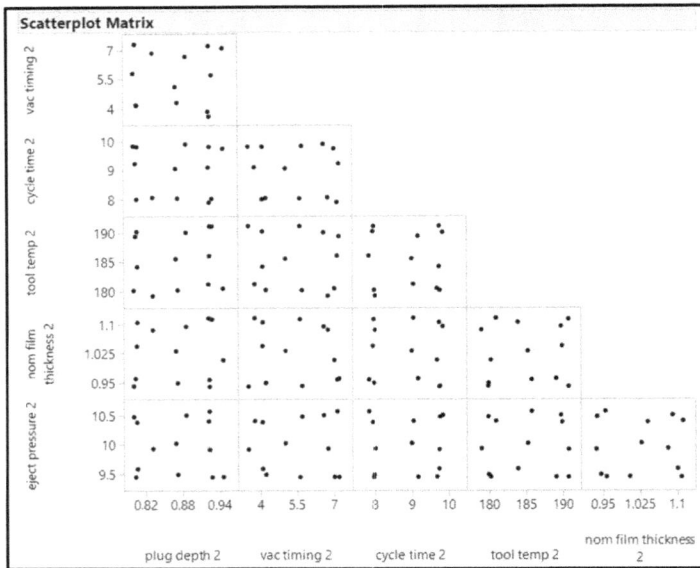

Figure 10.33: Scatterplot Matrix: Definitive Screening Design Space (Continuous Variables)

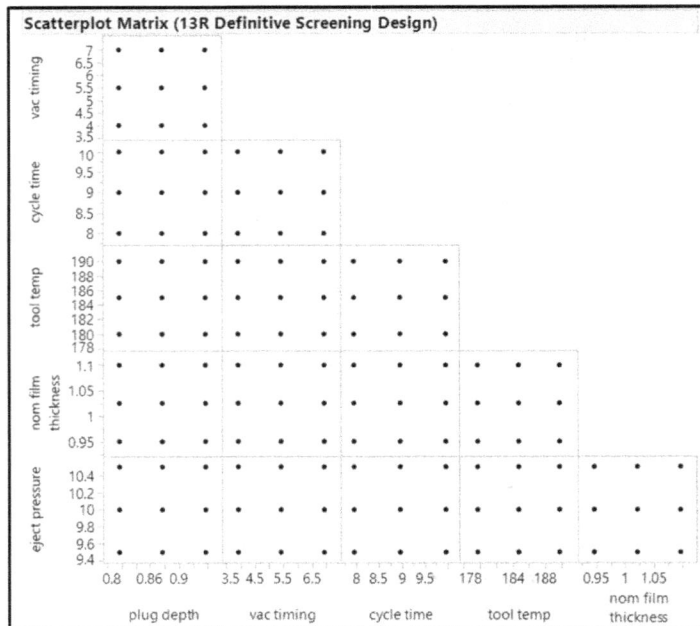

Figure 10.34 is the scatterplot matrix for the continuous variables of the custom design. The plot clearly illustrates that the corners of the design are covered, but no observations are located in the center of the space. An advantage of including the center point is the ability for the model to identify whether there might be model effects that include a rate of change (non-linear). The single, added run provides yet

another level of error detection and can greatly benefit the analysis by reducing the potential for making prediction errors. The DSD structure automatically includes a center point. However, you can specify a center point when you use the Custom Designer. It is good practice to re-evaluate model diagnostics; added center points will change things and some trial and error might be needed to find the optimal number of runs to meet experimental goals.

Figure 10.34: Scatterplot Matrix: Custom Design Space

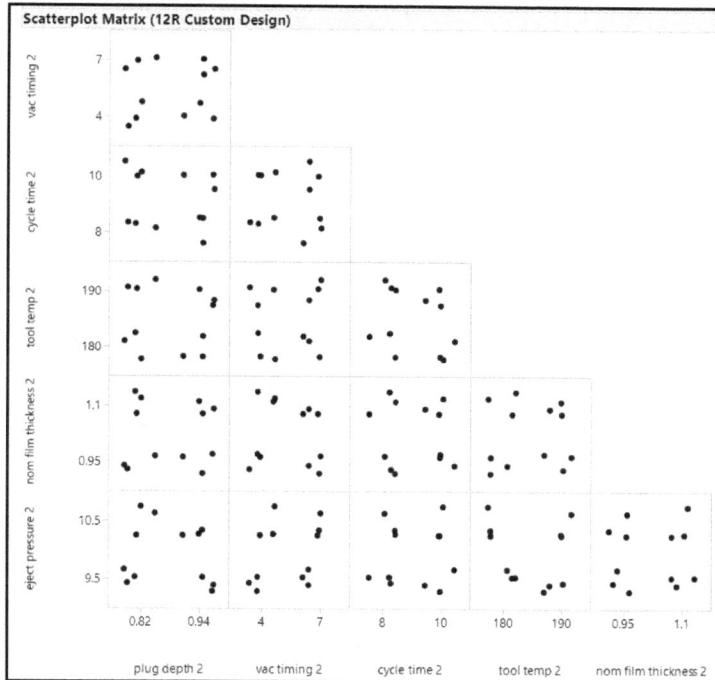

Michelyne could easily communicate with stakeholders throughout the design process through use of the excellent graphics and diagnostic tools in JMP. Good experimental design is typically iterative as the designer clarifies the need to strike the right balance between allocated resources (runs) and the information needed from the analysis. The project stakeholders are extremely pleased because the DSD has provided "clean" information about main effects and has also mitigated error by detecting the potential effects of interaction and curved terms.

Experimental Design for a Granulation Process with Multiple Outputs

Emily's team has been charged with developing a high-shear, wet granulation process that robustly meets five outputs of interest. She has had some experience visualizing data with JMP and knows that structured experiments can be easily designed with the excellent set of tools available in the DOE platform. JMP handles multiple outputs easily because you create a model for each, and there is a comprehensive profiler to determine whether the inputs can be optimized to meet the requirements for all outputs. There is no need to specify outputs in the design phase because you can add any number of outputs later, prior to the analysis of results. Outputs must be considered in discussions about experimental planning because the goals of experimentation result from the behavior that is seen in the outputs. This example involves the specification of outputs during the design phase; the outputs are well known to the stakeholders of the experiments.

JMP enables you to enter responses, specify the specific goals (including limit values), and set the importance of each. The detail entered for the responses will influence the type of model used for the analysis as well as provide structure for the profiler to determine the optimum settings for the inputs. This detail ensures that experimental goals can be met through predictions of the range of responses likely to represent the population of the end process. Start by selecting *DOE* ▶ *Custom Design*. Complete the following steps to enter the responses and inputs as shown in Figure 10.35:

1. Click *Add Response* for each response and specify the *Goal* of each.
2. Enter values for the *Upper Limit* and *Lower Limit*. Keep in mind that the *Minimize* and *Maximize* options are one-sided and use only the appropriate single limit for optimization.
3. The default value for *Importance* will result in the optimization of each response equally. You can specify the most important outputs (for example, CQAs) by entering values greater than 1. You can also minimize the importance of an output by entering values that are less than 1.
4. Keep in mind that you can leave the limits and importance blank during the design phase. The *Goal, Lower Limit, Upper Limit*, and *Importance* can be specified at the time of analysis of the results.
5. Enter 4 in the *Add N Factors* box and click *Add Factor*.
6. Choose *Continuous* to add factors *X1* through *X4* to the window.
7. Click on each factor to rename it and enter the limit *Values* as shown.
8. Click *Continue* to proceed to the model specification part of the design.

Figure 10.35: Custom Design

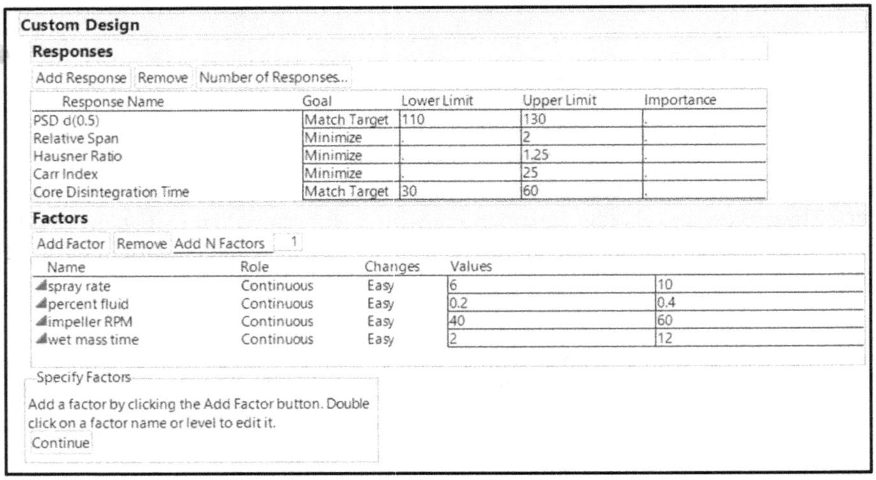

Custom Design

Responses

Add Response | Remove | Number of Responses...

Response Name	Goal	Lower Limit	Upper Limit	Importance
PSD d(0.5)	Match Target	110	130	.
Relative Span	Minimize	.	2	.
Hausner Ratio	Minimize	.	1.25	.
Carr Index	Minimize	.	25	.
Core Disintegration Time	Match Target	30	60	.

Factors

Add Factor | Remove | Add N Factors | 1

Name	Role	Changes	Values	
⊿spray rate	Continuous	Easy	6	10
⊿percent fluid	Continuous	Easy	0.2	0.4
⊿impeller RPM	Continuous	Easy	40	60
⊿wet mass time	Continuous	Easy	2	12

Specify Factors

Add a factor by clicking the Add Factor button. Double click on a factor name or level to edit it.

Continue

The purpose of the granulation experiments is to be able to accurately estimate the amount of effect that the inputs have on the multiple responses. The information needed for this design is much greater than the information required for the previous screening design. The stakeholders need to be prepared to allocate a greater amount of resources (runs) than what is required for a simple screening study. The example assumes that the need for information is high enough for the design to include the main effects and interactions, and the potential for curved terms. This response surface design can be easily included by using the RSM choice available in the model specifications.

1. In the *Custom Design* dialog box, click the *RSM* button located underneath the **Model** header to see all the individual inputs (main effects), combinations of 2 inputs (interactions), and each input crossed with the same input (squared terms) shown in Figure 10.36.

2. You can change *Estimability* from *Necessary* to *If Possible* so that fewer runs are required. However, the default value best meets the goals of experimentation.

3. Underneath the *Design Generation* header, change the *Number of Center Points* to 1 since the team believes that non-linear effects are possible and they want to add to the ability to detect them.

4. The *Default* number of runs is maintained at 21 for the design. You are encouraged to experiment with different numbers of runs to see how the design diagnostics change. However, this example uses the default number.

5. Click on the red triangle menu next to *Custom Design* header and select *Set Random Seed*.

6. Enter 2017 as the randomization seed, and click *OK*.

7. Click *Make Design* to get the design plan data table.

Figure 10.36: Custom Design: Model Specification and Design Generation

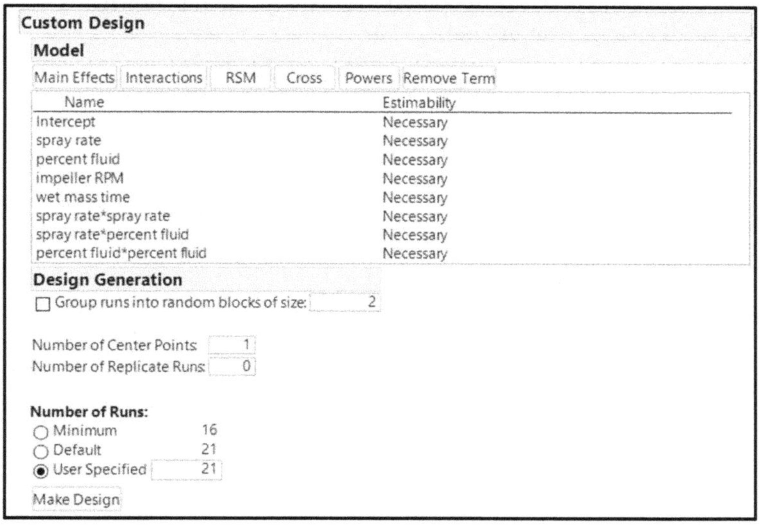

The 21-run design is shown in Figure 10.37. The high and low combinations of the four inputs are shown in the runs, as well as the center point value (run 21). There are also values between the high and low levels of the inputs that are necessary to be able to quantify the amount of change that can be expected from the inputs, interactions, and squared terms; that is the purpose of a response surface model.

Figure 10.37: Custom Design with 21 Runs and CP

Run	spray rate	percent fluid	impeller RPM	wet mass time
1	7.3	2	40	2
2	8	2	60	12
3	10	4	50	7
4	8	4	60	2
5	10	3	60	12
6	10	2	40	12
7	6	3	50	12
8	6	4	60	12
9	6	2	60	2
10	10	2	60	2
11	6	4	50	2
12	6	3	60	7
13	10	4	40	2
14	8	3	50	2
15	8	3	40	7
16	8	4	40	12
17	8	3	50	7
18	6	2	40	12
19	6	4	40	7
20	8	2	50	7
21	8	3	50	7

The diagnostics for a response surface model involve a different optimality algorithm than what is needed for the screening design. The power analysis is shown in Figure 10.38.

Figure 10.38: Custom Design: Power Diagnostics

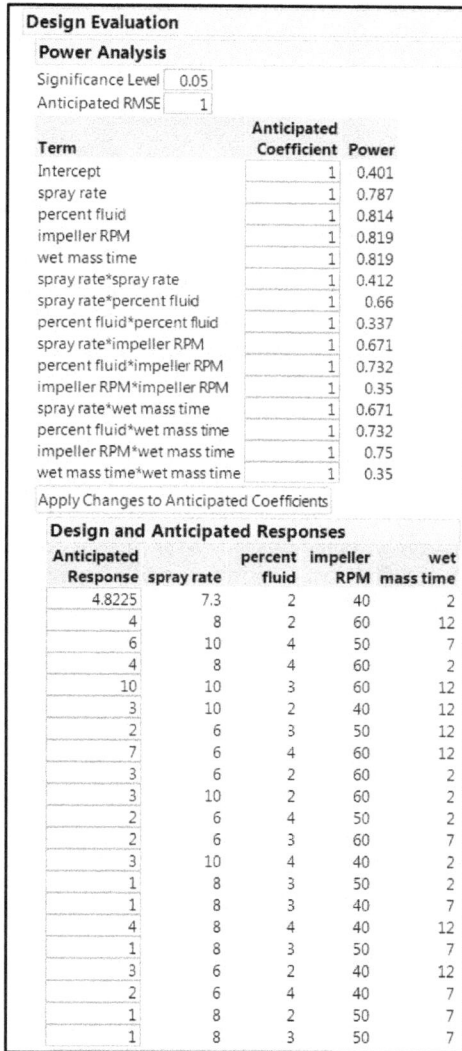

Design Evaluation

Power Analysis

Significance Level	0.05
Anticipated RMSE	1

Term	Anticipated Coefficient	Power
Intercept	1	0.401
spray rate	1	0.787
percent fluid	1	0.814
impeller RPM	1	0.819
wet mass time	1	0.819
spray rate*spray rate	1	0.412
spray rate*percent fluid	1	0.66
percent fluid*percent fluid	1	0.337
spray rate*impeller RPM	1	0.671
percent fluid*impeller RPM	1	0.732
impeller RPM*impeller RPM	1	0.35
spray rate*wet mass time	1	0.671
percent fluid*wet mass time	1	0.732
impeller RPM*wet mass time	1	0.75
wet mass time*wet mass time	1	0.35

Apply Changes to Anticipated Coefficients

Design and Anticipated Responses

Anticipated Response	spray rate	percent fluid	impeller RPM	wet mass time
4.8225	7.3	2	40	2
4	8	2	60	12
6	10	4	50	7
4	8	4	60	2
10	10	3	60	12
3	10	2	40	12
2	6	3	50	12
7	6	4	60	12
3	6	2	60	2
3	10	2	60	2
2	6	4	50	2
2	6	3	60	7
3	10	4	40	2
1	8	3	50	2
1	8	3	40	7
4	8	4	40	12
1	8	3	50	7
3	6	2	40	12
2	6	4	40	7
1	8	2	50	7
1	8	3	50	7

The power analysis continues to use the default values of 1 unit for RMSE and 1 unit for each of the anticipated responses. However, the model creates a table of anticipated responses listed by each input, including various values. The anticipated responses are typically not known unless there is reliable previous modeling that can be used to input the values. The estimated power is very good for main effects (~80%), good for the interactions (66% to 73%), and marginal for the squared terms (33% to 42%). The team might decide to go back and increase the estimated power by adding replicate runs, which can expand resources very quickly. In general, the estimated power seems adequate to move forward; actual power might increase if factors have stronger signals, and there is always the potential to augment a design to improve power.

The prediction variance trend is shown in the Fraction of Design Space Plot in Figure 10.39.

Figure 10.39: Custom Design: Variance Structure

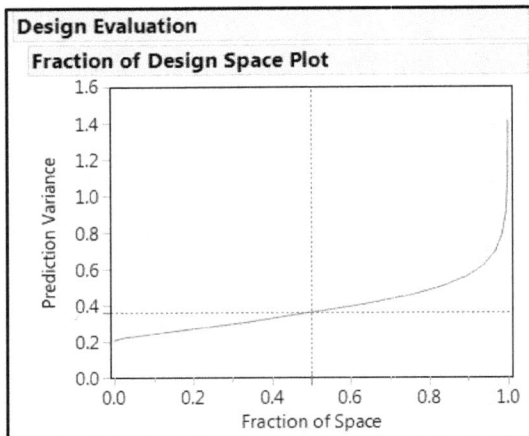

The Fraction of Design Space Plot illustrates a very flat prediction variance function over the design space. A flat function for variance from center point to the extreme levels indicates a minimal rate of increase in prediction variance over the design space, which is desirable. The added observations of levels within the extreme high and low levels really helps minimize prediction variance. This is one of the main reasons that an RSM is much better at making accurate predictions than is a screening design.

It is good practice to look at the prediction variance surface of the various two-way combinations to determine whether the allocation of runs for the RSM provides for symmetric error. Figure 10.40 shows the surface of prediction variance between the spray rate and percent fluid.

Figure 10.40: Custom Design: Variance Surface

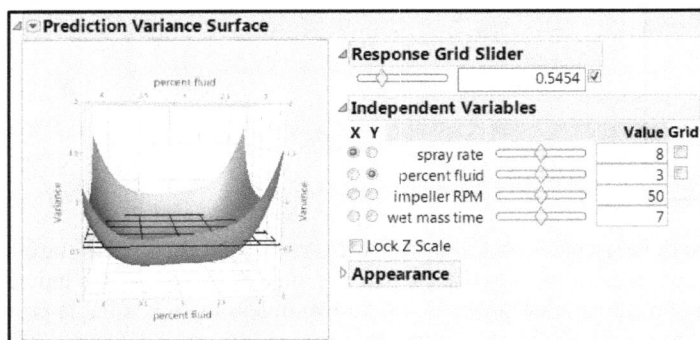

The 3-D surface was rotated to illustrate that the model does not perform as well for predictions made from the high level of fluid due to heightened prediction variance for the region. The options available to deal with this scenario include:

a. Go back to the custom model designer to add more runs to reduce the asymmetric prediction variance.

b. Maintain the model without change to accept regions of increased prediction variance. The risk associated with this option manifests if the results of the analysis indicate that the highest amount of fluid provides the best average results for the various outputs. The study stakeholders will need to be prepared to either augment the design or run multiple confirmation runs in order to ensure that the results are repeatable and robust.

The team is not prepared to ask for more resources than the 21 runs for this design and accepts the potential for increased prediction variance to mitigate resources. It is good practice to look at the prediction variance surfaces for all six of the combinations of the four inputs. This scenario is not included here for brevity.

The estimation efficiency and correlation map are shown in Figure 10.41.

Figure 10.41: Custom Design: Estimation Efficiency and Correlation Map

Design Evaluation

Estimation Efficiency

Term	Fractional Increase in CI Length	Relative Std Error of Estimate
Intercept	1.255	0.492
spray rate	0.384	0.302
percent fluid	0.336	0.291
impeller RPM	0.326	0.289
wet mass time	0.326	0.289
spray rate*spray rate	1.219	0.484
spray rate*percent fluid	0.614	0.352
percent fluid*percent fluid	1.51	0.548
spray rate*impeller RPM	0.593	0.348
percent fluid*impeller RPM	0.482	0.323
impeller RPM*impeller RPM	1.454	0.535
spray rate*wet mass time	0.593	0.348
percent fluid*wet mass time	0.482	0.323
impeller RPM*wet mass time	0.45	0.316
wet mass time*wet mass time	1.454	0.535

Color Map on Correlations

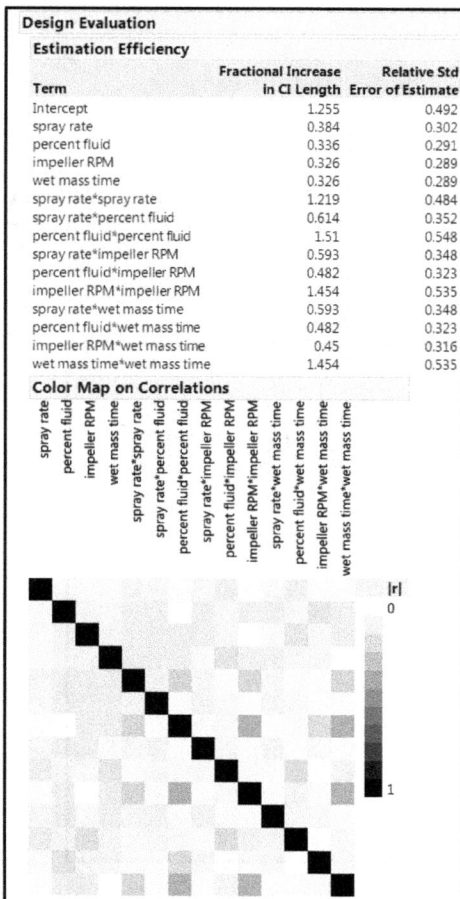

Estimation efficiency is often used to compare the merits of various designs. It can be used here to communicate to stakeholders the limits on information that will be gained from analysis to ensure that the best decision is made for the structured experiments plan. The correlation color map generally illustrates low correlation values for much of the space with light grey colored squares. There is a higher potential for the higher-order terms (interactions and squared terms) to correlate with each other. Typically, correlation among higher-order terms is not considered an issue unless the information from the analysis is expected to be extremely precise.

There are no needs voiced by stakeholders for an extremely high level of precision, and the team believes the model to be adequate to meet the goals of the project. The default values for the *Data Table Options*, shown in Figure 10.42, are used with the *Randomize* run order. Click *Make Table* to create the data table for the 21 runs of the experiment.

Figure 10.42: Output Options

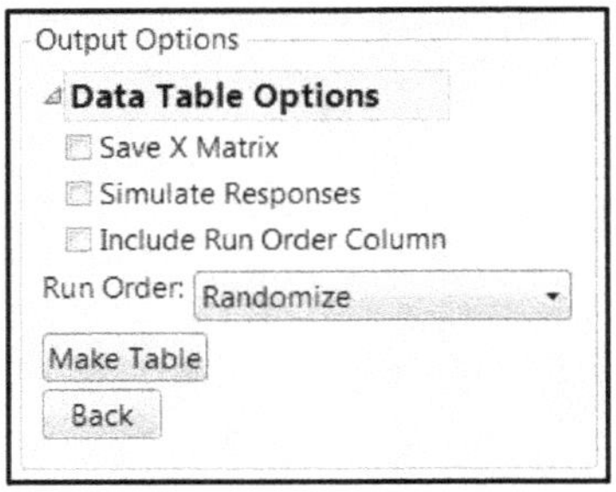

The data table shown in Figure 10.43 created in the Custom Designer illustrates a random allocation of the 21 runs of the I-Optimal set of experiments. Keep in mind that the allocation of runs is likely to change each time a table is created in the designer when runs are randomized.

Figure 10.43: Custom Design: Data Sheet

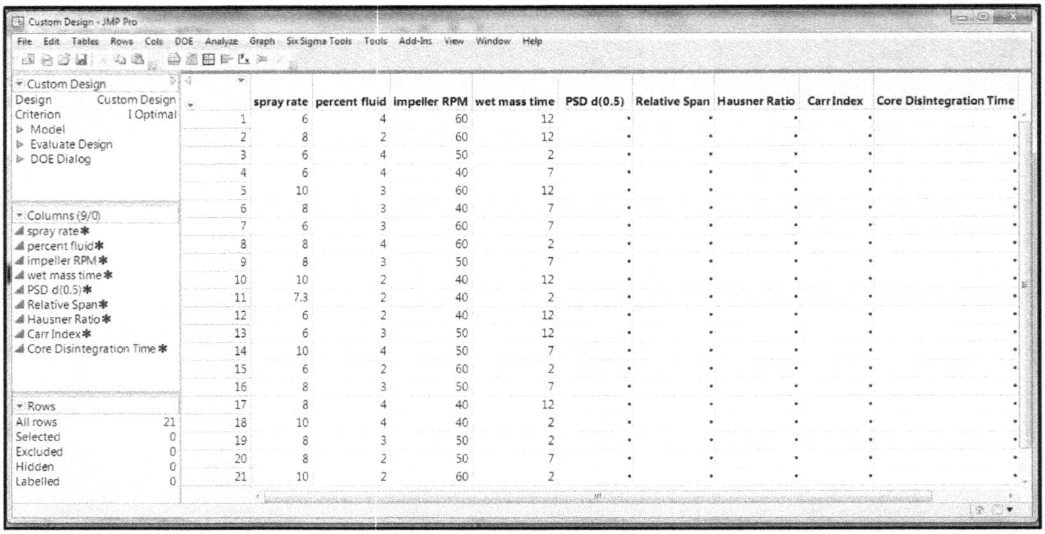

	spray rate	percent fluid	impeller RPM	wet mass time	PSD d(0.5)	Relative Span	Hausner Ratio	Carr Index	Core Disintegration Time
1	6	4	60	12	•	•	•	•	•
2	8	2	60	12	•	•	•	•	•
3	6	4	50	2	•	•	•	•	•
4	6	4	40	7	•	•	•	•	•
5	10	3	60	12	•	•	•	•	•
6	8	3	40	7	•	•	•	•	•
7	6	3	60	7	•	•	•	•	•
8	8	4	60	2	•	•	•	•	•
9	8	3	50	7	•	•	•	•	•
10	10	2	40	12	•	•	•	•	•
11	7.3	2	40	2	•	•	•	•	•
12	6	2	40	12	•	•	•	•	•
13	6	3	50	12	•	•	•	•	•
14	10	4	50	7	•	•	•	•	•
15	6	2	60	2	•	•	•	•	•
16	8	3	50	7	•	•	•	•	•
17	8	4	40	12	•	•	•	•	•
18	10	4	40	2	•	•	•	•	•
19	8	3	50	2	•	•	•	•	•
20	8	2	50	7	•	•	•	•	•
21	10	2	60	2	•	•	•	•	•

The inputs of the model were copied and changed to the nominal data type using the steps that produced figure 10.30 and the related scatterplot earlier in the chapter. A scatterplot matrix for the inputs for this example is shown in Figure 10.42. The I-optimal design utilized for the RSM saturates observations at both the outer corners and within the design space to allow for good quantifiable estimates of the effects of changes in the inputs. The more that the design space is represented by runs, the better the estimates will be that come from the model.

Figure 10.44: Scatterplot Matrix: View of Design Space

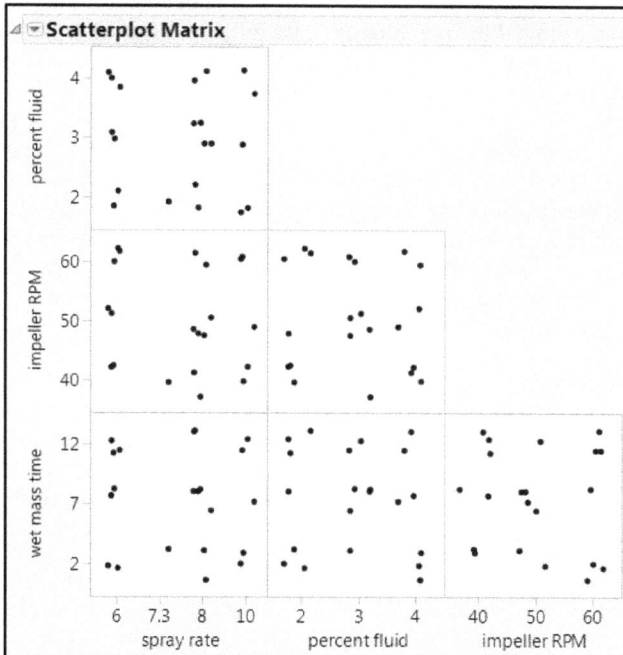

The project stakeholders are extremely pleased with the information presented regarding the model design. The tablets and plots provided from the JMP output enabled the team to understand the balance between available resources and the information desired from the analysis of the model. Everyone feels that the information that comes from the efforts is far better than utilizing OFAT designs to try to get acceptable results for each of the five outputs. The design is analyzed in a future chapter, that emphasizes the full value of using a structured experimental model to determine the optimal set of process controls to ensure that the goals for all outputs can be met simultaneously.

Practical Conclusions

Structured, multivariate experiments offer the most efficient method to answer questions about causational relationships between the inputs and outputs of a process. Planning starts by establishing the goal of the experiments. Screening for important inputs requires fewer resources than does establishing a response surface for accurately predict the amount of influence each input might have on outputs of a process. Teams might be reluctant to support the use of multivariate experimentation due to the amount of resources required. It is of the upmost importance to utilize the evaluation of designs to explain the quality of the information that will be gained from the number of runs that must be executed. JMP offers a wide variety of design evaluation tools to estimate statistical power, measure the amount of variance expected in predictions, and establish the robustness of the effects.

The DOE platform continues to expand as research continues and new models are added. The number of options available can be daunting for a novice working to find the right balance between the information provided and the resources required for a given model. Planning a set of experiments historically involved searching through DOE texts to find a classical model that seemed to fit the experimental scenario as closely as possible. A DOE practitioner needed to possess sufficient knowledge of statistics to guard against choosing an inadequate design. JMP has changed the planning process by incorporating the statistical knowledge within the DOE tools and supporting the user with easily accessible documentation integral to the software application. You can quickly create and evaluate multiple models by using the DOE options. The evaluation tools include both summary metrics and high-quality graphics to compliment the ability of non-statisticians to interpret the information. Stakeholders gain confidence in the methodology of choosing an appropriate model through the comparative design evaluation. This is due to quality of information that is likely gained from experimentation based on the resources that were planned for. The 13 runs required for the definitive screening design provide for robust experimentation to detect the important process inputs for the thermoforming process that ensures that surgical handle covers have adequate wall thickness. The 21 runs of the I-optimal response surface design set of experiments will work very well to define the design space of the wet granulation process.

You should exercise caution when designing and evaluating experimental designs. The evaluation activity is based on assumptions made about the degree of effects and amount of random variation expected in the model. Historical information can help teams to adjust the assumptions to be as accurate and precise as possible. Regardless of the effort expended to improve assumptions, the analysis of the actual results could erroneously indicate that an effect is significant. The default level of significance of 0.05 might need to be reduced if the process being studied is critical and the tolerance for error is low.

There is no substitute for careful planning of structured experiments. Teams must expend as much effort as possible to ensure that the variables studied are appropriate and all that other potential variables are either held as fixed controls or monitored and measured to later test for possible influence. Too much control can create a synthetic environment; too little control can result in a large amount of random variability. Experimental planning is somewhat of an art form because the set of experiments must be designed to represent the population of interest as closely as possible.

Exercises

E10.1—Use the results of the predictive model from exercise problem E8.1. Use treatment ranges that are slightly wider than the ranges determined from the sample data to create a set of structured, multivariate experiments. Management has allocated enough resources to be able to produce and test no more than 15 variants of the surgical tray sealing process. The team is most interested in a design that includes both the main effects and the interactions of the process. There is no expectation that terms more complex than interactions are present (curved terms).

1. Create your plan. You will likely want to have enough resources available to make a minimum of four confirmation runs, so plan wisely.
2. Create a report that includes a comparison of a good design with the best design, and be prepared to explain the balance between information and resource requirements.

E10.2—A set of structured experiments has been designed for the study of the tablet compression process involving four inputs. Leadership exerted a significant amount of pressure to keep the resource requirements at a minimal level, which means that a 6-run design has been created.

1. Open *Custom Process Design 4F 6R CP.jmp.*
2. Use the *Evaluate Design* script to go through the design evaluation.
3. Use the *DOE Dialog* script to open the design journal.
   - Click *Back* button at the bottom of the journal.
   - Create new designs with the design generator to add interactions to the model.
   - Use the minimum number of runs to create a new design with interactions.
4. Compare the initial 6-run design to the minimum-run design with interactions.
   - How did the new design affect expected statistical power?
   - How did the new design affect the prediction variance profile?
   - How did the new design affect correlation (use the color map to visualize)?
5. How would you present this information to leadership (you have a strong suspicion that interactions might exist)?

E10.3—You are working on an injection molding process used for critical components of a medical device that delivers a dose of an inhaled drug. There are a large number of process inputs including the following: back pressure, holding pressure, injection time, open mold time, shot size, clamping pressure, injection pressure, screw speed, and boost cut off. Several outputs will be measured from the parts, including part weight and key dimensions from physical features of the plastic part.

1. Select *DOE ▸ Custom Design* to create a design.
   - All factors are continuous (use default coded levels of -1, 1).
   - Include main effects and interactions.
   - Save the factors as a table with the design red triangle menu options.
   - Make a table with the minimum number of suggested runs.
   - Name the "Table 9F Custom Design."
2. Select *DOE ▸ Classical ▸ Full Factor* to create a design.
   - Load the factor table that you created in the custom design.
   - Make a table with the suggested number of runs.
3. Use *DOE ▸ Definitive Screening* to create a design.
   - Load the factor table created in the custom design.
   - Select the design option for adding blocks with additional runs to estimate quadratic effects.
   - Make a table with the suggested number of runs.
4. (JMP Pro Only) Select *DOE ▸ Design Diagnostics ▸ Compare Designs* to create output that compares the three design options.
5. How would you present the comparison, considering the balance between information provided and the resource requirements? Which option is the most feasible in the commercial operations environment of molding plastic parts?

Chapter 11: Analysis of Experimental Results

Overview

The designs of chapter 10 were formed through detailed collaboration and communication with project stakeholders and subject matter experts. The designs developed in JMP were carefully executed in compliance with an experimental protocol. The protocol was clearly documented, approved by leadership, and supported by resources provided by the project team and the operations group conducting the granulations. The randomization plans were followed in the exact order noted in the data collection plans. The output values entered in the data sheet were confirmed to be precise and accurate. This chapter explains how to use JMP to efficiently analyze the experimental models to provide evidence in support of the goals for each project. You will see the comparisons between the information provided by a simple screening design and a high-resolution response surface model with multiple outputs.

The Problems: A Thermoforming Process and a Granulation Process, Each in Need of Improvement

Michelyne's team is charged with quickly identifying the process inputs that influence the thinning of the material thickness of surgical handle covers so that customer complaints can be corrected as soon as possible. The project stakeholders prefer speed and decisive action over the obtainment of detailed knowledge of the thermoforming process. The definitive screening design (DSD) used for the experiments met the goal of the stakeholders and even provides a glimpse into the amount that various factors affect the handle thickness.

The granulation process being studied by Emily's team is more mature than the thermoforming process being studied by Michelyne's team. The project stakeholders have run enough prior experiments to identify the process inputs that have a significant effect on outputs. The problem involves multiple outputs that must be met robustly through control of four process inputs. The goal of the team is to quantify the effects of changing inputs so that the optimum settings can be identified. This information is crucial for the manufacturing team to have the best chance of robustly obtaining good product. The response surface model designed for the experiments provides a high resolution of information that will exceed the expectations of the team.

Execution of Experimental Designs

The team charged with addressing the incidence of cracked and loose surgical light handle covers created an experimental plan to screen for significant thermoforming process inputs in chapter 10. Michelyne found that the definitive screening design, exclusive to JMP, provides an extremely efficient set of experiments to screen to a limited number of process inputs. The main advantage of the DSD is a model that mitigates the typical correlation that is found between the individual inputs and the interactions. The small number of runs was very appealing to the project stakeholders because results can be obtained very quickly and with controlled expense. Additional advantages of the DSD is the detection of interactions and squared (non-linear) terms that might leverage the output.

The design of experiments (DOE) explained in chapter 10 is only a small part of the effort required to execute a set of multivariate structured experiments. The statistical analysis of an experimental model assumes that most of the variation in the output is due to changes in the inputs studied. All other potential sources of variation to the output must be discussed and appropriately controlled. The inclusion of controls to the process is easier said than done in most cases. The design team must include subject matter experts from the operations area to discuss the needed process controls before the experiments are executed. It is extremely important for the team to go into such meetings with healthy pessimism to ensure that all sources of potential variability are identified. These sources can include, but are not limited to, variations in the physical environment (temperature, humidity, air circulation), changes in people (operators, technicians, leadership), changes in the machines (warm up, continued wear and tear, automatic controls adjustments), and changes in materials (amount of incoming material, changes in physical characteristics, changes in lots).

Each potential influence should include a control as well as a team member responsible for ensuring that the control is in place as experimental runs are executed. There are some sources of variability that the team will decide to not control in order to ensure that the set of experimental runs adequately represent the population of process results being modeled. For instance, the operational facility might experience fluctuations in ambient temperature that range between 68 degrees F to 82 degrees F during typical processing. If the experimental environment were to be controlled to a rigid control of 70 degree F +/- 1 degree, the resulting model might not contain the amount of random error expected in ambient conditions and could provide misleading results. Operations would need to invest a significant expense to mimic a 70-degree F +/- 1-degree environment, which is highly unlikely.

There is a very high likelihood that the execution of a set of experimental runs will not go perfectly. An experimental model can extract the highest-value information when levels chosen for the design are pushed out as far as possible to the extreme low and high ends of the operational range. With wide process levels, a risk exists for the inability of the process to create a sufficient output with the most extreme combinations of input levels. This can be especially problematic if the process is unable to create an output several runs into the design plan. It is highly recommended that the design team discuss the plan with subject matter experts (SMEs) in order to identify any runs that are at high risk for an inability to create an output. For instance, a surgical handle might not adequately form with the shallowest plug depth (0.82), shortest vacuum time (4), and coolest tool temperature (180). The DSD plan from chapter 10 does not included a run with the noted extremes. However, one of the runs is very close. It is good practice to make the highest-risk run the first run executed in the plan. This violates the concept of a completely randomized design slightly. If the process is unable to produce a viable output, the design team can quickly adjust the levels in the design and add a run to replace the failed first run. Front-loading the highest-risk run can help maintain the intended number of runs. The movement of one run does not interfere with the assumption of randomness of runs enough to be practically relevant in most cases. If a failed processing attempt occurs late in the design plan, the team will need to exclude the run or scrap the effort and develop a new plan. Either option is very disruptive and costly, not to mention the loss of credibility that can occur with stakeholders of the project.

Analysis of a Screening Design

Michelyne assigned a member of the team to upload the outputs collected from the set of experiments to the design plan that they created in chapter 10. There are great advantages to using the original design plan created in JMP; scripts are created to make the job of analysis much easier for the analyst. Open *Thermoforming DSD Results.jmp* to view the data table shown in Figure 11.1.

Figure 11.1: Thermoforming Experiments Results

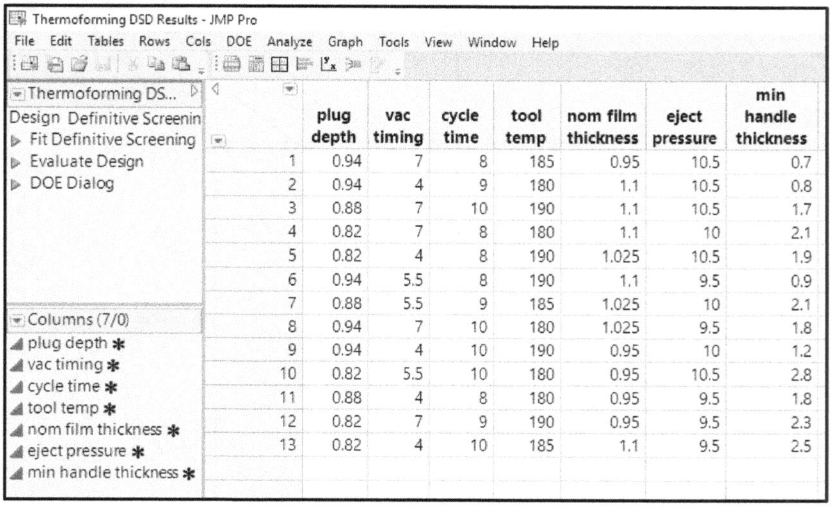

Thermoforming DSD Results - JMP Pro

File Edit Tables Rows Cols DOE Analyze Graph Tools View Window Help

Thermoforming DS...
Design Definitive Screenin
▷ Fit Definitive Screening
▷ Evaluate Design
▷ DOE Dialog

	plug depth	vac timing	cycle time	tool temp	nom film thickness	eject pressure	min handle thickness
1	0.94	7	8	185	0.95	10.5	0.7
2	0.94	4	9	180	1.1	10.5	0.8
3	0.88	7	10	190	1.1	10.5	1.7
4	0.82	7	8	180	1.1	10	2.1
5	0.82	4	8	190	1.025	10.5	1.9
6	0.94	5.5	8	190	1.1	9.5	0.9
7	0.88	5.5	9	185	1.025	10	2.1
8	0.94	7	10	180	1.025	9.5	1.8
9	0.94	4	10	190	0.95	10	1.2
10	0.82	5.5	10	180	0.95	10.5	2.8
11	0.88	4	8	180	0.95	9.5	1.8
12	0.82	7	9	190	0.95	9.5	2.3
13	0.82	4	10	185	1.1	9.5	2.5

Columns (7/0)
⬛ plug depth ✱
⬛ vac timing ✱
⬛ cycle time ✱
⬛ tool temp ✱
⬛ nom film thickness ✱
⬛ eject pressure ✱
⬛ min handle thickness ✱

The fit definitive screening script in Figure 11.1 will be used to execute the analysis. However, it is good practice to visualize the data at a high level before completing a detailed model analysis. Select *Analyze ▶ Distributions* and move the variables to the *Y, Columns* box with the response on top and the six inputs underneath, as shown in Figure 11.2. Click *OK* to get the output.

Figure 11.2: Distribution Setup Window

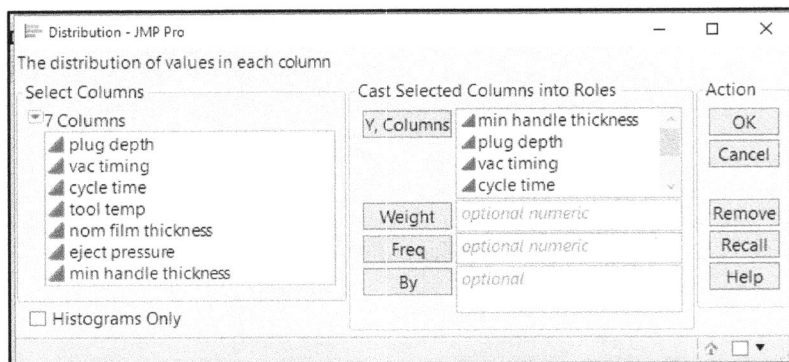

The output for the seven variables takes up a large amount of space and includes information that is not important. Press the Ctrl key and click the gray arrow next to the *Quantiles* header underneath *min handle thickness* to hide the quantiles summary for each variable. Press the Ctrl key and click the gray arrow next to the *Summary Statistics* header underneath *min handle thickness* to hide the summary statistics for each variable. Click on the gray arrow next to *Summary Statistics* underneath *min handle thickness* to unhide *Summary Statistics* for only the output. The summary statistics of the inputs are irrelevant since fixed levels were chosen when the model was designed. Lastly, use the red triangle menu next to *Distributions* to select *Arrange in Rows*, and enter 4 to display the output in two rows, as shown in Figure 11.3.

Figure 11.3: Distribution Output

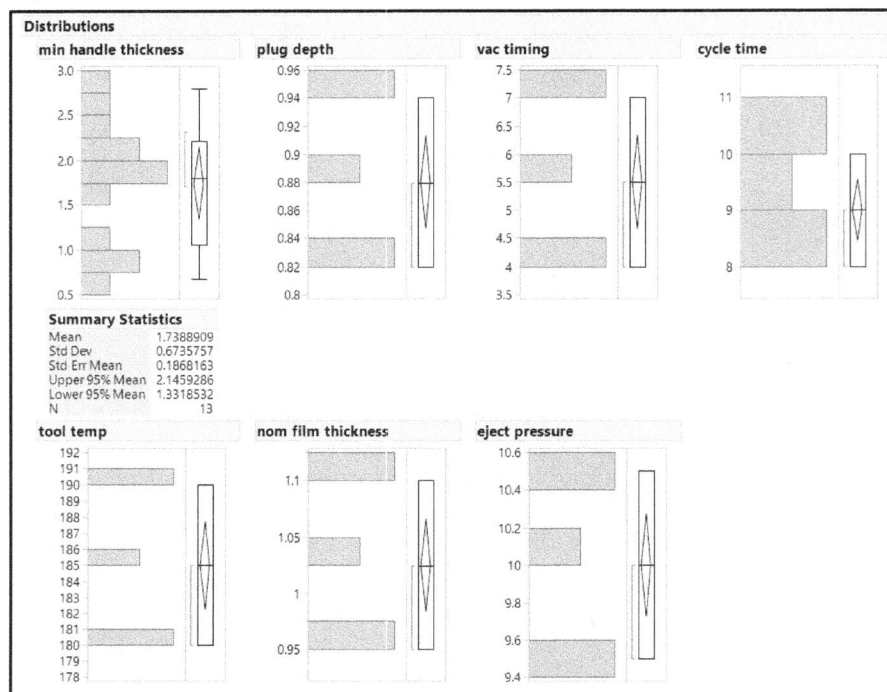

The average minimum handle thickness for the set of experiments is 1.74 mm and the standard deviation is 0.67 mm. The reasonable expectation for average minimal thickness for a population of handles made with the ranges of input changes is noted by the 95% confidence interval of 1.33 mm to 2.15 mm. The three levels explored for each process input are obvious since each distribution plot includes three bars. The dynamic functionality of distributions is used to determine whether any relationships are obvious. Notice that the *min handle thickness* plot seems to illustrate two groups of results. Press the Shift key and click on the lower group bars of *min handle thickness* to get the output shown in Figure 11.4.

Figure 11.4: Distribution Output with Dynamic Selection

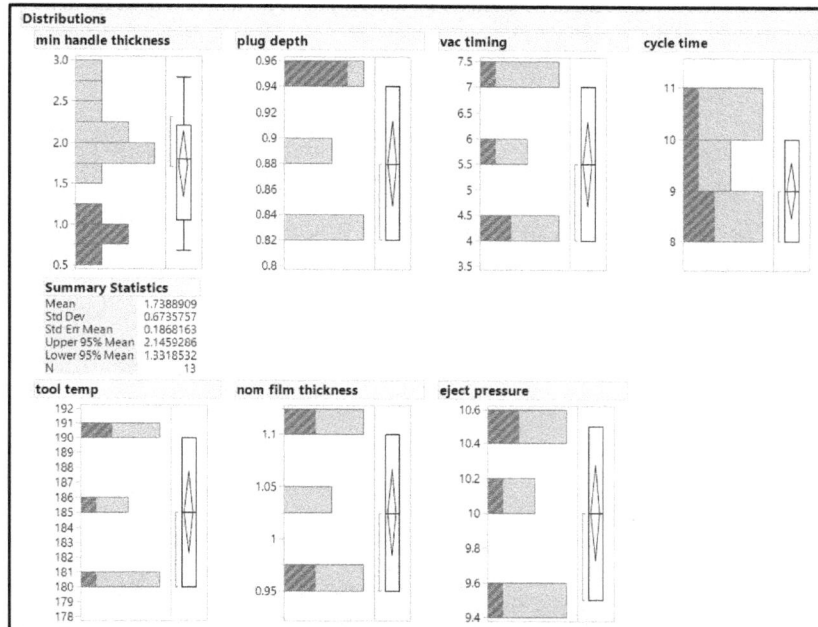

The lower group of *min handle thickness* results seem to relate to the high level of plug depth since the majority of the high bar is highlighted. Low cycle time has a larger portion of the bar highlighted, but is not as convincing a pattern as plug depth. The four other variables do not illustrate as much of a pattern. Try other dynamic selections to look for potential patterns in the model data. Patterns in distribution plots offer a high-level view of possible relationships. However, the analysis of the model includes detail needed to quantify the level of significance for process inputs that influence *min handle thickness*. Go back to the *Thermoform DSD Results.jmp* table shown in Figure 11.1, and click the green arrow beside the *Fit Definitive Screening* script to get the output shown in Figure 11.5.

Figure 11.6: Thermoforming DSD Initial Analyses

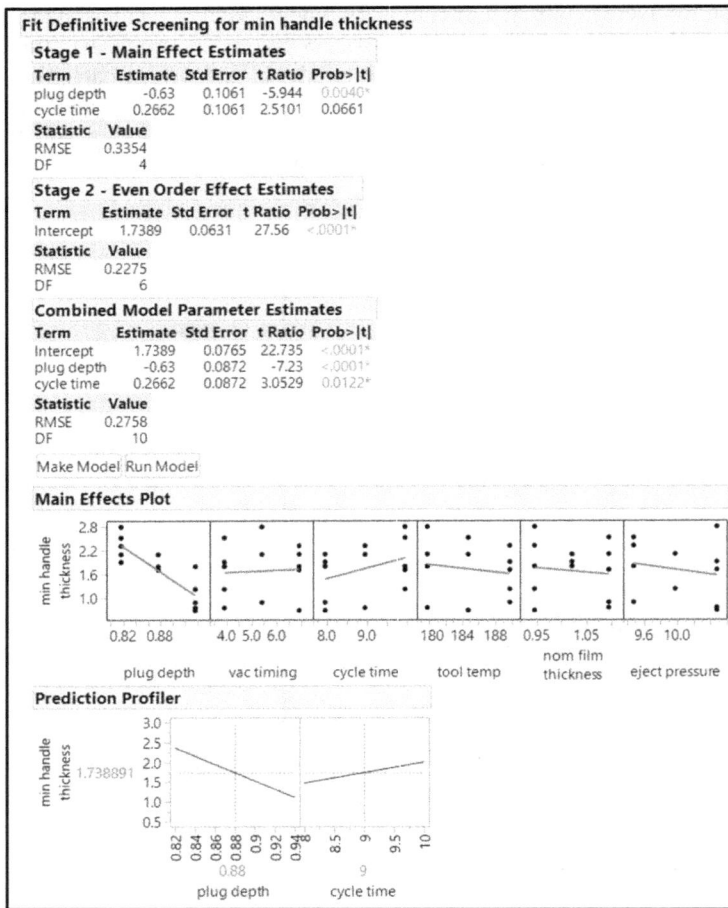

Fit Definitive Screening for min handle thickness

Stage 1 - Main Effect Estimates

Term	Estimate	Std Error	t Ratio	Prob>|t|
plug depth	-0.63	0.1061	-5.944	0.0040*
cycle time	0.2662	0.1061	2.5101	0.0661

Statistic	Value
RMSE	0.3354
DF	4

Stage 2 - Even Order Effect Estimates

Term	Estimate	Std Error	t Ratio	Prob>|t|
Intercept	1.7389	0.0631	27.56	<.0001*

Statistic	Value
RMSE	0.2275
DF	6

Combined Model Parameter Estimates

Term	Estimate	Std Error	t Ratio	Prob>|t|
Intercept	1.7389	0.0765	22.735	<.0001*
plug depth	-0.63	0.0872	-7.23	<.0001*
cycle time	0.2662	0.0872	3.0529	0.0122*

Statistic	Value
RMSE	0.2758
DF	10

Make Model Run Model

Main Effects Plot

Prediction Profiler

The analysis relies on a unique model selection algorithm, which detects for significant inputs in two stages (1-main effects, 2-interactions). The *Stage 1 – Main Effects Estimates* in the *Fit Definitive Screening for min handle thickness* window provides information about the inputs that have significant evidence of influence. The root mean square error of the model is 0.3334, which is relatively high compared to the overall range of thickness results of 0.7 mm to 2.8 mm. The 4 degrees of freedom relates to the number of inputs that were not selected as a main effect for a simple DSD without replicate center points or a blocking term.

The plug depth is highly significant (p=.004) in reducing the minimum thickness output. The estimate of the effect (-0.63) means that a 0.63 mm average reduction in thickness for every percent increase in plug depth can be expected. This estimate of change in minimum thickness due to plug depth is only 4/10ths of a percent likely to have come from random variation. The cycle time has a minimally significant evidence of influence on minimum thickness (p=0.066). This means that the estimate of a 0.27 mm average increase in thickness for every second of increase in cycle time identified by the model is 6.6% likely to have come from random variation. The DSD algorithm utilizes a p-value threshold for selecting main effects that is based on the degrees of freedom in the error term of the model. This is why it is possible for selected main effects to have a p-value higher than the 0.05 level of significance. (You can find additional technical details about selecting main effects for a DSD in the Design of Experiments Guide, available in the Help menu.). None of the remaining four inputs seem to have significant influence on minimum thickness.

The second stage is used to detect for significant interactions. The minimum resource DSD with no replicate runs, no blocks, and no extra center runs does not offer enough runs to gain the highest level of robustness in sensitivity to higher order factors. The results indicate a significant intercept, which means only that the response value of linear model is not zero when the explanatory value is at zero. There is no useful practical interpretation of the intercept value. The lack of any other terms in the Stage 2 analysis indicates that the influence on minimum thickness is limited to main effects.

Combined model parameter estimates are derived from the typical least squares analysis of the model. With the 10 degrees of freedom used for model selection, the significance of the plug depth and cycle time is stronger than the results of the specialized DSD main effects modeling. The slope of the blue model lines in the main effects plots of plug depth and cycle time indicate evidence of significant influence on *min handle thickness* because they are the two with the steepest slopes. In addition, the small distance between the observation points and the plug depth effect line also emphasize the strength of the significant relationship. The observations located about the cycle time effect line illustrate a similar pattern of a strong relationship. However, the greater distances between the points and the line illustrates the marginal significance of the effect.

The Prediction Profiler provides for a dynamic view of the effects of plug depth and cycle time on minimum thickness. Click on the vertical red segmented line of *plug depth*, and drag horizontally along the range of values to estimate the minimum thickness. Repeat the dynamic estimation of *min handle thickness* with *cycle time*. To estimate the minimum handle thickness with an exact numeric value, click on the red input value for *plug depth* and enter 0.9. Repeat with other values for plug depth and cycle time to see how the estimates change. It is clear that minor changes in plug depth have a larger effect on minimum thickness than changes in cycle time. Basically, plug depths that are 0.9 or greater estimate a minimum thickness that is less than the 1.5 mm low specification. In addition, excessive plug depth (>=90%) with longer cycle times exacerbate the risk of producing handles that have minimum wall thickness that is below the minimum specification.

Detailed Analysis of the DSD Model

The next step is to run the model to get additional details of the analysis. Click *Run Model* in the *Fit Definitive Screening for min handle thickness* window to get the model output shown in Figure 11.7.

Figure 11.7: Thermoforming Reduced Model

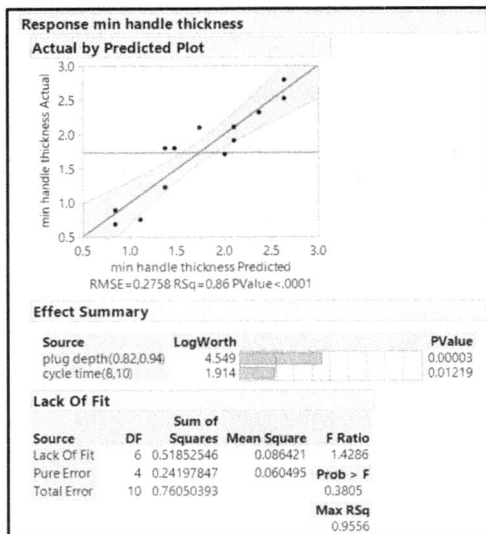

The Actual by Predicted plot enables you to see the relationship between actual and predicted values, indicated with a solid red line. The 95% confidence interval for the mean effect is illustrated with a shaded red area about the model line. The pattern of the predicted results by actual results of the individual runs is shown by the black circular markers. The relationship seems reasonable due to the evidence that the linear model is highly significant (p<0.0001). The model explains 86% of the changes in minimum thickness (r-square = 0.86), and has a relatively small amount of random error (RMSE=0.2758). The effect summary lists plug depth and cycle time as significant effects, (p=0.00003) and (p=0.01219) respectively.

The lack of fit tests are used to detect whether there are observations that have a poor fit to the model, even if the overall trend is of high significance. Notice that there are a limited number of observations that are outside of the shaded confidence interval region. If the observations were within the trend of the interval, the maximum r-square fit of 0.9556 would be achieved. The lack of evidence that poor fitting observations are within the experimental results is noted by the p-value of 0.3805. In general, the model for the two significant thermoforming process variables is very robust.

The use of residual analysis is shown in Figure 11.8.

Figure 11.8: Thermoforming Reduced Model Residual Diagnostics

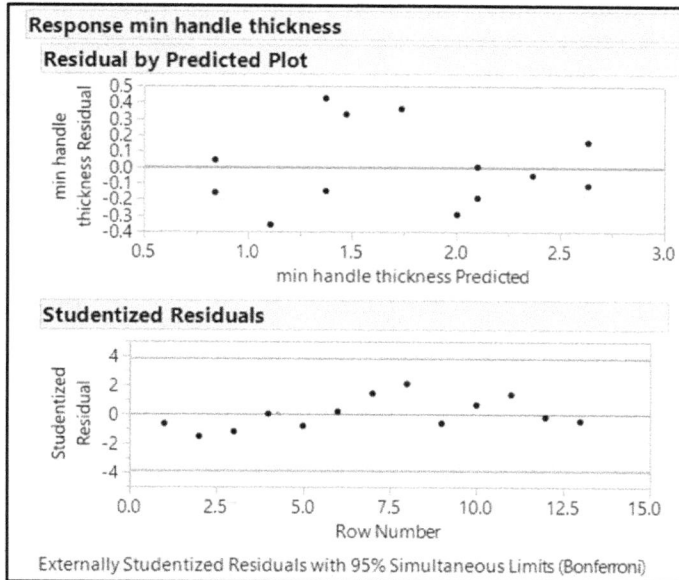

Residual analysis adds detail to the diagnostics of the robustness of the model. The random pattern of the residuals illustrated in the Residual by Predicted plot is highly desirable. If a cone pattern or other non-random pattern were evident in the plot, the conclusions of the model might be suspect due to the potential for error. The Studentized Residuals plot provides a reference for the user vis the red decision lines at the studentized residual values of +/- 4. None of the observations are beyond the decision lines. Therefore, the residual analysis provides further evidence of a robust model.

The output provided by JMP has justified that the model formed through the two significant process variables of plug depth and cycle time is robust. The goal of the set of experiments has been met. The analysis was used to narrow the improvement efforts in controlling the minimum handle thickness by 67% (two of the six inputs can be manipulated to better control the output). The set of structured, multivariate experiments enabled the team to achieve the goals quickly and efficiently, but even more useful information is available. Effects and predictions can be made from the model to provide additional direction for the team. Michelyne needs to use the information with caution because there is typically an inadequate number of runs included in a screening design to precisely quantify the effects. The parameter estimates are shown in Figure 11.9.

Figure 11.9: Thermoforming Reduced Model Analysis Interpretation

Response min handle thickness

Parameter Estimates

Term	Estimate	Std Error	t Ratio	Prob>|t|
Intercept	1.7388909	0.076485	22.73	<.0001*
plug depth(0.82,0.94)	-0.630489	0.087207	-7.23	<.0001*
cycle time(8,10)	0.2662302	0.087207	3.05	0.0122*

Effect Tests

Source	Nparm	DF	Sum of Squares	F Ratio	Prob > F
plug depth(0.82,0.94)	1	1	3.9751607	52.2701	<.0001*
cycle time(8,10)	1	1	0.7087854	9.3199	0.0122*

The parameter estimates indicate that an approximate 0.63 mm decrease in minimum thickness is expected as plug depth is increased one unit of standardized increase (standardized to the design space). This unit of increase is either between the low limit (0.82) and center point (0.88), or between the center point and the high limit (0.94). A one-unit of standardized increase in the design space of cycle time results in an approximate 0.27 mm increase in minimum thickness. The Prediction Profiler in Figure 11.10 is a dynamic illustration of how changes in the two process inputs of the model relate to changes in *min handle thickness*.

Figure 11.10: Thermoforming Reduced Model Prediction Profiler

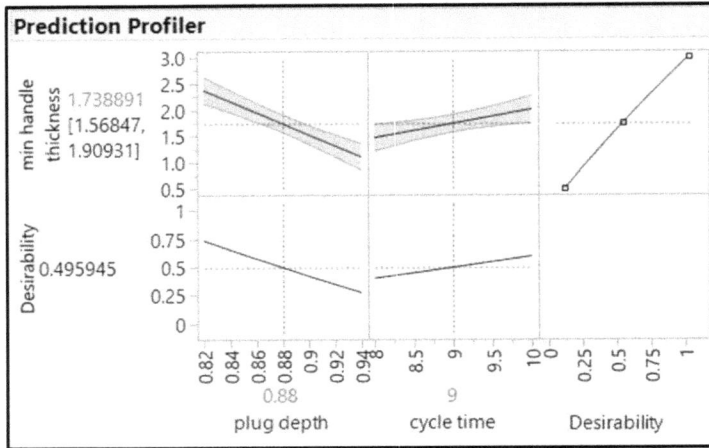

The Prediction Profiler provides a dynamic plot in which the analyst can try different values of plug depth and cycle time by sliding the vertical segmented red slider lines. You should consider general relationships because prediction accuracy might be lacking. The practical interpretation of the limits on prediction accuracy is evident in the 95% confidence limits of the minimum thickness prediction at the center point of the design space, as shown in the profiler. The average minimum thickness predicted is shown in red as 1.74 mm. However, the value could be as low as 1.57 mm or as high as 1.91 mm, as indicated by the black values for the confidence interval. Michelyne can guard against inaccurate conclusions made by project stakeholders by sticking with the interval estimates in all communication. The report on the analysis might include the statement "For the process set at a plug depth of 88% and a cycle time of 9 seconds, we can expect that the handles produced will have an average minimum thickness of between 1.57 and 1.91 mm". If the stakeholders want increased precision in the estimates, they will need to support the project with more resources for further study.

Use of the Fit Model Analysis Menu Option

The previous example went through the model analysis that is run from scripts that are automatically created when you make a table for a DSD model. There might be times when the data from the experiments was not collected via the JMP table, or the person doing the analysis has a JMP license that is a version prior to 13.0. In such cases, you can complete the analysis of experimental data from a DSD by using the Fit Model platform. Complete the following steps to set up the model analysis:

1. Open *Thermoform DSD Results.jmp*, and select *Analyze ▶ Fit Model*.
2. In the *Model Specification* window, make sure that *min handle thickness* is selected as *Y, Response*.
3. Select all six process inputs in the *Columns* box, click *Macros*, and then select *Response Surface*, from the options shown in Figure 11.11.

Figure 11.11: Thermoforming Full Model Creation

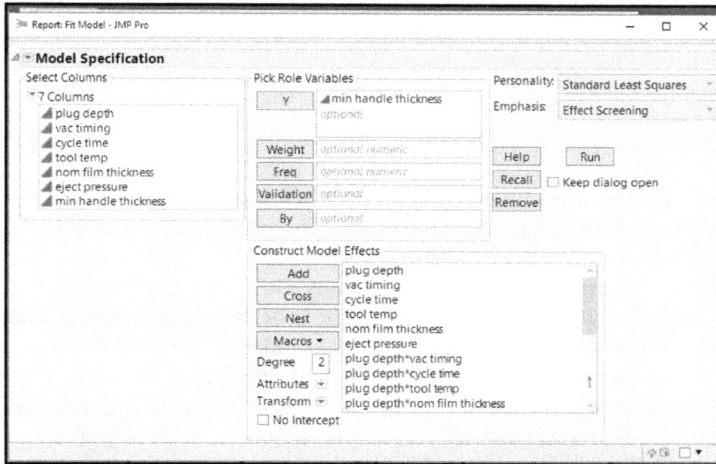

4. Make sure that the *Degree* box includes the default value 2 so that only the two-way interactions and squared terms are included in the model.
5. Click *Run* to get the model output.

Singularity

The thirteen runs of the model are not enough to be able to detect all six individual inputs, all interactions, and all squared terms. When there are more terms than runs in a model, the output includes singularity details. The singularity details in Figure 11.12 indicate that the response surface model does not have enough degrees of freedom for estimating the effects of the 21 factors.

Figure 11.12: Thermoforming Full Model Singularity

Singularity Details

plug depth*vac timing = - plug depth*cycle time - plug depth*tool temp - plug depth*nom film thickness - plug depth*eject pressure = - 1.16667*plug depth*cycle time - 0.5*plug depth*tool temp - 0.66667*plug depth*nom film thickness + 0.83333*vac timing*cycle time - 0.83333*vac timing*tool temp = 0.5*plug depth*cycle time + 2*plug depth*tool temp - 1.5*plug depth*nom film thickness + 2.5*vac timing*cycle time - 2.5*vac timing*nom film thickness = - 2*plug depth*cycle time + 0.33333*plug depth*tool temp - 2.33333*plug depth*nom film thickness + 1.66667*vac timing*cycle time - 1.66667*vac timing*eject pressure = - 0.57143*plug depth*cycle time - 0.14286*plug depth*tool temp - 1.14286*plug depth*nom film thickness + 0.71429*vac timing*cycle time - 0.71429*cycle time*tool temp = - 2*plug depth*cycle time - 1.33333*plug depth*tool temp - 2.33333*plug depth*nom film thickness + 1.66667*vac timing*cycle time - 1.66667*cycle time*nom film thickness = - 0.75*plug depth*cycle time + 0.75*plug depth*tool temp - 0.25*plug depth*nom film thickness + 1.25*vac timing*cycle time - 1.25*cycle time*eject pressure = 3*plug depth*cycle time - 3*plug depth*tool temp + plug depth*nom film thickness - 5*vac timing*cycle time + 5*tool temp*nom film thickness = - 0.33333*plug depth*cycle time - 0.5*plug depth*tool temp - 0.66667*plug depth*nom film thickness + 0.83333*vac timing*cycle time - 0.83333*nom film thickness*eject pressure
plug depth*nom film thickness = vac timing*cycle time - tool temp*eject pressure

The model must be updated to reduce terms down to the number of factors that can be estimated with the degrees of freedom available from 13 runs. The Effect Summary information in Figure 11.13 lists the factors with singularity without the LogWorth or PValue information. You could go back to the *Model Specification* window for a trial and error set of model selections. However, JMP provides an easy tool to reduce the model within the analysis output. Work from the bottom of the list upwards by selecting *plug depth*vac timing* and clicking *Remove* to eliminate it from the model. The model is reduced by the selected factor, and the analysis is redone automatically. Select multiple factors and click *Remove* until the remaining factors match the six that are included in Figure 11.14. (A more efficient method for reducing multiple factors from a model uses stepwise regression. Details about this method are more complex than the manual method. The topic is covered in chapter 13.)

Figure 11.13: Thermoforming Full Model Effect Summary

⊿ ▼ **Response min handle thickness**
⊿ **Actual by Predicted Plot**
⊿ **Effect Summary**

Source	LogWorth		PValue
plug depth(0.82,0.94)	0.926		0.11855
cycle time(8,10)	0.573		0.26717
eject pressure(9.5,10.5)	0.366		0.43084
tool temp(180,190)	0.303		0.49736
nom film thickness(0.95,1.1)	0.217		0.60699
vac timing(4,7)	0.103		0.78869
nom film thickness*eject pressure	.		.
tool temp*eject pressure			
tool temp*nom film thickness			
cycle time*eject pressure			
cycle time*nom film thickness			
cycle time*tool temp			
vac timing*eject pressure			
vac timing*nom film thickness			
vac timing*tool temp			
vac timing*cycle time			
plug depth*eject pressure			
plug depth*nom film thickness			
plug depth*tool temp			
plug depth*cycle time			
plug depth*vac timing			

Remove Add Edit ☐ FDR

Figure 11.14: Thermoforming Partially Reduced Model

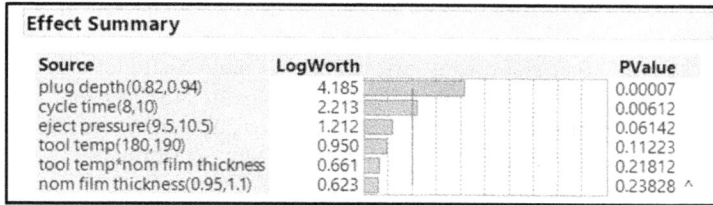

Effect Summary

Source	LogWorth		PValue
plug depth(0.82,0.94)	4.185		0.00007
cycle time(8,10)	2.213		0.00612
eject pressure(9.5,10.5)	1.212		0.06142
tool temp(180,190)	0.950		0.11223
tool temp*nom film thickness	0.661		0.21812
nom film thickness(0.95,1.1)	0.623		0.23828 ^

The Effect Summary includes a threshold value for detecting important factors, which is shown as a vertical blue line on the LogWorth horizontal bar chart. The effect summary horizontal bar chart illustrates that the plug depth and cycle time are the only factors that have enough evidence of influence to be considered significant to changes in minimum thickness. The model with the six factors also indicates that interactions are not likely to be of significant influence; the *tool temp*nominal film thickness* interaction has a LogWorth value that is very small (0.661) with an insignificant PValue of 0.218.

Analysis of a Partially Reduced Model

The model summary shown in Figure 11.15 includes additional factors as compared with the Fit Definitive Screening analysis.

Figure 11.15: Thermoforming Partially Reduced Model Fit

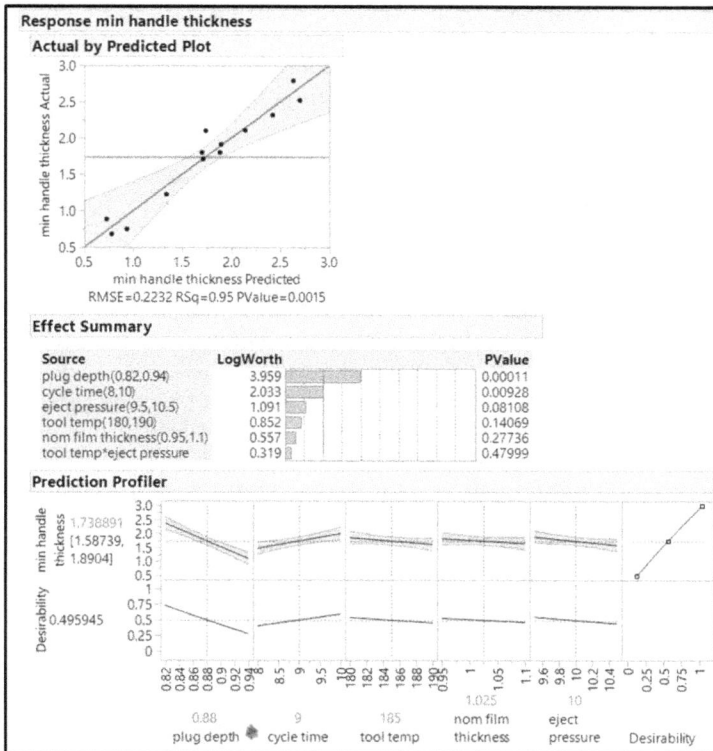

Response min handle thickness

Actual by Predicted Plot

RMSE=0.2232 RSq=0.95 PValue=0.0015

Effect Summary

Source	LogWorth		PValue
plug depth(0.82,0.94)	3.959		0.00011
cycle time(8,10)	2.033		0.00928
eject pressure(9.5,10.5)	1.091		0.08108
tool temp(180,190)	0.852		0.14069
nom film thickness(0.95,1.1)	0.557		0.27736
tool temp*eject pressure	0.319		0.47999

Prediction Profiler

Notice that the model fit of Rsquare =0.95 is better than the fit of the 2-factor DSD reduced model fit of Rsquare = 0.86. In addition, the amount of random error in the model decreased from an RMSE of 0.2758 to 0.2232 by including four additional factors. The basic conclusion has not changed; the plug depth is the most significant factor and cycle time the only other significant factor. The four additional factors explain more of the random error, but the signals are too weak to be significant. The analysis provides some hints that further study could offer additional value for the team to optimize the process. It is also possible that the model with additional factors is overfit, so adding runs might not provide useable information.

The team must keep in mind that the DSD provides the best results in defining which of the six process inputs has significant influence on *min handle thickness*. Even though the analysis tools allow for some interpretation of how much influence is exerted, the model lacks enough runs to provide robust estimates. Runs can be augmented easily for more detailed focus on plug depth and cycle time so that the amount of influence can be robustly quantified.

The results of the model analysis are enough for leadership to decide to take quick action. The process controls are changed so that plug depth does not exceed 90% and the maximum cycle time does not exceed 9.5 seconds. The action will immediately reduce the risk of producing thin handles, but the team will not be able to robustly estimate the effectiveness of the actions prospectively. Enhanced process monitoring must be initiated for a period that is sufficient to represent the population of all commercial production. The minimum thickness data collected from in-process checks is to be assessed for capability (chapter 3) and tested for significance to the distribution of data from parts produced prior to the improvement (chapter 4) to ensure that the changes have been effective.

Management has the option for further study of the process at any time to obtain additional improvement by augmenting the DSD model. Augmentation of an existing model requires that the process environment is equivalent to that which was in place for the initial experiments. Adding runs to the model can be a cost-effective way to improve the predictive nature of the model. The risk involved for augmentation is that more resources are added with no additional information extracted from the model.

Analysis of a Response Surface Model with Multiple Outputs

The stakeholders in the next set of experiments have a different set of goals. The high-shear granulation process, which is the subject of the study, has four process inputs that were found to exert significant influence on outputs of the process. Erica, the team leader, faces the challenge that the process must perform well for multiple outputs. JMP is an invaluable tool for the exploration of a process involving multiple inputs as well as multiple outputs. More resources are required for such experiments since the analysis will be used to quantify the levels of inputs that will be included in the manufacturing order protocols to get optimal results. Erica gained the support of leadership to include an adequate number of runs and robustly quantify the effects of the inputs. The response surface design from chapter 10 is complete and the data is ready for analysis. Open *Granulation Process Experiment Results.jmp*, shown in Figure 11.16.

Figure 11.16: Granulation Process RSM Data

	spray rate	percent fluid	impeller RPM	wet mass time	PSD d(0.5)	relative span	hausner ratio	carr index	core disintegration time
1	8	0.3	50	7	125.3	1.78	0.20	16.7	45.6
2	6	0.2	40	2	127.1	1.03	2.30	24.3	38.2
3	8	0.3	40	7	137.5	0.20	3.54	17.3	40.9
4	8	0.2	60	2	101.4	1.33	1.92	23.8	55.2
5	8	0.3	50	7	125.4	1.61	1.92	11.4	48.7
6	6	0.4	60	2	105.4	0.15	1.20	39.0	47.0
7	8	0.3	50	2	116.7	0.65	0.25	28.0	45.2
8	8	0.4	40	2	127.6	0.15	0.58	26.7	36.7
9	8	0.4	50	7	122.3	3.33	2.94	23.9	49.6
10	6	0.2	60	12	101.1	1.64	2.09	7.1	50.0
11	10	0.4	60	2	111.6	1.78	3.17	36.9	65.5
12	6	0.3	40	12	101.2	2.75	2.67	45.5	52.8
13	6	0.4	50	12	103.7	2.01	3.38	24.4	55.2
14	8	0.4	60	12	121.9	1.56	1.49	38.1	48.3
15	10	0.4	40	12	140.3	2.60	1.45	30.2	46.0
16	6	0.2	50	7	115.6	1.30	0.75	14.5	46.9
17	10	0.2	60	8.5	127.7	3.18	0.12	0.6	50.6
18	10	0.3	50	12	137.0	2.95	0.02	24.6	45.7
19	10	0.2	40	2	131.0	0.43	0.18	11.5	44.6
20	8	0.2	40	12	120.0	0.11	1.86	26.4	46.9
21	6	0.3	60	7	111.7	0.57	0.15	30.6	53.1

Prior to initiating the model analysis, a high-level look at the data using the distributions platform is appropriate. Select *Analyze* ▶ *Distributions* and move the variables to the *Y,Columns* box with the five responses on top and the four inputs underneath, as shown in Figure 11.17. Click *OK* to get the output.

Figure 11.17: Distribution Platform Window

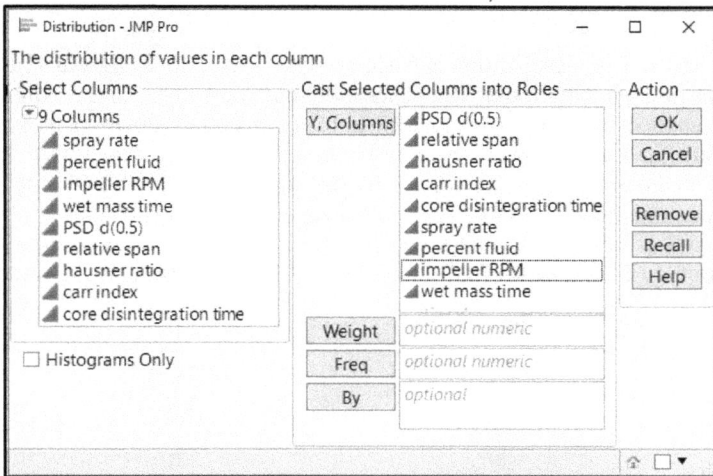

The output needs to be formatted for the best visualization of the experimental data. Press the Ctrl key and click on the gray arrow next to the *Quantiles* header underneath *PSD d(0.5)* to hide the quantiles summary for each variable. Click on the gray arrow next to each of the four inputs to hide the summary statistics. If capability studies are automatically added due to the response limits, use the red triangle menu of each to deselect the *Capability Analysis* option and remove the output. Lastly, use the red triangle menu next to *Distributions* to select *Arrange in Rows*. Enter 5 to arrange the output into two rows, as shown in Figure 11.18.

Figure 11.18: Distributions Output

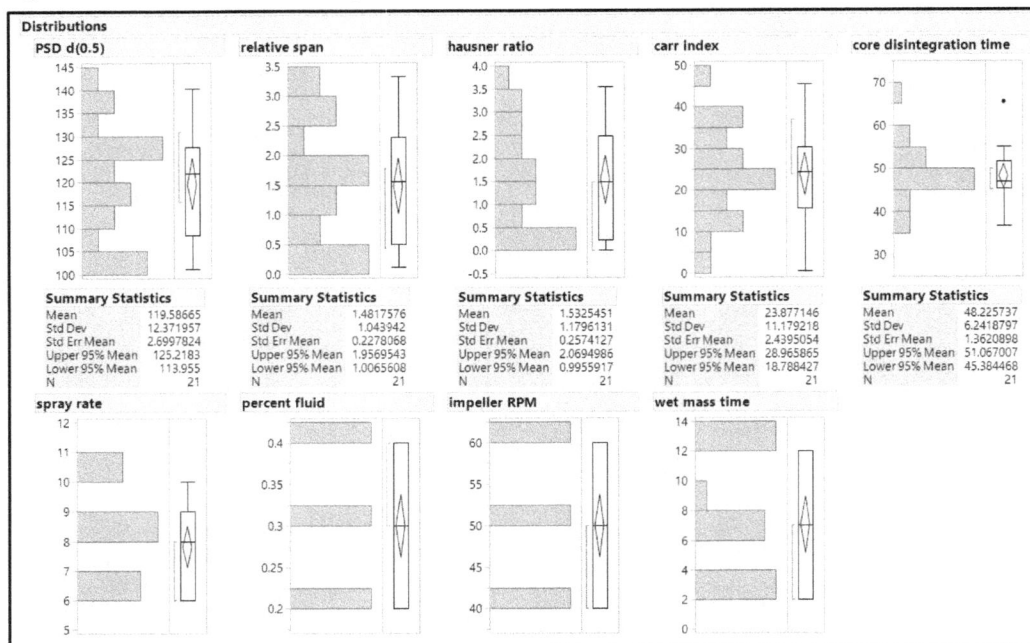

Use the dynamic features of the Distributions platform to look for high-level patterns of potential relationships among the inputs and outputs. The visualization is basic because interactions and squared terms cannot be viewed. Note any non-random pattern that you observe so that you can explore it with the detailed model analysis. Multivariate analysis can be used as an efficient way to look for relationships among several variables and is explained in a later chapter. With *Granulation Process Experiment Results.jmp* open and in view, click on the green arrow beside the *Model* script to run it. The *Model Specification* window appears, as shown in Figure 11.19. The column properties were previously set in the data table to identify the five outputs and to include response limits for each. Notice that the outputs are selected automatically and moved into the *Pick Role Variables* box in the *Model Specification* window due to the column property settings.

Figure 11.19: Model Specification for Granulation Process RSM

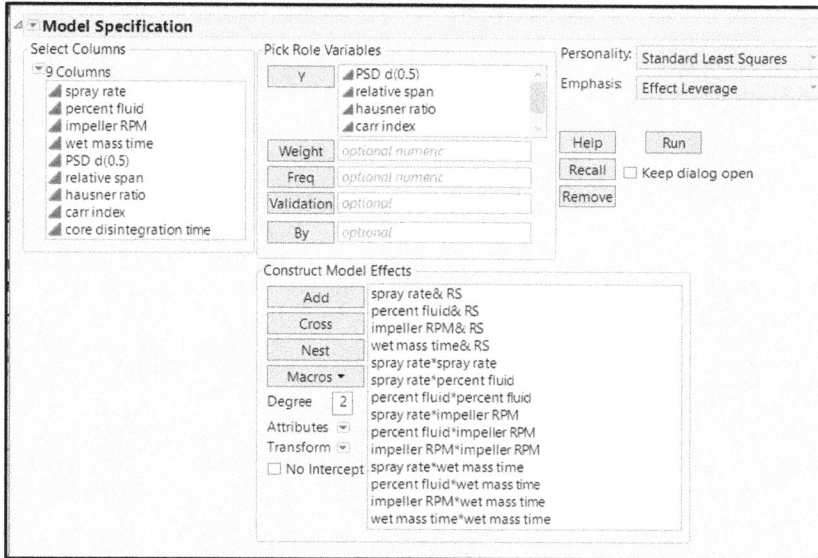

You can manually select the outputs if the data was not collected into the JMP design table or if you used a JMP version earlier than JMP 13:

1. Select *Analyze* ▶ *Fit Model* to open the *Fit Model* window.
2. Press the Shift key and select the five output variables. Drag the selections to the *Y* box in the *Pick Role Variables* section.
3. Press the Shift key, select the four input variables, and click *Macros* (keep the default value of 2 in the *Degree* box). Select *Response Surface* to create the model effects.
4. The *Construct Model Effects* section lists all individual inputs, two-way interactions, and squared terms, as shown in Figure 11.19

The *Model Specification* window is complete with the desired outputs and inputs to study a response surface model (RSM). The default *Personality* value *Standard Least Squares* is used. However, you should change the *Emphasis* to *Effect Leverage* for an RSM. Click *Run* to get the analyses output shown in Figure 11.20.

Figure 11.20: Effect Summary for Granulation Process RSM

Source	LogWorth		PValue
spray rate(6,10)	4.745		0.00002
impeller RPM(40,60)	3.781		0.00017
spray rate*wet mass time	3.683		0.00021
impeller RPM*wet mass time	3.252		0.00056
wet mass time*wet mass time	2.735		0.00184
percent fluid(0.2,0.4)	1.976		0.01058
spray rate*impeller RPM	1.116		0.07662
spray rate*spray rate	1.078		0.08361
percent fluid*impeller RPM	1.068		0.08541
wet mass time(2,12)	1.048		0.08953 ^
percent fluid*wet mass time	0.967		0.10791
impeller RPM*impeller RPM	0.800		0.15862
percent fluid*percent fluid	0.785		0.16398
spray rate*percent fluid	0.737		0.18319

The Effect Summary is at the top of the analysis output, which examines the significance of the factors regarding influence for all five outputs combined. A significance threshold represented by the vertical blue line adjusts for the number of factors being compared. Five experimental factors have strong enough influence to be significant for all outputs:

- *Spray rate* with a LogWorth of 4.745 and highly significant PValue of 0.00002
- *Impeller RPM* with a LogWorth of 3.781 and highly significant PValue of 0.00017
- The interaction of *spray rate*wet mass time* with a LogWorth of 3.683 and highly significant PValue of 0.00021
- The interaction of *impeller RPM*wet mass time* with a LogWorth of 3.252 and highly significant PValue of 0.00056
- The squared term of *wet mass time*wet mass time* with a LogWorth of 2.735 and significant PValue of 0.00184

It is clear that the team needs to ensure that the manufacturing limits for spray rate, impeller RPM, and wet mass time are set to obtain robust results. Other factors might be important to specific outputs and will be explored within the experimental models for each output. The analyses of the individual models must be completed next.

Examination of Fit Statistics for Individual Models

Figure 11.21 provides detailed information about the model of the physical attribute particle size d(0.5).

Figure 11.21: Model Fit for Granulation Process RSM

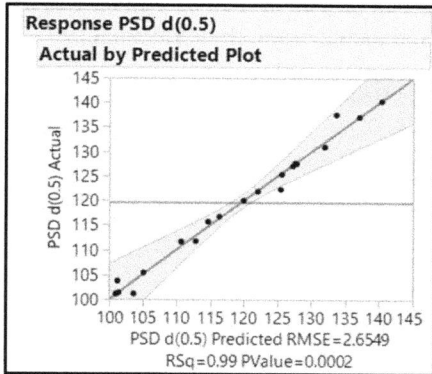

The RSM model exerts a high level of significant influence on particle size (p=0.0002). Changes in the factors of the RSM model explain 99% of the changes in particle size (r-square=0.99). The amount of random variation in the model is limited to 2.65 microns (RMSE=2.6549), which is very small relative to the scale of changes in the outputs approximately 45 microns. The pattern of observations in the predicted plot tightly follows the red model line, with a narrow 95% confidence interval indicated by the red shaded area about the model line.

The next evaluation is the lack of fit test, shown in Figure 11.22.

Figure 11.22: Lack of Fit for Granulation Process RSM

Response PSD d(0.5)

Lack Of Fit

Source	DF	Sum of Squares	Mean Square	F Ratio
Lack Of Fit	5	42.289390	8.45788	17624.66
Pure Error	1	0.000480	0.00048	Prob > F
Total Error	6	42.289870		0.0057*
				Max RSq
				1.0000

The test indicates that a significant lack of fit is present within the model (Prob > F = 0.0057). It is important to keep in mind that the fit of the model is extremely good. Therefore, individual observations that are a small distance from the model line can result in significant lack of fit. The Actual by Predicted plot in Figure 11.21 has one observation that is just below the 95% confidence interval at the approximate prediction of 126 and actual value of approximately 123. Residual analysis can help the analyst better visualize the points contributing to lack of fit.

Model Diagnostics through Residual Analysis

Figure 11.23 includes the residual plots used to further analyze and diagnose the fit of the model.

Figure 11.23: Model Residual Diagnostics Granulation Process RSM

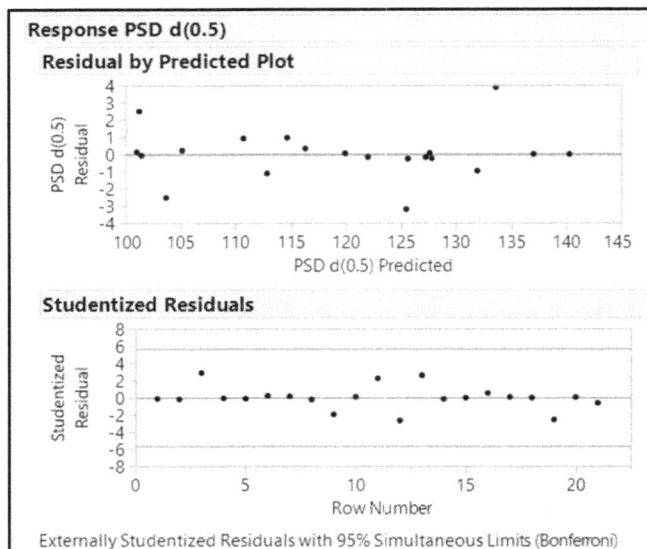

The Residual by Predicted plot illustrates a random pattern that is desired for a robust model. However, there are a few points that have residual values that are more extreme than the other observations. An observation with a predicted value of 126 microns has a residual of -3, indicating that the model overpredicted. An observation with a predicted value of 134 has a residual of 4, indicating that the model underpredicted.

The Studentized Residuals plot has no observations that are outside of the decision limits, which are illustrated by red horizontal lines above and below the studentized residual average of 0. The decision limits are adjusted for the number of factor comparisons included in the model. The Studentized Residual plot including all points that are well within the statistical limits provides evidence that the significant result for the lack of fit test is not likely to create prediction error large enough to be of practical relevance for the information gained.

If you are especially concerned about reductions in the precision of estimates due to lack of fit, you can filter to exclude the runs that contribute to lack of fit, and then run the model analysis with the remaining data. If the conclusions from the analysis of filtered data do not change enough to be practically relevant, the original model with the lack of fit is further supported. Running models with excluded observations should be done with caution. It is possible that the exclusions might not clear up the lack of fit, and the results might change the conclusions that come from the analysis. Other issues for filtered models include lack of power, changes in the structure that might increase correlation and confounding of factors, and reductions in statistical power. In general, it is typically best to note the lack of fit in assumptions for models that have a good overall fit to ensure that details are fully disclosed to the consumers of the information.

Parameter Estimates

The summary values of model fit and residual analysis help you determine that the model is adequate to explain changes that occur in PSD d(0.5). Parameter estimates shown in Figure 11.24 are the part of the analysis that provide a great deal of information about the design space.

Figure 11.24: Parameter Estimates for Granulation Process RSM

Response PSD d(0.5)				
Parameter Estimates				
Term	Estimate	Std Error	t Ratio	Prob>|t|
Intercept	125.63	1.309209	95.96	<.0001*
spray rate(6,10)	9.906923	0.808325	12.26	<.0001*
percent fluid(0.2,0.4)	1.4491105	0.764072	1.90	0.1067
impeller RPM(40,60)	-6.343936	0.764072	-8.30	0.0002*
wet mass time(2,12)	1.4187354	0.75404	1.88	0.1089
spray rate*spray rate	0.6152473	1.276877	0.48	0.6470
spray rate*percent fluid	1.3973299	0.928854	1.50	0.1832
percent fluid*percent fluid	-1.61861	1.425257	-1.14	0.2994
spray rate*impeller RPM	-1.186929	0.928854	-1.28	0.2485
percent fluid*impeller RPM	1.7606584	0.855945	2.06	0.0854
impeller RPM*impeller RPM	1.660268	1.425257	1.16	0.2883
spray rate*wet mass time	7.3441552	0.921117	7.97	0.0002*
percent fluid*wet mass time	0.2716136	0.851917	0.32	0.7607
impeller RPM*wet mass time	5.6633563	0.851917	6.65	0.0006*
wet mass time*wet mass time	-7.872113	1.486879	-5.29	0.0018*

The first term of the parameter estimate list is the intercept. There is significant evidence (p<0.0001) that the intercept of the model is not zero (intercept =125.63), which is information of no practical value for the context of the granulation study.

It is good practice to evaluate the effects of the complex factors before the main effects. There are two interactions and one squared term that have significant leverage on changes in PSD d(0.5):

- The interaction of *spray rate*wet mass time* is estimated to significantly increase PSD d(0.5) by an average of 7.3 microns for each unit increase of the factor (p=0.0002).
- The interaction of *impeller RPM*wet mass time* is estimated to significantly increase PSD d(0.5) by an average of 5.7 microns for each unit increase of the factor (p=0.0006).
- The curved effect of *wet mass time*wet mass time* is estimated to significantly decrease PSD d(0.5) by an average of 7.9 microns for each unit increase of the factor (p=0.0018).

The presence of highly significant two-way interactions and a curved effect provide evidence of a complex design space. The stakeholders of the developmental drug product could not have identified the real cause of changes in various responses without the use of the multivariate, structured experiments. The results of this example emphasize the quintessential reason why international regulatory agencies have emphasized Quality by Design (QbD). The knowledge of the cause and effect model relationships are used to set robust process controls. Such controls are not possible when teams use the one factor at a time (OFAT) methods.

The main effects are the last to be evaluated in an RSM. There are two significant main effects that also have a part in the interactions noted above:

- A unit increase in *spray rate* significantly increases PSD d(0.5) by 9.9 microns (p<0.0001).
- A unit increase in *impeller RPM* significantly decreases PSD d(0.5) by 6.3 microns (p=0.0002).

Refer to the model evaluation to determine the amount of correlation present between significant interactions and significant main effects to determine the practical relevance of the individual inputs. You can quickly explore the model evaluation by using the scripts that have been automatically added to the data table during the design phase. With *Granulation Process Experiment Results.jmp* open and in view, click on the green arrow beside the *Evaluate Design* script. *Color Map on Correlations* appears. Place your pointer over each cell of the matrix and move it slightly until a note appears with the amount of correlation present. Figure 11.25 shows the highest amount of correlation between a significant individual input and a significant two-way interaction. *Spray rate* is 12% correlated with the interaction of *spray rate*wet mass time*.

Figure 11.25: Color Map Analysis

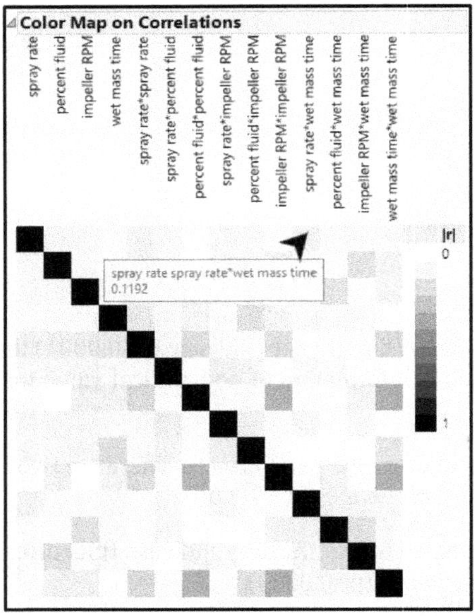

The low correlations of 6% to 12% are not of great concern. The evidence of a significant effect from the two individual inputs, two interactions, and the squared term can be included in the conclusions of the analysis. A mention of the low correlation between factors is advised to ensure that consumers of the analysis have all the important details.

Detailed Analyses of Significant Factors with Leverage Plots

The goal for the set of experiments is to obtain high-resolution information so that robust predictions can be made for how changes in the process inputs affect the outputs. The experimental design with a response surface model includes individual inputs as well as complex terms to aid the goal. You might have noticed that the model output for each response includes a set of Leverage plots, located to the left of the Prediction plot for the full model. Leverage plots isolate the effect of each factor on the output in the presence of the influences of all other factors. The Leverage plot shown in Figure 11.26 provides detail about the amount of leverage that results from the curved effect of wet mass time.

Figure 11.26: Leverage Plot of (Wet Mass Time)^2 for PSD d(0.5)

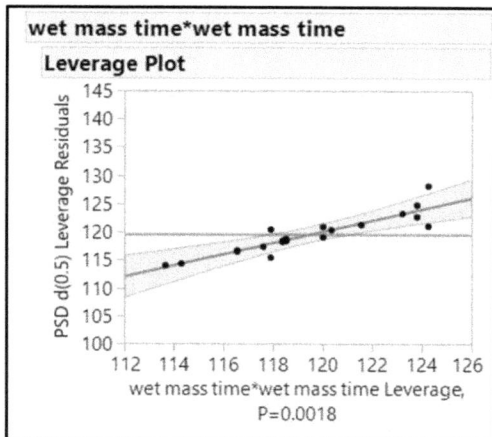

The squared term factor exerts highly significant (P=0.0018) leverage on PSD d(0.5). The power for the curved wet mass time factor exerting significant influence on PSD d(0.5) is 32.5%, as estimated during the design phase (Figure 11.27). The estimate is based on one unit of experimental error (RMSE) and a one-unit anticipated coefficient for the factor.

Figure 11.27: Estimated Model Power from the Experimental Design (from Chapter 10)

Design Evaluation		
Power Analysis		
Significance Level	0.05	
Anticipated RMSE	1	
	Anticipated	
Term	**Coefficient**	**Power**
Intercept	1	0.4
spray rate	1	0.781
percent fluid	1	0.823
impeller RPM	1	0.823
wet mass time	1	0.833
spray rate*spray rate	1	0.416
spray rate*percent fluid	1	0.666
percent fluid*percent fluid	1	0.348
spray rate*impeller RPM	1	0.666
percent fluid*impeller RPM	1	0.735
impeller RPM*impeller RPM	1	0.348
spray rate*wet mass time	1	0.673
percent fluid*wet mass time	1	0.738
impeller RPM*wet mass time	1	0.738
wet mass time*wet mass time	1	0.325

The analysis of the PSD d(0.5) model is complete. It is good practice to obtain the actual statistical power for the significant factors of the model and compare that with the estimates used to justify the design chosen. The power is easily calculated from the leverage plots by completing the following steps:

1. Click on the red triangle menu next to the *wet mass time*wet mass time* plot header.

2. Select the only available option for the plot, *Power Analysis*. The *Power Details* window shown in Figure 11.28 appears.

Figure 11.28: Power Details Window

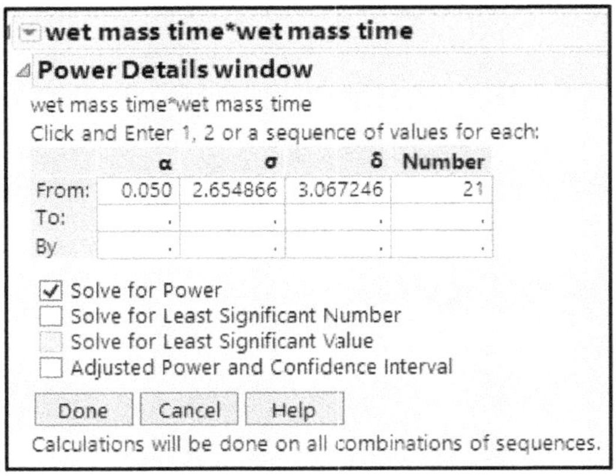

3. The significance value α, standard deviation σ, the size of effect to be detected δ, and sample size *Number* have been automatically included by JMP. There is no need to alter the given values for power to be calculated.

4. Select the *Solve for Power* check box.

5. Click *Done* to add the *Power Details* underneath the leverage plot, as shown in Figure 11.29.

Figure 11.29: Power of the (wet mass time)^2 Significant Leverage on PSD d(0.5)

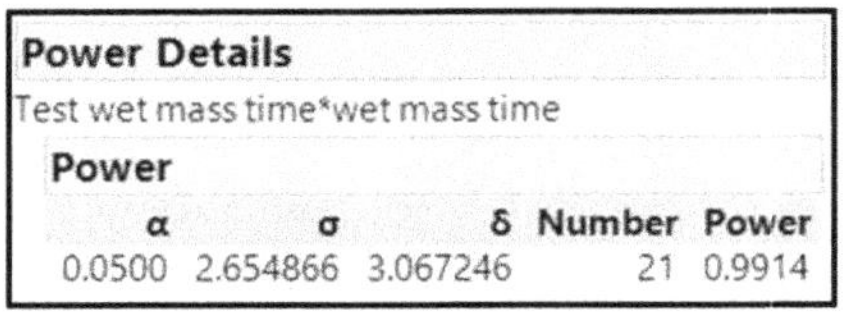

The results shown in Figure 11.29 indicate that the significant effect of *wet mass time*wet mass time* on PSD d(0.5) is more than 99% likely to be a real effect for the population of granulations represented by the design space explored by the set of experiments. Power increased dramatically from the design estimates largely because the magnitude of the amount of change in the output levered by the factor is larger than the random error represented by the standard deviation. Put simply, the signal is large as compared with the small amount of noise. The significance level of 0.05 and the sample size could affect power but neither changed between the design phase and the analysis of the experiments. There are four other significant effects that should be analyzed for statistical power and included somewhere in the details of an analysis report provided to the project stakeholders. There is no need to analyze power for the factors that lack evidence of significance; the definition of power is a value that explains the likelihood that a significant effect will be real for the population represented by the experimental sample.

Visualization of the Higher-Order Terms with the Interaction Plots

The analysis of the effects provides robust numeric evidence of how the factors affect the PSD d(0.5). Numeric estimates are very useful but interpretation is confusing with relationships that are known to be complex. JMP always includes excellent graphics to help people visualize results and interpret trends with ease. The Interaction Profiler is a tool that breaks down complex relationships for sound understanding, even for people who are not well versed in statistics. Click on the red triangle menu next to the *Response PSD d(0.5)* header in the analysis output. Select *Factor Profiling* ▶ *Interaction Plots* option to get the plot shown in Figure 11.30.

Figure 11.30: PSD d(0.5) Interaction Profiles

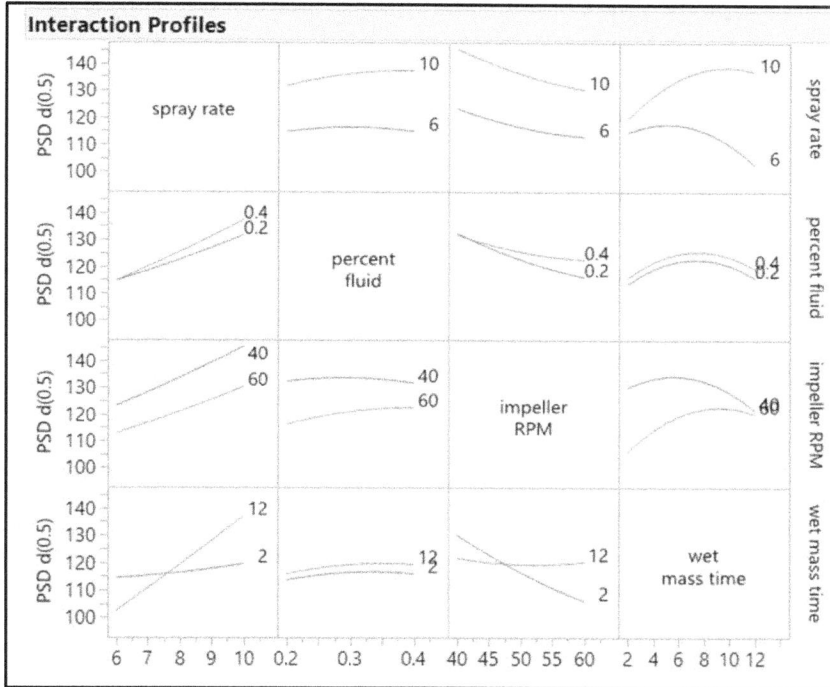

Trend lines are illustrated with blue and red colors for the noted levels, which is not evident in the plot converted to black and white for publishing. This text references the colors for ease of understanding.

The complex relationship for *spray rate* and *wet mass time* appears in the top right plot of the matrix. The curvature of the profiles illustrates the significance of the *wet mass time*wet mass time* factor. The blue profile shows that PSD d(0.5) increases at a steep rate of growth for the highest spray rate (10). The red profile illustrates that the PSD d(0.5) decreases at an increasing rate as *wet mass time* is increased and *spray rate* is at the lowest level (6). The differing profiles provide robust graphical evidence of the presence of the highly significant interaction. The plot in the lower left corner of the matrix illustrates the same interaction. However, *wet mass time* is shown in the fixed low and high values, with the *spray rate* indicated as the explaining variable along the X axis. The plot just below the *wet mass time*spray rate* plot examines the effect of *wet mass time*percent fluid*. Notice that the curved profiles are parallel to each other. The parallel profiles illustrate the lack of a significant interaction. There can be no argument that the model created from the set of experiments provides a great deal of useful information about the PSD d(0.5) response. Each of the five outputs will have a unique model included as outline headers in the analysis output. The model for the Hausner ratio is examined next.

Examination of an Insignificant Model

Figure 11.31 provides information to assess the fit of the model for the Hausner ratio.

Figure 11.31: Model Fit for Hausner Ratio

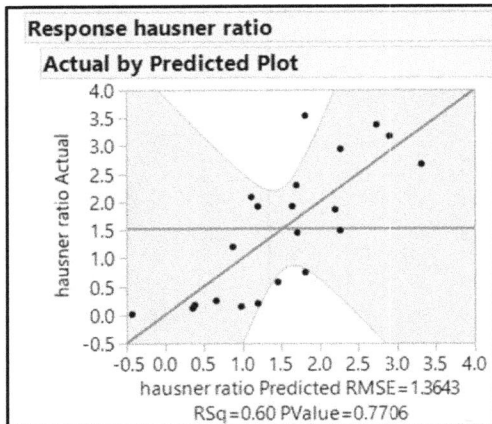

The model is insignificant for providing information about the Hausner ratio (PValue=0.77). Changes to the inputs of the model explain 60% of the changes in Hausner ratio (Rsquare = 0.60). The RMSE of 1.36 is compared with the overall range in output values of approximately 3.7 and is sizeable. The shaded area of 95% confidence interval widens dramatically about the center point of the predicted values, also known as centroid. The actual observations are scattered widely about the model prediction line

The information provided by the Actual by Predicted plot tells you that the effect on the Hausner ratio is very subtle or insignificant. Therefore, the detailed analysis of the Hausner ratio is pointless. JMP calculates values for residual analysis and model effects, but the detail is not needed for an insignificant model.

Dynamic Visualization of a Design Space with the Prediction Profiler

The detailed analysis of the remaining three outputs is left up to you to explore. The Prediction Profiler is an extremely useful tool for examining the dynamic relationships among all the inputs and the outputs simultaneously. This powerful tool typically becomes available by default when you run the model script. If the profiler is not initially visible, you can easily add it to the output. Click on the red triangle menu located to the left of the *Least Squares Fit* header, and select *Profilers* ▶ *Profiler* to add it to the bottom of the analysis output, as shown in Figure 11.32.

Figure 11.32: Granulation Process RSM Model Profiler

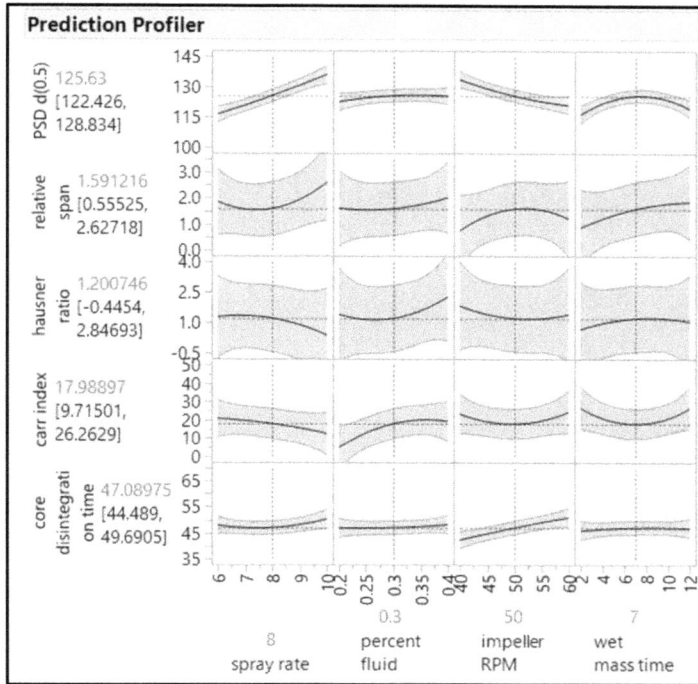

The profiler is a dynamic matrix plot of all inputs as columns and outputs as rows. Recall that the curved term of *wet mass time*wet mass time* is significant for several of the output models. The curvature of the model profile lines illustrates the non-linear responses of the model effects. The gray regions about the profiler lines show the 95% confidence interval for the average response and are illustrative of prediction error. The horizontal, segmented red lines indicate the average response for the chosen levels of the inputs. The vertical, segmented red lines are dynamic sliders that you can click and drag to explore the dynamic changes that take place in the outputs.

Notice that the slope of some profiles change as you manipulate the slider of an input between low and high values. Move the slider to represent changes in *wet mass time* to watch the slope change in *impeller RPM* for various outputs. Changing slopes indicate that significant interactions are present; vertical shifts indicate independence among inputs. You can also click on the red numeric value of an input to enter a specific value. The red numeric value for the output is the average prediction, and the black numeric values are the low and high 95% confidence limits for the prediction. It is best practice to focus on the 95% confidence intervals for the most precise predictions.

Recall that the responses modeled include both a goal and specified limits. The goal can be to match a target, or to minimize or maximize results. The profiler includes functionality to indicate how well the models meet the goals for all responses. The desirability function accounts for the goals of all responses. The range of desirability is between a minimum of 0 and a maximum of 1 to explain how well the models will meet goals. The higher the value, the more likely that all goals will be met satisfactorily. Click on the red triangle menu next to the *Prediction Profiler* header, and select *Optimization and Desirability* ▶ *Desirability Functions* to get the plot shown in Figure 11.33.

Figure 11.33: Granulation Process RSM Profiler with Desirability Function

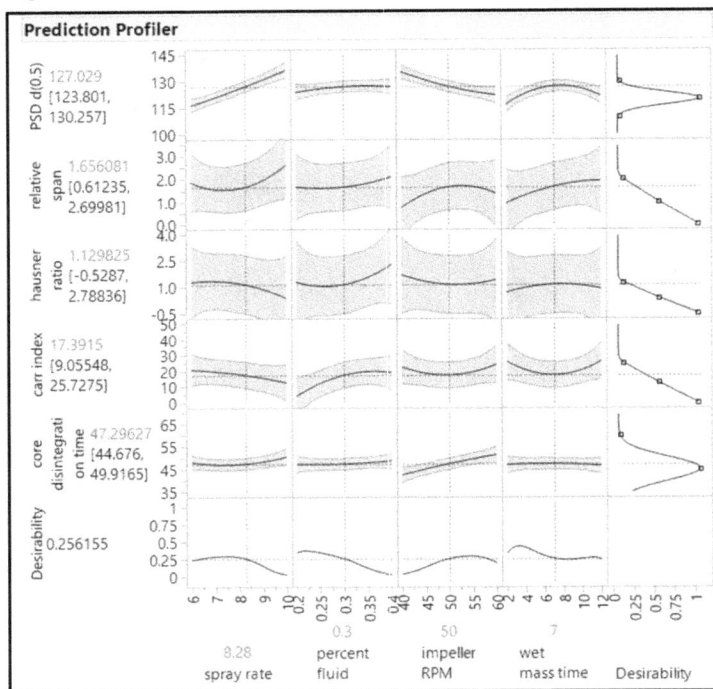

The peaked desirability for PSD d(0.5) and core disintegration time reflects the goal to match a target. The small squares located on the lower and higher tails of the function indicate the limits; the square on the peak is the target. The downward sloping desirability functions for relative span, Hausner ratio, and Carr index reflect the goal to minimize. The small square located at the highest point of the function is the upper limit, and the middle and lower squares reflect the median and minimum predicted values for the output. The mid-point settings for the four process inputs reflected in the profiler yield a relatively low desirability of 0.256. The analyst can manipulate each of the inputs to obtain higher desirability, but the function shape changes dynamically as inputs change. JMP provides an algorithm that quickly finds the optimum settings to get the highest possible level of desirability automatically. Click on the red triangle menu next to the *Prediction Profiler* header, and select *Optimization and Desirability ▶ Maximize Desirability* to get the plot shown in Figure 11.34.

Figure 11.34: Granulation Process RSM Profiler Optimized

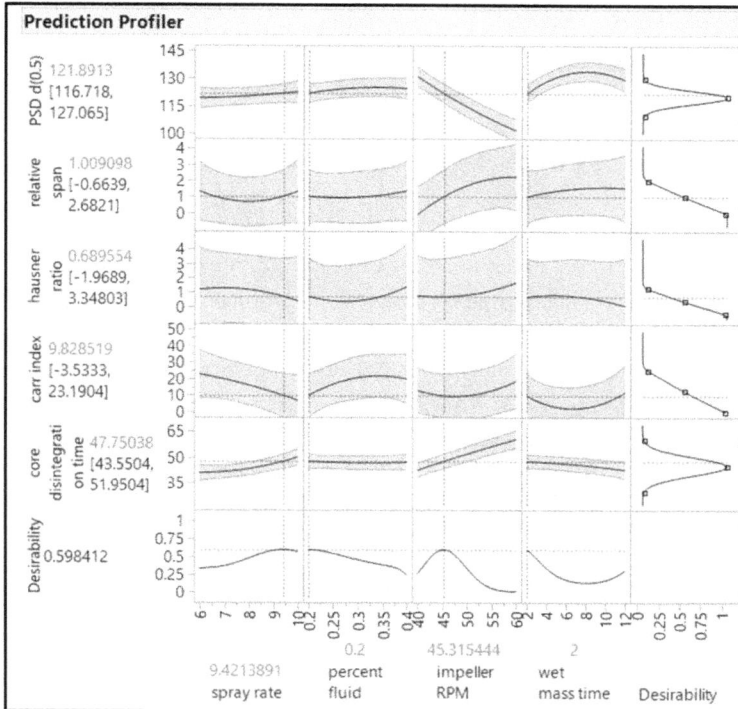

The highest possible desirability is 0.598, which is more than double the value obtained by the mid-point settings. The team can obtain the best possible results with a high spray rate (9.4), least amount of percent fluid (0.2), moderately low impeller RPM (45.3), and least amount of wet mass time (2). Stakeholders can expect the following outputs:

- PSD d(0.5) of between 116.7 and 127.1 microns
- Relative span maximum of 2.7*
- Hausner ratio maximum of 3.3*
- Carr index maximum of 23.2
- Core disintegration time of between 43.6 and 52.0 seconds
  *Model not significant

Elimination of Insignificant Models to Enhance Interpretation

It is very important to present a clear message to the project stakeholders. There is no requirement for the number of words in business; in fact, less is always more. There is little point in providing detailed results and predictions for models that you know to be insignificant. The analysis of all five outputs is an important step in the process of discovery. The analyst should save the analysis as a script so that it can be pulled up as needed at any time. Complete the following steps to create a new set of analyses limited to the significant models. The result provides the clarity needed to share results with the stakeholders.

1. Click on the red triangle menu next to the *Least Squares Fit* header.
2. Select *Save Script* ▶ *To data table* to save the 5-output script.
3. In the *Save Script As* window, click to the right of the *Fit Least Squares Name* and enter "5 outputs" so that the name of the script is clear.
4. Click *OK* to save the script to the data table.
5. Notice that other options exist for saving scripts. Explore these options to find the one the best meets the needs of your organization.
6. Click again on the red triangle menu next to the *Least Squares Fit* header.
7. Select *Redo* ▶ *Relaunch analysis* to get to the *Model Specification* dialog box.
8. Press the Ctrl key or the Shift key, and click *Relative Span* and *Hausner Ratio* to highlight them in blue in the *Pick Role Variables* box.
9. Click *Remove*, and then click *Run* to get the analysis output for three responses.
10. Click on the red triangle menu next to the *Least Squares Fit* header.
11. Select *Profilers* ▶ *Profiler* to add it to the bottom of the analysis output.

None of the models change. However, the effect summary shown in Figure 11.35 changes slightly because the LogWorth values are calculated for three models instead of for the original five.

Figure 11.35: Granulation Process RSM Profiler Optimized (Significant Models Only)

Effect Summary		
Source	**LogWorth**	**PValue**
spray rate(6,10)	4.745	0.00002
impeller RPM(40,60)	3.781	0.00017
spray rate*wet mass time	3.683	0.00021
impeller RPM*wet mass time	3.252	0.00056
wet mass time*wet mass time	2.735	0.00184
percent fluid(0.2,0.4)	1.976	0.01058
spray rate*spray rate	1.078	0.08361
percent fluid*impeller RPM	1.068	0.08541
percent fluid*wet mass time	0.967	0.10791
wet mass time(2,12)	0.963	0.10893 ^
spray rate*impeller RPM	0.863	0.13702
impeller RPM*impeller RPM	0.800	0.15862
percent fluid*percent fluid	0.785	0.16398
spray rate*percent fluid	0.737	0.18319

The profiler in Figure 11.36 illustrates an increased desirability for the inputs set to the mid-point values (desirability = 0.43). This makes sense because it is much easier to meet the goals for a reduced number of outputs (responses).

Figure 11.36: Granulation Process RSM Profiler Optimized (Significant Models Only)

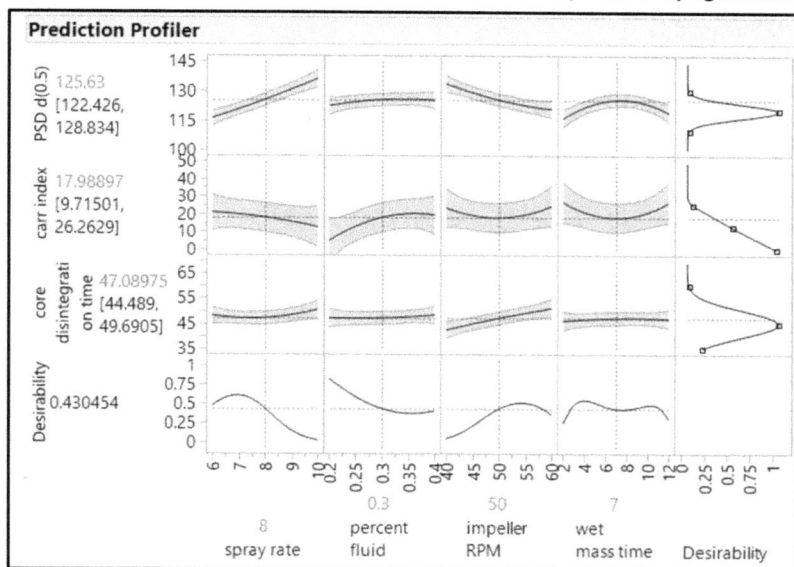

Use the red triangle menu options of the profiler to maximize the desirability of the three output models. Figure 11.37 indicates that the analysis can achieve an excellent desirability of 0.917.

Figure 11.37: Granulation Process RSM Profiler Reduced and Optimized

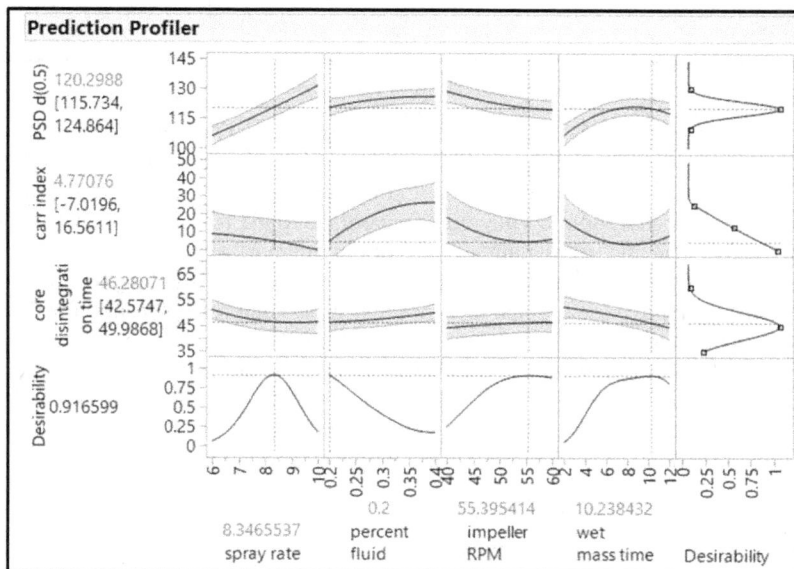

The settings listed in Table 11.1 are the most likely values needed for the granulation process to achieve all the experimental goals.

Table 11.1 Granulation Process RSM Profiler Reduced and Optimized

Input Settings	Outputs	95% Low	95% High
Spray rate at 8.3	PSD d(0.5)	115.7	125.0
Percent fluid at 0.2	Carr Index		16.6
Impeller RPM at 55.4	Core Disintegration Time	42.6	50.0
Wet mass time at 10.2			

The results for the RSM have provided a great deal of information. It is good practice to save the analyses for the reduced set of models to the data file as a final step. Click on the red triangle menu next to the *Least Squares Fit* header, and select *Save Script* ▶ *To data table* to save the 5-output script. In the *Save Script As* window, click to the right of the *Fit Least Squares Name* and enter "3 outputs" so that the name of the script is clear. Click *OK* to save it to the data table.

Practical Conclusions

The stakeholders for both projects enjoy a rich amount of information, providing detail about how the experimental goals have been met through the use of structured, multivariate experimentation. Michelyne's team tasked with the problems of surgical light handle covers with thin walls quickly found that they can focus on changing only two of the six process inputs. They quickly confirmed the results of the new input requirements with a few confirmation runs and immediately incorporated them into the manufacturing order process controls. Erica's team, studying the granulation process, used the detailed analysis outputs to create a set of process input controls that have the highest likelihood of producing robust results for three important outputs. The two outputs that had insignificant models are not critical. The team decided to monitor the uncontrolled outputs and determine whether further study might be necessary to determine whether controls are possible. Structured, multivariate experimentation provides the highest value information to incorporate as elements of QbD for the planned submission package in order to gain regulatory approval and achieve the highest level of quality in the drug product produced.

The resources required to design and execute the sets of experiments might have seemed daunting at first as compared with utilizing hierarchical or OFAT experimentation. However, the amount of information provides for great confidence in decisions made to ensure the highest level of robust output from each process. This chapter might paint a somewhat unrealistic picture, since a single set of experiments was executed for each problem. Use of these techniques can best be described as a journey of enlightenment about the process studied. There will be times when a single, well-designed set of experiments provides the information needed by the stakeholders. It is more likely that one set of experiments answers many questions yet initiates more detailed questions as layers of understanding are achieved through analysis of results. The wonderful aspect of the utilization of the techniques is the structure provided, which you can use throughout the lifespan of a process. If problems arise in commercial production, teams can use the models to focus efforts on the most likely contributing factors for unsatisfactory results. JMP provides a nearly infinite amount of design and analysis options within the DOE and Analysis platforms, which you can easily employ to facilitate the study of all types of processes.

Exercises

E11.1—The project to improve the film seal on the surgical trays is progressing quickly since the team started using JMP. Predictive modeling narrowed the scope from nine process inputs to three: line speed, dwell time, and head energy. Leadership has asked the team to define how each of the influential predictors is affecting the seal strength of the film. Each of the inputs can be easily changed for each run, and the project team was able to get agreement to study the process with a response surface design. The custom designer in the DOE platform included 16 runs as the default size of the experiment, which the team executed.

1. Open *Burst Testing Experiments 3F 16R CP.jmp*.
2. Use the *Model* script to launch the analysis. This script is available since the responses were added to the JMP data table that was created during the experimental design phase of the project.
3. In the *Model Specification* dialog box, change *Emphasis* to *Effect Leverage* before you run the analysis.
4. Summarize the results into a report , and share only the most important information Be sure to quantify the random error for the model.
   - Is the model fit strong enough to suggest a robust model?
   - Is the model significant?
   - Is there an issue with lack of fit?
   - Are the residuals indicating a random pattern?
   - Which are the significant parameter estimates and how much does each affect seal strength?
   - Use the red triangle menu options for each Leverage plot to get the actual power for the significant effects. Is there enough sample size for a robust model?
5. Use the red triangle menu next to the *Response burst pressure (in Hg)* and select *Factor Profiling ▶ Interaction Profiler* to visualize the interactions. Explain them in the report.
6. How important was it to include squared terms?
   - Go to the *Effect Summary* header near the top of the analysis and select the three squared terms to remove them from the model.
   - Compare and contrast the fit statistics, model significance, and random error to determine the importance of including the squared terms.
   - Click *Undo* to add the squared terms back to the model.
7. Use the red triangle menu next to the *Response burst pressure (in Hg)* and select *Factor Profiling ▶ Profiler* to use the model profiler.
   - Use the red triangle menu options to maximize the model.
   - Make a table of the settings required to get the best seal strength.

E11.2—A minimal screening study with six experimental runs was developed for the tablet compression process in the previous chapter. A pressing concern is to determine whether any of the process inputs are related to lowered tablet hardness and changes in variability. The technical operations team utilized pilot scale equipment to execute the set of six experiments and collected the results of mean tablet hardness (SCU) and the variability in tablet hardness (%RSD). The information is available in the Excel spreadsheet *tablet compression study results.xlsx.*

1. Open the Excel data sheet in JMP and utilize the effect screening emphasis of fit model to analyze the data.
2. Use the output to summarize the fit of the model and the evidence of model significance.
3. Utilize the residual plots to diagnose the health of the model.
4. What conclusion would you provide to project stakeholders?

E11.3—The injection molding process was modeled with a definitive screening design due to the large number of process inputs and the likelihood that influence on the outputs of part width and part weight is limited to a few of the ten inputs. The molded part is a trigger for the inhaler assembly and must have a width of between 11.75 mm and 12.25 mm in order to have the needed clearance to operate as designed. The weight of the molded trigger is designed to be between 3.05 g and 3.35 g in order for the return spring to work properly. Operations and engineering worked together to execute the set of experiments and collect the data from the parts made. The project stakeholders need to know the inputs that affect the outputs so that the team can quickly optimize the process and immediately improve the quality of parts made.

1. Open *inhaler molded component process study DSD 26R.jmp.*
2. Use the *Fit Definitive Screening Design* script to get a summary of the important factors.
3. Click *Run Design* to obtain the detailed analysis for each of the outputs.
4. Create a summary report with practical conclusions that you would present to the project stakeholders. Be sure to include a table of the optimized process with expected results in the summary report.
    -   What is the fit of each model? Do you expect that the models will product robust estimates for the two outputs?
    -   Which of the process inputs have the most influence? Is there a presence of combined effects (interactions) or complex (non-linear effects) of inputs?
    -   Are there any risks of parts that might not function properly for the combination of inputs studied?
    -   What can be expected for the average part width and average part weight of the population of all parts that are made at the optimized input settings?

E11.4—A materials study design was initiated in the chapter 9 exercises involving three materials inputs and one slack variable to make up the total weight of a dose. Mixes were made and test batches completed for the 12 runs of the experiments. The team collected data on six outputs: tablet assay, content uniformity (acceptance value of 10 tablets), average dissolution at 1 hour, average dissolution at 4 hours, and the variance within the six dissolution tablets tested for both 1 and 4 hours. Project stakeholders are interested in the effects of the changes in materials as well as any combined effects (interactions).

Goals for the outputs are as follows:

- Assay   90% to 110%
- AV10 (content uniformity)   NMT 15
- Dissolution at 1 hour   25% to 50%
- Dissolution at 4 hours   60% to 85%
- Variance in 1-hour dissolution   NMT 10
- Variance in 2-hour dissolution   NMT 10

1. Open *formulation materials study results.jmp*.
2. Run the analysis of the models for the six outputs.
3. Create a summary report of the analysis to present to the project stakeholders. Can the goals be met robustly?
   a. Be sure to comment on any limitations to the design.

Chapter 12: Getting Practical Value from Structured Experiments

Overview

The use of structured, multivariate experimentation is growing by leaps and bounds in the medical device and pharmaceutical industries. Analysis of a statistical model typically involves predictions made from specific point settings of the process inputs to better understand relationships and trends. The amount of resources required to execute a good set of structured experiments is not trivial. Therefore, the analysis of results should be utilized to extract the most useable information possible. This information enables teams to make optimal decisions and to illustrate quantifiable reductions in the risk of producing products that are inferior or that fail to meet customer demands. This chapter takes advantage of the excellent set of tools in the Prediction Profiler, which add detail to the model analysis and make the results practically applicable. You can think of the model analysis as the academic portion of the experimental journey, while the tools in this chapter help teams to apply the knowledge in order to gain the most possible value.

The Problems: Statistical Modeling Are Needed to Gain Detail About A Thermoforming Process and a Granulation Process

The results of the screening study for the thermoforming process are now available. Michelyne's team needs to quickly provide a new set of process controls that will reduce the risk of producing surgical light handle covers that are too thin. The Prediction Profiler in JMP can provide the set of tools necessary to visualize the model while working with the subject matter experts on updates to the production controls needed for robust product output.

Erica's team analyzed the response surface model for the granulation process. The analysis provides detailed information for optimizing the results of the three outputs. Her team is interested in digging deeper into the results so that they can accurately and precisely predict the population of batches made by the granulation process. The detailed analysis is needed to quantify the risks so that the product can move to the next phase of development. Addressing product risks, by development in accordance with international guidance on Quality by Design (QbD), will speed the approval process for the product submission.

Identification of a Control Space from the Thermoforming DSD

The modeling of the thermoforming process, utilizing a definitive screening design (DSD), provides clarity for two process inputs that have significant influence on the minimum thickness output. Michelyne used the results of the analysis to inform the project stakeholders about the adoption of a robust strategy to update the controls for plug depth and cycle time. The model profiler results from chapter 11 provide evidence indicating that a shallower plug depth and lengthened cycle time will mitigate excessive thinning of the surgical light handle covers. Thin areas of plastic are associated with cracked and loose handle covers reported by customers. Screening designs do not provide for the most precise predictions, but a good control space can be determined from the results of the model analysis. The team plans to define an adequate control space with the model and include confirmation runs to ensure that actual results fall within the intervals for the model predictions. Define the control space by completing the following steps:

1. Open *Thermoforming DSD Results with confirmation.jmp*.

2. Select rows 14 through 20 so that they are highlighted in blue. Right-click in the highlighted area, and select *Hide and Exclude* to filter out all but the DOE rows.

3. Run the *Fit Definitive Screening Design* script.

4. In the *Combined Model Parameter Estimates* window, click *Run*. Remember, if you are using a version of JMP earlier than JMP 13, you need to manually create the model by selecting *Analyze* ▶ *Fit Model*.

5. Use the red triangle menu next to the *Response min handle thickness* header, and select *Factor Profiling* ▶ *Contour Profiler* to get the output in Figure 12.1.

Figure 12.1: Contour Profiler

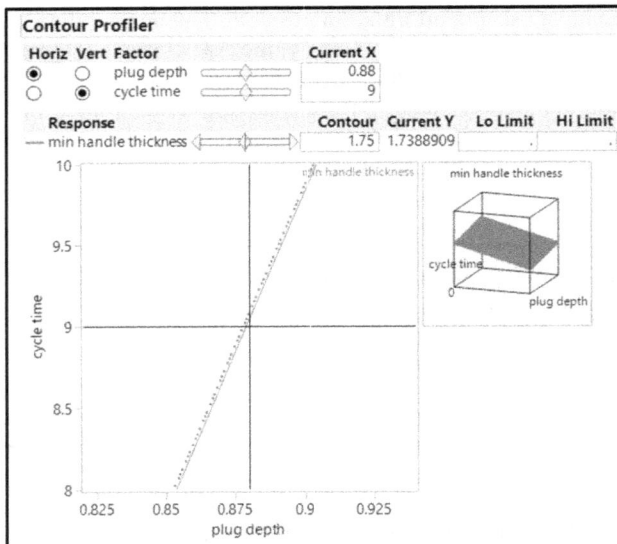

The contour profiler illustrates the relationship between the plug depth and cycle time inputs on *min handle thickness*. The midpoint of the *min handle thickness* output is automatically indicated by the red model line in the plot. The cube to the right of the plot illustrates the minimum thickness response surface plane from the two inputs. The red dots to the immediate left of the model line in the main contour plot indicate the upslope of the model response plane. The black cross hairs indicate specific levels of the two inputs, with values shown in the *Current X* boxes at the top of the output. You can dynamically click and drag the crosshairs to see the *Current Y* values for an infinite combination of input values. You can enter exact values for the inputs in the *Current X* boxes for specific predictions. You can also select different combinations of radio buttons to swap the inputs respective to the X and Y sides of the plot.

The midpoint model gives some information, but the team needs more information in order to determine a reasonable control space in which to operate the process. You can add gridlines to the plot to help visualize the design space with better resolution. Use the red triangle menu next to the *Contour Profiler* header, and select the *Contour Grid* option. The *Please Enter Values* dialog box appears with default values. Change *Low value* to 0.50, and change *Increment* to 0.25. There is no need to change *High value*. Click *OK* to get the output shown in Figure 12.2.

Figure 12.2: Contour Profiler with Grid Lines

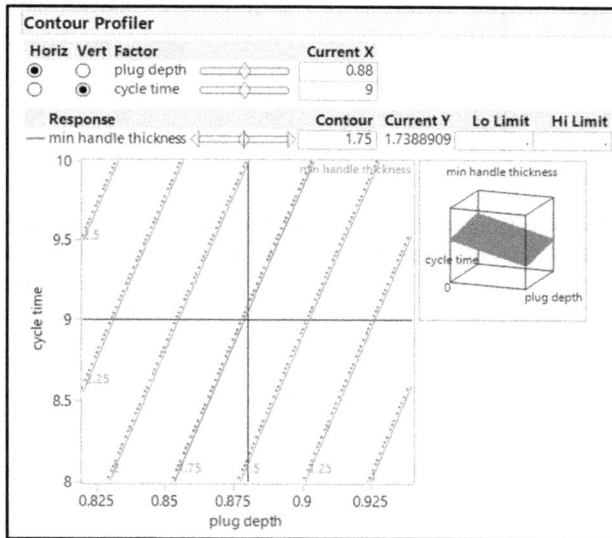

The settings create contour lines that start at 0.50 mm and cover the space for every 0.2 5mm change in thickness. Grid lines added to the plot provide a great deal of additional information. The minimum thickness clearly increases from the lower right corner of the contour plot up to the upper left corner. The even spacing of the gridlines indicates that the increases are at a constant rate.

The goal of the project is to ensure that the minimum handle thickness does not fall below 1.5 mm. Adding the value as a lower limit provides more clarity to the design space. Enter 1.5 into the *Lo Limit* box for the response to get the output shown in Figure 12.3.

Figure 12.3: Contour Profiler with Grid Lines and Lo Limit

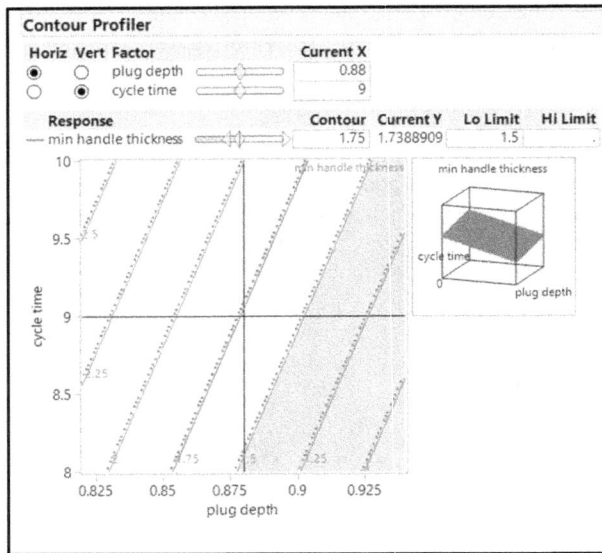

The contour profiler with gridlines and the shaded rejection region clearly show the operational space that must be maintained in the non-shaded region. The improvement team provides the plot to the subject matter experts in operations as a framework for determining the process control space that is satisfactory. The recommendation from the discussion is to control the plug depth from 0.825 to 0.875 and the cycle time from 8.5 to 9.5. Add a shaded box to the contour profiler by completing the following steps to provide visual representation of the control space:

1. Move the pointer to the shaded blue bar above the analysis output to show the main menu.
2. Select *Tools ▶ Simple Shape* in the main menu. Alternatively, click the ellipse icon on the tool bar.
3. The pointer changes to an ellipse with an arrow. Draw the shape by clicking on the plot close to the point on the plot with a plug depth of 0.825 and cycle time of 9.5. Drag the pointer down to the point with a plug depth of 0.875 and cycle time of 8.5. You can make final adjustments to the shape later.
4. A blue circle with an adjustment border is drawn on the profiler. Right-click inside the blue adjustment border (Hint: The pointer is an arrowed cross hair.), and select *Shape ▶ Rectangle* to change the shape.
5. Right-click in the blue adjustment border and select *Color*. Select the green color to get the output shown in Figure 12.4.

Figure 12.4: Contour Profiler with Grid Lines, Lo Limit, and Proposed Control Space

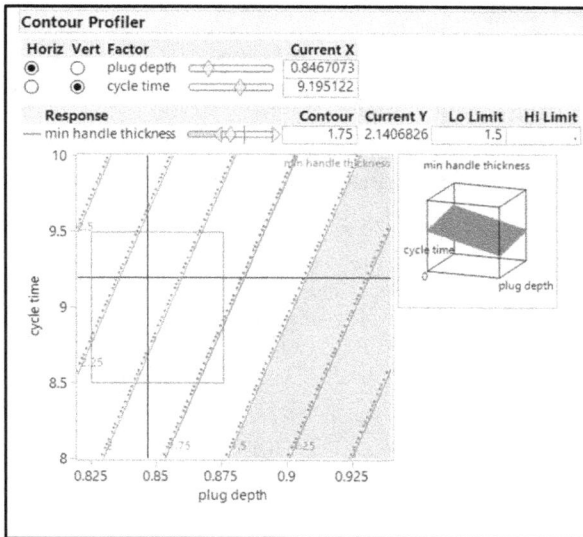

Verification of a Control Space with Individual Interval Estimates

The control space in the Contour Profiler seems to be reasonable for the team to protect against an average handle thickness of falling below the 1.5 mm limit. It is good practice to run a small number of confirmation runs based on the control space to ensure a high chance of success. Michelyne is concerned about the lower right corner of the proposed control space since it is close to the shaded region of results that are outside of the specification. She knows that the process includes variability and that the models provided information only about mean results. Will this space be robust enough to protect the customers from a few handles with thin spots that are less than 1.5 mm?

The model is used to understand the potential for extremely thinned individual parts by using a prediction interval for individual results. The small number of runs utilized by the screening model yield higher prediction error than what is obtained from a response surface model. Recall that additional resources are needed to create a response surface model. Knowing the limitations, the team defines a control space based on the individual interval limits. Go to the red triangle menu next to the *Response min handle thickness* header, and select *Save Columns ▶ Individual Confidence Limit Formula*, as shown in Figure 12.5.

Figure 12.5: Creating Individual Confidence Limits

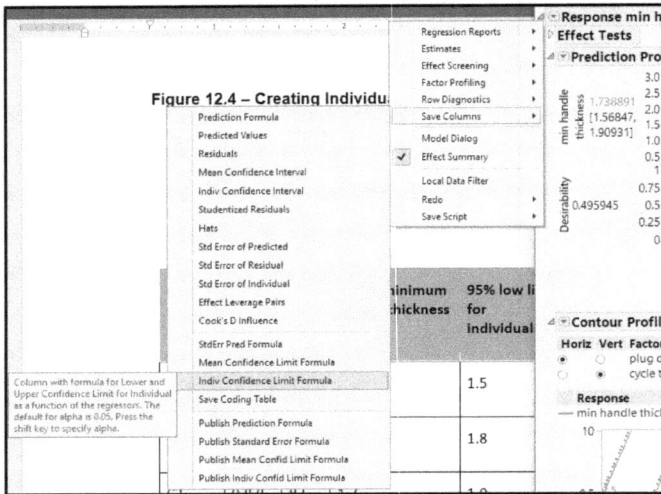

When you save the individual confidence limit formula, press the Shift key to specify the alpha value that you want (the default is an alpha of 0.05 for a 95% confidence level). Two columns are added to the data table, including the low and high individual confidence limits. Notice that the corners of the proposed design space are included as rows 14 through 17. The individual confidence limits for corners 3 and 4 include values that are indeed below the 1.5 mm specification limit.

Table 12.1: Proposed Control Space with Verification

control space corner	plug depth	cycle time	actual minimum handle thickness	95% low limit for individuals	95% high limit for individuals
C1	0.825	8.5	1.8	1.5	2.9
C2	0.825	9.5	2.0	1.8	3.1
C3	0.875	8.5	1.7	1.0	2.3
C4	0.875	9.5	1.9	1.3	2.6

The proposed control space will be a significant improvement, but the intervals indicate that there might still be some risk. Michelyne utilizes the individual interval formulas to determine how much to contract the control space so that all limits are greater than 1.5. Rows 18, 19, and 20 of the data table include the tightened control space. The tightened control space is shown in Figure 12.6.

Figure 12.6: Tightened Control Space Based on Individual Confidence Estimates

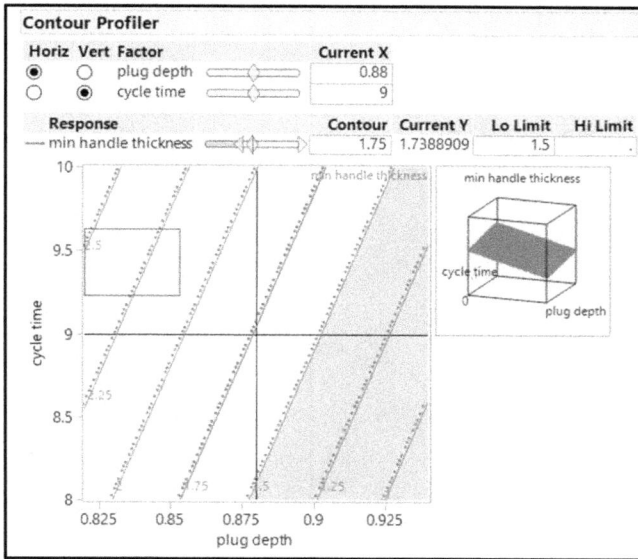

The formal control space that the team chooses will be determined from discussions with the subject matter experts who are responsible for the process. If the tolerance for risk is low, the operations team will need to maintain a tightened control space; moderate tolerance for risk allows for a widened control space. A wide control space is typically coupled with increased process monitoring. The samples from the process monitoring are used to assess the probability of producing a handle that is below the minimum thickness level of 1.5 mm. A capability study, explained in chapter 3, confirms that the space chosen is robust enough to produce handles of acceptable thickness.

Using Simulations to Model Input Variability for a Granulation RSM

The analysis of the response surface model for the granulation process created a model for the two significant inputs and three outputs. The Prediction Profiler can generate a large amount of detailed information from the statistical modeling. Open the *Granulation Process Experiment Results Simulation.jmp* file, and run the *Fit least squares 3 outputs* script to open the model analysis output. Utilize the red triangle menu next to the *Least Squares Fit* header, and select *Profilers ▶ Profiler* to get the output shown in Figure 12.7.

284 Pharmaceutical Quality by Design Using JMP

Figure 12.7: Granulation Experiments Prediction Profiler

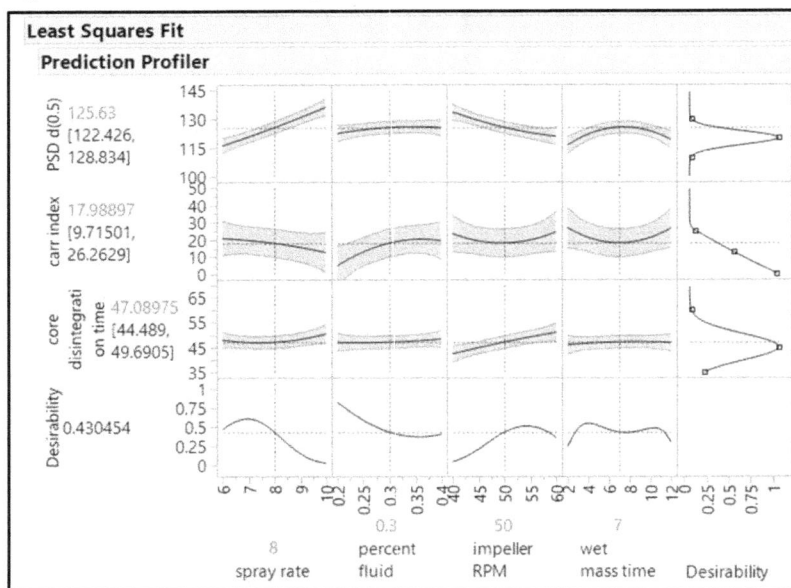

The Prediction Profiler illustrates the trends of influence that each of the inputs has on the outputs as well as the band of random error that is expected. The profiler is an excellent, dynamic tool for optimizing a process to achieve goals for multiple outputs.

The optimization completed with the *Prediction Profiler* is based on the assumption that the inputs are controlled to a single value. It is clear to subject matter experts that each of the inputs is likely to vary randomly about a setpoint. You can set the spray rate to 8 but the amount of spray measured from run to run, and even within each run, varies. The actual value is between 7.8 and 8.1. The actual values for a spray rate set to 8 form a distribution about the target setpoint. It might be unlikely that the subject matter experts will know the process settings with such detail as to know the distribution of values. Including this variation in the analysis makes the results much more representative of the actual population of batches that will be produced from the process. Representative simulations with the expected variation from the process inputs provides practical predictions on how well that the process goals will be met for the real process. Use the red triangle menu next to the *Prediction Profiler* header to select *Simulator* and obtain the output shown in Figure 12.8.

Figure 12.8: Granulation Experiments Prediction Profiler with Simulator

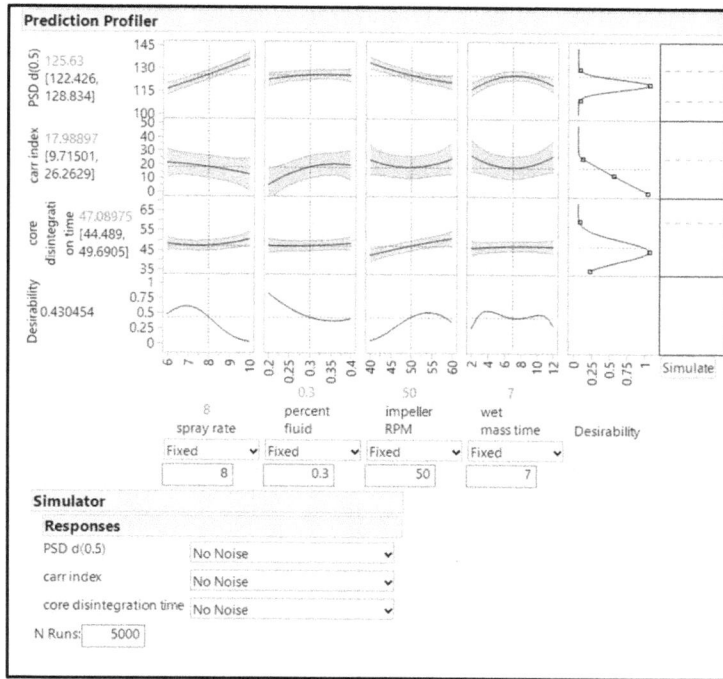

The default for the simulator function is 5,000 runs, which is adequate to estimate populations of drug products and medical devices. The simulator changes the profiler output by adding boxes to the right of the desirability plots and including a *Simulate* button. Notice that choice boxes appear underneath each of the inputs as well. The default setting models an input as a fixed value, as shown in Figure 12.8.

Creating variability for the process inputs can be as detailed as you want. However, a great deal of information will be needed on patterns of variability as the complexity increases. This example assumes that each of the inputs has a normal distribution of variability. The subject matter experts provide the range of values for each of the inputs based on the monitoring of the process, the technical information from the equipment manufacturer, or both.

The spray rate has been identified to have a variability of +/- 0.6 for the range of settings studied by the experiments. A simple way to include this variability in the model estimates is to take the full range of variability (1.2) and divide by 6 to get an estimate for standard deviation (0.2). Recall that the empirical rule states that 99.7% of values from a normal distribution is captured with 6 standard deviations. The assumption made for each of the inputs is that the highest frequency of actual values is very close to the set point, with low frequency of values tailing in either direction forming a normal distribution.

There are more than 20 options within the random function used to simulate variability for each input. You can also simulate the variability by using an expression or through multivariate estimates; however, this example uses the random function for simplicity. Use the function boxes underneath each of the first three inputs to select *Random ▶ Normal* option and enter the values shown in Figure 12.9. Use the red triangle menu next to the *Simulator* header to select the *Spec Limits* option. Enter the goals of the optimization in the *LSL* (lower spec limit) and *USL* (upper spec limit) cells of the table, and click *Save*.

Figure 12.9: Granulation Experiments Prediction Profiler with Simulator Input Values

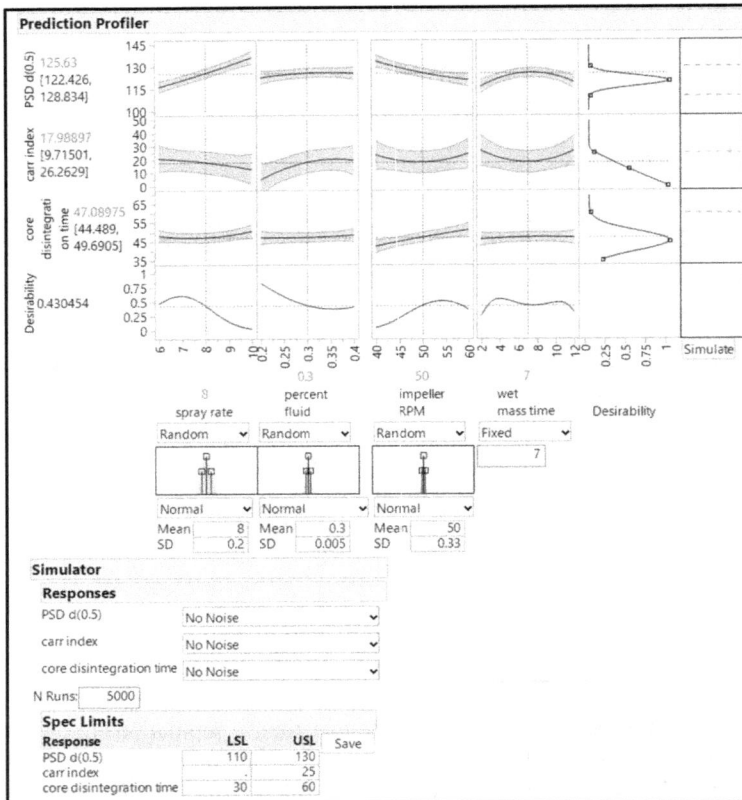

The model includes the expected variability for each of the inputs to compare to the specification limits and determine the percentage of the simulated population (n=5000) that might be outside the requirements. Make sure that the Prediction Profiler is set to the center values for each input shown in Figure 12.9. Click *Simulate,* located in the middle far right of the output, to get the simulated population of results shown in Figure 12.10.

Figure 12.10: Prediction Profiler with 5000 Simulated Runs

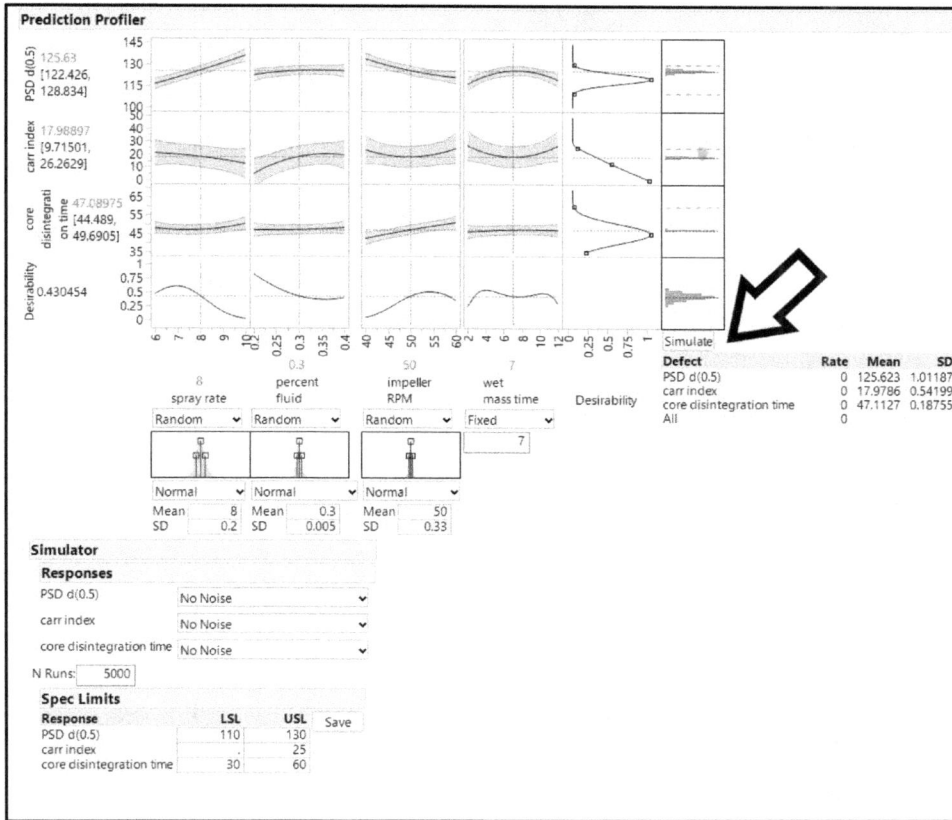

In Figure 12.10, the location of the *Simulate* button is highlighted with a black outlined arrow. The boxes above the button now show the distribution of the 5000 simulated runs, which includes variation in three of four inputs of the process. The segmented horizontal lines illustrate the specification limits, and the green histograms illustrate the expected results. Below the *Simulate* button is a table of summary results. With the process running at the center targets for all inputs, the likely percentage defect is 0. This confirms that the team created a process that is properly targeted. The mean results of the simulation are nearly identical to the profiler mean results for the outputs for the fixed input settings, which are the red values to the left of the profiler matrix.

The simulator can be run for an infinite number of adjustment settings for the inputs. Figure 12.11 illustrates the results after *spray rate, percent fluid*, and *wet mass* are adjusted lower, and *impeller RPM* is adjusted higher. This might be a case when the team is trying to shorten process time and conserve materials to reduce processing costs. Click the *Simulate* button after you adjust the profiler settings. A new set of 5000 runs is created "behind the scenes" by JMP for analysis.

Figure 12.11: Prediction Profiler with 5000 Simulated Runs and Adjusted Inputs

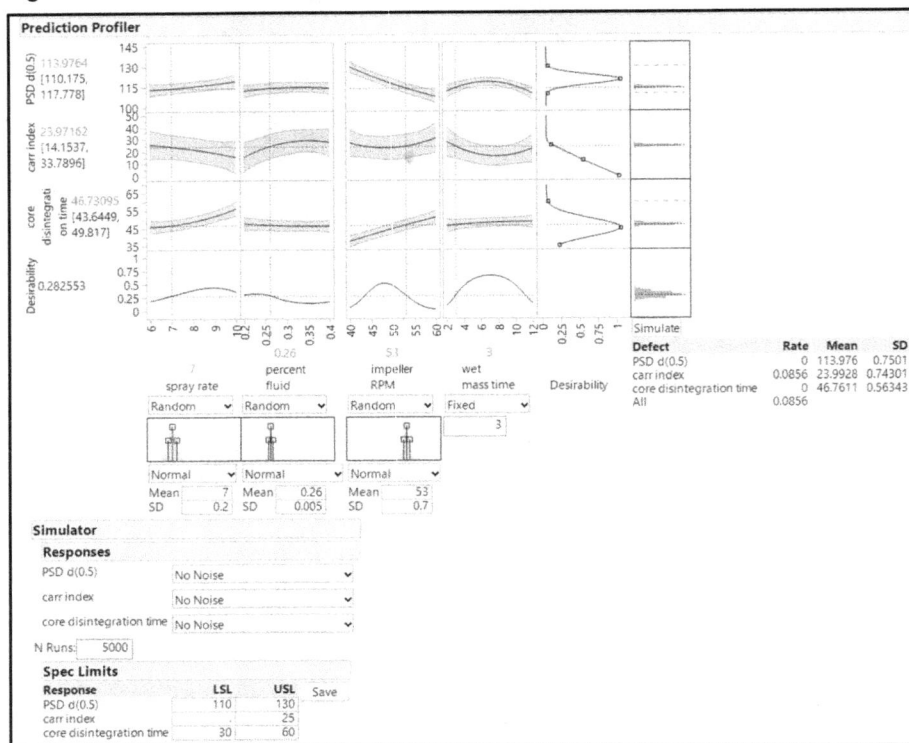

The more aggressive process settings now hold the potential to yield a defect rate of around 8.6%. The Carr Index, a value that estimates the potential compressibility of the mix going through a tablet press, is the lowest performing output of the process. Since the index is not a critical quality attribute (CQA) and is not necessarily definitive of potential compressibility problems, the team might choose the more aggressive settings to minimize production costs. You can study an unlimited number of process scenarios with simulations to predict the risks compared to the practical value of the results. The profiler with simulations is an extremely valuable tool for assessing and quantifying risks, which is the essence of QbD.

There is another important aspect of prediction profiler simulations. The model might indicate process settings that exceed the normal operating range for the processing equipment. The impeller speed is a great example since high speeds might be possible but are not recommended for extended production. The high speeds consume excess energy and accelerate the wear of the equipment. The profiler enables you to fix one or more inputs and optimize the model for the remaining variable inputs.

Complete the following steps to fix an input and redo the model optimization:

1. In the *Prediction Profiler*, enter 53 for the *impeller RPM* setting.
2. Press the Ctrl key, and right-click on the desirability plot for *impeller RPM*. The *Factor Settings* dialog box appears.

3. Select the *Lock Factor Settings* check box shown in Figure 12.12, and click *OK*.

Figure 12.12: Locking Factor Settings

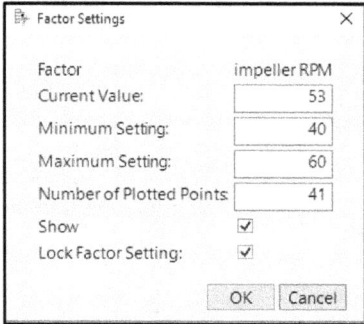

4. Click *Simulate,* located in the middle far right of the output, to get a new simulated population of results for the three variable process inputs.

5. Use the red triangle menu next to the *Prediction Profiler* header, and select *Optimization and Desirability* ▶ *Maximize Desirability* to get the output shown in Figure 12.13.

Figure 12.13: Profiler Simulation Maximized with a Fixed Factor

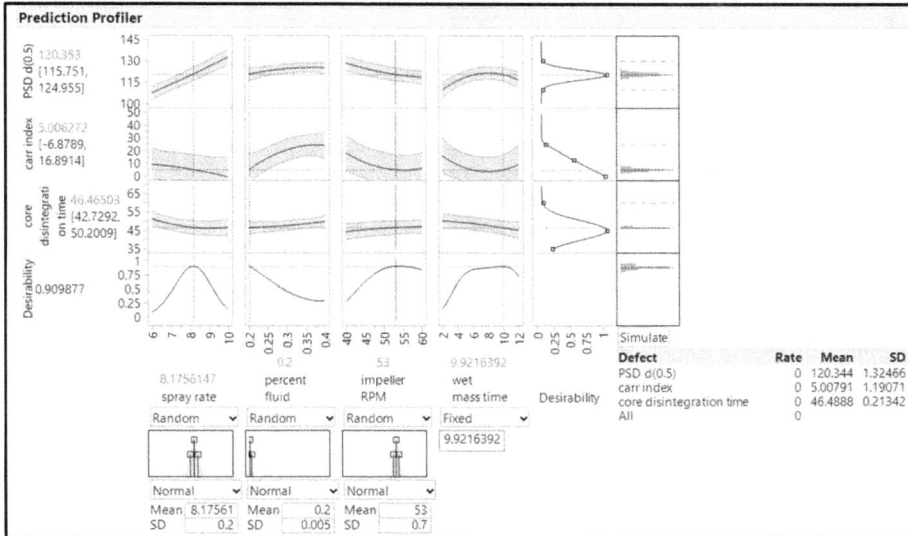

Using a fixed factor provides the application of an additional level of reality to the results. The profiler with the simulated results indicates that *spray rate* is best at the target level, *percent fluid* is best at the lowest setting, and *wet mass time* is best at around 10 minutes. The likely results of the process are excellent since the expected rate of defect for the granulation process is 0.

Including Variations in Responses Within RSM Simulations

The team used the simulator to evaluate the effect of variability of the input settings on the expected population of process results. Simulation modeling gives a realistic picture of risks that are inherent to the process, but one aspect of variability is missing from the simulated population. You know that the measurement process is not perfect and includes variability. Analytical methods are developed with rigor and evaluated for precision before a test method is released for quality control. The measurement process for physical attributes might not be viewed as critical as analytical methods, yet it should be analyzed for measurement uncertainty. The expected amount of measurement error can be included as variation in the responses if it is quantified. It is highly recommended to explore the measurement processes with the methods noted in chapter 7 at minimum to quantify the measurement error and include it in the simulation process to gain the most representative results from modeling.

The team identified that PSD d(0.5) has a measurement standard deviation of 0.05 and that the Carr Index has a measurement standard error of 0.08. They complete the following steps to include the known measurement error for simulations:

1. Go to the *Responses* header within the *Simulator* tools.
2. On the *PSD d(0.5)* line, click the *No Noise* drop-down menu and change the value to *Add Random Noise*.
3. Enter 0.05 as the *Std Dev*.
4. Do the same for the *Carr index* and enter 0.08 as the *Std Dev*.
5. There is no data on the measurement error for core disintegration time so leave that as the default value *No Noise*, as shown in Figure 12.14.
6. Click *Simulator* to update results.

Figure 12.14: Profiler Simulation with Response Noise Added

Simulator			
Responses			
PSD d(0.5)	Add Random Noise	Std Dev:	0.05
carr index	Add Random Noise	Std Dev:	0.08
core disintegration time	No Noise		

The random noise option is the simplest way to add the expected measurement error to the simulated results. You might have noticed that weighted random noise and multivariate noise options are available; you should explore these options to determine how different values affect the results. The basic options are the easiest for explaining results to high-level decision makers. Complex options need to be fully researched to ensure that you can answer all potential questions. It is good practice to compare a simulation run with basic random noise options to more complex options in order to mitigate error and better understand the value of the added complexity. If the higher level of complexity offers little practical value to the results, it is best to default to the simplest options for ease of understanding.

Making Detailed Practical Estimations of Process Performance with a Table of Simulated Modeling Data

The simulation has the options required to offer a realistic picture of what can be expected from the granulation process used by the operations team. The summary results for the simulation offer excellent insight to the risks for producing defects. The team might want an additional level of detail to be able to fully communicate expected risks to the quality team in terms that they understand. It is time to bring the simulated data out from behind the scenes and create a data table. Click on the arrow to the left of the *Simulate to Table* header and click *Make Table* to create a new data table including 5000 simulated runs (shown in Figure 12.15). The simulated data shown in this chapter is different from your results since your output is created with a different seed from a random number generator within JMP. The summary results of the simulated data are not significantly different, so the differences in simulated data are not of great concern.

Figure 12.15: Simulated Process Data Table

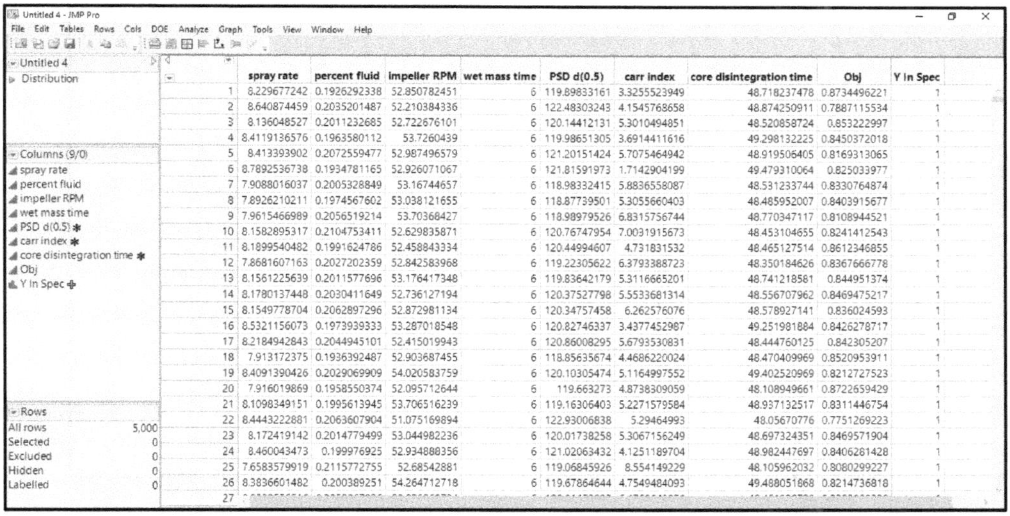

The table of 5000 hypothetical runs of the process includes the variation in three of the inputs and the random noise for two of the three outputs. The data set can now be used for the typical analyses that would be done on real production data to gain more detail on process performance. The capability studies shown in chapter 3 are to be applied to the data to illustrate the quality aspects of the proposed process.

1. Save the open table as *granulation process simulation.jpg* in a file location of your choice.
2. With the simulated data table open, select *Analyze* ▶ *Quality and Process* ▶ *Process Capability*.
3. Select all three outputs and move them to the *Y, Process* box in the *Process Capability* dialog box. Click *OK* to launch the capability output.
4. Use the red triangle menu options to select *Summary Reports* ▶ *Overall Sigma Summary Report*.
5. Right-click in the body of the *Overall Sigma Capability Summary Report* and select the *Columns* option.

6. Deselect the columns that are not in the output shown in Figure 12.16. You need to repeat the previous step and this step until all undesired columns are deselected.

7. Change the view options by hiding output headers to match the output shown.

Figure 12.16: Simulated Process Data Capability Studies

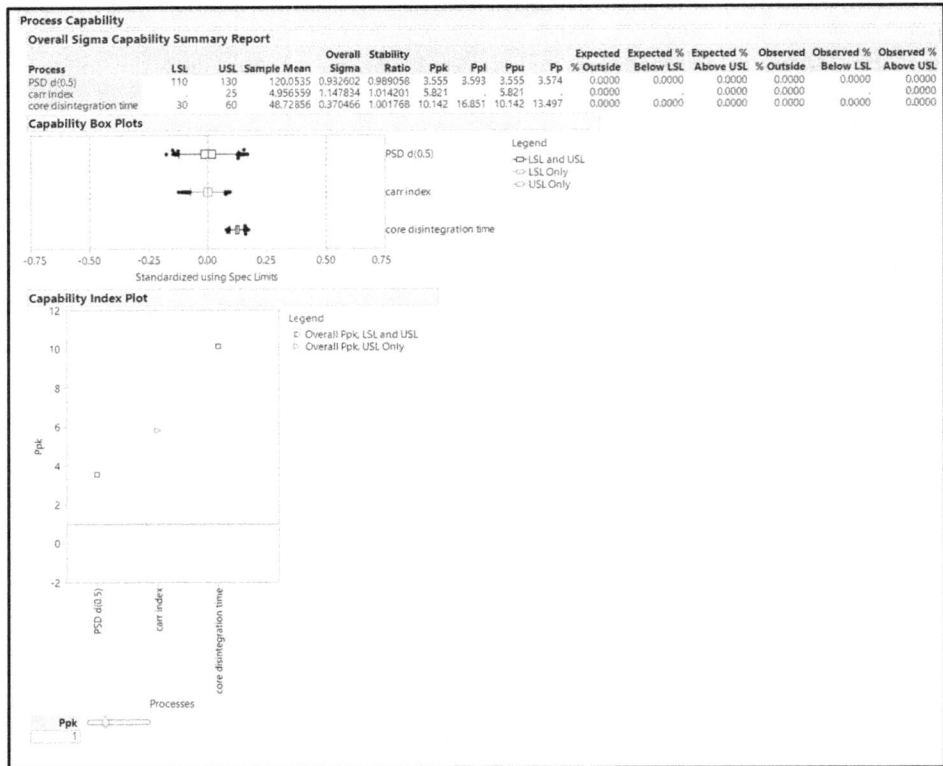

Creating a PowerPoint Presentation from JMP Results

The condensed capability report provides the information in the format that is the most pleasing to the quality group that Erica is working with. She can use an infinite number of options to ensure that the output to be shared contains the desired amount of information. Erica knows that the quality team responds well to information provided with Microsoft PowerPoint slides, although she is not sure why this is the case. Regardless, she decides to save the output as a PowerPoint file so that she can create a presentation of the results. With the output of capability studies open, position your pointer on the bar just underneath the window header until it is highlighted in blue. Click on the highlighted blue bar to see the base menu bar, and select *File* ▶ *Save As*. In the *Save Report As* dialog box, change *File Name* to "Granulation Potential Process Capability" and change *Save As Type* to *PowerPoint Presentation (*.pptx)*. Specify a file location and click *Save* to execute.

Figure 12.17: Save as a PowerPoint Presentation

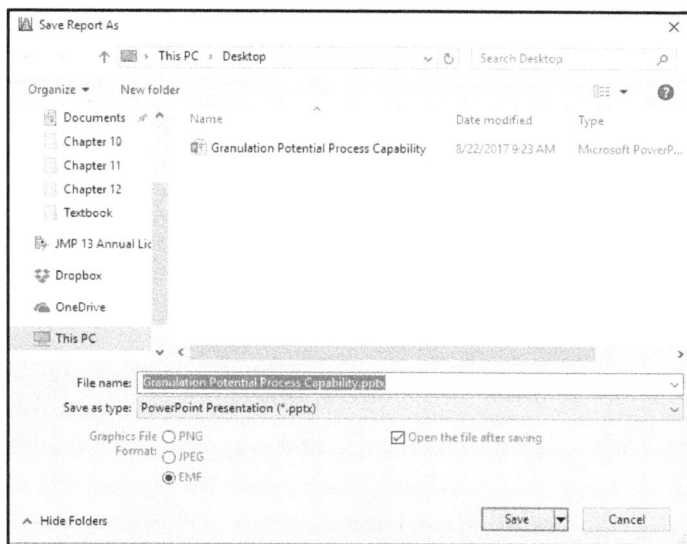

The output converts to an opened PowerPoint presentation. The graphics differ from the format seen in JMP because they have been converted to the format used by Microsoft Office. This example uses the Office 365 version of PowerPoint; your results might differ based on the version of Microsoft Office that you are using. The final version requires some format changes regarding font size, colors, and sizes of outputs.

Practical Conclusions

The Prediction Profiler provides a large amount of practical information that is extracted from a set of experiments. The analyst should consider the fit of the model, the expected power, and the structure of the prediction variance to understand the limitations that exist. The topics covered in this chapter come from models that are robust to the diagnostics used, so you can be confident that error is mitigated. Simulation greatly expands the usefulness of a model. The practical realities of variation in process inputs as well as measurement error of outputs are incorporated to increase the precision of modeling, making it more likely to represent real trends in processing.

Another potential benefit to the modeling of input variability is the ability to analyze the potential effect of equipment upgrades. Subject matter experts must convert technical information to the range of values that are expected for process set points. Once the tightened variability from upgraded equipment is plugged into the modeling simulation, the stakeholders can obtain a realistic estimate of the benefits that might result from the investment. A reduction in defect rates that resulted from the simulation can be quickly converted to annualized savings for a robust projection of the return on the investment.

The amount of information that can be extracted from modeling is limited only by the quality of the model, the accuracy of the information gained on input and output variability, and the creativity and skill of the JMP user. Astute users will catalog the value provided by the JMP tools demonstrated in this chapter, ensuring that high-level decision makers are continuously sold on the value of structured, multivariate experimentation.

Exercises

E12.1—Open the file *Granulation Process Experiment Results Simulation.jmp* used in this chapter and create the least squares model for the three process inputs. The goal of the analysis is to define a process that provides a PSD d(0.5) of 110 to 130 microns, a Carr index less than 25, and a core disintegration time between 30 and 60 seconds. Utilize overlaid contour plots to define an appropriate control space for the granulation process.

E12.2—The definitive screening design for the injection molding process of the plastic trigger for an inhaler helped the team to narrow 10 process inputs to four individual inputs, one interaction, and one squared effect, including the following process inputs:

- Injection time
- Clamping force
- Cooling time

The project stakeholders are very interested in determining a control space for the process to ensure that the goals for part width and part weight are met.

1. Open *inhaler molded component process study DSD 26R.jmp*.
2. Select *Analyze* ▶ *Fit Model* to create a response surface model for the three inputs noted above, with two-factor interactions and squared terms.
3. In the *Effect Summary* section of the model output, remove the unimportant factors until the models are reduced to include only the following: *injection time*, *clamping force*, *injection time*clamping force*, and *clamping force*clamping force*.
4. Add the contour profiler to the output, and add appropriate contour lines. Enter the part width limits of 11.9 to 12.1 and part weight requirements of 3.1 and 3.3. Define a robust control space for the process for the combinations of the two process inputs.
5. Confirm that the points used to create the control space will likely produce good parts.
6. Use the red triangle menu options to save the model analysis as the script "reduced model with 3 inputs."

E12.3—Utilize the techniques covered in this chapter to define an adequate control space for study results of the surgical tray sealing process that are captured in *Burst Testing Experiments 3F 16R CP.jmp*. The process validation reports indicate that the actual line speed varies normally about the input setting with a standard deviation of 1.7, and head energy varies normally about the setting with a standard deviation of 0.8.

1. Use the simulator red triangle menu option in the *Prediction Profiler* to estimate the settings needed to ensure that the sealed trays resist at least 20 inHg before bursting.
2. Create a set of PowerPoint slides that you can use to present the information to the project stakeholders.

E12.4—The process engineers monitored the process records for the plastic injection molding of the inhaler trigger. Their findings indicate that the actual injection time varies normally about the set point with a standard deviation of 0.03, and actual clamping force and cooling time have respective standard deviations of 3 and 0.5. The stakeholders of the medical device project are interested in the robustness of the process set to the optimal input levels.

1. Open *inhaler molded component process study DSD 26R.jmp.*
2. Run the *reduced model with 3 inputs* script to get to the model output.
3. Utilize the profiler simulator to estimate the robustness of the process for meeting the part width and weight requirements noted in exercise E12.2 for the optimized settings, as well as the edges of the control space defined.
4. How would you summarize this information to the project stakeholders?

Chapter 13: Advanced Modeling Techniques

Overview

We live in a period when the practice of data analysis is exploding. Businesses maintain and enhance success by collecting data from operations for analysis to increase the level of intelligence for all processes. This intelligence is practically applied as continuous improvement to meet increasing demands on the business. The pharmaceutical and medical device industries are late comers to the use of analytics, but changes are occurring rapidly in each industry. Organizations find it increasingly difficult to remain competitive without embracing analytics. JMP and JMP Pro are leaders in the field for providing predictive modeling techniques that are easy to use with graphics that are easy to interpret. Basic predictive modeling techniques were introduced in chapter 8 to help narrow down a large number of potential process inputs for the design of multivariate, structured experimentation. This chapter provides some review of predictive modeling techniques and expands on the concepts to further illustrate the value offered.

The Problem: A Shift in Tablet Dissolution

Sudhir's team is faced with the reality that a shift in dissolution values has been noticed for recent batches made for a commercial extended release tablet product. The team is charged with determining the influences on dissolution trends so that the technical team can find ways to mitigate the shift. There is a great deal of processing data available, which the team analyzes with predictive modeling techniques in order to detect the inputs that have the greatest influence on changes in dissolution results.

Preparing a Data Table to Enhance Modeling Efficiency

The road from raw materials to tablets is complex for the subject product. Raw materials are mixed and passed through a high-shear, wet-granulation process and dried in a fluid bed dryer. A high-shear granulator is a type of industrial mixer that incorporates a solution spray during mixing to build the size of particles and blend materials together. The equipment is available in various sizes and configurations.

A cross section of a high high-shear granulator with an impeller at the bottom of the powder bed is shown in Figure 13.1. The granulation process includes a dry bed of powder materials charged into the unit. A large impeller rotates to move the powder while a spray of solution is added over the powder bed. A small chopper is located underneath the spray to declump the material and spread the fluid equally throughout the charge of material.

Figure 13.1: High High-Shear Granulator Cross- Section

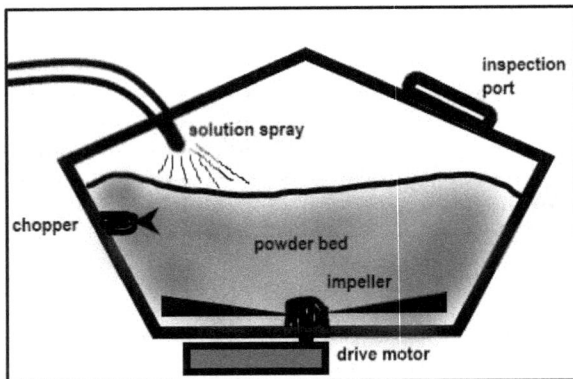

Once the wet granulation is completed, the charge of material is placed in a dryer until the material is at the desired moisture level. The type of unit used in this example is a fluid bed dryer, which pumps hot air from underneath the charge of material so that it fluidizes and dries. Figure 13.2 is a cross-sectional schematic of the drying process.

Figure 13.2: Fluid Bed Dryer Cross Section

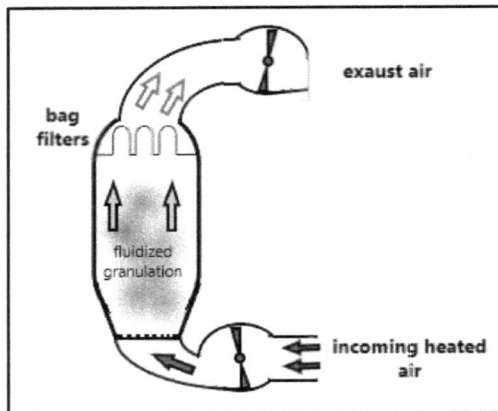

The dried granulation is typically milled to the final size and added to a final blend of materials for table compression. The inputs to the final blend include two granulations, multiple lots of active pharmaceutical ingredient (API) and polymer, and various physical properties of the materials involved. Table 13.1 includes descriptions of the input variables:

Table 13.1: Input Variables

Variable	Details
Moisture of granulation 1 after drying	% moisture sampled from powder bed
Moisture of granulation 2 after drying	% moisture sampled from powder bed

Variable	Details
Hold minutes for granulation 1	start time – end time of granulation process
Hold minutes for granulation 2	start time – end time of granulation process
API age 1	DOM-time of use for first lot of API used
API age 2	DOM-time of use for second lot of API used
API weighted avg of d(0.1)	Weighted average based on % use of d(0.1)
API weighted avg of d(0.5)	Weighted average based on % use of d(0.5)
API weighted avg of d(0.9)	Weighted average based on % use of d(0.9)
Weighted avg of polymer A curve @ target time	Weighted average based on % use of polymer A lots for rheology value at target time
Weighted avg age of polymer A	Weighted average based on % use of age DOM - time of use for first lot of API used
Weighted avg age of polymer B	Weighted average based on % use of age DOM - time of use for first lot of API used

The 12 process inputs are studied to determine whether any have influence on changes in the dissolution profile. The raw data set shown in Figure 13.3 includes all 12 variables, outputs, and other metadata collected from the processing records.

Figure 13.3: Raw Data Table

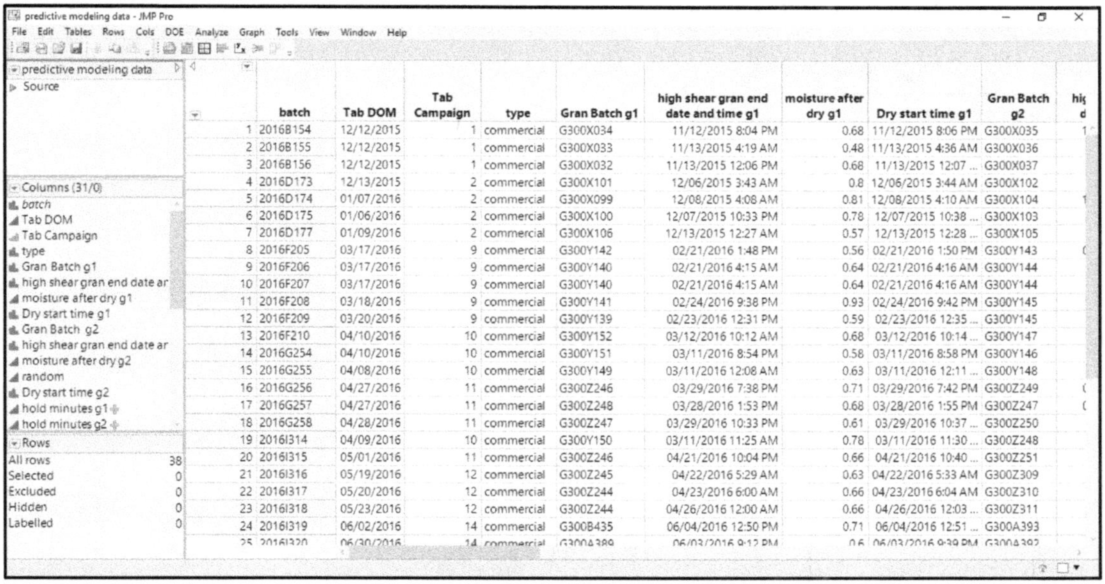

	batch	Tab DOM	Tab Campaign	type	Gran Batch g1	high shear gran end date and time g1	moisture after dry g1	Dry start time g1	Gran Batch g2
1	2016B154	12/12/2015	1	commercial	G300X034	11/12/2015 8:04 PM	0.68	11/12/2015 8:06 PM	G300X035
2	2016B155	12/12/2015	1	commercial	G300X033	11/13/2015 4:19 AM	0.48	11/13/2015 4:36 AM	G300X036
3	2016B156	12/12/2015	1	commercial	G300X032	11/13/2015 12:06 PM	0.68	11/13/2015 12:07 ...	G300X037
4	2016D173	12/13/2015	2	commercial	G300X101	12/06/2015 3:43 AM	0.8	12/06/2015 3:44 AM	G300X102
5	2016D174	01/07/2016	2	commercial	G300X099	12/08/2015 4:08 AM	0.81	12/08/2015 4:10 AM	G300X104
6	2016D175	01/06/2016	2	commercial	G300X100	12/07/2015 10:33 PM	0.78	12/07/2015 10:38 ...	G300X103
7	2016D177	01/09/2016	2	commercial	G300X106	12/13/2015 12:27 AM	0.57	12/13/2015 12:28 ...	G300X105
8	2016F205	03/17/2016	9	commercial	G300Y142	02/21/2016 1:48 PM	0.56	02/21/2016 1:50 PM	G300Y143
9	2016F206	03/17/2016	9	commercial	G300Y140	02/21/2016 4:15 AM	0.64	02/21/2016 4:16 AM	G300Y144
10	2016F207	03/17/2016	9	commercial	G300Y140	02/21/2016 4:15 AM	0.64	02/21/2016 4:16 AM	G300Y144
11	2016F208	03/18/2016	9	commercial	G300Y141	02/24/2016 9:38 PM	0.93	02/24/2016 9:42 PM	G300Y145
12	2016F209	03/20/2016	9	commercial	G300Y139	02/23/2016 12:31 PM	0.59	02/23/2016 12:35 ...	G300Y145
13	2016F210	04/10/2016	10	commercial	G300Y152	03/12/2016 10:12 AM	0.68	03/12/2016 10:14 ...	G300Y147
14	2016G254	04/10/2016	10	commercial	G300Y151	03/11/2016 8:54 PM	0.58	03/11/2016 8:58 PM	G300Y146
15	2016G255	04/08/2016	10	commercial	G300Y149	03/11/2016 12:08 AM	0.63	03/11/2016 12:11 ...	G300Y148
16	2016G256	04/27/2016	11	commercial	G300Z246	03/29/2016 7:38 PM	0.71	03/29/2016 7:42 PM	G300Z249
17	2016G257	04/27/2016	11	commercial	G300Z248	03/28/2016 1:53 PM	0.68	03/28/2016 1:55 PM	G300Z247
18	2016G258	04/28/2016	11	commercial	G300Z247	03/29/2016 10:33 PM	0.61	03/29/2016 10:37 ...	G300Z250
19	2016I314	04/09/2016	10	commercial	G300Y150	03/11/2016 11:25 AM	0.78	03/11/2016 11:30 ...	G300Z248
20	2016I315	05/01/2016	11	commercial	G300Z246	04/21/2016 10:04 PM	0.66	04/21/2016 10:40 ...	G300Z251
21	2016I316	05/19/2016	12	commercial	G300Z245	04/22/2016 5:29 AM	0.63	04/22/2016 5:33 AM	G300Z309
22	2016I317	05/20/2016	12	commercial	G300Z244	04/23/2016 6:00 AM	0.66	04/23/2016 6:04 AM	G300Z310
23	2016I318	05/23/2016	12	commercial	G300Z244	04/26/2016 12:00 AM	0.66	04/26/2016 12:03 ...	G300Z311
24	2016I319	06/02/2016	14	commercial	G300B435	06/04/2016 12:50 PM	0.71	06/04/2016 12:51 ...	G300A393
25	2016I320	06/30/2016	14	commercial	G300A389	06/03/2016 9:12 PM	0.6	06/03/2016 9:39 PM	G300A392

Multiple modeling techniques are to be explored in this chapter. The grouping of the 12 input variables from Table 13.1 will create efficiency for the analysis. Press the Ctrl key and select the 12 column variables for the inputs of interest, as shown in Figure 13.4.

Figure 13.4: Input Variable Selection

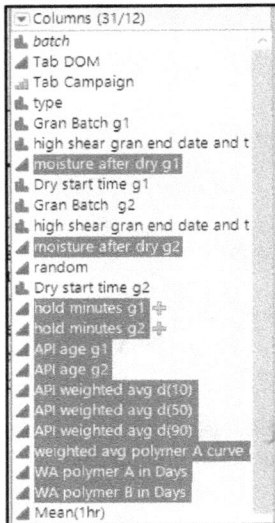

Select *Cols* ▶ *Group Columns* to complete the grouping of the columns, as shown in Figure 13.5.

Figure 13.5: Grouping Columns

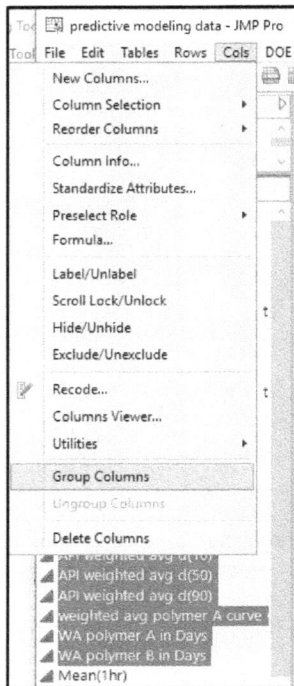

The data table contains an outlined arrow in the *Columns* list for the group, as shown and highlighted in Figure 13.6.

Figure 13.6: Columns with a Group

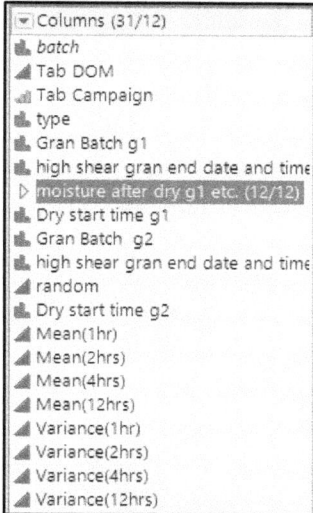

With the 12 inputs of interest grouped, it is easy to try different modeling techniques without the possibility of inadvertently missing an input.

Partition Modeling

In chapter 8, predictive modeling was utilized to identify potential influential inputs from the surgical handle cover processing data that suggest further study via structured experimentation. The partitioning platform proved to be a simple and effective method for modeling a data set with multiple inputs. The advantage of predictive modeling is that you don't have to assess the assumptions for the type of distribution that results from the output. To investigate the partitioning technique, open *predictive modeling data.jmp*, and select *Analyze* ▶ *Predictive Modeling* ▶ *Partition*. The Partition dialog box appears, as shown in Figure 13.7.

Figure 13.7: Partition Dialog Box

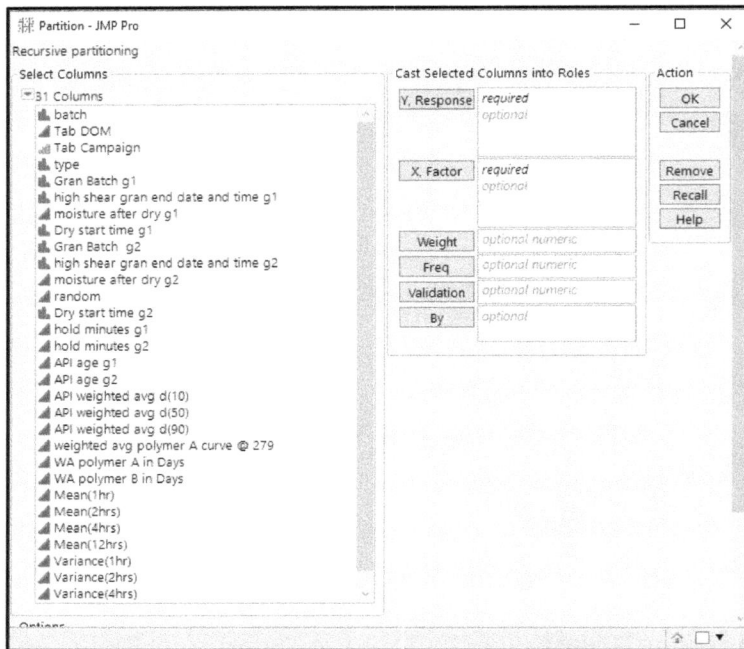

In the *Partition* dialog box, move *Mean(1hr)* to the *Y, Response* box and move *moisture after dry g1 etc.* (grouped variables) to the *X, Factor* box, as shown in Figure 13.8. Notice that the individual input variables appear in the *X, Factor* box when you choose a group.

Figure 13.8: Partition Dialog Box

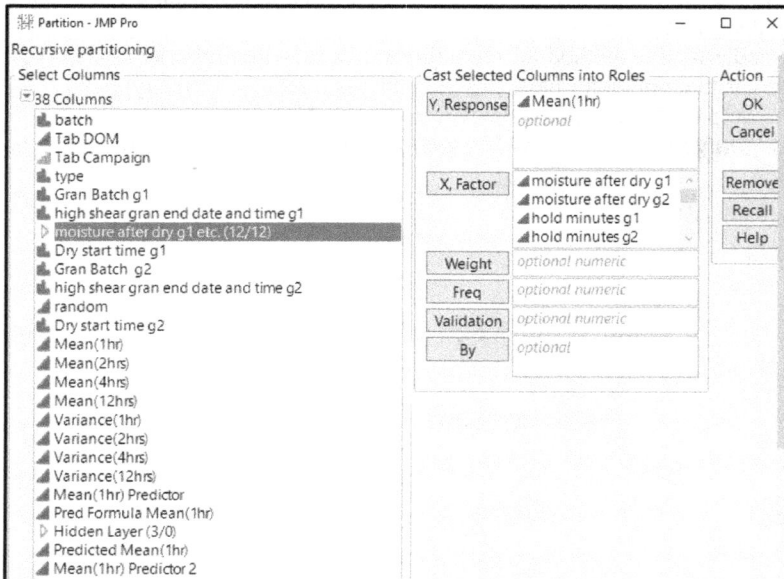

Click *OK* to get the output shown in Figure 13.9.

Figure 13.9: Partition Initial Output

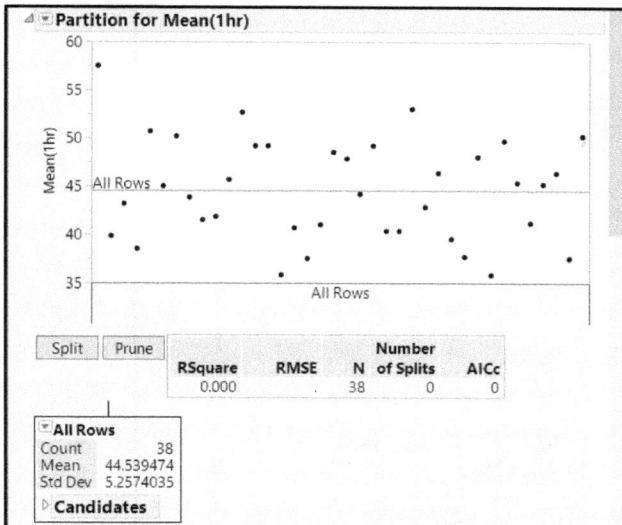

The basic partition plot is the starting point in the modeling journey. Predictive modeling includes the risk of overfitting the model to random variation in results. A great way to mitigate this risk is to include the K-fold cross-validation technique. This technique breaks up the set of data into a finite number (K) of subsets known as folds. The first fold is a small random selection held back from the data. A model is created using the remaining observations, and the model predictions are checked against the fold to determine accuracy. The remaining folding process repeats with a new randomly selected fold, a model made from the remaining data, and model predictions checked with the fold. The results for predictability of the folds are averaged to evaluate the model fit (Rsquare) and compare with the fit of the original model. A stopping rule is built into the algorithm to stop optimization of the fold modeling when the improvement in fit is minimal. Complete the following steps to run partitioning with K-fold cross-validation:

1. Use the red triangle menu next to the *Partition for Mean (1hr)* header and select *K Fold Crossvalidation* from the list of options.

2. In the dialog box, you request the number of folds. This example uses the default value 5. Click *OK* to set the folding parameter.

3. Note that the partitioning control buttons located underneath the dot plot now include a *Go* button next to the *Split* and *Prune* buttons.

4. Click *Go* to get the output shown in Figure 13.10. (Alternatively, you can iteratively click *Split* until the best model fit (R square) is obtained. This example uses the quicker *Go* button.)

Figure 13.10: Partition Output with Cross-Validation

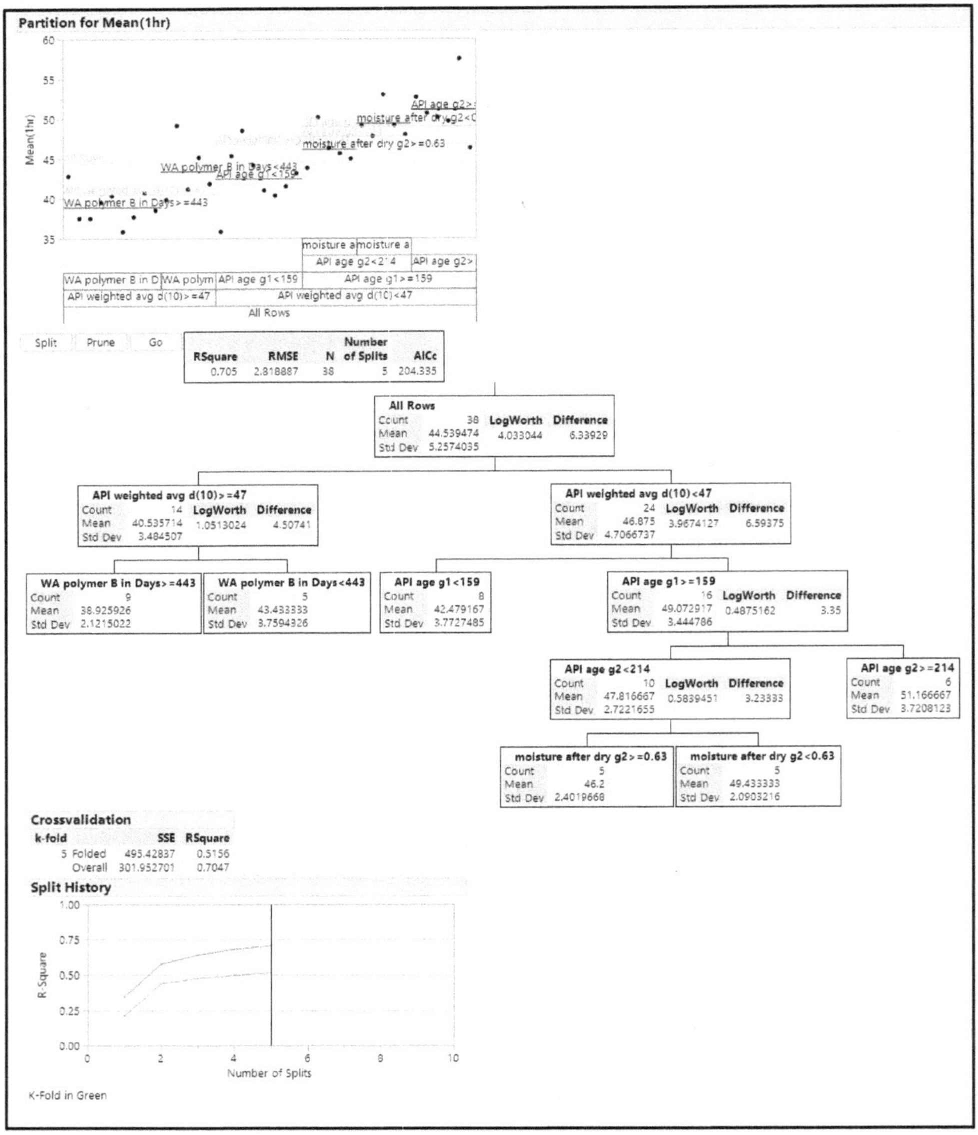

The partitioning with 5-fold crossvalidation resulted in five splits of the data. The split history illustrates that the crossvalidation model obtains the maximum Rsquare fit of 52% at the stopping point. The overall model has an Rsquare fit of 70% fit at the stopping point. The difference between the cross-validated model and the overall model indicates that overfitting is present. As the fit between the overall and folded model diverges, more risk of overfitting is present. The K-fold cross-validation algorithm stops the splits when additional splits of the cross-validated model add minimally to the Rsquare model fit.

The details in the split tree explain the potential relationships that exist between the various inputs and the 1-hour average dissolution results. The first split indicates that the batches with API d(0.10) smaller than 47 microns tend to have 1-hour dissolution results that are more than six percentage points higher than batches with a larger API d(0.5). The second split indicates that batches with API d(0.10) less than 47 microns yield 1-hour dissolution results even greater for API lots that are 159 days or older at the time of tablet manufacturing. An efficient summary of the inputs that might have the most influence is obtained as a table of column contributions. Use the red triangle menu next to the *Partition for Mean (1hr)* header, and select *Column Contributions* from the list of options. The plot illustrating the contribution of each column in order of the amount of influence on the result is shown in Figure 13.11.

Figure 13.11: Column Contributions

Column Contributions				
Term	Number of Splits	SS		Portion
API weighted avg d(10)	1	355.333647		0.4930
API age g1	1	231.880208		0.3217
WA polymer B in Days	1	65.3037478		0.0906
API age g2	1	42.084375		0.0584
moisture after dry g2	1	26.1361111		0.0363
moisture after dry g1	0	0		0.0000
hold minutes g1	0	0		0.0000
hold minutes g2	0	0		0.0000
API weighted avg d(50)	0	0		0.0000
API weighted avg d(90)	0	0		0.0000
weighted avg polymer A curve @ 279	0	0		0.0000
WA polymer A in Days	0	0		0.0000

The column contributions table ranks the portion of potential influence determined through the sum of squares difference (SS). The *API weighted average d(0.10)* has the highest portion of influence at 49.3%. Keep in mind that the amount of influence is a portion of the Rsquare fit of the entire model. Using the conservative estimate provided by the folded model of 52%, you can conclude that the *API weighted average d(0.10)* has approximately 26% total influence on the 1-hour average dissolution; 0.52 total model fit multiplied by 0.493 portion from the API weighted avg d(10) column. The age of the API used in the first granulation makes up a 32.2% portion and 16% of total influence. The weighted average of the age of polymer B has a 9% portion and 4% influence on the dissolution results.

The amount of influence the three input variables have might not seem to be very high. However, you can expect that there is a large amount of random noise in the model due to the weighted averaging of measures from multiple granulations and multiple lots. It is possible that the influence is greater than what the predictive modeling can provide.

You can save model predictions as formula variables and use them to make estimations of the output. The predictions will provide value later in the chapter when you compare models. Use the red triangle menu next to the *Fit Group* header, and select *Save Columns ▶ Save Prediction Formula*. A column of model predictions is added to the data table, as shown in Figure 13.12.

Figure 13.12: Predictive Modeling Data with Partitioning Predictor

	batch	rs)	Mean(12hrs)	Variance(1hr)	Variance(2hrs)	Variance(4hrs)	Variance(12hrs)	Mean(1hr) Predictor
1	2016B154	77.2	100.0	25.0	21.6	14.2	0.4	49.072916667
2	2016B155	78.2	99.0	19.9	18.6	12.6	0.8	49.072916667
3	2016B156	77.3	99.2	1.9	1.9	1.1	0.6	49.072916667
4	2016D173	82.3	100.3	35.5	31.5	16.3	0.7	49.072916667
5	2016D174	78.5	101.3	31.9	29.0	18.7	0.3	49.072916667
6	20^6D175	76.8	101.3	5.8	5.8	3.8	0.3	43.433333333
7	2016D177	76.0	102.3	5.5	5.1	2.8	1.5	49.072916667
8	2016F205	74.3	100.8	3.9	4.7	4.7	1.4	49.072916667
9	2016F206	76.3	102.0	3.8	3.1	3.1	0.0	49.072916667
10	2016F207	76.0	101.3	2.8	3.5	2.0	0.3	49.072916667
11	2016F208	77.2	100.8	11.0	10.2	5.4	0.2	49.072916667
12	2016F209	72.3	101.0	3.0	1.9	1.9	0.4	49.072916667
13	2016F210	76.3	100.8	5.0	28.3	3.5	0.2	49.072916667
14	2016G254	79.3	101.3	6.8	5.4	3.5	0.7	49.072916667
15	2016G255	70.5	100.5	29.9	30.4	25.9	0.7	42.479166667
16	2016G256	74.0	102.7	5.6	6.3	4.8	0.3	49.072916667
17	2016G257	74.7	102.2	10.3	10.3	6.7	0.2	49.072916667
18	2016G258	73.5	101.7	12.9	9.2	7.5	0.2	43.433333333
19	2016I314	78.0	102.3	1.8	3.9	2.4	0.3	49.072916667
20	2016I315	71.0	100.5	3.0	2.3	2.4	0.3	43.433333333
21	2016I316	70.0	100.3	10.2	9.1	4.4	0.3	43.433333333
22	2016I317	70.0	100.8	2.2	1.5	1.6	0.2	43.433333333

Left panel:
- predictive modeling data - JMP Pro
- File Edit Tables Rows Cols DOE Analyze Graph Tools View Window Help
- predictive modeling data ▸ Source
- Columns (32/12)
  - high shear gran end date and time g2
  - random
  - Dry start time g2
  - Mean(1hr)
  - Mean(2hrs)
  - Mean(4hrs)
  - Mean(12hrs)
  - Variance(1hr)
  - Variance(2hrs)
  - Variance(4hrs)
  - Variance(12hrs)
  - Mean(1hr) Predictor ✦ ✱
- Rows
  - All rows 38
  - Selected 0
  - Excluded 0
  - Hidden 0
  - Labelled 0

The study is observational by design, which offers much more limited information than what structured, randomized experimentation can provide. The partition technique provides one limited view into potential relationships. Predictive modeling tends to work best when the analyst employs different techniques to look for inputs that are found to have common influence. Save the data table with the prediction columns to ensure that it is available for model comparisons. Select *File ▸ Save As* and save it as "granulation process models with predictions.JMP" to a location of your choice. You don't need to close the table since you will use it later in the chapter.

Another technique to try is the stepwise selection technique that is available in the fit model platform.

Stepwise Models

The stepwise technique is located in the Fit Model platform. Stepwise modeling involves the repeated creation of models using different numbers of factors, based on a set of rules. One option is to start with one factor and build the number of factors until the best model fit is attained; this technique is known as forward selection. Backward selection starts with a model that includes all factors. Then, you repeatedly create models, reducing the number of factors each time, until you obtain the best model fit. The platform even allows for a combination of forward and backward selection. The data table *granulation process models with predictions.jmp* is used again for this technique. Open that data table, and select *Analyze ▸ Fit Model* to get to the *Model Specification* dialog box, shown in Figure 13.13.

Figure 13.13: Model Specification Dialog Box

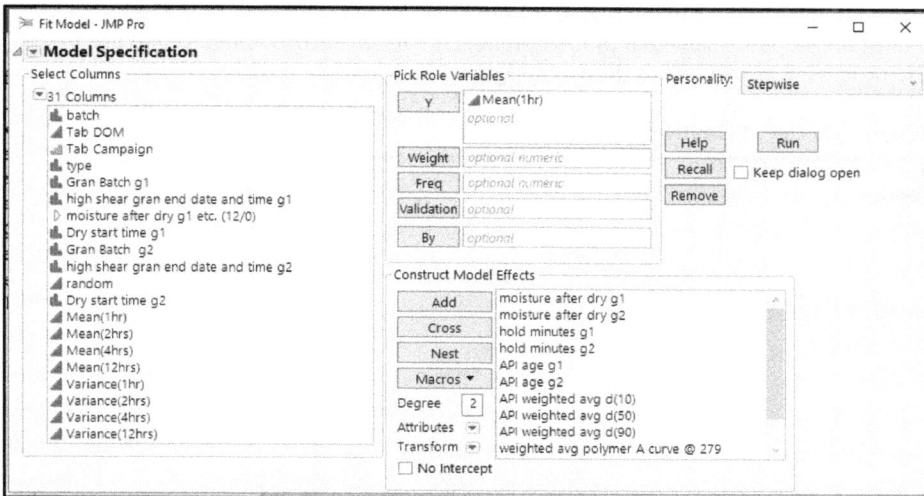

Move *Mean(1hr)* to the *Y* box in the *Pick Role Variable* section, and move *moisture after dry g1 etc.* (a group) to the *Construct Model Effects* box. Change the *Personality* to *Stepwise*, and then click *Run* to get the output shown in Figure 13.14.

Figure 13.14: Stepwise Fit Dialog Box

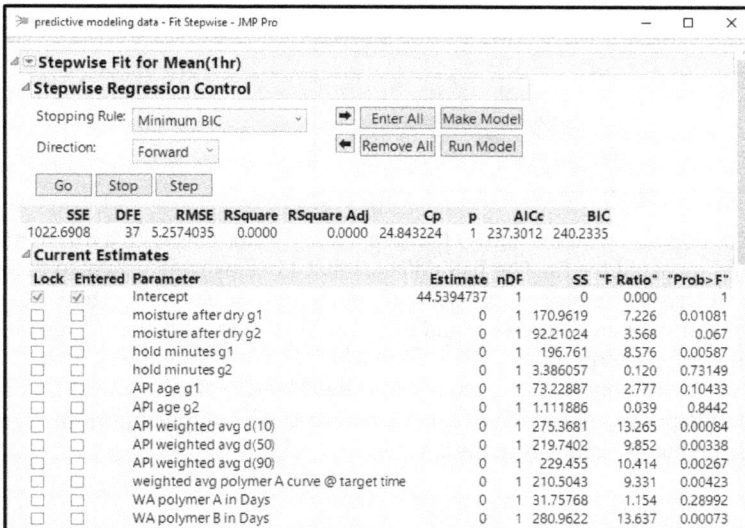

As noted previously, the stepwise technique includes options for stopping as well as for the direction of the steps. By default, the algorithm uses forward selection building factors until the Bayes Information Criterion (BIC) is minimized, indicating an optimum model fit. Up to five options for the stopping rule are available. The Help menu within JMP provides detail about each rule and the considerations for each. You can use forward, backward, or mixed direction. This example uses the default values, but you should try different combinations to compare results.

Like partitioning, the stepwise technique enables you to manually step through the process to see the iterative changes in the model. This example uses the automated stepwise technique; the steps automatically run until the stopping rule goal of the minimized BIC is reached. Click *Go* to use the automated stepwise process. The results are shown in Figure 13.15.

Figure 13.15: Stepwise Model at Best Condition

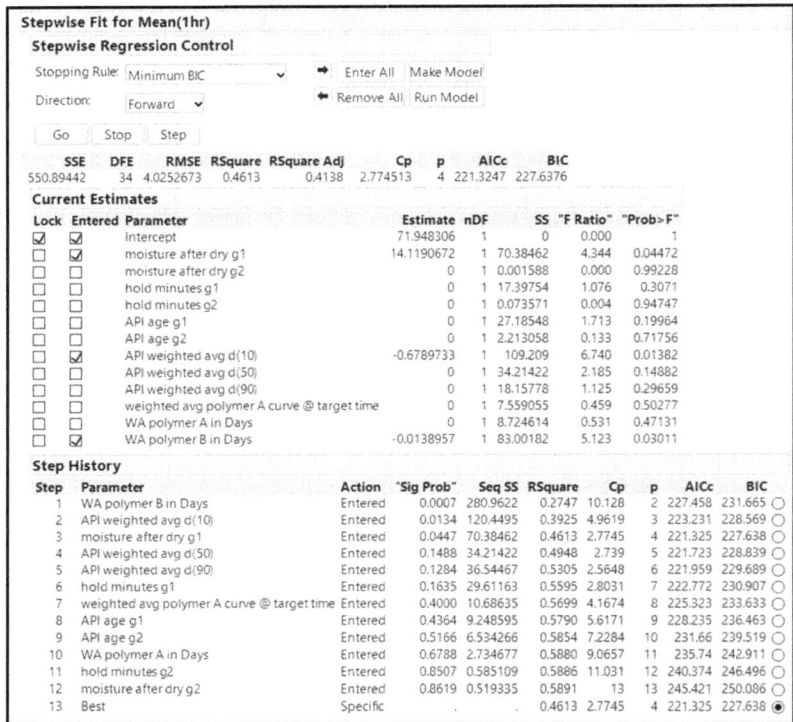

Stepwise Fit for Mean(1hr)

Stepwise Regression Control

Stopping Rule: Minimum BIC ⌄ → Enter All   Make Model

Direction: Forward ⌄ ← Remove All   Run Model

Go   Stop   Step

SSE	DFE	RMSE	RSquare	RSquare Adj	Cp	p	AICc	BIC
550.89442	34	4.0252673	0.4613	0.4138	2.774513	4	221.3247	227.6376

Current Estimates

Lock	Entered	Parameter	Estimate	nDF	SS	"F Ratio"	"Prob>F"
☑	☑	Intercept	71.948306	1	0	0.000	1
☐	☑	moisture after dry g1	14.1190672	1	70.38462	4.344	0.04472
☐	☐	moisture after dry g2	0	1	0.001588	0.000	0.99228
☐	☐	hold minutes g1	0	1	17.39754	1.076	0.3071
☐	☐	hold minutes g2	0	1	0.073571	0.004	0.94747
☐	☐	API age g1	0	1	27.18548	1.713	0.19964
☐	☐	API age g2	0	1	2.213058	0.133	0.71756
☐	☑	API weighted avg d(10)	-0.6789733	1	109.209	6.740	0.01382
☐	☐	API weighted avg d(50)	0	1	34.21422	2.185	0.14882
☐	☐	API weighted avg d(90)	0	1	18.15778	1.125	0.29659
☐	☐	weighted avg polymer A curve @ target time	0	1	7.559055	0.459	0.50277
☐	☐	WA polymer A in Days	0	1	8.724614	0.531	0.47131
☐	☑	WA polymer B in Days	-0.0138957	1	83.00182	5.123	0.03011

Step History

Step	Parameter	Action	"Sig Prob"	Seq SS	RSquare	Cp	p	AICc	BIC	
1	WA polymer B in Days	Entered	0.0007	280.9622	0.2747	10.128	2	227.458	231.665	○
2	API weighted avg d(10)	Entered	0.0134	120.4495	0.3925	4.9619	3	223.231	228.569	○
3	moisture after dry g1	Entered	0.0447	70.38462	0.4613	2.7745	4	221.325	227.638	○
4	API weighted avg d(50)	Entered	0.1488	34.21422	0.4948	2.739	5	221.723	228.839	○
5	API weighted avg d(90)	Entered	0.1284	36.54467	0.5305	2.5648	6	221.959	229.689	○
6	hold minutes g1	Entered	0.1635	29.61163	0.5595	2.8031	7	222.772	230.907	○
7	weighted avg polymer A curve @ target time	Entered	0.4000	10.68635	0.5699	4.1674	8	225.323	233.633	○
8	API age g1	Entered	0.4364	9.248595	0.5790	5.6171	9	228.235	236.463	○
9	API age g2	Entered	0.5166	6.534266	0.5854	7.2284	10	231.66	239.519	○
10	WA polymer A in Days	Entered	0.6788	2.734677	0.5880	9.0657	11	235.74	242.911	○
11	hold minutes g2	Entered	0.8507	0.585109	0.5886	11.031	12	240.374	246.496	○
12	moisture after dry g2	Entered	0.8619	0.519335	0.5891	13	13	245.421	250.086	○
13	Best	Specific	.	.	0.4613	2.7745	4	221.325	227.638	●

The *Stepwise Fit for Mean(1hr)* dialog box window shows the step history as models with a different number of predictors are created and assessed for fit. The list indicates that 11 steps were executed to reduce the BIC to 227.638. The third step of the process created a model with four parameters and the best fit of all. The four parameters consist of the model intercept and the three significant factors. It is useful to run the model to summarize the results for interpretation. Click *Run Model* switch from the stepwise output to a least squares model of the best fit option. You can also use the *Make Model* to add additional customization before you run the model; this is not required for this example. The output shown in Figure 13.16 has been manipulated to hide all but the analysis of variance table.

Figure 13.16: Model Fit Results

Fit Group

Response Mean(1hr)

Analysis of Variance

Source	DF	Sum of Squares	Mean Square	F Ratio
Model	3	471.7964	157.265	9.7061
Error	34	550.8944	16.203	**Prob > F**
C. Total	37	1022.6908		<.0001*

Prior to evaluation of the model, the robustness of the three input model is assessed with a review of the residuals. The Residual by Predicted plot is an easily interpreted graphic used to assess the model. Use the red triangle menu next to the *Response Mean(1hr)* header, and select *Row Diagnostic* ▶ *Plot Residual by Predicted* to add the plot to the output as shown in Figure 13.17.

Figure 13.17: Residual by Predicted (QQ) Plot

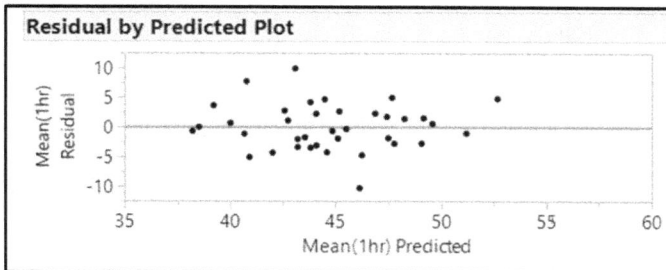

The plot illustrates a random pattern of residuals across the Mean(1hr) Predicted axis. Therefore, the assumptions of equal variance have been met to ensure minimal error in the results. If a curved or cone pattern was evident in the plot, using this model would be risky due to the high potential for statistical error. Residual diagnostics indicate that assumptions for modeling are met. The output shown in Figure 13.18 has been manipulated to hide all but the lack of fit and summary of fit results.

Figure 13.18: Reduced Model Results

Fit Group

Response Mean(1hr)

Lack Of Fit

Source	DF	Sum of Squares	Mean Square	F Ratio
Lack Of Fit	33	550.89442	16.6938	.
Pure Error	1	0.00000	0.0000	**Prob > F**
Total Error	34	550.89442		.
				Max RSq
				1.0000

Summary of Fit

RSquare	0.461328
RSquare Adj	0.413799
Root Mean Square Error	4.025267
Mean of Response	44.53947
Observations (or Sum Wgts)	38

The lack of fit detail for the reduced model created by stepwise regression indicates no evidence that observations of extreme value outside of the model trend are present. The Rsquare adjusted statistic of 0.414 is interpreted as the model explaining just over 41% of the variability in average 1-hour dissolution. The adjusted value must be used when you use multiple modeling techniques in an attempt to find the best possible fit. The fit statistic is adjusted for the number of inputs included in the model, which might differ depending on technique. The model analysis indicates that the technique has produced a reasonable model. Therefore, the effect summary shown in Figure 13.19 is likely to provide information that is not highly biased.

Figure 13.19: Reduced Model Results

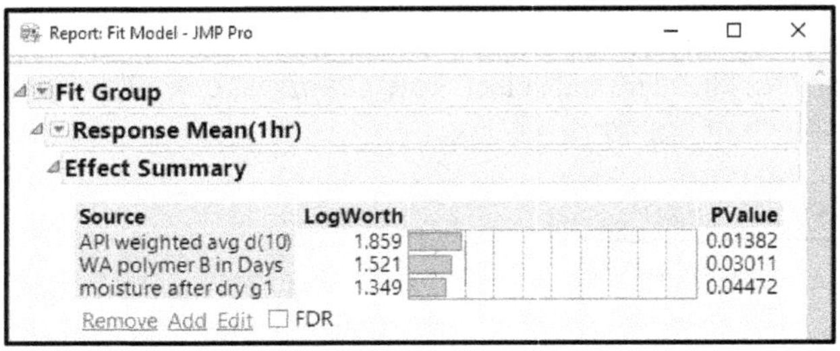

Source	LogWorth	PValue
API weighted avg d(10)	1.859	0.01382
WA polymer B in Days	1.521	0.03011
moisture after dry g1	1.349	0.04472

The Effect Summary for the reduced model is reasonable for explaining differences in the average 1-hour dissolution. The inputs of the API weighted average d(0.10) and the weighted average age of polymer B are like the partitioning results. Moisture after drying is an input with potential influence that was not detected by partitioning.

Continue the good practice of adding the model predictions to the set of data. Use the red triangle menu next to the *Fit Group* header, and select *Save Columns* ▶ *Save Prediction Formula*. A column of model predictions is added to the data table, as shown in Figure 13.20.

Figure 13.20: Predictive Modeling Data with Stepwise Predictor

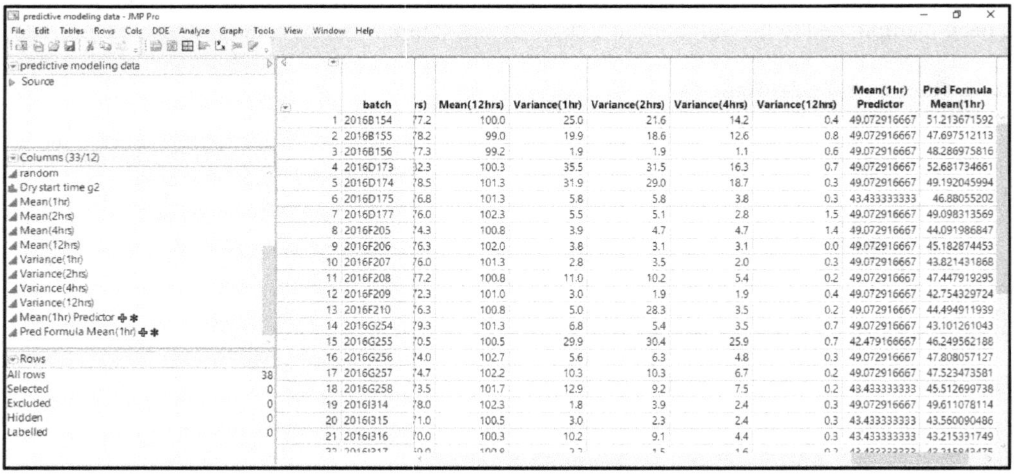

	batch	rs)	Mean(12hrs)	Variance(1hr)	Variance(2hrs)	Variance(4hrs)	Variance(12hrs)	Mean(1hr) Predictor	Pred Formula Mean(1hr)
1	2016B154	77.2	100.0	25.0	21.6	14.2	0.4	49.072916667	51.213671592
2	2016B155	78.2	99.0	19.9	18.6	12.6	0.8	49.072916667	47.697512113
3	2016B156	77.3	99.2	1.9	1.9	1.1	0.6	49.072916667	48.286975816
4	2016D173	82.3	100.3	35.5	31.5	16.3	0.7	49.072916667	52.681734661
5	2016D174	78.5	101.3	31.9	29.0	18.7	0.3	49.072916667	49.192045994
6	2016D175	76.8	101.3	5.8	5.8	3.8	0.3	43.433333333	46.88055202
7	2016D177	76.0	102.3	5.5	5.1	2.8	1.5	49.072916667	49.098313569
8	2016F205	74.3	100.8	3.9	4.7	4.7	1.4	49.072916667	44.091986847
9	2016F206	76.3	102.0	3.8	3.1	3.1	0.0	49.072916667	45.182874453
10	2016F207	76.0	101.3	2.8	3.5	2.0	0.3	49.072916667	43.821431868
11	2016F208	77.2	100.8	11.0	10.2	5.4	0.2	49.072916667	47.447919295
12	2016F209	72.3	101.0	3.0	1.9	1.9	0.4	49.072916667	42.754329724
13	2016F210	76.3	100.8	5.0	28.3	3.5	0.2	49.072916667	44.494911939
14	2016G254	79.3	101.3	6.8	5.4	3.5	0.7	49.072916667	43.101261043
15	2016G255	70.5	100.5	29.9	30.4	25.9	0.7	42.479166667	46.249562188
16	2016G256	74.0	102.7	5.6	6.3	4.8	0.3	49.072916667	47.808057127
17	2016G257	74.7	102.2	10.3	10.3	6.7	0.2	49.072916667	47.523473581
18	2016G258	73.5	101.7	12.9	9.2	7.5	0.2	43.433333333	45.512699738
19	2016I314	78.0	102.3	1.8	3.9	2.4	0.3	49.072916667	49.611078114
20	2016I315	71.0	100.5	3.0	2.3	2.4	0.3	43.433333333	43.560090486
21	2016I316	70.0	100.3	10.2	9.1	4.4	0.3	43.433333333	43.215331749

Save the data table with the prediction columns to ensure that it is available for model comparisons. Select *File* ▶ *Save* to ensure that the predictions are retained. Don't close the table; you will use it later in the chapter.

The next modeling technique to explore is the neural network model.

Neural Network Models

A neural network model is much more complex than the partition or stepwise regression techniques. The model involves a hidden set of flexible nonlinear functions that are set up in a network of up to two layers. Neural networks work well for detecting important predictors when you are not interested in describing the relationship between the inputs and outputs. The neural network is more fully explained in the *Predictive and Specialized Modeling* book, which is available through the Help menu. Users of JMP should familiarize yourself with the details before using the technique.

The following example is a very basic use of a neural network using JMP. Neural network modeling is available in JMP Pro, which includes many ways to customize the model. Make sure that *granulation process models with predictions.jmp* is open. Get started by selecting *Analyze* ▶ *Predictive Modeling* ▶ *Neural* to open the *Neural* window shown in Figure 13.21.

Figure 13.21: Neural Network Selection Dialog Box

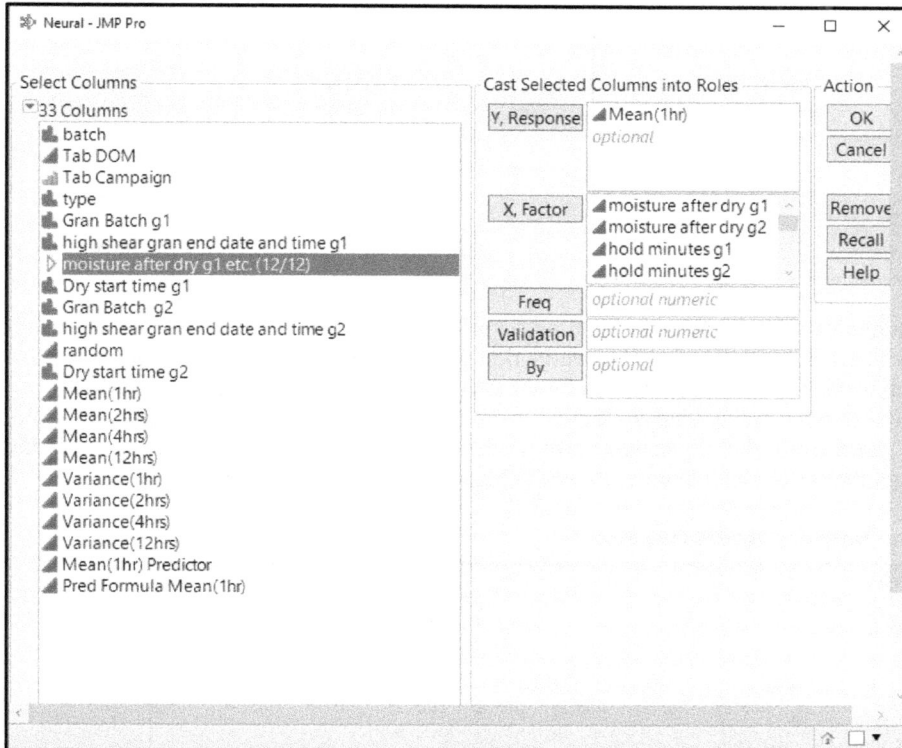

Move *Mean(1hr)* to the *Y,Response* box, and move *moisture after dry g1 etc.* (a group) to the *X,Factor* box. Click *OK* to get the dialog box shown in Figure 13.22.

Figure 13.22: Neural Network Model Launch

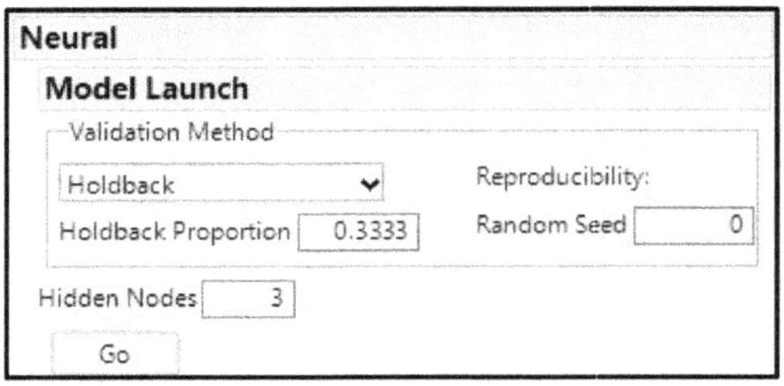

Available for neural network models include the percentage of data that is to be held back for model validation, the ability to set a randomization seed, and the ability to specify the number of hidden nodes. For simplicity, the default options are used for this example. Click *OK* to launch the model and get the results shown in Figure 13.23.

Figure 13.23: Neural Network Model Results

Neural
Validation: Random Holdback
Model Launch
Model NTanH(3)

Training: Mean(1hr) Measures	Value	Validation: Mean(1hr) Measures	Value
RSquare	0.5671704	RSquare	0.6239019
RMSE	3.5550845	RMSE	2.8493853
Mean Abs Dev	2.7676416	Mean Abs Dev	2.4935143
-LogLikelihood	67.182934	-LogLikelihood	32.058544
SSE	315.96564	SSE	105.54696
Sum Freq	25	Sum Freq	13

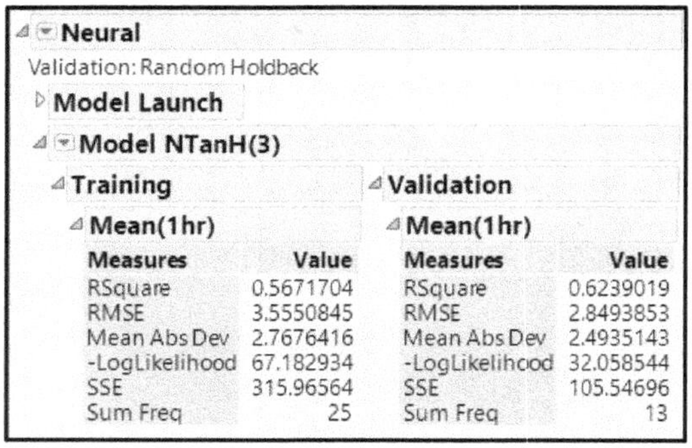

This neural network model has decent fit; the training set model explains 57% (RSquare = 0.567) of the variability in an average 1-hour dissolution. The validation set model explains a similar amount; 62% of the variability in the output. The actual fit is likely to be somewhere between the two fit values. Model diagnostics obtained through residual analysis can help you determine whether the assumptions for model use have been met. Use the red triangle menu next to the *Model NTanH(3)* header to select *Plot Residual by Predicted* and get the set of residual plots shown in Figure 13.24.

Figure 13.24: Neural Network Model Results

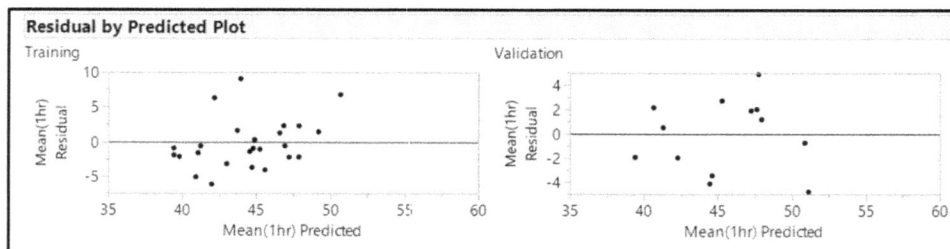

The Residual by Predicted (QQ) plot is added to the output. The random pattern of residuals across the Mean(1hr) Predicted axis for both models indicates that the assumption of equal variance is met. The next step is to evaluate the inputs in order to detect the ones with evidence of leverage on the average 1-hour dissolution; the Profiler plot is a visual representation of leverage. Click on the red triangle menu next to the *Model NTanH(3)* header to select *Profiler* and get the output shown in Figure 13.25.

Figure 13.25: Neural Network Prediction Profiler

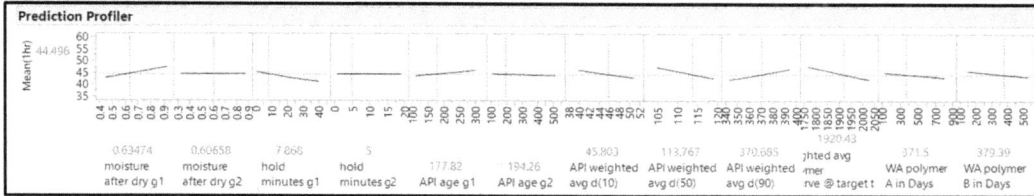

Each of the twelve model inputs has a profile for the average 1-hour dissolution. The profiles with the steepest model lines exert more influence than the others. The following inputs have the most potential influence:

- API weighted average d(0.50)*
- API weighted average d(0.90)
- API weighted average d(0.10)
- weighted average polymer A curve @ target time
- moisture after dry of granulation 1*

Noted in other predictive models

The neural network model detected more potential leveraging inputs than did the previous techniques. Save the predictions from this model to the set of data. Use the red triangle menu next to the *Model NTanH(3)* header and select *Save Formulas*. A column of model predictions is added to the data table, shown in Figure 13.26.

Figure 13.26: Predictive Modeling Data with Neural Network Formulas

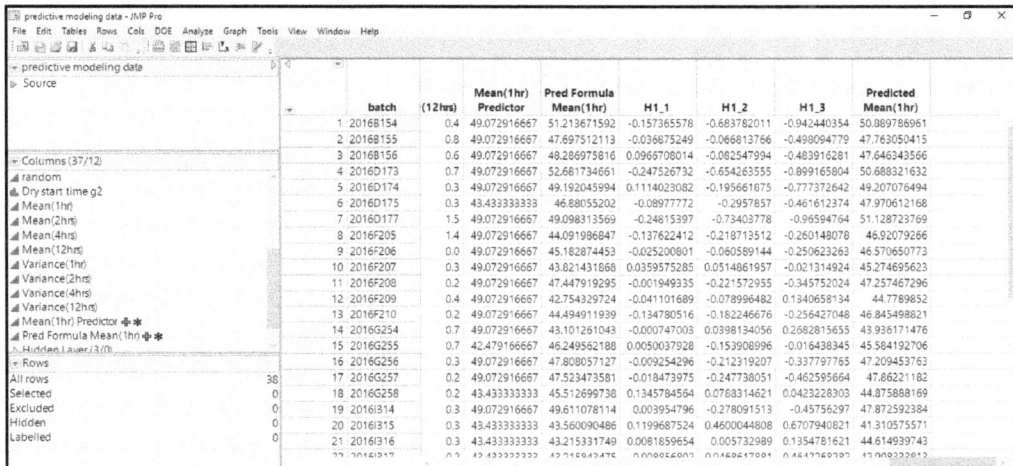

The neural network includes formulas for the portions of the network as well as a predicted value column. Select *File* ▶ *Save* to ensure that the predictions are not lost. You don't need to close the table because you will use it later in the chapter.

The next modeling technique to explore is the bootstrap forest model, which is available only in JMP Pro.

Advanced Predictive Modeling Techniques (Bootstrap Forest) (JMP Pro Only)

One of the most important advantages to having a JMP Pro license is the ability to run cutting-edge predictive models that include statistical simulations. Advanced modeling involves algorithms that can perform complex, recursive functions that offer increased sensitivity to inputs of moderate influence. This section includes the popular bootstrap forest model. By default, the bootstrap forest model creates 100 trees using the 12 process inputs. Essentially, several random selections from the data run partitioning with at least 10 splits. The 100 partition models are averaged together to form a comprehensive model. Users of JMP should read about bootstrap modeling in the *Predictive and Specialized Modeling* book, which is available through the Help menu, before using the technique.

Make sure that *granulation process models with predictions.jmp* is open. Select *Analyze* ▶ *Predictive Modeling* ▶ *Bootstrap Forest* to get the dialog box shown in Figure 13.27.

Figure 13.27: Bootstrap Forest Model Specification

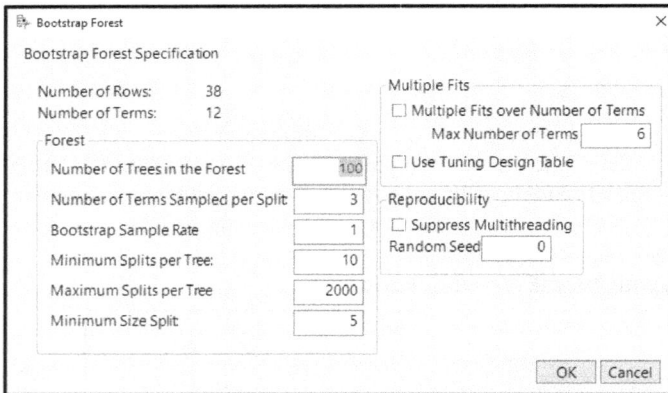

You can explore the many options that are available to tune the model for best results. This example uses the default settings. Click *OK* to get the output shown in Figure 13.28. Keep in mind that the bootstrap forest model is randomized every time you use it. Therefore, your results will not match the output shown in this example.

Figure 13.28: Bootstrap Forest Model

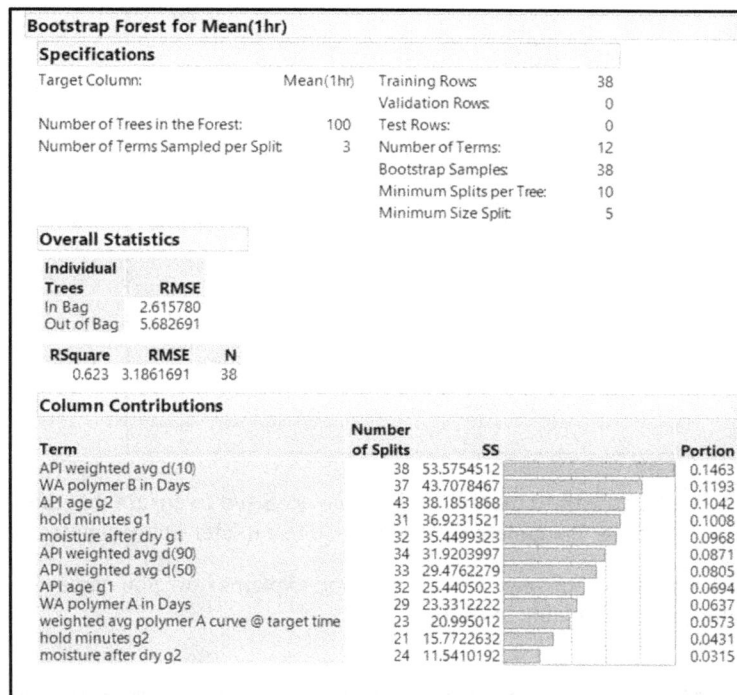

Bootstrap Forest for Mean(1hr)

Specifications

Target Column:	Mean(1hr)	Training Rows:	38
		Validation Rows:	0
Number of Trees in the Forest:	100	Test Rows:	0
Number of Terms Sampled per Split:	3	Number of Terms:	12
		Bootstrap Samples:	38
		Minimum Splits per Tree:	10
		Minimum Size Split:	5

Overall Statistics

Individual Trees	RMSE
In Bag	2.615780
Out of Bag	5.682691

RSquare	RMSE	N
0.623	3.1861691	38

Column Contributions

Term	Number of Splits	SS		Portion
API weighted avg d(10)	38	53.5754512		0.1463
WA polymer B in Days	37	43.7078467		0.1193
API age g2	43	38.1851868		0.1042
hold minutes g1	31	36.9231521		0.1008
moisture after dry g1	32	35.4499323		0.0968
API weighted avg d(90)	34	31.9203997		0.0871
API weighted avg d(50)	33	29.4762279		0.0805
API age g1	32	25.4405023		0.0694
WA polymer A in Days	29	23.3312222		0.0637
weighted avg polymer A curve @ target time	23	20.995012		0.0573
hold minutes g2	21	15.7722632		0.0431
moisture after dry g2	24	11.5410192		0.0315

The results indicate a good fit (Rsquare = 0.623) to explain variation in the average 1-hour dissolution results. The random variability present in the model is relatively low compared with the average results (RMSE = 3.19). Column contributions verify the increased sensitivity offered by advanced modeling. The five inputs with the highest potential to influence an average 1-hour dissolution include:

- API weighted average d(0.10) [14.6%]

- weighted average of polymer B age [11.9%]

- API age from granulation 2 [10.4%]

- hold minutes for granulation 1 [10.1%]

- moisture after dry for granulation 1 [9.7%]

The technique yields influence of inputs that is subtler and shared among additional inputs. Use the red triangle menu next to the *Bootstrap Forest for Mean(1hr)* header and select *Save Columns ▶ Save Prediction Formula*. A column of model predictions is added to the data table, shown in Figure 13.29.

Figure 13.29: Predictive Modeling Data with Bootstrap Forest

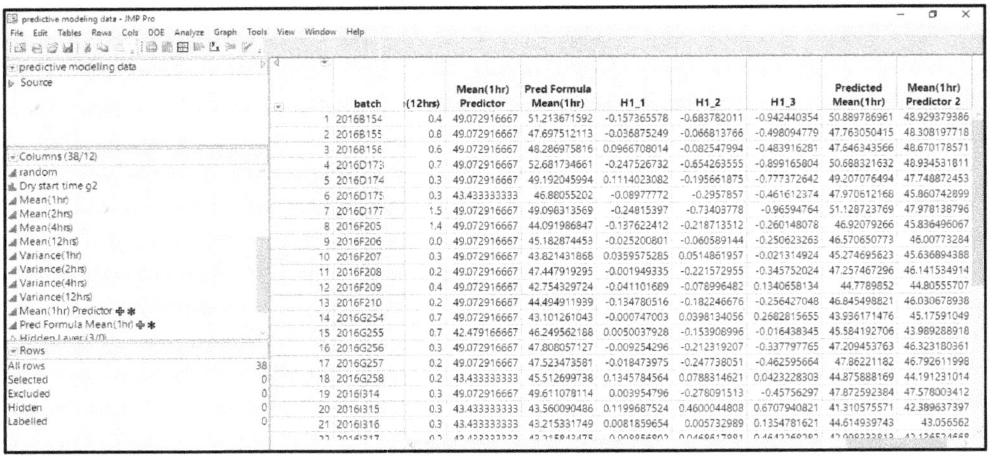

	batch	(12hrs)	Mean(1hr) Predictor	Pred Formula Mean(1hr)	H1_1	H1_2	H1_3	Predicted Mean(1hr)	Mean(1hr) Predictor 2
1	2016B154	0.4	49.072916667	51.213671592	-0.157365578	-0.683782011	-0.942440354	50.889786961	48.929379386
2	2016B155	0.8	49.072916667	47.697512113	-0.036875249	-0.066813766	-0.498094779	47.763050415	48.308197718
3	2016B156	0.6	49.072916667	48.286975816	0.0966708014	-0.082547994	-0.483916281	47.646343566	48.670178571
4	2016D173	0.7	49.072916667	52.681734661	-0.247526732	-0.654263555	-0.899165804	50.688321632	48.934531811
5	2016D174	0.3	49.072916667	49.192045994	0.1114023082	-0.195661875	-0.777372642	49.207076494	47.748872453
6	2016D175	0.3	43.433333333	46.88055202	-0.08977772	-0.2957857	-0.461612374	47.970612168	45.860742899
7	2016D177	1.5	49.072916667	49.098313569	-0.24815397	-0.73403778	-0.96594764	51.128723769	47.978138796
8	2016F205	1.4	49.072916667	44.091986847	-0.137622412	-0.218713512	-0.260148078	46.92079266	45.836496067
9	2016F206	0.0	49.072916667	45.182874453	-0.025200801	-0.060589144	-0.250623263	46.570650773	46.00773284
10	2016F207	0.3	49.072916667	43.821431868	0.0359575285	0.0514861957	-0.021314924	45.274695623	45.636894388
11	2016F208	0.2	49.072916667	47.447919295	-0.001949335	-0.221572955	-0.345752024	47.257467296	46.141534914
12	2016F209	0.4	49.072916667	42.754329724	-0.041101689	-0.078996482	0.1340658134	44.7789852	44.80555707
13	2016F210	0.2	49.072916667	44.494911939	-0.134780516	-0.182246676	-0.256427048	46.845498821	46.030678938
14	2016G254	0.7	49.072916667	43.101261043	-0.000747003	0.0398134056	0.2682815655	43.936171476	45.17591049
15	2016G255	0.7	42.479166667	46.249562188	0.0050037928	-0.153908996	-0.016438345	45.584192706	43.989288918
16	2016G256	0.3	49.072916667	47.808057127	-0.009254296	-0.212319207	-0.337797765	47.209453763	46.323180361
17	2016G257	0.2	49.072916667	47.523473581	-0.018473975	-0.247738051	-0.462595664	47.86221182	46.792611998
18	2016G258	0.2	43.433333333	45.512699738	0.1345784564	0.0788314621	0.0423228303	44.875888169	44.191231014
19	2016I314	0.3	49.072916667	49.611078114	0.003954796	-0.278091513	-0.45756297	47.872592384	47.578003412
20	2016I315	0.3	43.433333333	43.560090486	0.1199687524	0.4600044808	0.6707940821	41.310575571	42.389637397
21	2016I316	0.3	43.433333333	43.215331749	0.0081859654	0.005732989	0.1354781621	44.614939743	43.056562

The bootstrap forest data set includes predicted value columns. Select *File* ▶ *Save* to ensure that the predictions are retained. You don't need to close the table because you will use it later in the chapter.

You have created several models and evaluated each one. The next section explains how you can evaluate all models together with a model comparison, available in JMP Pro.

Model Comparison (JMP Pro only)

This chapter has explained four predictive modeling techniques used to detect important predictors of average 1-hour dissolution from 12 potential process inputs. The data table *granulation process models with predictions.jmp* includes predictions from the following modeling techniques:

- Partitioning (five important inputs detected)
- Stepwise Regression (three important inputs detected)
- Neural Network (five important inputs detected with other subtle inputs)
- Bootstrap Forest (five important inputs detected with other subtle inputs)

You can access the mathematical details for each technique by selecting the column to use for a specific prediction and opening the formula in the column properties. However, the complex techniques of neural networking and bootstrap forest are very difficult to interpret from the formulas that are included.

Make sure that *granulation process models with predictions.jmp* is open in order to use the model comparison tool. Select *Analyze* ▶ *Predictive Modeling* ▶ *Model Comparison* to get the dialog box shown in Figure 13.30.

Figure 13.30: Model Comparison Setup

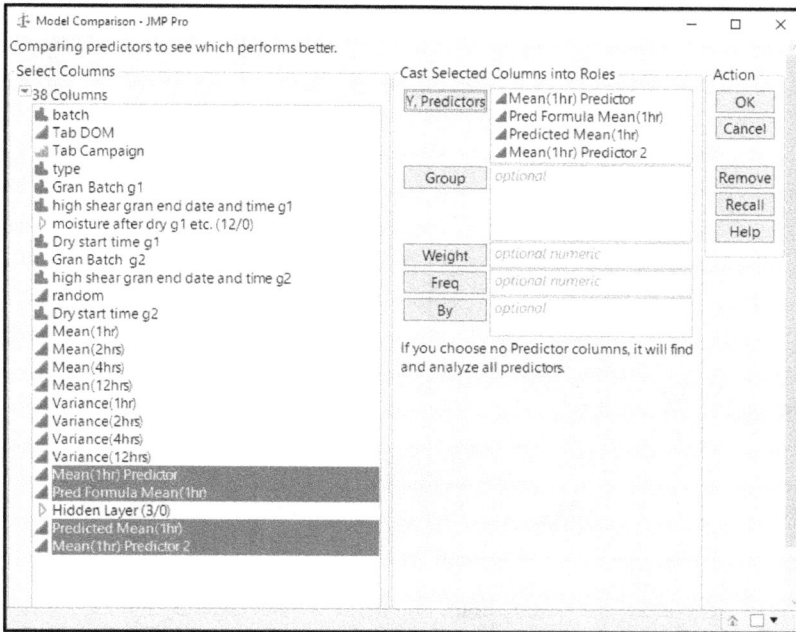

Move each of the predictor columns for the four modeling techniques to the *Y, Predictors* box, and click *OK* to get the output shown in Figure 13.31.

Figure 13.31: Model Comparison Output

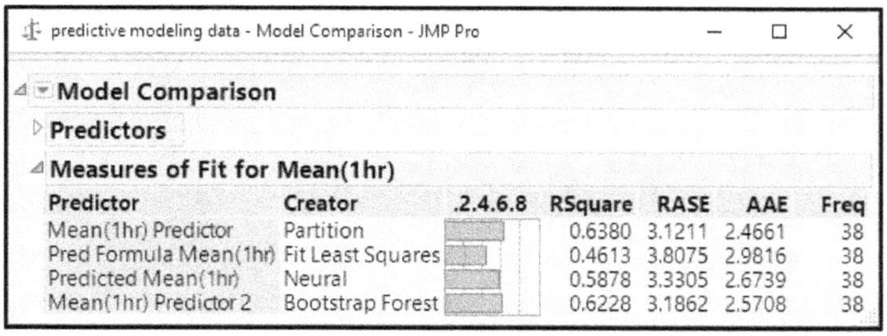

The Measures of Fit for Mean(1hr) summary provides comparisons with multiple fit statistics. The Rsquare values noted are adjusted for the number of inputs that were included in each model and are evaluated first. The highest Rsquare fit of 63.8% was obtained by the partition technique. The square root of the mean square prediction error (RASE=3.12) is lowest for the partition model, which is evidence of increased precision in estimates from reduced random variability. The average absolute error (AAE=2.47) is also lowest for the partition model and is evidence of increased precision.

Next, look for the random spread of points needed to meet modeling assumptions by looking at residual plots. Use the red triangle menu next to the *Model Comparison* header and select *Plot Actual by Predicted* and *Plot Residual by Row* to get the output in Figure 13.32.

Figure 13.32: Model Comparison Diagnostics

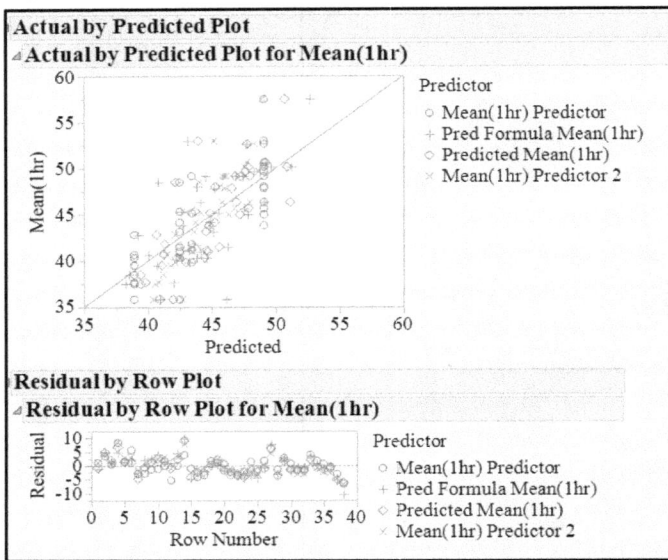

In the residual plots, all four techniques are identified by color and symbol as defined by the legend to the right of the plots. To highlight the residual pattern for a specific technique, click on the marker in the legend that corresponds to that technique. The selected technique shows up in full color, with the markers for all other techniques muted. The pattern of residuals for all techniques does not indicate non-random patterns. Therefore, residual analysis cannot be used to deselect specific models.

Now, evaluate the prediction profiler for the four models. Click on the red triangle menu next to the *Model Comparison* header and select *Profiler* to get the output in Figure 13.33.

Figure 13.33: Model Comparison Profiler

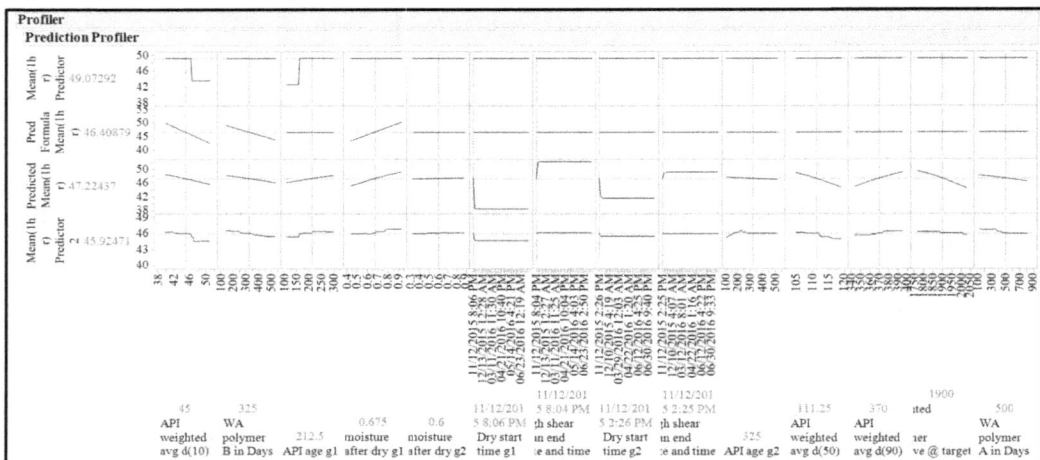

The model profiler offers a view of the potential influence for each of the inputs across the four techniques. The inputs with influence across all techniques include the weighted average of API d(0.10) and the weighted average of age of polymer B. Moisture after drying of granulation 1 indicates potential influence across three techniques. Other inputs have been detected as important by specific models and are not commonly shared. Shared influence among multiple modeling techniques adds to the robustness of conclusions that are made from the analysis.

Practical Conclusions

Predictive modeling provides a robust method to analyze observational data that includes several inputs. The technique does not require the data to be from a specific distribution shape and there can be more input variables than the number of observations (which causes significant issues for typical least squares modeling). A significant challenge for predictive modeling is the potential for overfitting. Overfitting is when the technique is so sensitive that random variability falsely indicates that an input is important. Validation techniques available in JMP help mitigate overfitting as well as support the use of multiple techniques. Another method that mitigates overfitting is to obtain new data for the process and repeat modeling techniques to determine whether similar results are obtained. Model results that are free of overfitting enable the analyst to make appropriate conclusions about important predictors for the output.

Sudhir utilized four different techniques and found that there are between two and three inputs that have a high potential for influence on changes in average 1-hour dissolution. This is great news since the team started by needing to consider 12 input variable that might have influence. Focusing on a small number of high-potential contributing factors greatly enhances the efficiency of the process of investigating trends and determining preventive and corrective action.

It is clear from the results that efforts need to be put into evaluating the API source, specifically to looking for changes in the particle size distribution. The team presented the information to the API source and learned that changes in the process were implemented recently to improve yield. The changes to the process were believed to be minor and were not formally reported. However, the modeling results now provide strong evidence of a related change in a critical quality attribute for the end tablet product.

The age of polymer B is another input that required further investigation. The team found that the procurement group negotiated a deal with the supplier, reducing costs by increasing the size of the containers used and amount shipped. Sudhir's team worked with the analytical group to determine that the rate of activation of the polymer after water is added slows as the age of the material is increased. The team worked with the quality assurance team to shorten the retest period for stored lots to increase monitoring for changes in activation.

Exercises

E13.1 — A new tablet formula has been developed and scaled up to be ready for commercial production. Unfortunately, the hardness of the tablets has dropped in recent batches and is becoming a quality concern. Process data has been compiled for several batches of the two tablet products with formulations that differ only by flavor. You have been asked to analyze the information for relationships between the inputs and the average tablet hardness and range in tablet hardness outputs.

1. Open *tablet production data.jmp*.
2. Select the *columns product, type of batch, batch size, lube time, mix room temp (F), mix room RH%, tote used, hold time (days), press speed (RPM), weight range (g), pre-compress thick*, and *side of press* to group them together.
3. Use the techniques described in this chapter to create predictive models for average tablet hardness and tablet hardness range.
4. Create a summary to explain the chosen model, why you feel it is robust, and which inputs should be focused on to make improvements in tablet hardness outputs.

E13.2 — Basic predictive modeling was executed in chapter 8 to determine which process inputs have the highest potential of a relationship to the variation in the outputs. Run the predictive models again with the new techniques noted in this chapter and determine whether the advanced predictive models provide more reliable information than the basic models.

1. Open *burst testing with process factors.jmp*.
2. Run the advanced predictive modeling techniques noted in this chapter.
3. Compare the results to the model analysis from chapter 8. How might you update the report of results made to the project stakeholders?

E13.3 — Repeat predictive modeling for the data in *mix and compression process data.jmp* by using the neural network modeling technique.

1. What is the AICc statistic? Is there evidence that the neural network model is better than the basic partitioning (smaller value for AICc)?
2. Is the list of most important inputs the same as the basic partitioning?
3. Does the neural network model add any robustness to specific inputs?
4. How would you update the project stakeholders with this information?

Chapter 14: Basic Mixture Designs for Materials Experiments

Overview

A simple materials experiment was executed in chapter 9 that involved the study of materials variables as independent factors and that used a slack variable to maintain a fixed amount of total materials. Such designs are easy to design, analyze, and interpret. However, the chance for error is increased because you can never know whether the changes are due to the model or the varying amounts of the slack variable. Mixture designs offer an approach that deals with materials variables as proportions of the total mix. A change of the proportional amount of one variable requires changes in other variables to maintain the proportional materials total of 100%.

Variables in mixture designs are named mixture components because dependency among the components is assumed. The simplest of mixtures involves three mixture components, each allowed to vary from 0% to 100%. Such designs are rare in the pharmaceutical industry because each material tends to contribute a specific property to a formula, so materials can rarely be allowed to vary the full range. This chapter explains a design that includes limits on the minimum and maximum proportional levels for each mixture component. Such designs are referred to as extreme vertices designs and are relatively easy to set up and execute in JMP.

The Problem: Precipitants in a Liquid Drug Solution

The consumer's affairs group of a pharmaceutical manufacturer has growing concern over a liquid drug solution product. Customer complaints have been received regarding solid precipitants forming on the bottom of the bottle. The problem seems to be at its worst after being refrigerated; the product is supposed to be stored at room temperature but apparently some people refrigerate it. Jon is the manager of a group of formulations scientists assigned to reformulate the product in order to address the problem. Based on years of experience and subject matter expertise, the scientific team narrowed down a three material solvent system that meets all requirements of the product. Each of the three materials of the solvent system are known to affect the amount of active ingredient that remains in solution, which is measured with assay values. Jon knows that the total of the three ingredients must be fixed per unit volume since the dose amount (in ml) is to remain constant. The team will create solutions that have the maximum amount of dissolved active ingredient pharmaceutical ingredient (API) in solution and then evaluate changes in the assay values, which are expressed as a percentage of the theoretical full amount. A mixture design will offer a powerful analysis to determine how proportional differences between the three materials might be related to changes in assay for product stored at both room temperature and in refrigerated environments for 24 hours.

Design of Mixture Experiments

Previous chapters provided step-by-step instructions for how to create an experimental design via the custom designer in the design of experiments (DOE) menu in JMP. The team must define the outputs and inputs to the study before they create the experimental design.

The outputs of interest for the set of experiments include the following:

assay RT
the percentage amount of active ingredient measured from the drug solution kept at real-time conditions (temperature of 72° F/~22° C), with the goal of maximizing the amount

assay cool
the percentage amount of active ingredient measured from the drug solution kept at refrigerated conditions (temperature of 38° F/~3° C), with the goal of maximizing the amount

density
the density of the solution in mg/ml, with the goal of matching a target amount

The materials inputs, known as mixture components when using mixture designs, for the set of experiments are listed below. The viable ranges are based on subject matter expertise and experience of the scientific team.

purified water
the amount can be varied between 0% and 75% of the mixture

solubilizer
the amount can be varied between 0% and 65% of the mixture

surfactant
the amount can be varied between 0% and 35% of the mixture

There is one special consideration for two of the materials:

design constraint
the combined amount of water and surfactant must be no less than 75% of the mixture for the active ingredient to go into solution properly

You have learned that controls are very important to successfully execute a set of structured experiments. The list of controls can be extensive, but the following are the two most important controls:

process settings
an equivalent process must be in place for all runs

lots of materials
all mixes come from the same lot of the API, and from the same lot of each of the mixture components

The outputs, inputs, and controls for the project have been defined. The JMP Home window must be open to get started. Select *DOE* ▶ *Classical* ▶ *Mixture Design* and enter the responses and the factors shown in Figure 14.1.

Figure 14.1: Mixture Design Dialog Box

Response Name	Goal	Lower Limit	Upper Limit	Importance
assay RT	Maximize	.	.	.
assay cool	Maximize	.	.	.
density	Match Target	.	.	.

Factors — Add [1] Mixture / Remove Selected

Name	Role	Values	
purified water	Mixture	0	0.75
solubilizer	Mixture	0	0.65
surfactant	Mixture	0	0.35

Specify Factors

Specify desired number of factors. Double click on a factor name or setting to edit it.

Continue

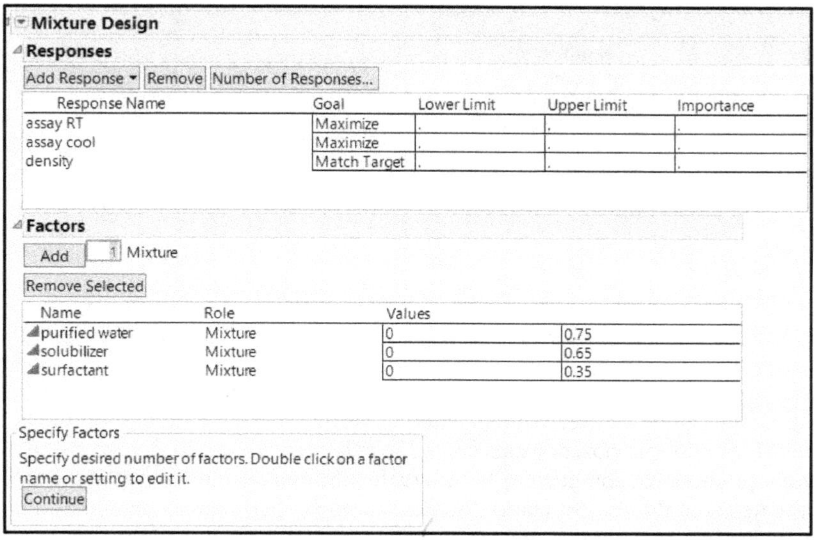

Each of the responses have a target value of interest and can vary above and below the target. The limits are not known at the time of the design and will be added during the analysis of results. The factors (mixture components) are the defined amounts included in Figure 14.1.

It is good practice to save the responses and factors as data files in case you need to explore more model options due to feedback on the design from subject matter experts. Use the red triangle menu next to the *Mixture Design* header and select *Save Responses* to create a data file. Select *File* ▶ *Save As*, name the file "liquid mixture responses," and save it to a location of your choice. Use the red triangle menu next to the *Mixture Design* header and select *Save Factors* to create a data file. Select *File* ▶ *Save As* and name the file "liquid mixture factors". Go back to the *Mixture Design* dialog box and click *Continue* to add the design type choices to the output, as shown in Figure 14.2.

Figure 14.2: Choose Design Type

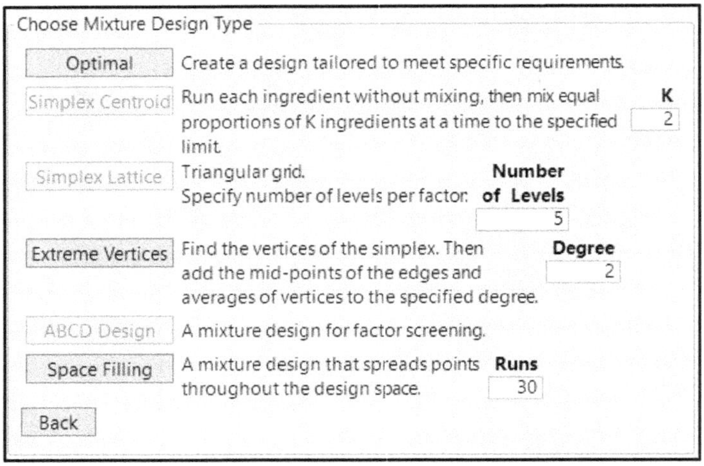

The output includes three available options for the example mixture design. The limits to proportional changes for the mixture components can be studied with an extreme vertices design. Another option is a space filling design, which involves multiple observations spread evenly throughout the design space in order to make the most accurate predictions. Jon decides to utilize the optimal design since it requires the fewest resources and meets the needs of the project team. The goal for the experiments is to determine whether a difference exists between the product solutions stored in room temperature and those in the refrigerated environments. Click the *Optimal* button in the *Choose Mixture Design Type* area of the dialog box, shown in Figure 14.2.

Recall the special consideration of the total amount of water and surfactant, which must be at least 75% of the total mix. Such a consideration is added to the model design as a linear constraint. Select the radio button option *Specify Linear Constraints*, and click *Add*. Enter 1 in both the *purified water* and *surfactant* boxes. Click ≥ and enter 0.75, as shown in Figure 14.3.

Figure 14.3: Design Factor Constraints

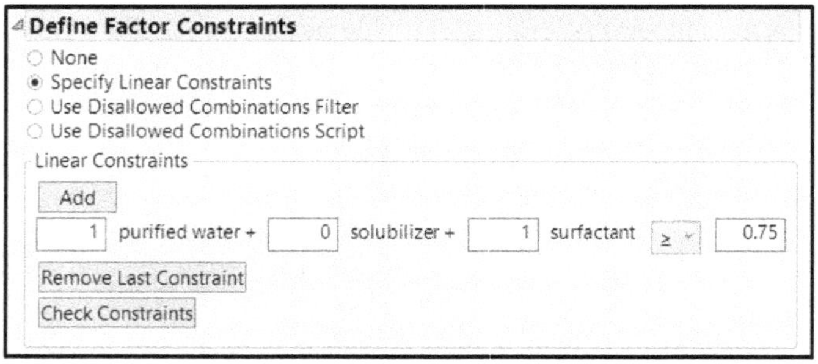

The constraints have been entered to the model designer output, it is good practice to save them in case future designs must be considered. Use the red triangle menu next to the Mixture Design header and select Save Constraints to create a data file. Choose *File ▶ Save As* and name the file "liquid mixture constraints" saved to a desired location.

The total number of effects for the model must be determined. The simplest form is given as the default, which includes the three main effects of the mixture components. Recall that interactions for an experiment with independent factors, studied in Chapter 13, are included as factors to check for dependent relationships among two or more inputs. The inputs of mixture models are defined as dependent by design; therefore, the interactions are not needed to check for dependency. Interactions of a mixture model identify complex, non-linear trends in outputs. Significant mixture interactions are illustrated in the profiler plots as curved contours, since the response surfaces are non-planer. Jon has reason to believe that curved contours might be present in the response surfaces. Curved contours represent interactions of materials in the model that are significant. In the *Model* box shown in Figure 14.4, click *Interactions*, and select *2nd* to add them to the model.

Figure 14.4: Model Designation

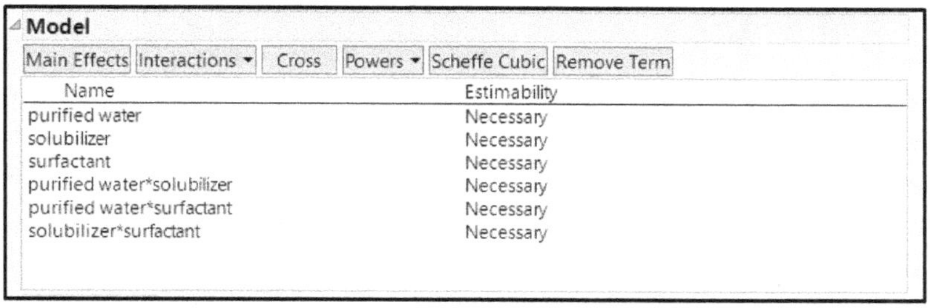

The model is well-defined and ready for generation. The optimal design provides resource options that are similar to those in the custom designer. The team decides that they want a center point for the model and that the 12-run default number of runs is acceptable for the resources available. Enter 1 in the *Number of Centerpoints* box in the *Design Generation* dialog box, and click *Make Design*.

Figure 14.5: Design Generation

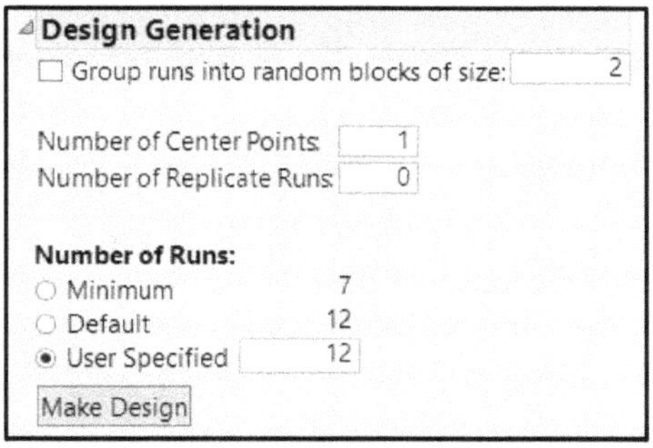

JMP runs a model optimization algorithm behind the scenes to create a randomized model for the three mixture components that include the least amount of experimental error. A dialog box opens as the designer is running. The amount of computer processing time required by the optimization algorithm to create a model depends upon the complexity of the model. When the analysis is complete, the model of 12 runs displays within the designer journal. The last thing to do is to review the design and create the data table needed to execute the experimentation.

Independent factor models typically involve the review of design diagnostics to help you fine-tune the design to ensure that goals are likely to be met without correlation and excessive error. Mixture designs are unique and cannot be evaluated via typical design diagnostics. Statistical power has little meaning and checking for correlations makes no sense for the dependent mixture components. The prediction error is the only diagnostic worth reviewing in the diagnostic results to ensure the minimum amount possible.

The amount of available resources and desire for information mitigates the potential to fine-tune with other modeling options to reduce prediction error. The design will work for the team and needs to be generated. Use the red triangle menu next to the *Custom Design* header, select *Set Random Seed*, and enter 2018 so that the data table produced matches the table shown Figure 14.6. Leave the default value *Randomize* in the *Run Order* drop-down options, select the *Include Run Order Column* check box, and click *Make Table* to get the data table shown in Figure 14.6.

Figure 14.6: Mixture Design Data Table

purified water	solubilizer	surfactant	assay RT	assay cool	density	Run Order
0.75	0.25	0	•	•	•	1
0.75	0	0.25	•	•	•	2
0.575	0.25	0.175	•	•	•	3
0.75	0.119363065	0.130636935	•	•	•	4
0.75	0.25	0	•	•	•	5
0.4	0.25	0.35	•	•	•	6
0.65	0	0.35	•	•	•	7
0.75	0	0.25	•	•	•	8
0.6260546448	0.1484784801	0.2254668752	•	•	•	9
0.575	0.25	0.175	•	•	•	10
0.4	0.25	0.35	•	•	•	11
0.5439582855	0.1060417145	0.35	•	•	•	12

The data table includes a randomized set of 12 experimental runs, which should match your results since you entered the randomization seed of 2018.

Ternary Plots for Model Diagnosis

Once the team has created the model table, they can create a visualization to evaluate the specialized design space of a mixture model. Recall that independent factor designs include a cubic three-dimensional design space. The mixture design differs in shape due to the dependency among the inputs and is rendered as a triangular design space. If there are no restrictions on proportional amounts of the mixture components (0% to 100%), the three points represent runs that include 100% of each mixture component. Models with restrictions in the proportional levels of materials include more than three points and are referred to as extreme vertices designs.

The project is limited to three materials inputs, so the design space includes one triangle. The plot of the triangular design space is known as a ternary plot. A design might include more than three mixture components. Such designs have multiple ternary plots to illustrate two of the mixture components with all others as the three sides of the triangle. It is very easy to create by using the Graphs menu options in JMP. Select *Graphs* ▶ *Ternary Plot* to get the dialog box shown in Figure 14.7.

Figure 14.7: Ternary Plot Set Up

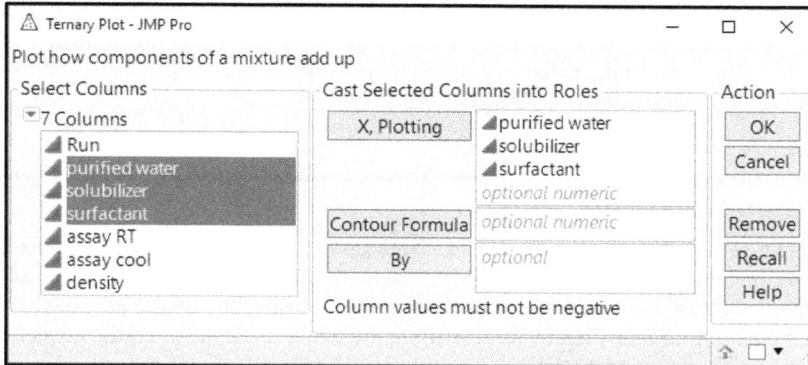

Move the mixture components *purified water*, *solubilizer*, and *surfactant* to the *X, Plotting* box, and then click *OK* to get the output shown in Figure 14.8.

Figure 14.8: Ternary Plot (Design Space)

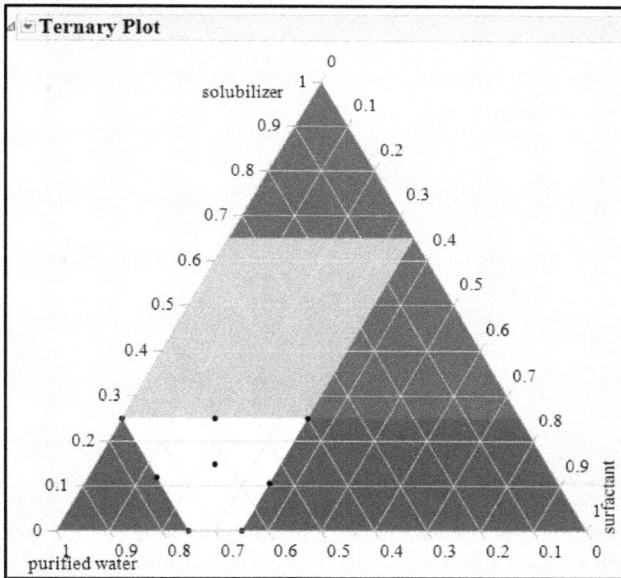

The ternary plot contains a space that includes four vertices within the triangular simplex. The proportional amount of purified water is the axis across the base of the triangle, with 0% at the right vertex and 100% toward the left. The solubilizer is the axis on the left face of the triangle, with 0% at the base and 100% at the upper vertex. Surfactant makes up the right axis, with 0% at the top vertex and 100% at the right base vertex. The restrictions do not allow for any solutions to be made with 100% of any of the three ingredients, so the dark gray areas of the simplex are not relevant. The linear constraint of water and surfactant totaling more than 75% of the mix is shown in light gray. The unshaded design space includes the dark black dot markers representing each run studied in the design.

Eight markers are shown on the plot, which means that replicate runs are included in the 12 runs. Replicates work to minimize the amount of random error included within the model. The space is defined with at least one run at each of the vertices. The optimizer placed the runs per the algorithm shown, with two edges of the space that include a point in the middle. The team decides that the 12-run design is likely to be adequate and moves on the execution portion of the project.

Analysis of Mixture Design Results

The 12 runs of the design have been created and the testing has been done to get the room temperature and refrigerated assay results as well as the density of each. Jon determined that the relative cost of each of the solutions would be easy to calculate based on the price of each of the three materials by the proportional weight for each run. Adding a cost output helps teams to determine the differences between the costs of a current formulation and the potential cost of an optimized formulation. The results from the experimental campaign have been added to the end of the data table. Open *liquid mixture design 12R results.jmp* to view the design with results, as shown in Figure 14.9.

Figure 14.9: Mixture Design Results Table

Run Order	purified water	solubilizer	surfactant	assay RT	assay cool	density	cost function
1	0.575	0.250	0.175	78.8	79.7	0.83	1.28
2	0.575	0.250	0.175	70.1	90.2	0.85	1.44
3	0.650	0.000	0.350	74.8	39.0	0.91	1.34
4	0.626	0.148	0.226	94.1	79.0	0.92	1.32
5	0.543	0.107	0.350	87.0	63.2	0.91	1.39
6	0.400	0.250	0.350	72.0	54.0	0.81	1.37
7	0.750	0.000	0.250	74.8	56.4	0.92	1.40
8	0.750	0.119	0.131	89.8	58.9	0.92	1.31
9	0.750	0.000	0.250	72.9	55.2	0.90	1.42
10	0.750	0.250	0.000	96.1	60.3	0.87	1.37
11	0.400	0.250	0.350	57.0	50.8	0.84	1.43
12	0.750	0.250	0.000	89.4	70.6	0.93	1.33

Tables of results that come from the designs that you create with the DOE platform include a model script to easily kick off the analysis. The cost function column has been added to the data table to include as an output of the mixture model. It is a good idea to check the *Column Properties* ▸ *Response Limits* of *cost function* to make sure that *Output Goal* is set to *Minimize*. Run the *Model* script in the upper left area of the data table by clicking on the green arrow next to it. The *Model Specification* dialog box includes all but the *cost function* response, which you must add to the *Pick Role Variables* box. The completed *Model Specification* dialog box is shown in Figure 14.10.

Figure 14.10: Mixture Model Specifications

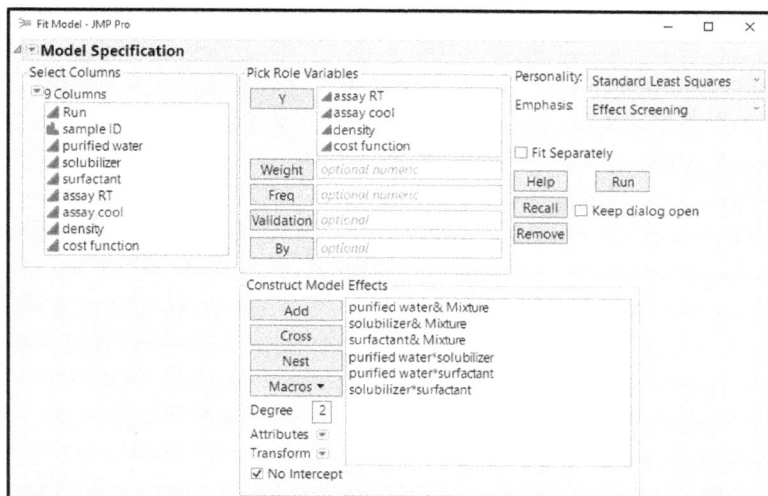

You can analyze data that is collected and provided without the use of the JMP design table by selecting *Analyze ▶ Fit Model* to get the *Model Specification* dialog box. Move the *assay RT*, *assay cool*, *density*, and *cost function* role variables (outputs) to the *Y* box. Click the mixture components *purified water*, *solubilizer*, and *surfactant* and drag them to the *Select Columns* box. Select all three so that they are highlighted in blue, click *Macros*, and select *Factorial to degree*. The result will be the same as in Figure 14.10.

The model specifications have been completed and the analysis is ready to run. Click *Run* to get the analysis output.

The model analysis output includes multiple outline headers. The analysis sections of greatest influence are shown individually to allow for a detailed explanation of each figure. The first section to interpret is the Effect Summary, shown in Figure 14.11.

Figure 14.11: Effects of the Mixture Models

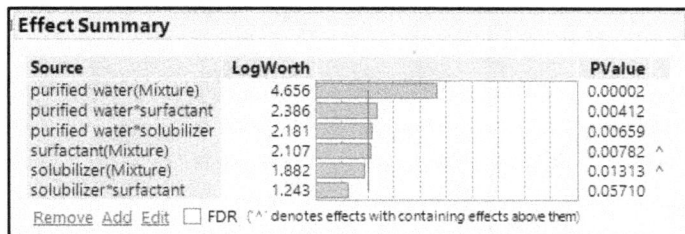

The summary ranks the factors in top to bottom order by the amount of influence they have on all the responses. The vertical blue line is a decision line for determining the evidence of significance for the inputs, which is adjusted by the number of inputs included in the models. Four of the six factors listed have significant influence on all outputs, noted by the bars that pass to the right of the decision line. The next step of the analyses is to evaluate the model detail for each of the outputs. For brevity's sake, the assay for room temperature solutions is explained based on the Actual by Predicted plot, shown in Figure 14.12.

Figure 14.12: Plot of the Assay RT Model

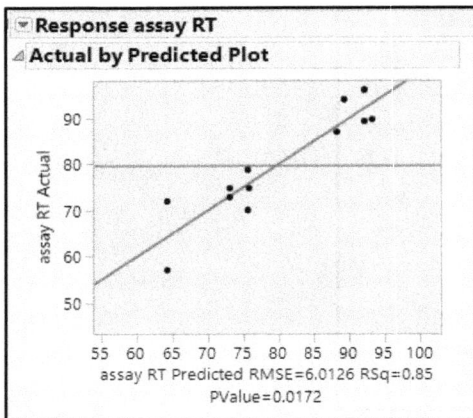

The plot of actual values of assay of room temperature solutions by model predictions contains points that generally follow the model prediction line. The fit of the model is very good, shown by the Rsquare of 0.85. There is evidence that the mixture model has significant influence on assay RT, noted by the Pvalue of 0.0172. Basically, there is only a 2% chance that changes in assay RT are due to random variation. The relatively small amount of variability in results is defined by the RMSE of 6.0126. The overall average result in assay RT of 80 is illustrated by the horizontal blue line.

The next detail to review is the set of residual plots to investigate for non-random patterns, shown in Figure 14.13.

Figure 14.13: Residual Plots

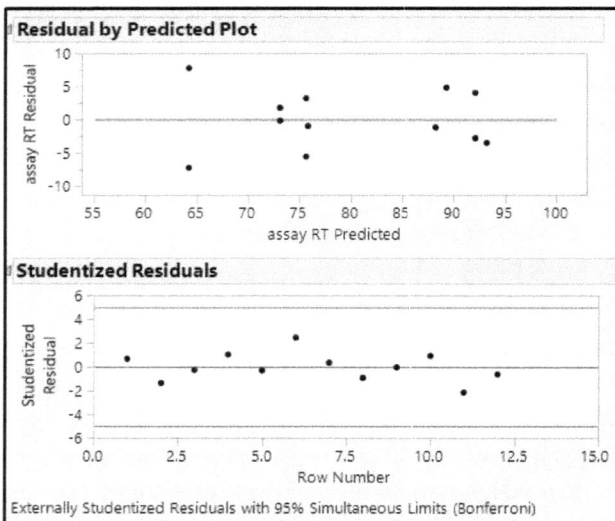

Residuals indicate a random pattern for both plots. No cone shaped or curved patterns are evident. Therefore, the variability is expected to be constant across the predicted values, which meets assumptions for modeling.

Parameter estimates are shown in Figure 14.14. They indicate which of the model effects have significant influence on the assay RT results.

Figure 14.14: Estimates of Model Parameters

Parameter Estimates				
Term	Estimate	Std Error	t Ratio	Prob>|t|
purified water(Mixture)	73.6054	20.7473	3.55	0.0121*
solubilizer(Mixture)	-703.7543	202.1987	-3.48	0.0131*
surfactant(Mixture)	134.25951	87.28821	1.54	0.1749
purified water*solubilizer	1135.3257	279.0911	4.07	0.0066*
purified water*surfactant	-83.61097	161.8575	-0.52	0.6239
solubilizer*surfactant	707.94472	301.3197	2.35	0.0571

The mixture components of purified water (Prob>\|t\|<0.0121) and solubilizer (Prob>\|t\|=0.0131) have highly significant influence on assay RT. The interaction of purified water*solubilizer shows strong evidence of significance (Prob>\|t\|<0.0066). The input with the highest estimate of influence on assay RT is the interaction of purified water and solubilizer, with an estimate of 1135.3. None of the remaining estimates have significant influence on assay RT.

The remaining models need to be analyzed and interpreted, with special attention given to comparing differences in significance and estimates. You should do this on your own in order to understand the individual models before you practically apply the information to the formulation with the model profilers.

Model Profiler

A Prediction Profiler for the model is at the bottom of the model analyses output. Use the matrix of plots to visualize the influence of the mixture components for each of the four outputs, as shown in Figure 14.15.

Figure 14.15: Mixture Model Prediction Profiler

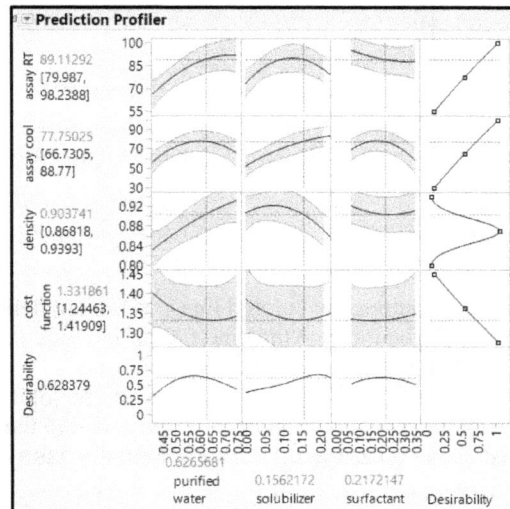

The influence of mixture components is illustrated by the steepness of the profile shown for each cell plot. The complex, curved shape of the profiler plots is indicative of the significant mixture interaction term noted previously. The default for the profiler includes all mixture components set to near the middle of the proportional range. Explore the plot by clicking on the vertical red segmented line of a mixture component and manipulating it to lower and higher proportions. The other two mixture components will move in concert with the manipulation of one component due to the requirement that the total must always be 100%.

Previous chapters about structured experiments provided detail about the function of the prediction profiler and interpretation of results. Mixture designs offer a unique profiler that is based on the ternary plot. Use the red triangle menu next to the *Least Squares Fit* output header to select *Profilers* ▶ *Mixture Profiler* from the list of options to get the output in Figure 14.16.

Figure 14.16: Mixture Design Profiler

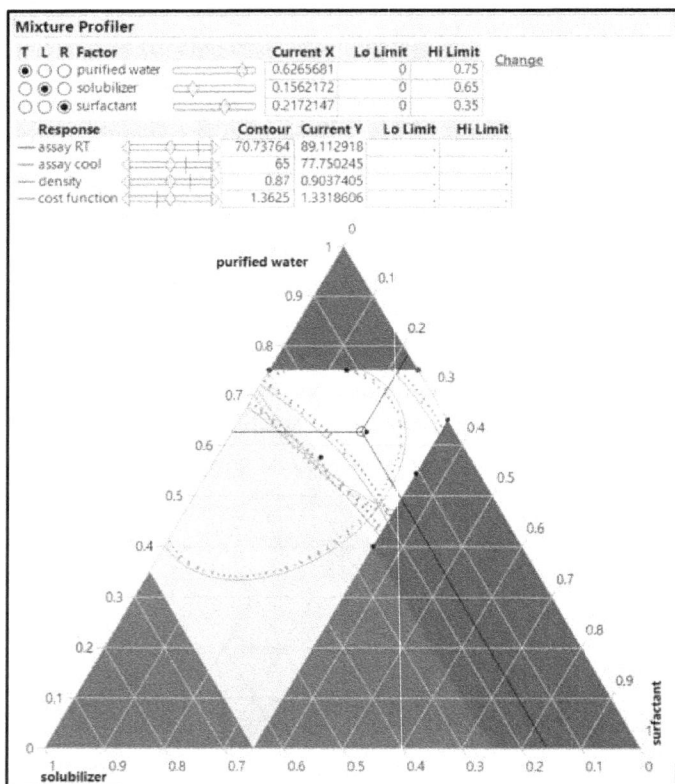

Response surface curved contours are shown for the four model outputs by colored lines, which are defined in the *Response* legend above the plot. The dots next to the profile indicate the direction of increase for the output being modeled. Subject matter expertise indicates that the assay values below 55% are undesirable because they indicate the point at which it is possible for the API to have fallen out of solution as precipitants that collect at the bottom of the product container. The low limit needs to be added to the profiler so that the desirable design space can be illustrated. Enter 55 as the *Lo Limit* for both the **assay RT** and *assay cool* outputs.

Figure 14.17: Mixture Design Profiler (with Limits)

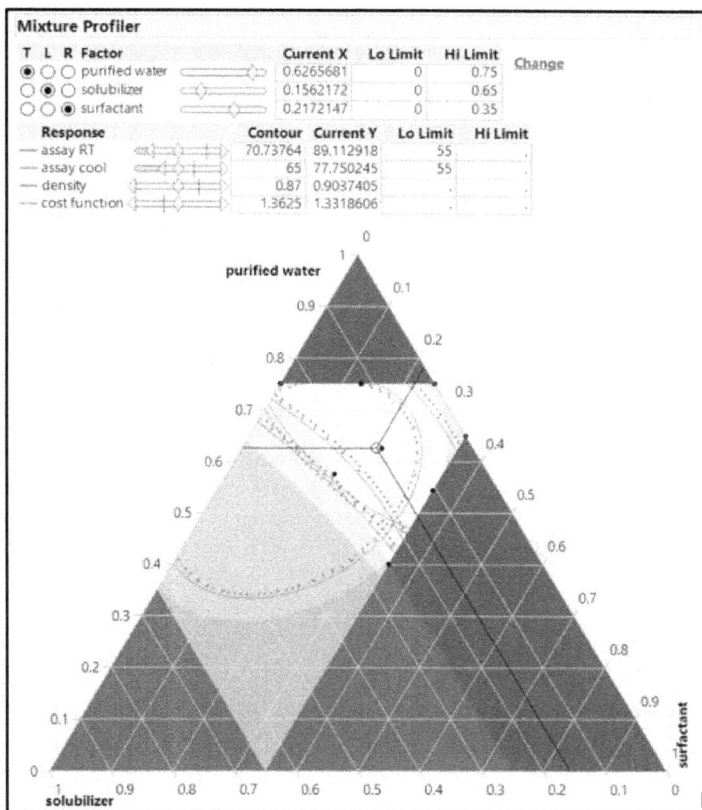

The shading at the lower left of the design space shown in Figure 14.17 reflects the results that are less than the 55% lower limit for assay RT. It is light gray because it is overlaid with the shading of the results less than 55% for assay cool. The density and cost function outputs have no specifications and are included in the model for additional information.

More detail for the two assay outputs is available by defining the contour grid lines for each. Use the red triangle menu next to the *Mixture Profiler* header to select *Contour Grid*. Select *assay RT* as shown in Figure 14.18. Utilize the option to modify the contour grid lines 3-6 to change *assay cool* to the same gridline settings.

Figure 14.18: Gridlines: Select Response

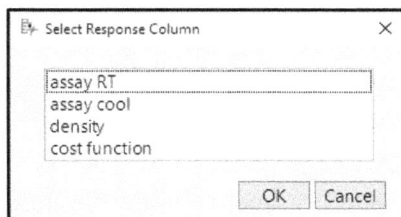

Enter 55 as the *Low value*, and change *Increment* to 10, as shown in Figure 14.19. Click *OK* to add the gridlines for *assay RT* to the mixture profiler.

Figure 14.19: Gridlines: Enter Values

Repeat the process of adding contour gridlines for *assay cool*. Enter 40 as the *Low value*, change *Increment* to 10, and click *OK* to add the gridlines for *assay cool* to the mixture profiler shown in Figure 14.20.

Figure 14.20: Mixture Profiler (with Limits and Gridline Detail)

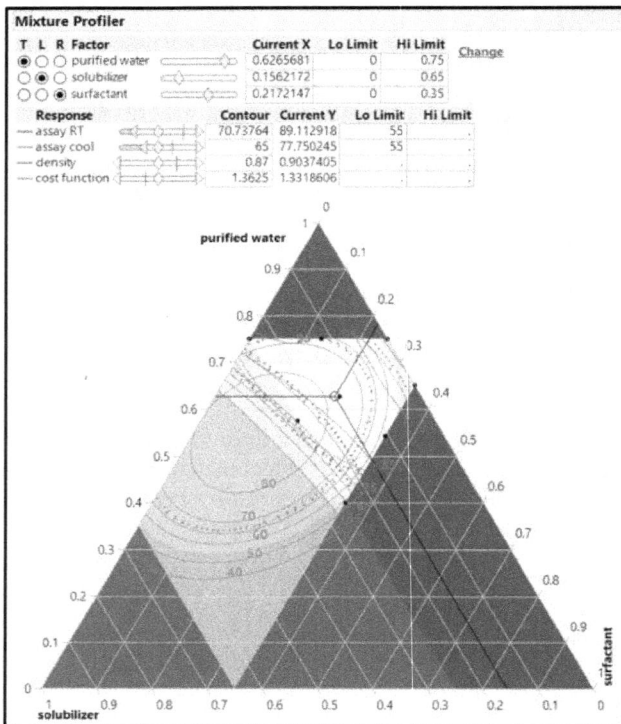

The Mixture Profiler with limits and gridlines for the most important outputs provides a great deal of practical information. It is very clear that the changes in assay for the room temperature and refrigerated conditions follow very different response surfaces. It is also clear that the refrigerated solutions are much more likely to create precipitants than the room temperature solutions. The best results for assay tend to be at the mid-top of the design space, indicating that a purified water proportion of around 65% and solubilizer proportion of around 20% provide the best functional results. The Mixture Profiler is great for visualizing the trends from the models and determining a design space. The Prediction Profiler provides additional value because the models can be maximized for best results.

The Practical Application of Profiler Optimization

The Prediction Profiler provides a great deal of information about how proportional changes in the mixture components affect changes in all the responses. The next set of steps enables the team to convert the significant experimental results to practical applications that have the greatest potential for value. Recall that the non-linear profiles shown in Figure 14.21 are due to the inclusion of mixture interactions in the model.

Figure 14.21: Prediction Profiler

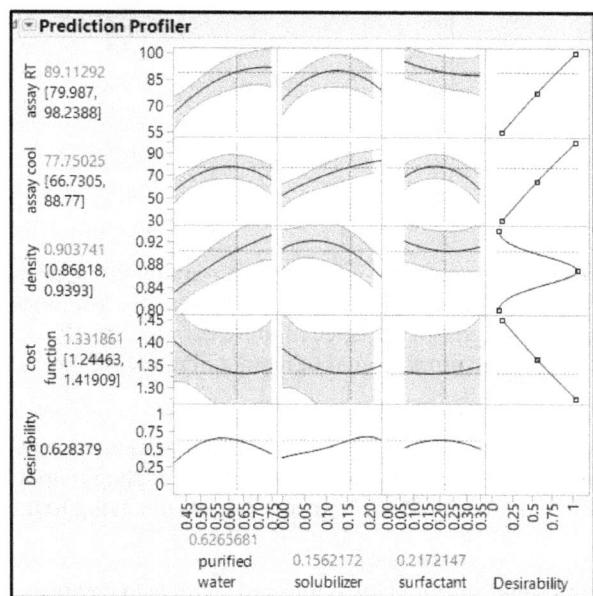

Use the red triangle menu next to the *Prediction Profiler* header to select *Optimization and Desirability*
▶ *Maximize Desirability* and get the output shown in Figure 14.22

Figure 14.22: Prediction Profiler (with Desirability Function Maximized)

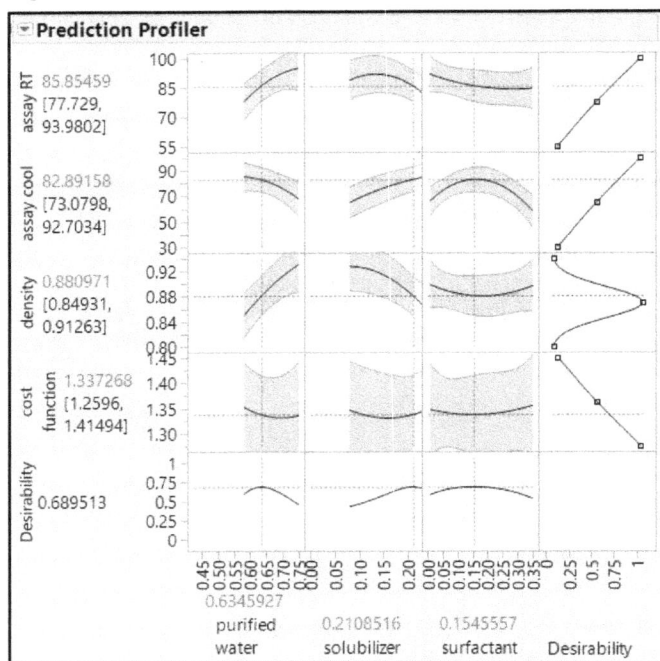

The Prediction Profiler indicates that solutions are likely to have the highest concentration with 63.5% water and the maximum amount of solubilizer of 15.4%. The cost of the solution is relatively low since water makes up the majority of the solution and is the cheapest of the three mixture components. It is possible to add water and reduce the cost minimally, but the primary goal of maintaining the API in solution would be compromised.

The modeling can be adjusted to add weight to certain outputs as well as to include specification limits and run simulations by using the steps explained in chapter 13. The dynamic nature of the modeling output provided by JMP allows for an unlimited amount of exploration of a design space without having to make actual batches and submit them for analytical results.

It is good practice to use the profilers to find mixtures that surround the optimized settings to create a control space. A control space for a campaign of process experiments was explained in chapter 13. The mix of materials used to formulate a drug product are typically expressed as fixed amounts, not ranges that are typical for process settings. Slight variations in the actual amount of each material per batch will occur in the commercial population since it is impossible to include exactly the same amount of material for every batch. The team makes three or four confirmation batches from material input amounts that surround the optimum mix and compares them with the model predictions. Go back to the output for the model analysis to create prediction variables for each output. Use the red triangle menu next to *assay RT* to select *Save Columns ▶ Prediction Formula*, as shown in Figure 14.23.

Figure 14.23: Saving a Prediction Formula

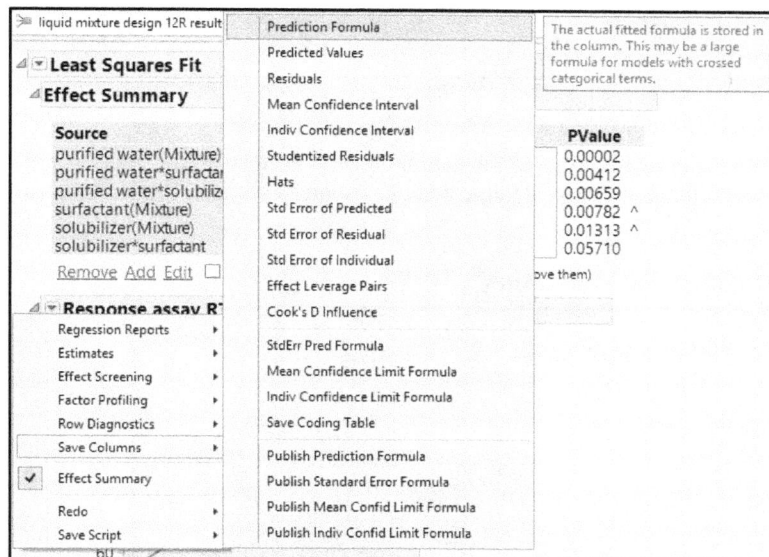

liquid mixture design 12R result	Prediction Formula	The actual fitted formula is stored in
	Predicted Values	the column. This may be a large
◢ ▾ **Least Squares Fit**	Residuals	formula for models with crossed categorical terms.
◢ **Effect Summary**	Mean Confidence Interval	
	Indiv Confidence Interval	**PValue**
Source	Studentized Residuals	0.00002
purified water(Mixture)		0.00412
purified water*surfactar	Hats	0.00659
purified water*solubiliz	Std Error of Predicted	0.00782 ^
surfactant(Mixture)	Std Error of Residual	0.01313 ^
solubilizer(Mixture)		0.05710
solubilizer*surfactant	Std Error of Individual	
Remove Add Edit ☐	Effect Leverage Pairs	ove them)
	Cook's D Influence	
◢ ▾ **Response assay R**		
Regression Reports ▸	StdErr Pred Formula	
Estimates	Mean Confidence Limit Formula	
Effect Screening ▸	Indiv Confidence Limit Formula	
Factor Profiling ▸	Save Coding Table	
Row Diagnostics ▸		
Save Columns ▸	Publish Prediction Formula	
✓ Effect Summary	Publish Standard Error Formula	
	Publish Mean Confid Limit Formula	
Redo ▸	Publish Indiv Confid Limit Formula	
Save Script ▸		

Repeat the process of saving a prediction formula for the three other outputs: *assay cool, density,* and *cost function.* The data table *liquid mixture 12R results.jmp* shown in Figure 14.24 contains the four newly created model prediction columns. Select *File* ▸ *Save As* to save the updated data table as "liquid mixture 12R results with predictions".

Figure 14.24: Data Table with Prediction Formulas

purified water	solubilizer	surfactant	assay RT	assay cool	density	cost function	Pred Formula assay RT	Pred Formula assay cool	Pred Formula density	Pred Formula cost function
0.575	0.250	0.175	78.8	79.7	0.83	1.28	75.642237665	85.547639...	0.8423966...	1.3561358648
0.575	0.250	0.175	70.1	90.2	0.85	1.44	75.642237665	85.547639...	0.8423966...	1.3561358648
0.650	0.000	0.350	74.8	39.0	0.91	1.34	75.812843913	47.019555...	0.9108244...	1.378294873
0.626	0.148	0.226	94.1	79.0	0.92	1.32	89.287586755	76.636967...	0.9063431...	1.3320015085
0.543	0.107	0.350	87.0	63.2	0.91	1.39	88.243014879	57.430730...	0.9181195...	1.3457286571
0.400	0.250	0.350	72.0	54.0	0.81	1.37	64.26660894	53.062247...	0.8268974...	1.4080546889
0.750	0.000	0.250	74.8	56.4	0.92	1.40	73.091871503	51.866100...	0.9066586...	1.3869655128
0.750	0.119	0.131	89.8	58.9	0.92	1.31	93.236214518	66.776293...	0.9318095...	1.338475046
0.750	0.000	0.250	72.9	55.2	0.90	1.42	73.091871503	51.866100...	0.9066586...	1.3869655128
0.750	0.250	0.000	96.1	60.3	0.87	1.37	92.139037998	64.195722...	0.8997701...	1.3486490267
0.400	0.250	0.350	57.0	50.8	0.84	1.43	64.26660894	53.062247...	0.8268974...	1.4080546889
0.750	0.250	0.000	89.4	70.6	0.93	1.33	92.139037998	64.195722...	0.8997701...	1.3486490267

The team used the dynamic profilers to create three confirmation runs shown in Table 14.1.

Table 14.1: Confirmation Runs

				actual results		
run	purified water	solubilizer	surfactant	assay RT	assay cool	density
13	0.635	0.211	0.154	84.8	79.3	0.88

				actual results		
14	0.620	0.207	0.173	85.3	82.9	0.89
15	0.645	0.193	0.162	88.1	81.4	0.87

Enter the input settings of the three confirmation runs to the data table with prediction formulas as runs 13, 14, and 15. The formulas of the predictions will have a value added as soon as all three inputs have values entered.

Figure 14.25: Data Table with Confirmation Predictions

Run Order	purified water	solubilizer	surfactant	assay RT	assay cool	density	cost function	Pred Formula assay RT	Pred Formula assay cool	Pred Formula density	Pred Formula cost function
1	0.575	0.250	0.175	78.8	79.7	0.83	1.28	75.642237665	85.547639...	0.8423966...	1.3561358648
2	0.575	0.250	0.175	70.1	90.2	0.85	1.44	75.642237665	85.547639...	0.8423966...	1.3561358648
3	0.650	0.000	0.350	74.8	39.0	0.91	1.34	75.812843913	47.019555...	0.9108244...	1.378294873
4	0.626	0.148	0.226	94.1	79.0	0.92	1.32	89.287586755	76.636967...	0.9063431...	1.3320015085
5	0.543	0.107	0.350	87.0	63.2	0.91	1.39	88.243014879	57.430730...	0.9181195...	1.3457286571
6	0.400	0.250	0.350	72.0	54.0	0.81	1.37	64.26660894	53.062247...	0.8268974...	1.4080546889
7	0.750	0.000	0.250	74.8	56.4	0.92	1.40	73.091871503	51.866100...	0.9066586...	1.3869655128
8	0.750	0.119	0.131	89.8	58.9	0.92	1.31	93.236214518	66.776293...	0.9318095...	1.338475046
9	0.750	0.000	0.250	72.9	55.2	0.90	1.42	73.091871503	51.866100...	0.9066586...	1.3869655128
10	0.750	0.250	0.000	96.1	60.3	0.87	1.37	92.139037998	64.195722...	0.8997701...	1.3486490267
11	0.400	0.250	0.350	57.0	50.8	0.84	1.43	64.26660894	53.062247...	0.8268974...	1.4080546889
12	0.750	0.250	0.000	89.4	70.6	0.93	1.33	92.139037998	64.195722...	0.8997701...	1.3486490267
13	0.635	0.211	0.154	.	.	.	.	85.867485665	82.885512...	0.8809827...	1.3372707197
14	0.620	0.207	0.173	.	.	.	.	85.276895967	82.939143...	0.8799786...	1.3376550123
15	0.645	0.193	0.162	.	.	.	.	88.130049672	81.357587...	0.8920301...	1.3334608973

The comparison of the predicted to the actual values indicates that the model slightly overpredicted the assay RT by approximately 0.8. Assy cool values were overpredicted by approximately 2.5. The density values have no practically relevant difference since variation in density of +/-0.01 is minimal.

Practical Conclusions

Jon and the formulation team were smart to utilize a mixture design as the search for the best solution to rectify the issues of API precipitants falling out of solution in cold conditions. The mixture profiler and prediction profiler are invaluable tools for the team to simulate an infinite amount of variations in the model. Work in the virtual space of JMP models is an extremely efficient way to create robust products.

Three confirmation runs yield actual values that are very close to the actual analytical results. Jon's team utilizes the model results with confirmation to explain to the project stakeholders and assure them that the optimal settings will robustly produce product.

Jon intentionally held back the actual materials values for the current product from the scientists who ran the model analysis to mitigate the potential for bias. The actual mix of 50% purified water, 16% solubilizer, and 34% surfactant was added to the data prediction table. The problem of precipitating API in cool storage is evident with the predicted assay cool of under 61%. The predicted assay RT of 85% illustrates why the problem was not evident in the batch retains that are stored in room temperature conditions.

The product formula is updated to the optimized settings of purified water at 63.5%, solubilizer of 21.1%, and surfactant of 15.4% and submitted for approval. The customer complaints have been eliminated with the new knowledge of the changes in assay likely once the product is cooled by refrigeration. Adding to the value of the structured experimentation is the fact that the new formulation includes 30% more water resulting in reduced costs significantly.

Exercises

E14.1—You are working on a team involved in formulating a new liquid solution for a drug product. Through risk analysis it has been determined that the proportional mix of the 3-component base is likely to have an effect on the CQA of soluble solids. The base includes water, sorbitol, and glycerine; which have no restrictions on the amount of each in the base.

1. Select *DOE ▶ Classical ▶ Mixture Designs* to create a simplex mixture design.
2. Add the three components and use the red triangle menu options to save a factor table named "base factors."
3. Create a data collection plan table with the *Optimal* mixture design type. Use the default number of runs, and save the table as "optimal base mixture."
4. Repeat steps a-c to create a *Simplex Centroid* and a *Space Filling* design with the default number of factors. Save them as "simplex centroid base" and "space filling base", respectively.
5. For each of the design plans, select the three mix component columns. Right-click inside the columns, and select *Standardize Attributes.* Change the *Modeling Type* to *Nominal.*
6. Make ternary plots of each by selecting *Graphs ▶ Ternary Plot.*
7. Make scatterplot matrix plots of each by selecting *Graphs ▶ Scatterplot Matrix.*
8. How would you summarize the robustness of each of the models and the potential for useable information by comparing the plots in a report to the stakeholders?

E14.2—The simplex centroid design was executed on the mix factors of the liquid base obtained in exercise E14.1. Analyze the design to determine the relationship of the mix components to the variability in soluble solids.

1. Open *simplex centroid base.jmp.*
2. Select *Analyze ▶ Fit Model* to create output for a least squares statistical model for the main effects.
3. Create a report using the output options that are described in this chapter to present the results to the project stakeholders.
4. What is recommended for the proportional mix of the three materials to achieve the soluble solids of 33% to 39%?

E14.3—The Quality by Design process of evaluating risk has identified that the soluble solids have high risk for maintaining the API within the liquid solution. The project team is justified for allocating more resources to the experimentation on the base to get the most precise information possible. This precise information enables the team to have the best chance at hitting the optimum mix of materials with the liquid base. A space filling design has been executed and requires analysis.

1. Open *space filling base.jmp*.
2. Select *Analyze ▶ Fit Model* to create output for a least squares statistical model for the main effects as well as the input interactions.
3. Create a report using the output options that are described in this chapter to present the results to the project stakeholders.
4. What is recommended for the proportional mix of the three materials to achieve the soluble solids of 38% to 46%?
5. Compare and contrast the space filling analysis report with the simplex centroid analysis report to prepare for questions from the project stakeholders. Was the allocation of additional resources justified?

Chapter 15: Analyzing Data with Non-linear Trends

Overview

Technical professionals working in the pharmaceutical industry regularly deal with data that have trends that are not linear. These trends involve rates of change, and they create challenges in the interpretation of results. It might be possible to transform the data or the axis used to plot the data as a linear trend. However, you must remember to use the right formula to decode results for practical application. JMP provides excellent tools for detailed exploration of trends modeled by functions that are more complex than linear models. This chapter provides an overview of some of the more common aspects in the study of non-linear models common to the pharmaceutical industry.

The Problems: Comparing Drug Dissolution Profiles and Comparing Particle Size Distributions

Heinz is a Senior Scientist working on a solid drug formulation for an extended release formula that is a generic equivalent to a branded product. The development team must have a drug release profile that is like the target drug. Candidates for the final formula include tablets made from two unique granulations that include materials intended to form a polymer matrix for control of the release. The final blends were compressed into tablets of various hardness levels. Six sample tablets were pulled at random from each developmental batch and tested in discriminatory media over a 14-hour period, with results compiled for 12 time segments. The brand target has been established by using equivalent methods. It is critical that the team provide robust evidence that a developmental candidate is equivalent to the target dissolution profile.

Sudhir is managing a team that has been utilizing structured experimentation to optimize a granulation. The team has five lots of granulates; three that are known to have desirable physical characteristics, and two are considered at risk. The project is nearing commercial production, and the development team must develop process monitoring so that teams can determine whether commercial granulations will have acceptable particle size profiles. The instrument used for manufacturing is a Ro-Tap Sieve Shaker; it is easy to use and provides relatively repeatable results. A stack of 12 sieves made up of screen mesh that moves from coarse to fine are used to capture percentages of a 100-gram sample and characterize the particle size profile. Nonlinear modeling will work exceptionally well for the team to narrow down to the best target screens for determining good granulations from those that might be at risk due to particle size trends.

Formatting Data for Non-linear Modeling

The dissolution testing has been completed and checked with the results posted. The JMP file was prepared from the posted data and represents the stacked data format needed to efficiently run analyses. Open *Stacked disso data.jmp* to see the data table shown in Figure 15.1.

Figure 15.1: Analytical Report Dissolution Data File

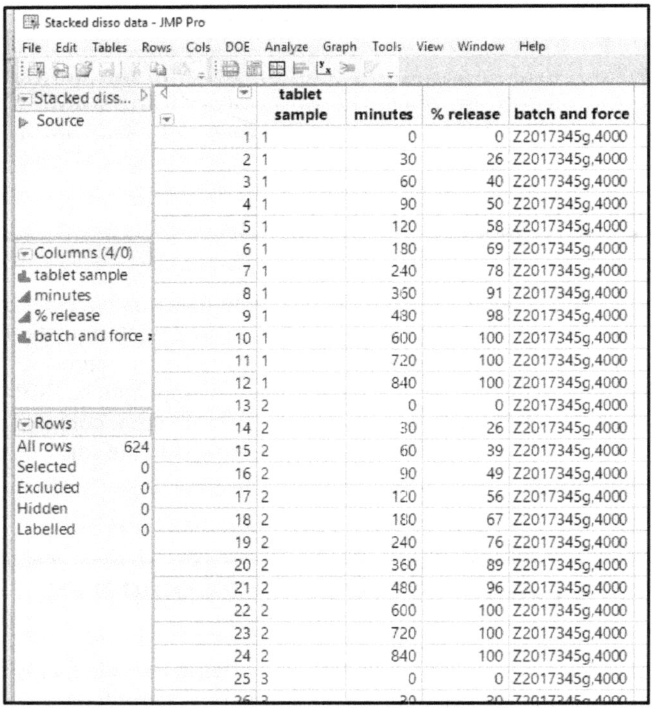

	tablet sample	minutes	% release	batch and force
1	1	0	0	Z20173459,4000
2	1	30	26	Z20173459,4000
3	1	60	40	Z20173459,4000
4	1	90	50	Z20173459,4000
5	1	120	58	Z20173459,4000
6	1	180	69	Z20173459,4000
7	1	240	78	Z20173459,4000
8	1	360	91	Z20173459,4000
9	1	480	98	Z20173459,4000
10	1	600	100	Z20173459,4000
11	1	720	100	Z20173459,4000
12	1	840	100	Z20173459,4000
13	2	0	0	Z20173459,4000
14	2	30	26	Z20173459,4000
15	2	60	39	Z20173459,4000
16	2	90	49	Z20173459,4000
17	2	120	56	Z20173459,4000
18	2	180	67	Z20173459,4000
19	2	240	76	Z20173459,4000
20	2	360	89	Z20173459,4000
21	2	480	96	Z20173459,4000
22	2	600	100	Z20173459,4000
23	2	720	100	Z20173459,4000
24	2	840	100	Z20173459,4000
25	3	0	0	Z20173459,4000

Notice that the data includes a time of zero with *% release* output of zero for each of six tablets. We do not expect any amount of drug to be released into the media at the exact time that the tablet enters a dissolution well. However, the zero values will help anchor the mathematical function used for non-linear modeling. The data includes a column for the number of tablets tested, the elapsed minutes for the 12 times that the medium in the dissolution well is sampled, the test result in *% release* for each row, and the grouping column that includes the number of the developmental batch and the amount in pounds force used for main compression of the tablets.

The target drug is noted as rld, which means reference listed drug. An rld is typically the original brand drug used as a comparator for over the counter and generic drug development. The dissolution data for the target drug was tested several weeks prior and is in the separate data file *rld dissolution data.jmp*, which is in an unstacked format. Open *rld dissolution data.jmp* to see the file shown in Figure 15.2.

Figure 15.2: Dissolution Data File for Target Drug

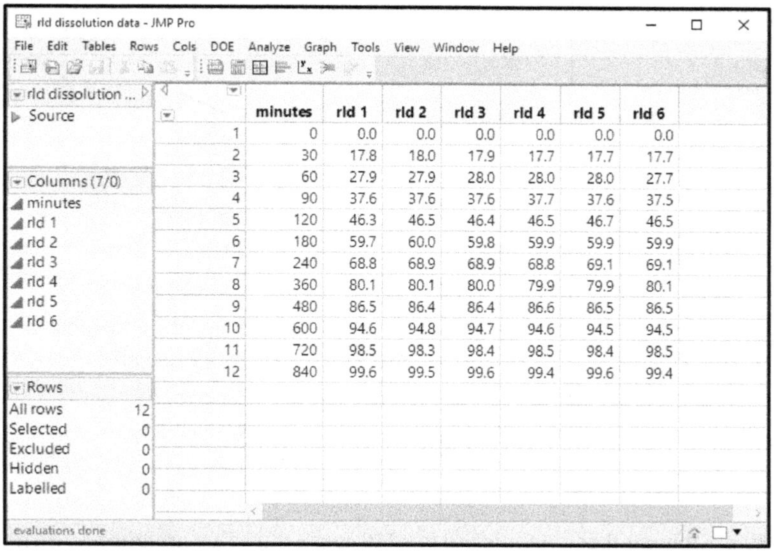

An unstacked format arranges the results in multiple columns that identify the column name by group. The rld example data groups results by tablet. The rld results need to be added to the batch test data file *Stacked disso data.jmp*, but the format must be changed into a stacked format. With *rld dissolution data.jmp* as the active view, select *Tables* ▶ *Stack* to get the dialog box shown in Figure 15.3. Select the columns *rld 1* to *rld 6* and click *Stack Columns*. Enter *% release* for the name of the *Stacked Data Column* and *tablet sample* for the name of the *Source Label Column*. Deselect the *Stack by Row* check box and click *OK* to execute.

Figure 15.3: Stack Table Dialog Box

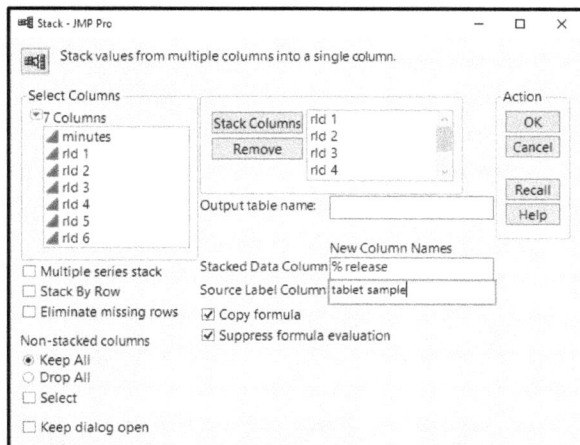

The new data file is shown in Figure 15.4.

Figure 15.4: Stacked Target Dissolution Data

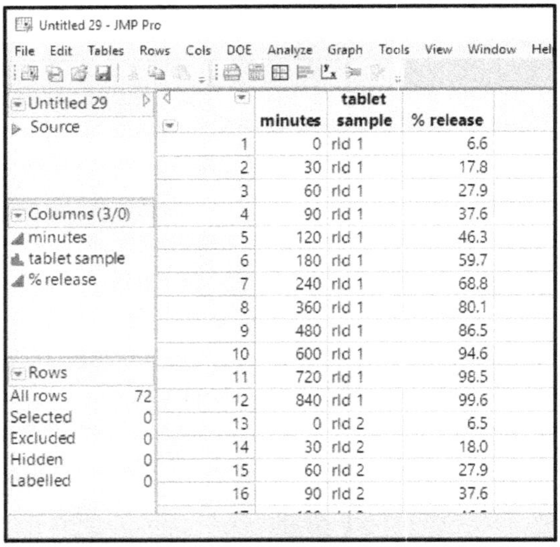

	minutes	tablet sample	% release
1	0	rld 1	6.6
2	30	rld 1	17.8
3	60	rld 1	27.9
4	90	rld 1	37.6
5	120	rld 1	46.3
6	180	rld 1	59.7
7	240	rld 1	68.8
8	360	rld 1	80.1
9	480	rld 1	86.5
10	600	rld 1	94.6
11	720	rld 1	98.5
12	840	rld 1	99.6
13	0	rld 2	6.5
14	30	rld 2	18.0
15	60	rld 2	27.9
16	90	rld 2	37.6

The file is generated as “Untitled##”; the number symbols are used because the value assigned by JMP varies from what is shown in in this example. The name of the file is changed. You don’t need to change the name since it will be added to the test data in *Stacked disso data.jmp*.

The table for the rld is now in the correct format, but the tablet sample column includes both the group identity and the tablet number. Use the column utilities to split the information into two columns. Select *Cols* ► *Utilities* ► *Text to Columns*, as shown in Figure 15.15.

Figure 15.5: Re-code the Tablet Sample Column

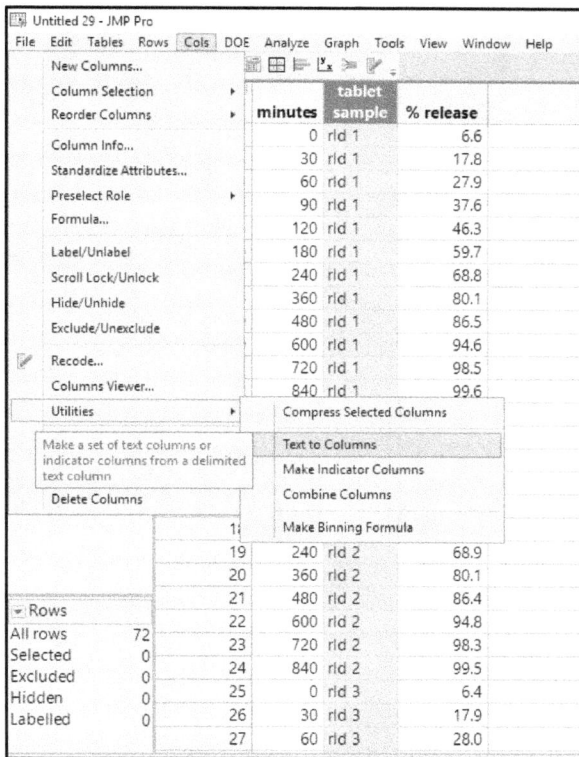

The *Text to Columns* dialog box is shown in Figure 15.6. Type open quotation marks, enter a space, and type closed quotation marks in the *Delimiter* box. The delimiter separates the group of "rld" and number of the tablet sample into separate variables. Click *OK* to see the output shown in Figure 15.7.

Figure 15.6: Options for Text to Columns

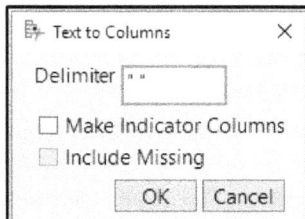

Figure 15.7: Data Table with New Columns

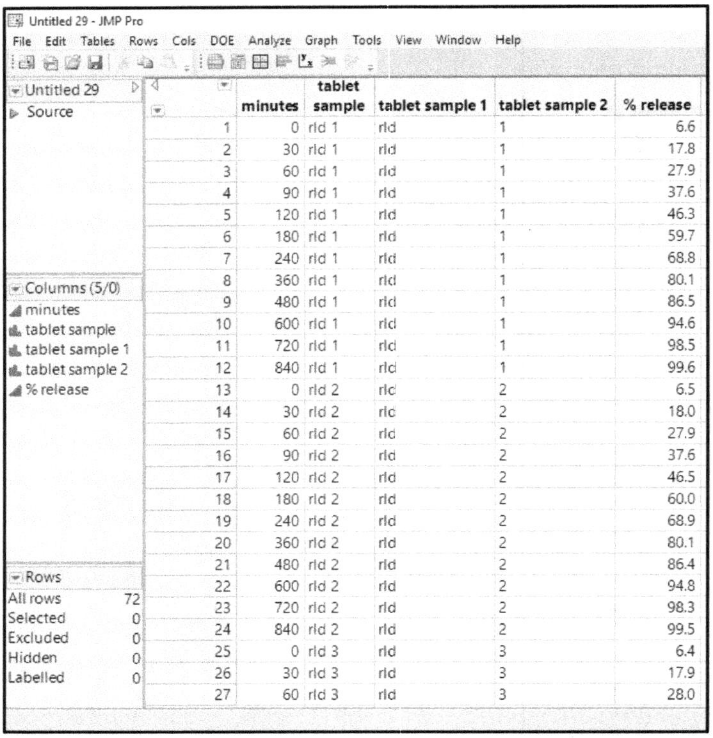

	minutes	tablet sample	tablet sample 1	tablet sample 2	% release
1	0	rld 1	rld	1	6.6
2	30	rld 1	rld	1	17.8
3	60	rld 1	rld	1	27.9
4	90	rld 1	rld	1	37.6
5	120	rld 1	rld	1	46.3
6	180	rld 1	rld	1	59.7
7	240	rld 1	rld	1	68.8
8	360	rld 1	rld	1	80.1
9	480	rld 1	rld	1	86.5
10	600	rld 1	rld	1	94.6
11	720	rld 1	rld	1	98.5
12	840	rld 1	rld	1	99.6
13	0	rld 2	rld	2	6.5
14	30	rld 2	rld	2	18.0
15	60	rld 2	rld	2	27.9
16	90	rld 2	rld	2	37.6
17	120	rld 2	rld	2	46.5
18	180	rld 2	rld	2	60.0
19	240	rld 2	rld	2	68.9
20	360	rld 2	rld	2	80.1
21	480	rld 2	rld	2	86.4
22	600	rld 2	rld	2	94.8
23	720	rld 2	rld	2	98.3
24	840	rld 2	rld	2	99.5
25	0	rld 3	rld	3	6.4
26	30	rld 3	rld	3	17.9
27	60	rld 3	rld	3	28.0

Untitled 29 - JMP Pro
File Edit Tables Rows Cols DOE Analyze Graph Tools View Window Help

Untitled 29
Source

Columns (5/0)
minutes
tablet sample
tablet sample 1
tablet sample 2
% release

Rows
All rows 72
Selected 0
Excluded 0
Hidden 0
Labelled 0

The table is almost ready to add to the stacked rld data table. However, more work is needed to clean up the table by completing the following steps:

1. Click to highlight the *tablet sample* column.
2. Right-click in the column and select *Delete Column*.
3. Right-click in the *tablet sample 1* column and select *Column Info*. Change the *column name* to "batch and force."
4. Right-click in the *tablet sample 2* column and select *Column Info*. Delete "2" from the column name so that the name is "tablet sample."
5. Select *tablet sample* and select *Cols* ▶ *Reorder Columns* ▶ *Move Selected Columns* to make the columns the first columns in the table.
6. Select *batch and force* and select *Cols* ▶ *Reorder Columns* ▶ *Move Selected Columns* so that the columns are located to the right of *tablet sample*, matching the table format shown in Figure 15.8.

Figure 15.8: Correct Format and Order of Target Dissolution Data

The target data is stacked with the order of columns matching *stacked disso data.jmp*. The *Concatenate* option in the *Tables* menu enables you to quickly combine the tables. Select *Stacked disso data.jmp* to be in view and select *Tables ▶ Concatenate*. The *Concatenate* dialog box is shown in Figure 15.9.

Figure 15.9: Concatenate Tables

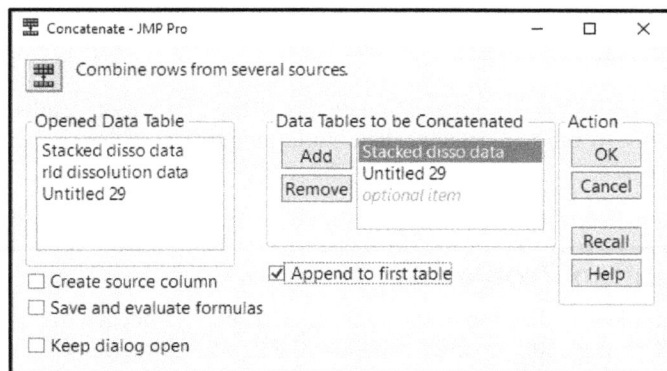

In the *Concatenate* dialog box, select *Untitled ##* (recall that the numerical value in your dialog box will not match the one shown in this example) in the *Opened Data Table* box so that it is highlighted. Click *Add* to add the stacked target data table. Select the *Append to first table* check box and click *OK* to execute.

Figure 15.10: Concatenated Stacked Disso Data Table

	tablet sample	minutes	% release	batch and force
619	4	240	61.0	X2017135d,4500
620	4	360	73.0	X2017135d,4500
621	4	480	83.0	X2017135d,4500
622	4	600	90.0	X2017135d,4500
623	4	720	95.0	X2017135d,4500
624	4	840	100.0	X2017135d,4500
625	1	0	0.0	rld
626	1	30	17.8	rld
627	1	60	27.9	rld
628	1	90	37.6	rld
629	1	120	46.3	rld
630	1	180	59.7	rld
631	1	240	68.8	rld
632	1	360	80.1	rld
633	1	480	86.5	rld
634	1	600	94.6	rld
635	1	720	98.5	rld
636	1	840	99.6	rld
637	2	0	0.0	rld
638	2	30	18.0	rld
639	2	60	27.9	rld
640	2	90	37.6	rld
641	2	120	46.5	rld

Stacked disso data - JMP Pro
File Edit Tables Rows Cols DOE Analyze Graph Tools View Window Help

Stacked diss...
Source

Columns (4/0)
tablet sample
minutes
% release
batch and force

Rows
All rows   696
Selected   0
Excluded   0
Hidden   0
Labelled   0

evaluations done

The stacked disso data table is now in the proper stacked format and includes all the dissolution results for the development batches and the target. It is now time to visualize the dissolution profiles to determine whether any of the candidates are like the target.

Making a Simple Plot of Dissolution Profiles

The theme of this book is to start with a simple visualization of the data and add detail as needed to extract information. Heinz plans to start with the Graph Builder in order to picture the trends of drug release over time for the various groups. Complete the following steps:

1. Make sure that *Stacked disso data.jmp* is open and select *Graph* ▶ *Graph Builder.*
2. Drag *% release* to the main drop zone in the *Graph Builder* window.
3. Drag *minutes* to the *X* drop zone.
4. Drag *batch and force* to the *Overlay* box.
5. Click *Done* to get the output shown in Figure 15.11.

Figure 15.11: Concatenated Stacked Disso Data Table

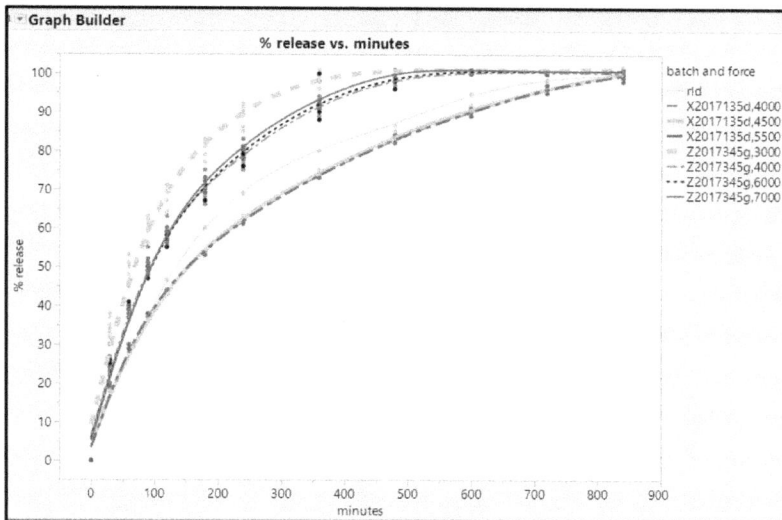

The plot provides the first visualization of the various dissolution trends. Each group is assigned a color automatically by JMP; the example plot was modified to differentiate the trends by gray-scale coloration and line style. The rld target dissolution profile is shown with a thin gray line, which is the profile at the lower middle of the plot. The target profile does not stand out from the seven other profiles. Using the graphics options to highlight the target profile will help the stakeholders of the project see the comparisons better. The example run that you create will use different colors used for each batch and force, but the instructions in the following steps will work as explained. Right-click on the line segment next to *rld*, and select *Line Color* to get the color pallet shown in Figure 15.12.

Figure 15.12: Graph Builder Color Pallet

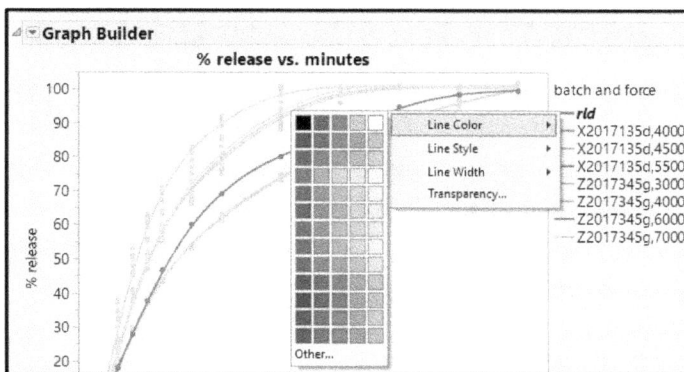

Click the black box in the upper left corner of the line color matrix. Right-click on the line segment next to *rld* again, select Line *Width*, and then select *6*, as shown in Figure 15.13.

Figure 15.13: Graph Builder Line Width

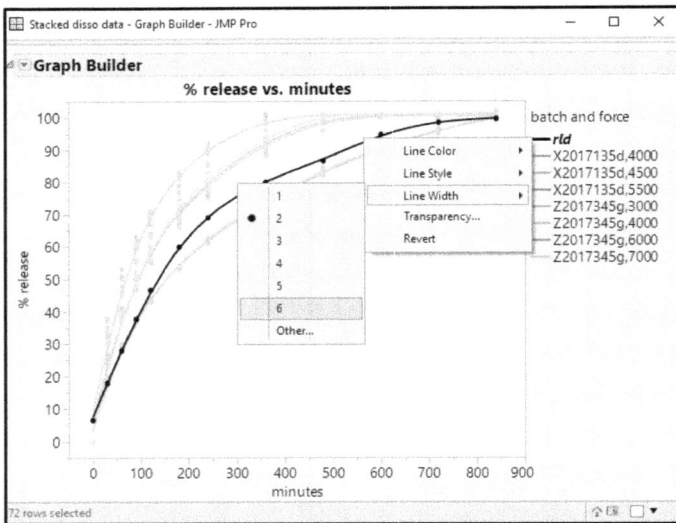

The line is now thicker so that it stands out as the target profile, as shown in Figure 15.14.

Figure 15.14: Dissolution Plots with Highlighted Target Profile

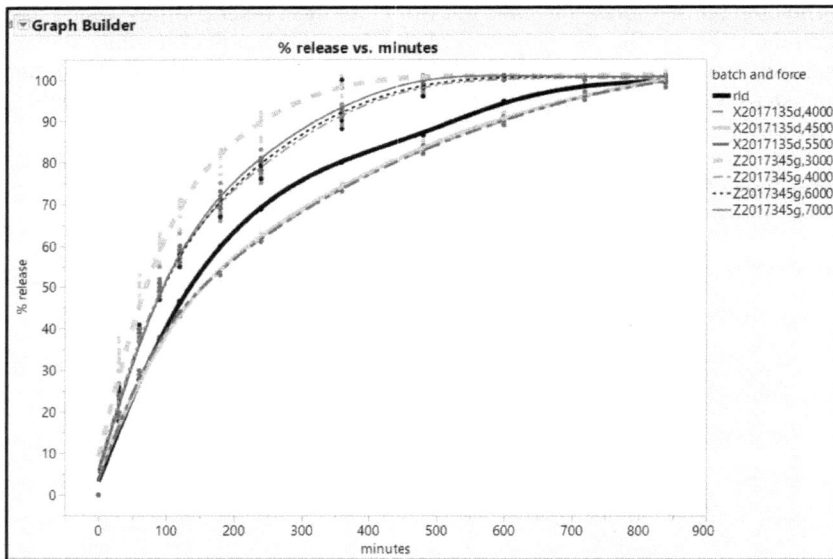

The results of the changes in the graphics options provide a clear comparison of the dissolution profiles for the various candidates with the target profile. There are clear differences among the candidates. At least two of them look to be very like the target. However, the overlap of profiles makes it a bit tough to determine which candidate is best. You can use the line options to change color, thickness, style, or transparency, which help you differentiate the profiles. In addition, the visualization is limited to a subjective comparison. Heinz needs a more detailed comparison with objective measures to justify the choice of a final formula to the stakeholders and for product filing purposes. Non-linear modeling will provide everything he needs to gain objective evidence.

Creating a Non-linear Model of Dissolution Profiles

The Graph Builder started our quest to find a candidate with a granulation and tablet hardness that creates a similar dissolution profile to the target. The expectation is that a similar in vitro dissolution profile will reflect in vivo extended drug release that is bioequivalent to the rld target. Evaluation of profiles via a non-linear function is one approach to obtain objective evidence of similarity. A model-based approach is not necessarily the most popular for the pharmaceutical industry due to a regulatory guideline that favors the similarity criterion (F2) approach. The use of non-linear modeling includes diagnostics to determine the fit of the dissolution results and how well assumptions are met.

Heinz decides to try both the non-linear modeling and similarity criterion approach in parallel to compare the information provided by each. With Stacked disso data.jmp open, select *Analyze* ▶ *Specialized Modeling* ▶ *Nonlinear* to get the *Nonlinear* setup window shown in Figure 15.15.

Figure 15.15: Nonlinear Dialog Box

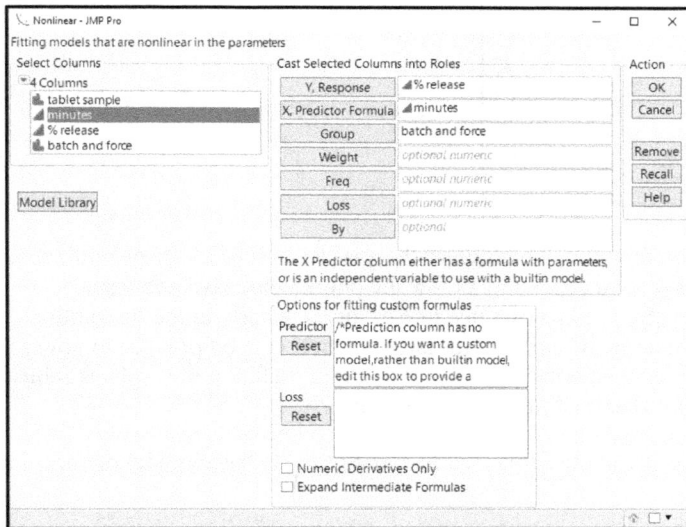

Move *% release* to *Y, Response*; *minutes* to *X, Predictor Formula*; and *batch and force* to *Group*. Click *OK* to get the output shown in Figure 15.16

Figure 15.16: Fit Curve Output

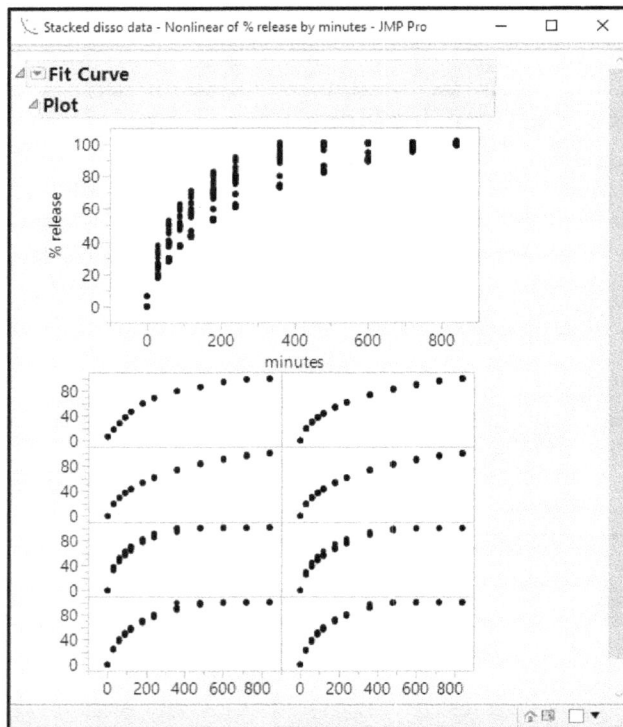

The pooled observations are shown in the first plot, which gives the analyst an idea of the general pattern for the dissolution profile. Each group is plotted separately below the pooled plot, so some visual comparisons can be made. The general pattern indicates a fast rate of increase early in the dissolution profile with an inflection to a more gradual rate of increase for later times. It is good practice to always start with the simplest function and add complexity sparingly to ensure that interpretation of results is robust for all stakeholders. A linear model is the most basic of models, so try that first.

1. Use the red triangle menu to the left of the *Fit Curve* header to see the basic model options that are available.
2. Select *Polynomials* ▶ *Fit Linear* to get the output shown in Figure 15.17.

Figure 15.17: Linear Model Detail

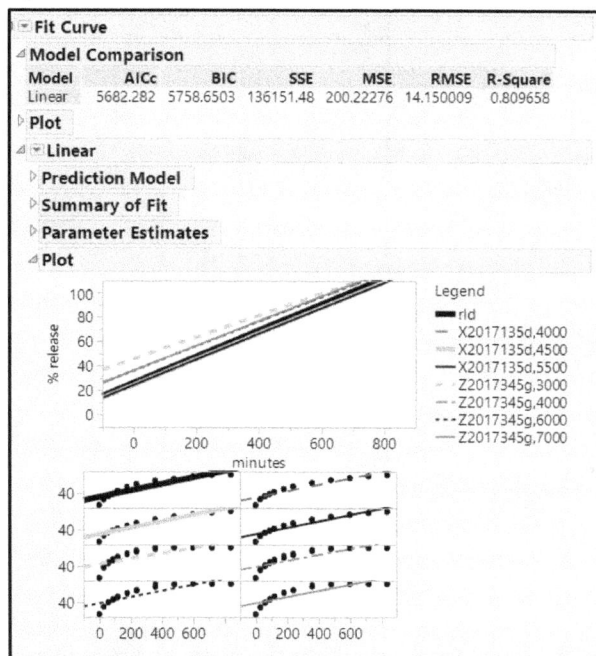

The linear model has a moderate fit noted by an R-square of 0.81. The interpretation of the fit is that 81% of the variation in dissolution values can be explained by the model lines. The overlaid model lines are plotted with a legend to define the profiles, which is located to the right of the plot. A visual indicator that a linear model might not offer the best fit is the y-intercepts of each model that are significantly greater than 0. We do not expect that the method will be able to detect any of the active ingredient in the media solution until the tablet starts to disintegrate, which happens after time 0. The eight plots below the overlaid model plot further illustrate how many observations do not follow the linear model. Estimations made from linear modeling include more error than desired due to the poor fit of linear models. Therefore, a more complex model function is needed.

Many of the trends in the testing of pharmaceutical products are known. Dissolution profiles of extended release formulas tend to show the % release building up rapidly at early time points, with an inflection to slower growth at later times. This makes sense because the tablet formulation is designed to build to a therapeutic level as quickly as possible in a patient, with slower release needed to maintain the level over time as the body metabolizes the drug. A logistic function or exponential function typically fit dissolution profiles very well.

Exponential functions tend to be a bit better for in vitro testing since the very early release of a drug must reach a level that can be detected in solution (limit of detection), which nullifies the initial inflection of a logistic function. Use the red triangle menu next to the *Fit Curve* header and select *Exponential Growth and Decay* ▶ *Fit Exponential 3P* to get the output shown in Figure 15.18.

Figure 15.18: Three-Parameter Exponential Model Detail

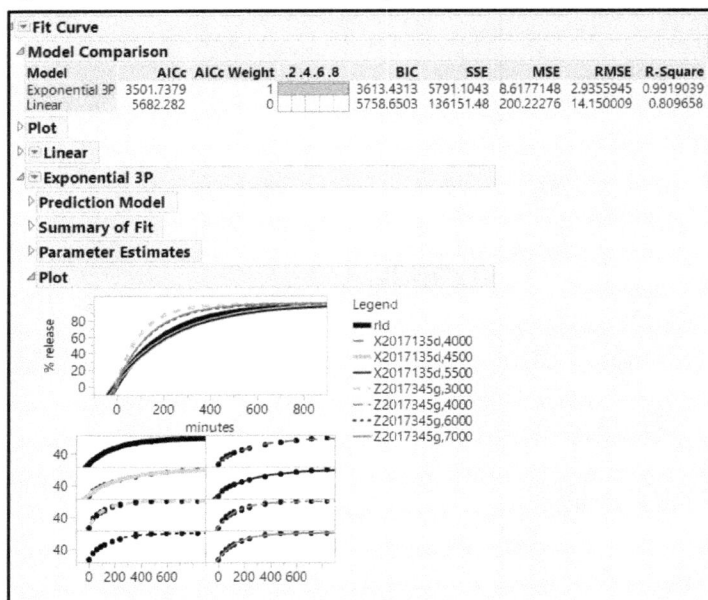

The output includes the overlay plot of the three-parameter exponential function, with the linear plot hidden for clarity. The *Model Comparison* at the top of the output provides summary statistics for the linear and exponential models. The fit of the three-parameter exponential function is much better than the linear function, as noted by an R-square of 0.99. The RMSE for the Exponential 3P model is five times less than the linear model. The model profiles fit the actual observations extremely well. Each model originates very near 0% release at time zero, which makes sense for actual trends expected for dissolution. The added complexity of the three-parameter exponential function adds a great deal of practical value for modeling the dissolution profiles.

It is easy to double-check how well the model performs by plotting the actual values against the predicted. Use the red triangle menu next to the *Exponential 3P* header and select *Plot Actual by Predicted* to get the output shown in Figure 15.19.

Figure 15.19: Exponential 3P Actual by Predicted Plot

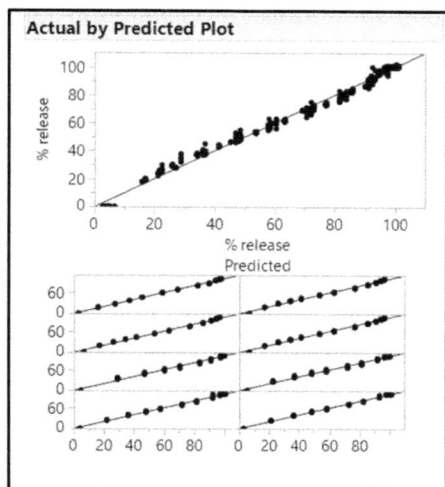

The actual dissolution values create a tight linear pattern along the prediction model line, which is highly desired. There is a group of points at zero that do not follow the plot. However, the 0 entered for each group to anchor the model lines are not real values and should not be considered for the robustness of the fit.

This exponential model has three parameters. It is helpful to understand what each of the parameters represent. Figure 15.20 illustrates the model parameters of Asymptote, Scale, and Growth Rate.

Figure 15.20: Exponential 3P Parameters

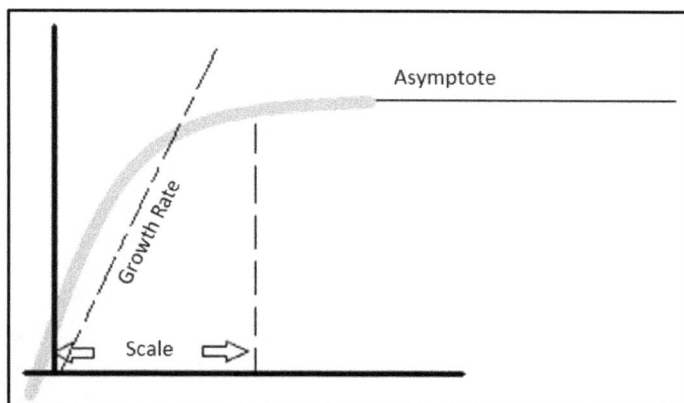

Asymptote: This is where the model terminates at and beyond the maximum time of observations. Drug dissolution typically ends up with complete release, which is approximately 100%.

Scale: Scale is the magnitude of how much the function stretches along the X axis

Growth Rate: The growth rate is the approximate slope set before the trend inflects and reaches the asymptote.

The next step in the analysis is to evaluate the candidates by each of the model parameters. Use the red triangle menu next to the *Exponential 3P* header and select *Compare Parameter Estimates* to get the output shown in Figure 15.21.

Figure 15.21: Parameter Comparison for Exponential 3P Modeling

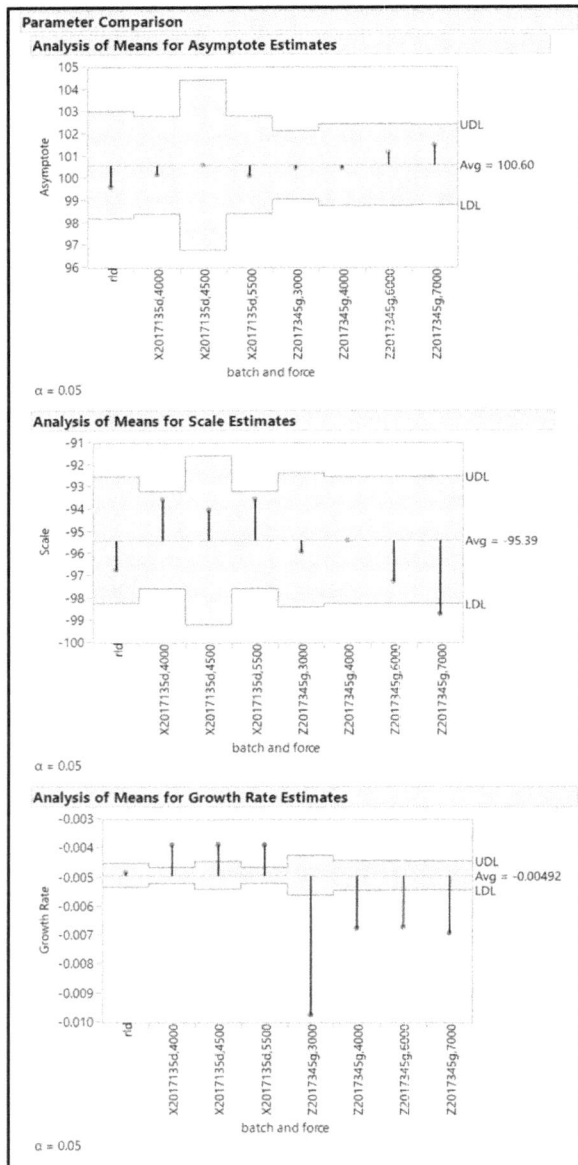

Parameter plots include a pooled average parameter and a blue-shaded zone that represents the random variability expected due to the variation in observations. Non-random differences are highlighted with a red point to illustrate a significant difference from the pooled average.

The rld group illustrates the parameters for the target dissolution profile. The goal is to find a candidate group that has parameter estimates that are closest to the rld parameter estimates. When analyzing parameters for dissolution profiles, the asymptote typically offers information of practical value, unless candidate groups are not fully releasing active ingredient to 100%. The estimates for the asymptote in the example include very little variation.

The scale estimate is important because it is an analog for the length of time that the tablet continues to dissolve, and that drug product is releasing active into the media. The rld target has a scale that is smaller than the pooled average. Two of the candidate groups have a scale estimate that is like that of the rld target: Z2017345g, 6000 and Z2017345g, 7000.

The growth rate parameter is the most important of the three for dissolution profile comparisons since it describes the speed of release. There is much more differentiation for the estimates of growth rate. The target profile can be considered statistically similar to the pooled average since it is basically in the middle of the plot. There are three candidate groups with growth rate estimates that are close to the target: X2017135d, 4000; X2017135d, 4500; and X2017135d, 5500. The analyst can look more deeply into parameter estimates. Use the red triangle menu next to the *Exponential 3P* header and select *Make Parameter Table* if actual numerical estimates are desired to further support the graphical evidence.

Equivalence Testing of Dissolution Profiles

The problem with comparing parameter estimates is that all are compared to a pooled average. Heinz' team needs to know which of the candidate profiles is the closest match to the target profile. JMP includes a red triangle menu option for an equivalence test, which enables you to choose the target profile as the basis of parameter comparisons. Use the red triangle menu next to the *Exponential 3P* header, select the *Equivalence Test* option, and then select *rld* as the reference group, as shown in Figure 15.22. Then, click *OK*.

Figure 15.22: Reference Group Selection

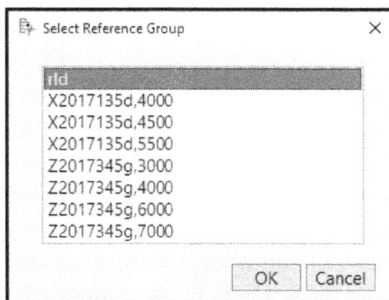

Seven Equivalence plots are added to the output in order to compare the rld to each candidate batch, including a ratio comparison to the rld target profile estimates. The plots of the best and worst candidates are shown in Figures 15.23 and Figure 15.24 respectively.

Figure 15.23: Equivalence Test (Best Candidate Profile)

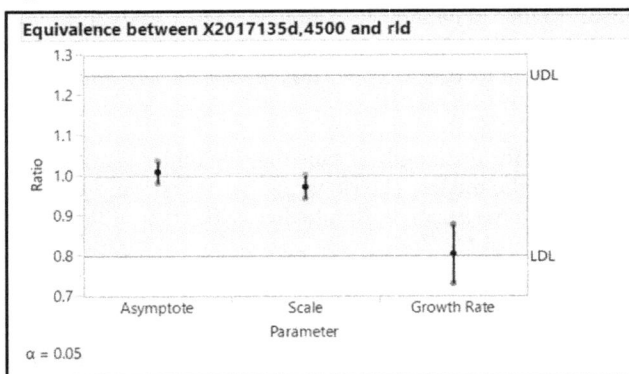

The equivalence plot for the best candidate shows the equivalence of the group X2017135d, 3500. The asymptote and scale parameters have the ratio value of approximately 1.0 with very tight confidence intervals, indicating robust equivalence to the target. The growth rate is less than the target, with the lower portion of the confidence interval just outside of the blue-shaded zone of random variation. The slight departure of the growth rate might cause some concern. However, you must keep in mind that testing for parameter equivalence tends to be conservative.

Figure 15.24: Equivalence Test (Worst Candidate Profile)

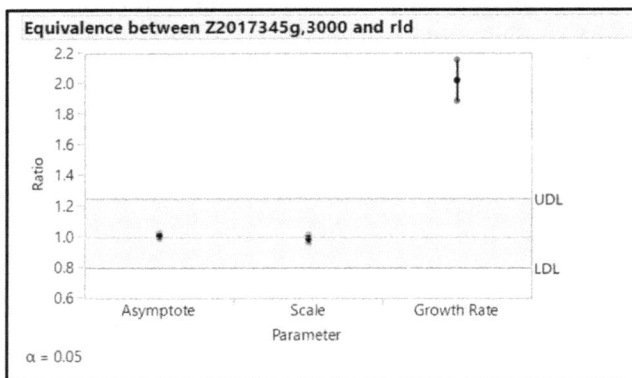

The candidate group Z2017345g, 3000 is the worst regarding equivalence to the target profile. The asymptote and scale parameters are equivalent to the target, but the growth rate is greater to a high degree of statistical significance. The entire confidence interval is significantly distant from the decision zone of random variation. Go back to the Exponential 3P overlay plot in figure 15.18 to see the functional difference in growth rate and compare the candidates with the best and worst equivalence results.

Comparisons of Dissolution Profiles with the F2 Similarity Criterion

The comparison of dissolution profiles via model-based methods requires that stakeholders understand the mathematical function used and the meaning of the parameters. You cannot explain the relationship between the candidate groups without using multiple parameter comparisons, which can be confusing. A method was developed that simplifies comparisons into a single value. This value is known as the F2 similarity criterion and is used by regulatory agencies to justify the choice of a candidate dissolution profile. The F2 value is calculated by a log-based function that is calculated from the sum of squares difference between the candidate and target for each time point.

The goal is an F2 value of at least 50, which signifies a 10% average difference between the candidate and target profiles. The data you have worked with in this chapter includes six individual tablet values for both the target profile and the candidate groups. The application of F2 involves mean values for both the target and the candidate groups, so the data set must be summarized. The data table has been prepared by summarizing each group and formatting the data back to an unstacked table. Open mean disso.jmp to see the summary data table shown in Figure 15.25.

Figure 15.25: Summary Unstacked Dissolution Data

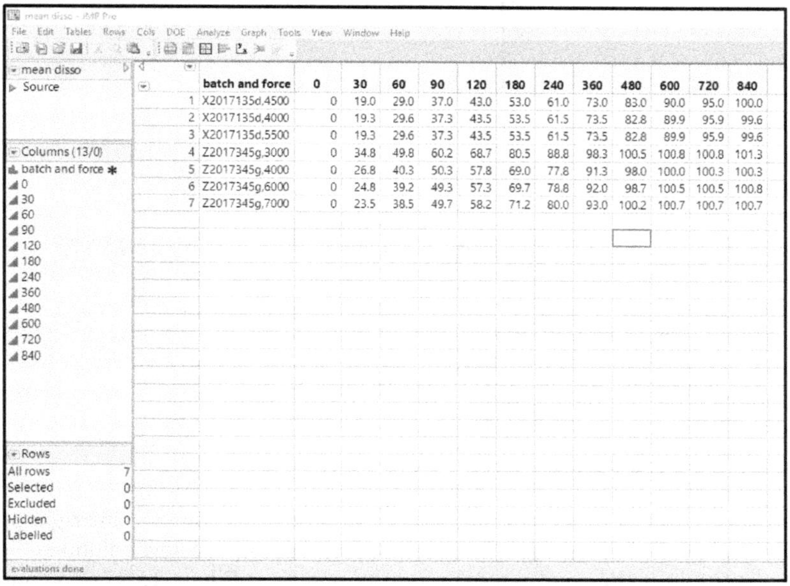

	batch and force	0	30	60	90	120	180	240	360	480	600	720	840
1	X2017135d,4500	0	19.0	29.0	37.0	43.0	53.0	61.0	73.0	83.0	90.0	95.0	100.0
2	X2017135d,4000	0	19.3	29.6	37.3	43.5	53.5	61.5	73.5	82.8	89.9	95.9	99.6
3	X2017135d,5500	0	19.3	29.6	37.3	43.5	53.5	61.5	73.5	82.8	89.9	95.9	99.6
4	Z2017345g,3000	0	34.8	49.8	60.2	68.7	80.5	88.8	98.3	100.5	100.8	100.8	101.3
5	Z2017345g,4000	0	26.8	40.3	50.3	57.8	69.0	77.8	91.3	98.0	100.0	100.3	100.3
6	Z2017345g,6000	0	24.8	39.2	49.3	57.3	69.7	78.8	92.0	98.7	100.5	100.5	100.8
7	Z2017345g,7000	0	23.5	38.5	49.7	58.2	71.2	80.0	93.0	100.2	100.7	100.7	100.7

Summary target values include the average % release of the rld group for each time point. The summary target values are used to calculate the sum of squares difference for all time points. The F2 formula involves the squared difference (SSQD) between the % released and the average % release of eight time points from 30 to 480 minutes. Complete the following steps to add the calculated SSQD column to the data table:

1. Create a new column to the right of the *840* column by selecting *Cols* ▶ *New Columns*.
2. Enter SSQD as the name of the new column.
3. Click *Column Properties* and select the *Formula* option.
4. Enter the formula shown in Figure 15.26, and click *OK* to execute.

Figure 15.26: Sum of Squares Difference Formula (Comparing to "rld")

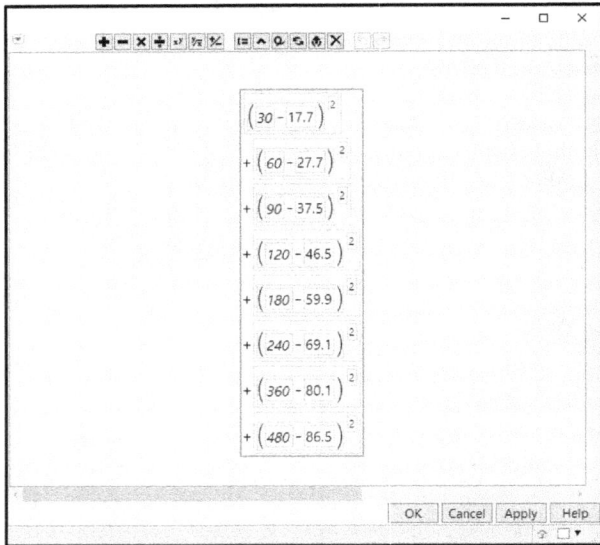

The SSQD variable is the variable portion of the F2 similarity calculation, shown in Figure 15.27.

Figure 15.27: F2 Similarity Criterion

$$50 \cdot \text{Log10}\left(\left(1 + 0.2 \cdot SSQD\right)^{-0.5} \cdot 100\right)$$

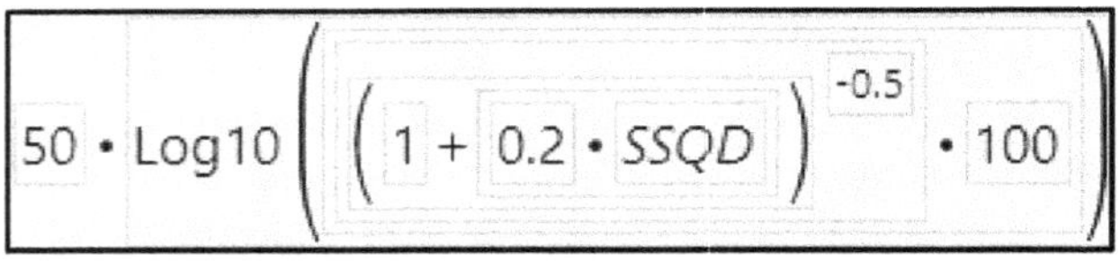

Add another column to the data table for the F2 similarity criterion. Create a new column to the right of the *SSQD* column by selecting *Cols>New Columns* and entering F2 as the name of the new column. Click *Column Properties*, select the *Formula* option, and enter the formula shown in Figure 15.27. You must use the Transcendental functions within the formula editor to get Log 10 into the formula. Click *OK* to execute and get the data table results shown in Figure 15.28.

Figure 15.28: Dissolution Profile Data Table with F2

	batch and force	0	30	60	90	120	180	240	360	480	600	720	840	SSQD	F2
1	X2017135d,4500	0	19.0	29.0	37.0	43.0	53.0	61.0	73.0	83.0	90.0	95.0	100.0	191.76	60
2	X2017135d,4000	0	19.3	29.6	37.3	43.5	53.5	61.5	73.5	82.8	89.9	95.9	99.6	171.00166667	61
3	X2017135d,5500	0	19.3	29.6	37.3	43.5	53.5	61.5	73.5	82.8	89.9	95.9	99.6	171.00166667	61
4	Z2017345g,3000	0	34.8	49.8	60.2	68.7	80.5	88.8	98.3	100.5	100.8	100.8	101.3	3130.7933333	30
5	Z2017345g,4000	0	26.8	40.3	50.3	57.8	69.0	77.8	91.3	98.0	100.0	100.3	100.3	953.67666667	43
6	Z2017345g,6000	0	24.8	39.2	49.3	57.3	69.7	78.8	92.0	98.7	100.5	100.5	100.8	919.52111111	43
7	Z2017345g,7000	0	23.5	38.5	49.7	58.2	71.2	80.0	93.0	100.2	100.7	100.7	100.7	1033.3544444	42

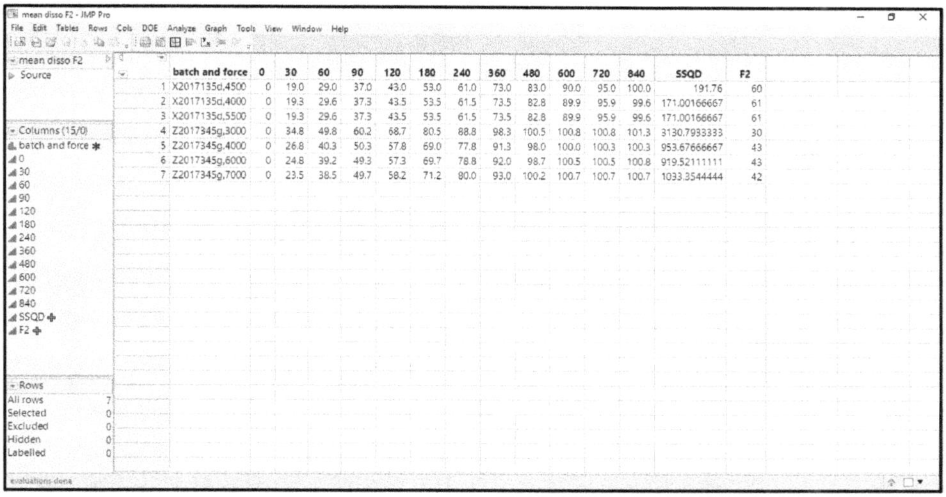

The F2 similarity criterion values are added to the end of the dissolution data table. It is not clear that there are three candidate groups with an F2 that is above the acceptance level of no less than 50. Since a data table is not the best way to visualize data, use a simple Graph Builder plot. Select *Graph* ▶ *Graph Builder* to get to the *Graph Builder* dialog box. Drag *F2* to the main drop zone, and drag *batch and force* to the *X* drop zone. Click *Done* to get the plot in Figure 15.29.

Figure 15.29: F2 vs. batch and force plot

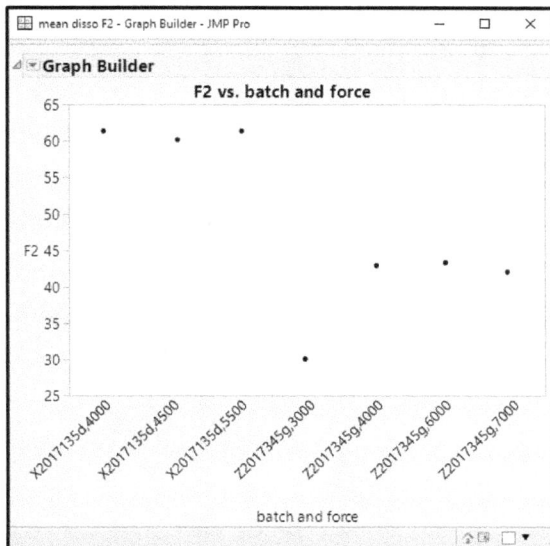

The plot now provides visual evidence of the candidate groups that have acceptable similarity to the target profile. Use the axis settings to clarify the plot by shading the undesirable zone for F2. Complete the following steps:

1. Right-click on the *F2* axis to get to the Y axis settings shown in Figure 15.30.
2. In the *Reference Lines* area of the settings, select the *Allow Ranges* check box.
3. Enter 20 for the *Min Value*, and enter 50 for *Max Value*.
4. Click on the black *Color* field and select deep red from the color pallet.
5. Click *Add*. A preview of the shaded plot appears to the right of the dialog box.
6. Click *OK* to execute the shaded area of the plot.

Figure 15.30: Y Axis Settings

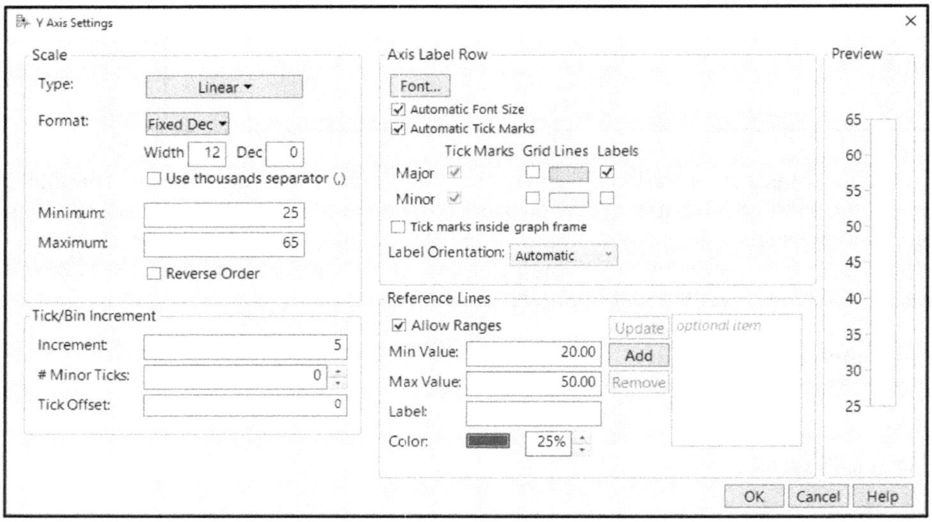

Figure 15.31: F2 Plot with Shaded Zone of Undesirability

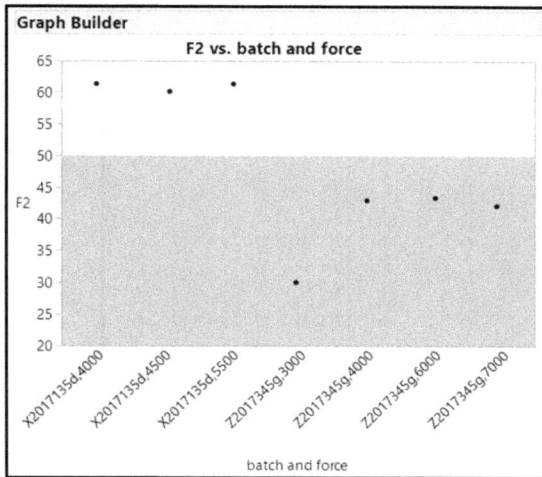

The result provides simple, graphical evidence that three of the seven candidate groups have F2 values that are between 60 and 61. A valid critique of using F2 for comparing dissolution profiles to target is that the single value has no predictive value. Since you know that variability is inherent to analytical testing, you cannot be sure that an F2 value of 60 will not be 50 or below on a repeat dissolution test of the same group. The model-based testing has a big advantage since the parameter estimates are predictive by nature.

Making F2 Similarity Predictive

You can make F2 similarity predictive. Each tablet in a group is dissolved in a unique well of an analytical instrument. Therefore, the results over time are dependent on the individual tablet. You can use a table of individual tablet values, with the same calculations, to get F2 values. Use JMP to plot the groups of F2 values with statistical error to provide stakeholders with a visualization of similarity.

1. Open *individual disso.jmp* to see results for each tablet over time in an unstacked format.
2. Go back to the *mean disso.jmp* table that includes the *SSQD* and *F2* columns.
3. Right-click on the *SSQD* column, and select *Formula*. Right-click on the formula and select *Copy*.
4. Go to *individual disso.jmp*.
5. Create a new column to the right on the *840* column by selecting *Cols* ▶ *New Columns*.
6. Enter *SSQD* as the name of the new column
7. Click *Column Properties* and select the *Formula* option.
8. Paste the copied formula, and click *OK* to execute.
9. Repeat steps 2–8 to create an *F2* column and to copy and paste the formula. The resulting table is shown in Figure 15.32.

Figure 15.32: Individual Disso Table with F2

	tablet sample	batch and force	0	30	60	90	120	180	240	360	480	600	720	840	SSQD	F2
1	1	Z2017345g,4000	0	26	40	50	58	69	78	91	98	100	100	100	921.76	43
2	2	Z2017345g,4000	0	26	39	49	56	67	76	89	96	100	100	100	686.56	46
3	3	Z2017345g,4000	0	30	45	55	63	75	83	94	100	100	101	101	1825.76	36
4	4	Z2017345g,4000	0	27	40	50	57	68	77	88	100	100	100	100	876.96	44
5	5	Z2017345g,4000	0	27	40	50	58	69	78	92	97	100	101	100	940.16	43
6	6	Z2017345g,4000	0	25	38	48	55	66	75	94	98	100	100	100	739.36	46
7	1	Z2017345g,3000	0	35	50	60	69	81	90	100	101	101	101	101	3297.36	30
8	2	Z2017345g,3000	0	38	53	63	71	83	92	100	100	100	100	101	3938.96	28
9	3	Z2017345g,3000	0	33	48	59	67	79	86	96	101	101	101	102	2642.16	32
10	4	Z2017345g,3000	0	32	46	56	64	77	85	94	100	101	101	101	2108.56	34
11	5	Z2017345g,3000	0	34	51	62	71	81	89	99	100	101	101	101	3389.76	29
12	6	Z2017345g,3000	0	37	51	61	70	82	91	101	101	101	101	102	3634.96	28
13	1	Z2017345g,6000	0	25	40	50	58	70	79	91	99	100	100	101	968.16	43
14	2	Z2017345g,6000	0	25	39	49	57	70	79	91	99	100	100	101	898.56	44
15	3	Z2017345g,6000	0	24	37	47	55	67	76	88	96	101	101	101	539.36	49
16	4	Z2017345g,6000	0	25	39	49	57	69	78	90	98	101	101	101	815.76	45
17	5	Z2017345g,6000	0	24	39	50	58	71	81	100	100	100	100	100	1298.96	40
18	6	Z2017345g,6000	0	26	41	51	59	71	80	92	100	101	101	101	1150.16	41
19	1	Z2017345g,7000	0	23	37	48	57	70	80	92	100	101	101	101	879.76	44
20	2	Z2017345g,7000	0	24	38	49	57	71	80	93	100	100	100	100	978.96	43
21	3	Z2017345g,7000	0	22	37	48	56	69	78	91	101	101	101	101	796.56	45
22	4	Z2017345g,7000	0	24	39	51	60	73	81	98	101	101	101	101	1375.76	39
23	5	Z2017345g,7000	0	24	40	50	59	72	81	92	99	100	100	100	1089.36	41
24	6	Z2017345g,7000	0	24	40	52	60	72	80	92	100	101	101	101	1172.56	41

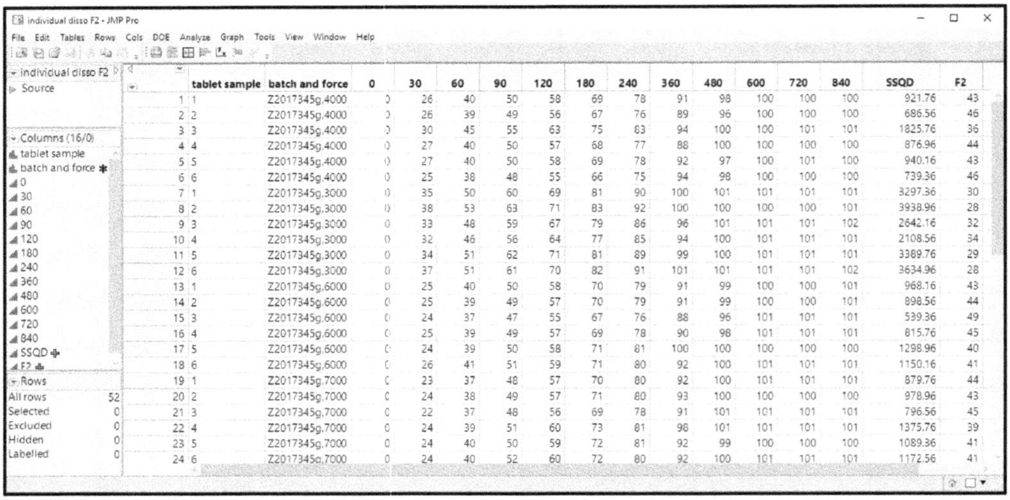

The individual disso table now contains 52 F2 values for each tablet tested. The Graph Builder is an important tool for the team to get a summary view of the results. Complete the same steps as you used for the mean dissolution to get the graph shown in Figure 15.33.

Figure 15.33: Graph Builder for F2 Individuals by batch and force

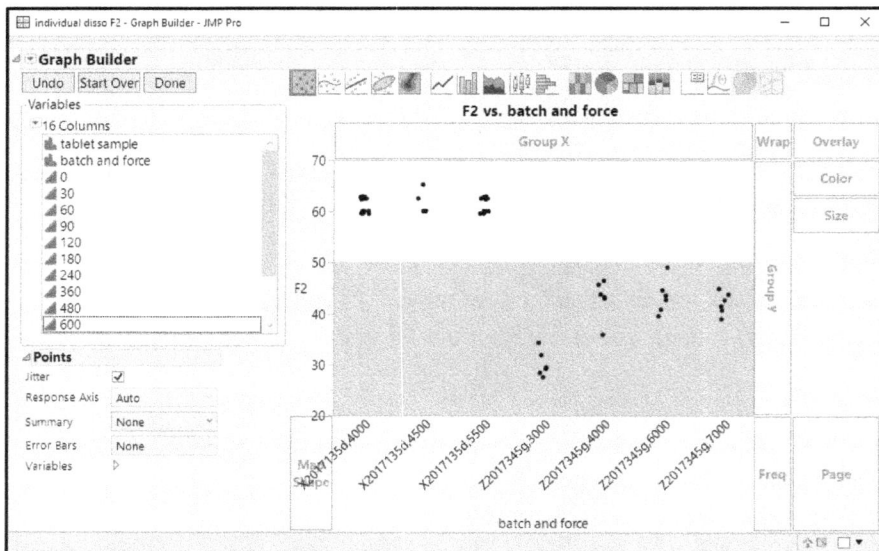

The markers indicate the individual F2 results of each batch. A summary view is better to show the range predicted from the data. Under the *Points* header, change the *Summary* to *Mean*, and *Error Bars* to *Confidence Interval*, as shown in Figure 15.34.

Figure 15.34: Graph Builder Points Edits

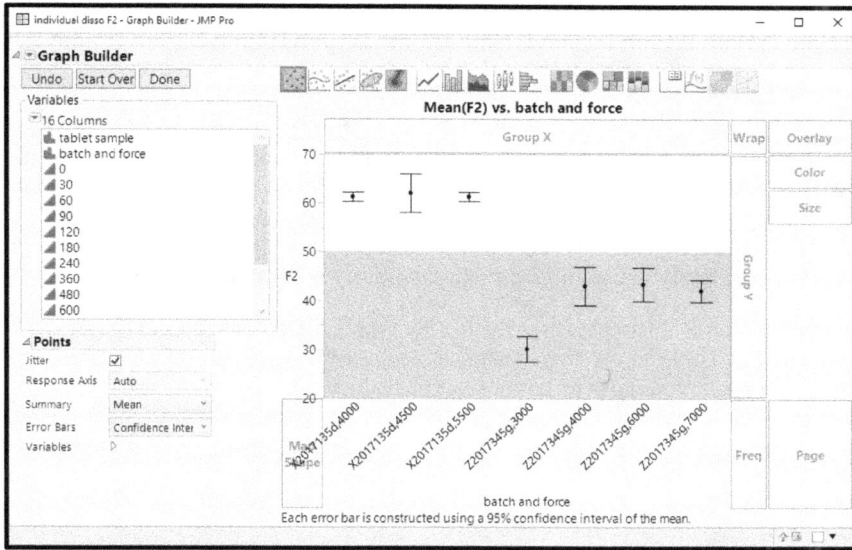

Click *Done* to get the plot in Figure 15.35.

Figure 15.35: Graph of Predictive F2

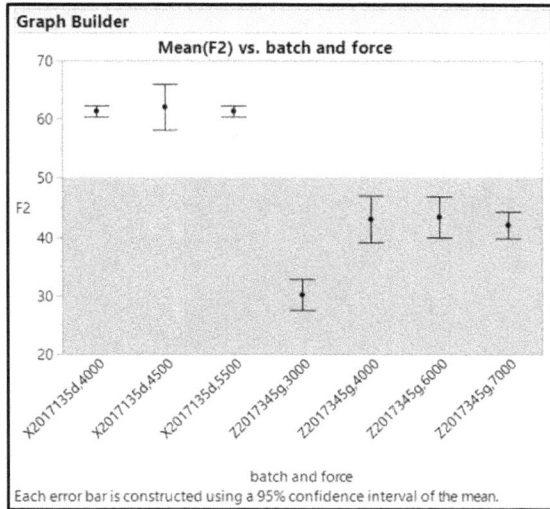

The F2 calculations for individual values provide some inference as to the robustness of the similarity of candidate groups to the target profile. It is clear in the plot that the X2017135d formula is the best for being like the target dissolution profile regardless of the tablet hardness (noted by compression force in the groups). There was more variability for tablets produced at 4500 pounds of force. However, this could be due to something other than the tablet compression process. The plot suggests that the population of tablets made for any of the three candidates will have no less that an F2 of 58. Heinz can guide the team to make a robust choice for the formula and compression forces needed to have the best possible chance of having an equivalent drug release to the target product.

Using Non-linear Models for Mesh Testing of Particle-Size Trends

Non-linear modeling is well suited to the study of cumulative distributions. Particle-size study of solid materials used in pharmaceutical applications typically involves the analysis of the amount of material retained over a stack of increasingly dense mesh screens. The percentage of total weight of material collected on each screen creates a particle-size distribution. Mesh screens are used regularly because that method does not require specialized analytical equipment or a laboratory environment.

The pilot plant technicians completed Ro-Tap particle-size testing with a 12-screen stack for five developmental granulation batches. Three of the five batches have desirable particle size characteristics. The goal of the team is to have a limited number of target screens that can be used to effectively differentiate particle size trends. The screens will be used for the in-process testing of future granulation batches to identify extremes. The data has been compiled in the stacked table format required to run non-linear models. Open *particle.jmp* to see the data shown in Figure 15.36.

Figure 15.36: Particle Data File

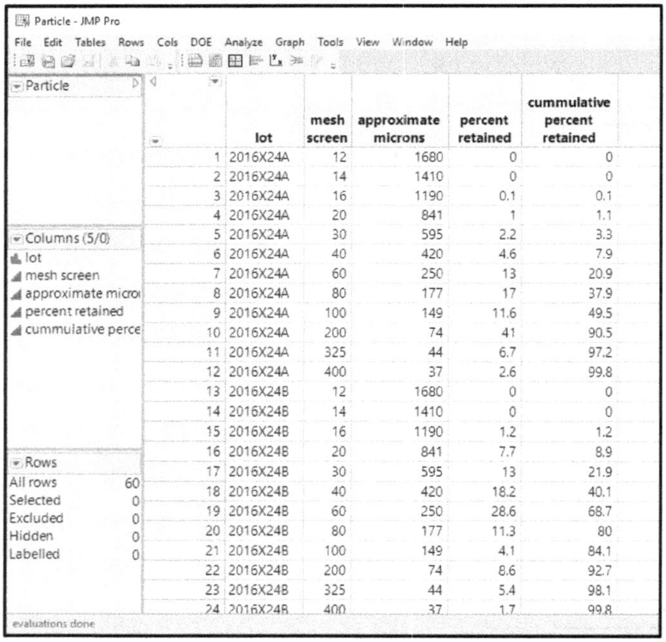

	lot	mesh screen	approximate microns	percent retained	cummulative percent retained
1	2016X24A	12	1680	0	0
2	2016X24A	14	1410	0	0
3	2016X24A	16	1190	0.1	0.1
4	2016X24A	20	841	1	1.1
5	2016X24A	30	595	2.2	3.3
6	2016X24A	40	420	4.6	7.9
7	2016X24A	60	250	13	20.9
8	2016X24A	80	177	17	37.9
9	2016X24A	100	149	11.6	49.5
10	2016X24A	200	74	41	90.5
11	2016X24A	325	44	6.7	97.2
12	2016X24A	400	37	2.6	99.8
13	2016X24B	12	1680	0	0
14	2016X24B	14	1410	0	0
15	2016X24B	16	1190	1.2	1.2
16	2016X24B	20	841	7.7	8.9
17	2016X24B	30	595	13	21.9
18	2016X24B	40	420	18.2	40.1
19	2016X24B	60	250	28.6	68.7
20	2016X24B	80	177	11.3	80
21	2016X24B	100	149	4.1	84.1
22	2016X24B	200	74	8.6	92.7
23	2016X24B	325	44	5.4	98.1
24	2016X24B	400	37	1.7	99.8

Non-linear modeling provides an excellent overlaid plot to visualize trends. Select *Analyze* ► *Specialized Modeling* ► *Nonlinear* to open the *Nonlinear* dialog box shown in Figure 15.37.

Figure 15.37: Nonlinear Dialog Box

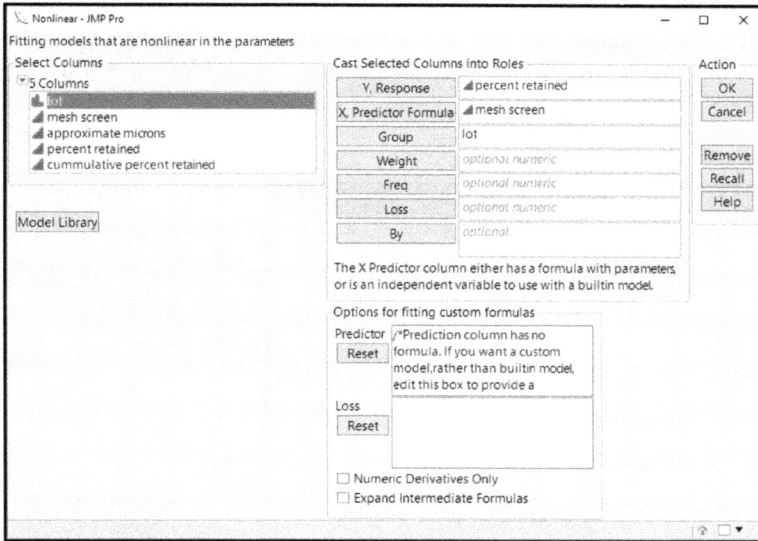

In the *Nonlinear* dialog box, move *cumulative percent retained* to the *Y, Response* box, *mesh screen* to the *X, Predictor Formula* box, and *lot* to the *Group* box. Click *OK* to get the output shown in Figure 15.38.

Figure 15.38: Fit Curve Output

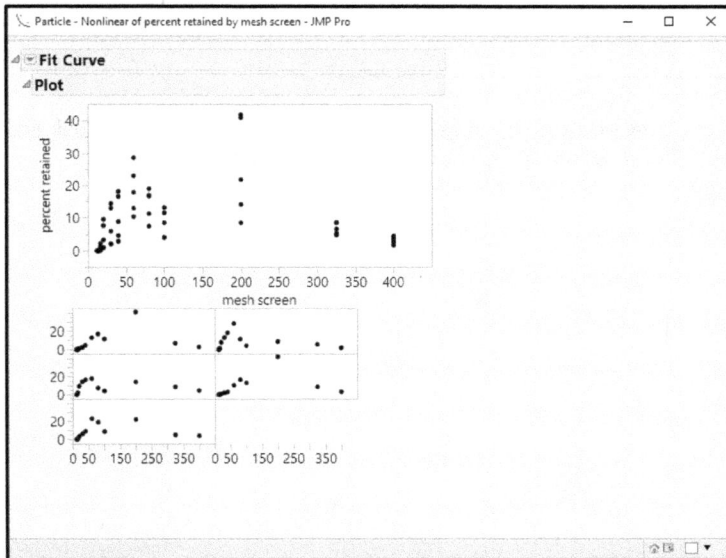

The pooled and individual group patterns in the plots follow a peaked pattern. There are two options for peaked models within the list of nonlinear modeling options: Gaussian Peak (Normal Bell Curve) and Lorentzian Peak. Both options need to be evaluated for fit to determine whether additional analysis is appropriate. Use the red triangle menu next to the *Fit Curve* header to select *Peak Models* ▶ *Fit Gaussian Peak*, as shown in Figure 15.39.

Figure 15.39: Fit Curve Red Triangle Menu List of Models

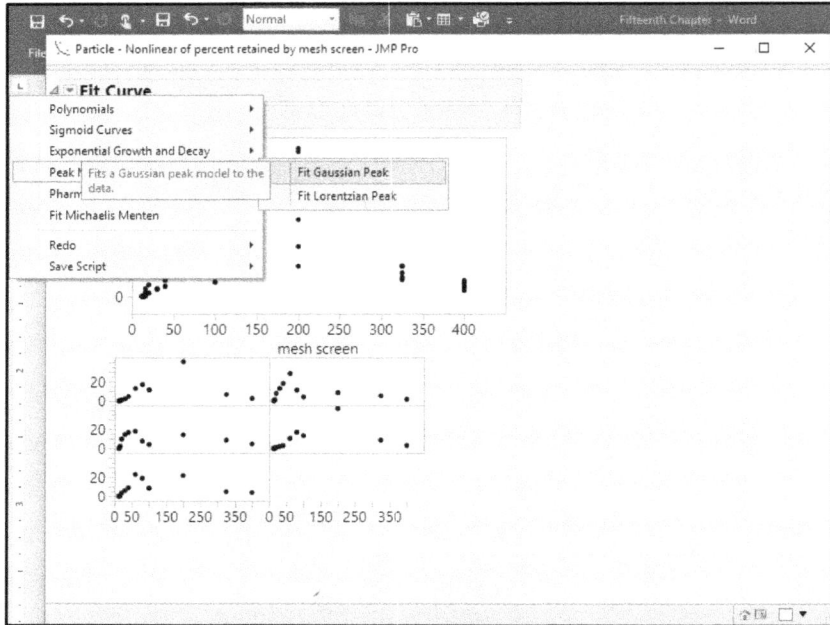

Use the red triangle menu again to select *Peak Models* ▶ *Fit Lorentzian Peak.* The Gaussian Peak and Lorentian Peak models are added to the *Fit Curve* output, as shown in Figure 15.40.

Figure 15.40: Peaked Non-linear Models

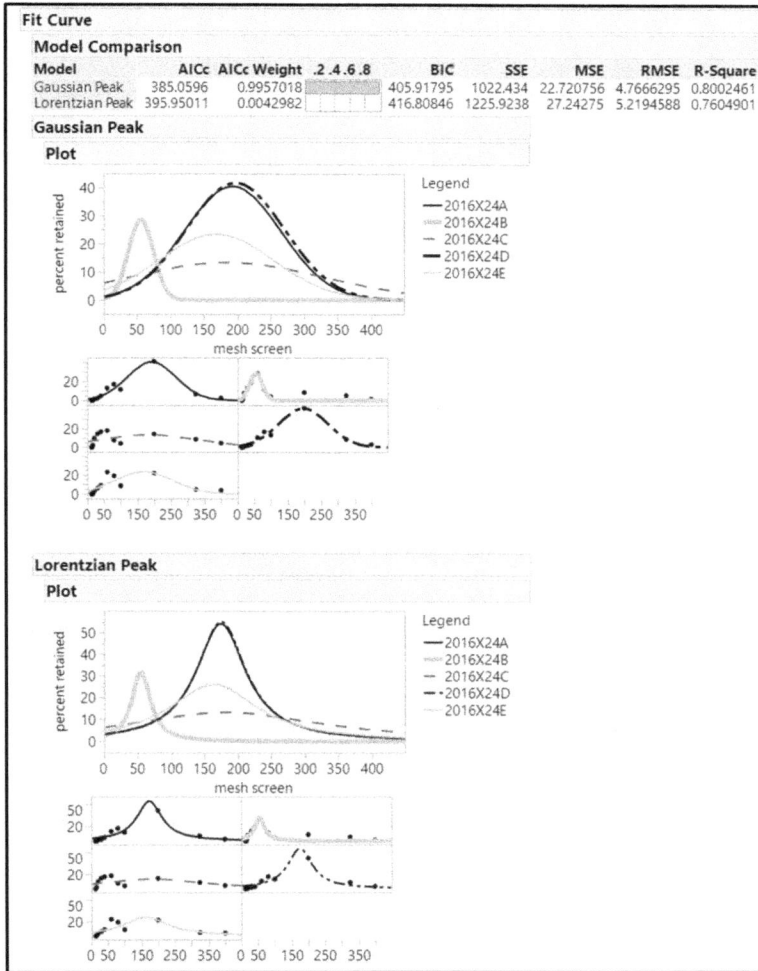

Each plot of the peak models illustrates a moderate fit to the percent retained observations. The models explain roughly 80% of the variability in percent retained results, and some of the lots are not well represented by a peak model (especially 2016V24C and 2016V24E).

Modeling the cumulative particle size output might yield better fits. The cumulative model is more representative of how target screens are used during operations, which makes interpretation of results practically applicable. Use the red triangle menu next to the *Fit Curve* header to select *Redo* ▶ *Relaunch Analysis*. Remove *percent retained* from the *Y, Response* box and replace with *cumulative percent retained*. Click *OK* to get the output shown in Figure 15.41.

Figure 15.41: Fit Curve for Cumulative Percent Retained

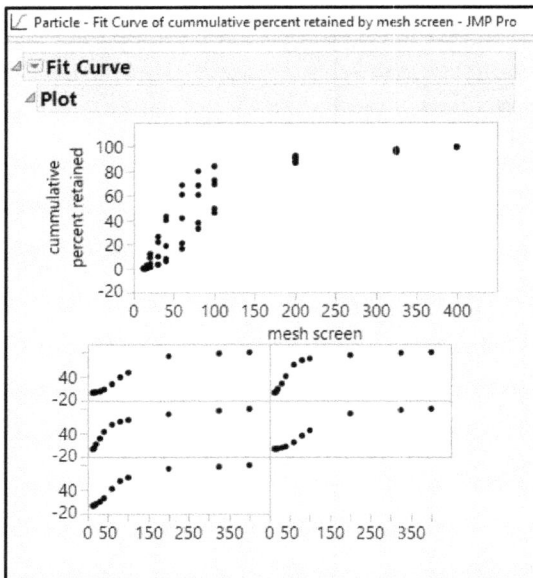

The plot of cumulative percent retained includes trends of observations that start out flat initially, inflect to a steep growth rate, and inflect again to an asymptote for the larger number screens. The pattern described seems to be of a logistic function, which is one of the model choices in the Nonlinear platform. Use the red triangle menu next to the *Fit Curve* header to select *Sigmoid Curves* ▶ *Logistic Curves* ▶ *Fit Logistic 3P* to get the output shown in Figure 15.42.

Figure 15.42: Fit Curve for Three-Parameter Logistic Model

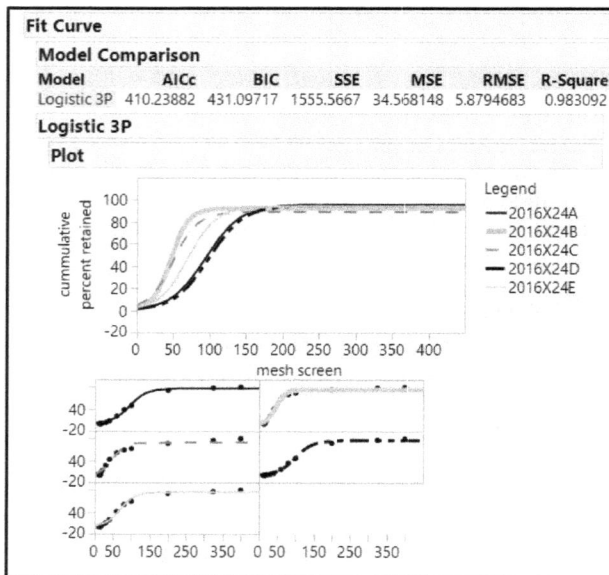

The three-parameter logistic models have an excellent fit to the cumulative screen results data, with the R-square of 0.983. The three-parameter logistic model explains 98.3% of the variability in the screen testing observations. You should try logistic models with additional parameters to see the differences in results. It will become quite clear that the simple three-parameter model fits the data very well and is the easiest to interpret and explain.

Comparing model parameters provides insight into where the lots vary significantly. Use the red triangle menu next to the *Logistic 3P* header to select *Compare Parameter Estimates* and add the parameter comparison to the output, as shown in Figure 15.43.

Figure 15.43: Fit Curve for Three-Parameter Logistic Model

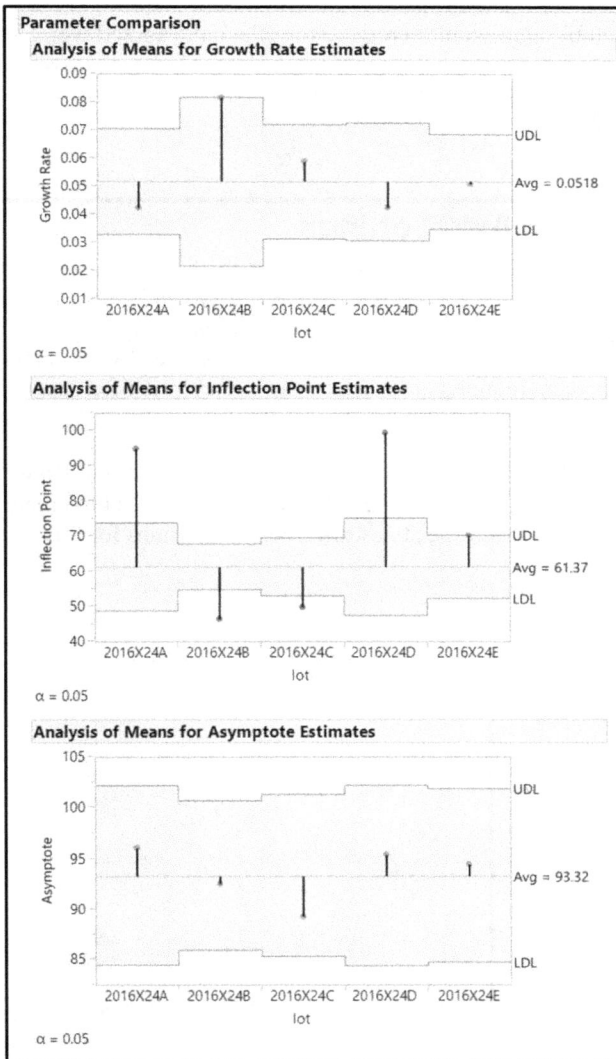

Significant differences in the inflection points between the batches are present. Batches 2016X24A and 2016X24D have inflection points that are located at much higher number mesh screens than the other three batches. The overlay plot for the logistic 3P models illustrates a function for the two batches that is much flatter. A flat logistic function for particle size indicates high variability in sizes throughout the distribution.

The growth rate of batch 2016X24B differs significantly from the other batches. Fast growth of a cumulative particle-size distribution can be interpreted as one that has less spread in size amounts than others. Significant differences in the parameters of non-linear functions for physical properties can be a clue to differences in the function of the granulations. It is wise for teams to track trends in physical properties and relate them to differences in analytical results in order to better understand the effects of the process.

Differences in the asymptote are not expected to be significant. Screen test results have a fixed weight of sample that is placed onto the top screen of the test instrument. The cumulative total of all screens should always be close to 100%. A significant difference in the asymptote of a batch might indicate testing error and loss of portions of the sample during testing.

Augmenting Non-linear Plots by Using Axis Settings

The rate of increase of the logistic 3P function for the cumulative particle size of the batches is steepest at the inflection point. A minor shift in the inflection point on the mesh screen axis between batches results in a big change on the percent retained axis. Target screens proposed for in-process monitoring of granulations must be located away from the inflection point to be the most discriminative. Target screens chosen within this zone of high variability for percent retained are not likely to detect true differences in particle size profiles.

Add two mesh screens within this zone near the inflection point to the plot to illustrate the point. Right-click on the mesh screen axis, and select *Axis Settings*. Add the detail shown in Figure 15.44 to create lines for the *60M* and *80M* screen mesh sizes on the plots. Line styles differentiate the lines for this example, but colors are typically used as the default.

Figure 15.44: Axis Settings

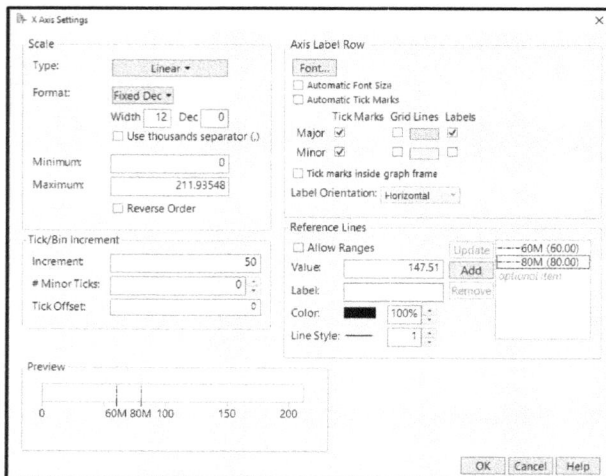

Leave all other options unchanged, and click *OK* to get the plot shown in Figure 15.45.

Figure 15.45: Fit Curve with 60M and 80M Screens

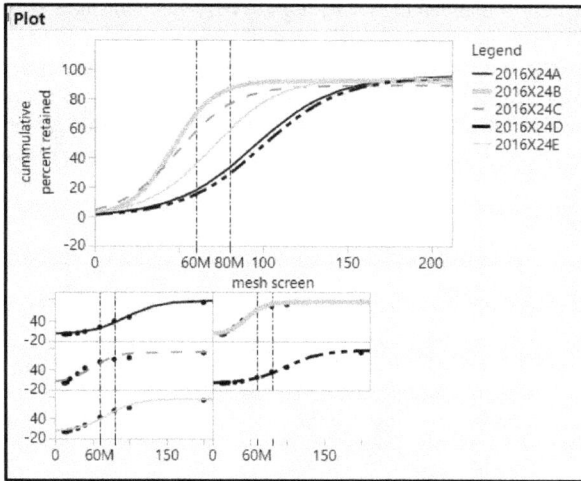

The overlay plots with reference lines illustrate large differences in cumulative percent retained among all five lots as vertical shifts of the functions. The amount of random variability creates great difficulty for the team to attempt to set specifications that will be utilized by operations as acceptance criteria for granulation particle size. The team needs to pick screens that are further away from the inflection point without including the asymptotes on either end of the function to effectively control particle size for granulations. Right-click on the mesh screen axis again, and select *Axis Settings*. Select each of the *60M* and *80M* reference lines, and enter the values *40M* and *120M*, respectively. Click *Update* for each change, and then click *OK* to get the plot shown in Figure 15.46.

Figure 15.46: Fit Curve with 40M and 120M Screens

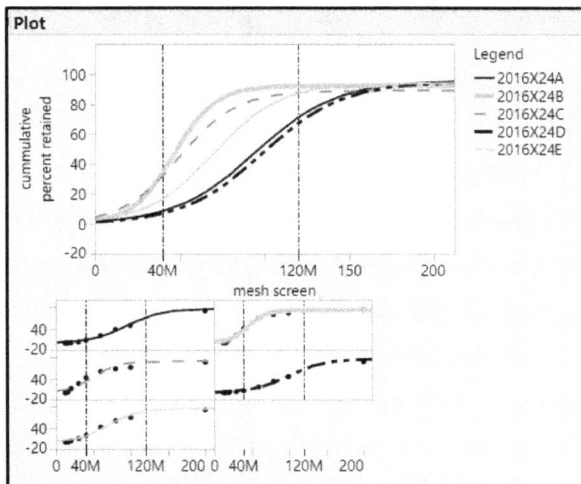

Using the 40 M and 120 M screens as targets for in-process monitoring of particle size is more reasonable than using the 60 M and 80 M screens. The lots known to have good physical characteristics (2016X24A, 2016X24D, and 2016X24E) are differentiated from the two that are known to be less desirable due to excessive variation in particle size. The 120 M screen is located closer to the curve, representing a slowing growth rate for the function, which yields reduced random variability. One challenge is that a 120 M mesh screen was not included in the stack used to create the data. The next section makes specific predictions of %retained values for a 120 M screen. The predictions help justify the obtainment of a 120 M mesh screen for in-process testing. Reduced variability is also shown for the 40 M screen located close to the curve, representing a quickening growth rate. The best choice for target screens justified by the overlay plot are the 40 M and the 120 M screens.

Making Predictions with Non-linear Models

The focus of this analysis turns to creating reasonable specifications for material that indicate good physical characteristics. Having a good fitting functional model for the cumulative particle-size distribution is extremely beneficial since you can use the model to make robust predictions. The predictions enable you to study a large number of scenarios without the need to make physical batches of granulations to get actual results from particle-size testing.

Save the prediction formula variable by using the model options and use it to provide the needed information. Use the red triangle menu next to the *Logistic 3P* header to select *Save Formula* ▶ *Save Prediction Formula*. A new column is added to the end of the table. This column is automatically named *cumulative percent retained Predictor* and is shown in the data table in Figure 15.47.

Figure 15.47: Particle Table with Predictions

	lot	mesh screen	approximate microns	percent retained	cummulative percent retained	cummulative percent retained Predictor
47	2016X24D	325	44	8.7	96.8	95.403805337
48	2016X24D	400	37	3.2	100	95.410054636
49	2016X24E	12	1680	0	0	4.6267257869
50	2016X24E	14	1410	0.1	0.1	5.0962971834
51	2016X24E	16	1190	0.4	0.5	5.6105493962
52	2016X24E	20	841	3.3	3.8	6.7877652566
53	2016X24E	30	595	6	9.8	10.783359932
54	2016X24E	40	420	8.9	18.7	16.683373498
55	2016X24E	60	250	23	41.7	35.211297502
56	2016X24E	80	177	19	60.7	58.767851064
57	2016X24E	100	149	8.6	69.3	77.475724425
58	2016X24E	200	74	21.9	91.2	94.331886904
59	2016X24E	325	44	4.8	96	94.458332058
60	2016X24E	400	37	3.9	99.9	94.45854465
61	2016X24A	120	.	.	.	71.338234437
62	2016X24B	120	.	.	.	92.33123645
63	2016X24C	120	.	.	.	87.902716411
64	2016X24D	120	.	.	.	67.3183543
65	2016X24E	120	.	.	.	87.533134142

Once you have added a prediction formula variable to a data table, you can create new predictions by adding rows with the desired input values. Scroll down the table to row 60, which is the last value of the original table. Right-click in the cell for the 60th row, and select *Add Rows*. Enter 5 for the number of rows to be added after row 60, and click *OK*. Rows 61–65 are added to the table. Type the identifying labels shown in Figure 15.48 for each of the five new lots in the empty cells of *lot*. Enter 120 in the empty cells in *mesh* for the 5 new rows. The predicted values fill in automatically. A subset of the table including only the *40M* and *120M* results is shown in Figure 15.48.

Figure 15.48: Subset of Particle-Size Table with Predictions

	lot	mesh screen	approximate microns	percent retained	cummulative percent retained	cummulative percent retained Predictor
1	2016X24A	40	420	4.6	7.9	8.5303796816
2	2016X24B	40	420	18.2	40.1	34.376156532
3	2016X24C	40	420	16.6	43.1	32.143787213
4	2016X24D	40	420	2.9	5.9	7.052879172
5	2016X24E	40	420	8.9	18.7	16.683373498
6	2016X24A	120	.	.	.	71.338234437
7	2016X24B	120	.	.	.	92.33123645
8	2016X24C	120	.	.	.	87.902716411
9	2016X24D	120	.	.	.	67.3183543
10	2016X24E	120	.	.	.	87.533134142

(JMP Pro window — Subset of Particle data table. Columns (6/0): lot, mesh screen, approximate microns, percent retained, cummulative percent retained, cummulative percent retained. Rows — All rows 10, Selected 0, Excluded 0, Hidden 0, Labelled 0.)

The granulations of greatest concern (2016X24B and 2016X24C) have 40 M values that are greater than 32%. The highest value for the three desirable batches is approximately 17%. A specification of no more than 25% retained on the 40 M screen seems to be reasonable for detecting granulations with an excessive amount of coarse particles.

A calculation is needed to determine the amount retained on the 120 M screen. The 40 M portion will be removed from the predicted amounts of % retained since the 40 mesh screen traps coarse particles with finer particles passing through to the 120 M. The calculations are noted in Table 15.1.

Table 15.1: 120 M %Retained Calculations

lot	120M cumulative % retained	40M % retained	120M % retained
2016X24A	71.3	8.5	62.8
2016X24B	92.3	34.4	57.9
2016X24C	87.9	32.1	55.8
2016X24D	67.3	7.1	60.2
2016X24E	87.5	16.7	70.8

The 120 M screen is predicted to contain no less than 60% retained material during in-process testing using the target screens. The variability of the 120 M predicted results from the study is minimal, so the

specification for the 120 M might not be as robust as the 40 M. Additional monitoring of results from the 120 M screen is warranted to ensure that it works as a target screen for in-process particle-size testing.

Practical Conclusions

The development team led by Heinz provides a strong justification for the candidate formula that was used as the exhibit batch for expensive clinical trials. There are healthy debates among the technical experts regarding the technique of choice used to compare dissolution profiles: model-based tools or the F2 similarity criterion. Non-linear modeling and visualizing F2 results using JMP led to the same formulation (X2017135d) as the best choice for moving forward. A minimal effect of compression force on dissolution is a desired result for new formulations. The compression force is not related to differences in dissolution shown in the analysis, which adds robustness to the exhibit batch choice. The model-based analysis identifies the growth rate as the parameter that differs the most between the target and candidate dissolution profiles, which is very good to know. The growth rate parameter should be an output for continued fine-tuning of the formula and developing the granulation process. The graphs and tables provided by JMP help the development team effectively explain their work to the project stakeholders and gain concurrence on the strategy for moving forward.

The operations team has manufacturing orders that specify the amount of retained material for two target screens (40 M and 120 M). The process of choosing the target screens is robustly supported by the non-linear modeling executed in JMP. The plots with the model predictions are easy for the stakeholders to interpret, and they offer robust justification to regulatory agencies. Leadership is exceptionally pleased at the amount of detail that is now available regarding the process of picking operational monitoring specifications for estimating the particle-size trends of granulations.

Exercises

E15.1—A tablet formula is in the late stages of development, and scientists are studying the release profile in OGD media. OGD refers to a specific mix of chemicals used to dissolve a drug tablet during dissolution testing. There is a suspicion that the hardness of tablets might have an effect on the dissolution profile. The team has tablet samples that were produced to two different hardness levels (20 Scu and 24 Scu) and has submitted 12 of each group for dissolution testing.

1. Open *OGD media dissolution comparision.jmp* and stack the results for analyses.
2. Rename the *Data* column to *% released* and the *Label* column to *minutes*.
3. Recode the *minutes* column to include numbers only, and change the column properties to numeric, continuous data.
4. Select *Analyze ▶ Specialized Modeling ▶ Nonlinear* with a three-parameter exponential function. Is the fit of the function appropriate for further analyses?
5. Use the red triangle menu next to the *Exponential 3P* header to compare parameter estimates. Do you see significant evidence of a difference? Which parameter (if any) differs? What does this mean?
6. Use the red triangle menu next to the *Exponential 3P* header to test for equivalence, using the 20 Scu batch as the reference group. How would you use this information to report to the project stake holders?

E15.2—There is a set of data for both batches that you worked with in exercise E15.1, with F2 calculations to the exhibit batch reference dissolution profile.

1. Open *OGD media dissolution comparison with F2 to reference.jmp*.

2. Use the Graph Builder to create box plots in order to compare the F2 similarity criterion of both batches.

3. How would you summarize the sensitivity of the formula to the minor difference in tablet hardness?

E15.3—Early development of a new tablet formula is underway so that the team can understand the effect of a high-shear granulation process. The team is evaluating the particle-size distribution with a stack of mesh screens and a Ro-Tap device. The cumulative percent retained on the mesh screens is to be analyzed with a non-linear model so that differences among the 12 batches of the set of structured experiments can be determined.

1. Open *newformulagrans.xlsx* in JMP using the Excel Wizard.

2. Use the Excel Wizard to start the data on row 6 with headers included.

3. Save the table as "newformulagrans unstacked actual.jmp."

4. The non-linear model of interest requires that the data be converted to cumulative results. Create five new columns by using the *Cols* menu options.

5. Add a new column named "0 mesh" that specifies 0 for all rows. Use the *Cols* menu to move the column to be between the *run* and *Sieve 30* columns. This column will be used to anchor the intercept of the non-linear model.

6. The table must be in stacked form in order to perform the analysis. Select *Tables* ▶ *Stack* to stack the eight columns of sieve results.
    -   Be sure that the default setting *Stack by Row* is selected.
    -   Use the default radio button to keep all non-stacked columns.
    -   Change *New Column Name* for *Data* to *% retained*, and change the *Label* column to *Sieve*.
    -   Add the *Output table name* "newformulagrans stacked."
    -   Click *OK* to create the table.

7. Use the *Cols* menu to recode the *sieve* column by splitting down to retain the last word (sieve number); change the *Pan* value to 300 to make it numeric. Change the default of the column recode to save the sieve numbers *In Place* to replace the values of the column.

8. Change the column properties of *sieve* to *Numeric, Continuous* data.

9. Create a new column with the name "cumulative % retained."

10. Calculate the cumulative results from the sieves 0 to 300 for each batch in the *cumulative % retained* column.

11. Utilize the tools in this chapter to analyze the *cumulative % retained* response (Y) with the *sieve* regressor (X) grouped by *batch*. Cumulative distribution functions for particle size tend to fit the three-parameter logistic function very well since it is expected that the PSD cannot be less than 0 and will tend to asymptote at 100%.
    -   Use the graphics options to size the plot so that you can see where the function for each batch reaches the asymptote.
    -   Does the plot make practical sense?

- The development team utilized a standard stack of eight sieves to collect data. Do you have evidence to support the standard stack… or not?

12. You have likely come to the conclusion that the standard sieve stack used to collect data was too coarse (the openings in the screens are too large) to provide much value. A big clue is the large amount collected in the pan (after particles passed through all sieves). How would you summarize this finding to the project stakeholders?

E15.4—The developmental scientists involved in the high-shear granulation experiments studied in exercise E15.2 collected more samples from the batches and passed them through a new sieve stack that included finer mesh. The data is saved in *newformulagrans fine screens stacked.jmp*.

1. Utilize the techniques explained in this chapter to create a set of non-linear models of *cumulative % retained* response (Y) with the *sieve* regressor (X) grouped by *batch* using a three-parameter logistic function. What is the fit of the function? Is it appropriate for further analysis?

2. Part of the development process includes the choice of target screens to be used in the commercial processing environment to detect differences in granulations. The upper and lower inflection points for changes in the rates of increase or decrease in the logistic function tend to be good areas for target screens because variability in such areas tends to differentiate particle-size distributions. The steep growth of the curve between the rate inflection points is a bad area for a target screen due to the high amount of random variability. Use the plot to determine the best target screens for detecting coarse particles (low sieve numbers) and fine particles (high sieve numbers).

3. Run a parameter comparison by using the red triangle menu options. Is there evidence of a significant difference among the batches with regard to particle size?

4. Relaunch the analyses utilizing *microns* as the regressor (X) with a three-parameter logistic function. What range of particle sizes will be difficult to detect due to the amount of variation?

5. How would you summarize this analysis to the project stakeholders?

Chapter 16: Using Statistics to Support Analytical Method Development

Overview

The United States Pharmacopeia (USP) has issued guidance regarding the use of statistics in support of analytical methods as one more leg of the Quality by Design (QbD) initiative. The proposed new General USP Chapter <1220> and the Analytical Procedure Lifecycle notes the goal to identify an analytical target profile (ATP). The ATP is being developed to ensure that an analytical method is robust for the entire lifecycle of use. The proposed General USP Chapter <1210> provides additional details on statistical validation techniques that are appropriate for the method so that necessary steps are taken to mitigate measurement uncertainty. This chapter provides a basic example of an analytical method to illustrate the incorporation of statistical methods to assess accuracy and precision. The example involves a few of many possible options for incorporating statistical techniques.

The Problem: A Robust Test Method Must Be Developed

An analytical research and development team is tasked to create a dissolution method for an immediate release tablet. Callie is the manager of the team and has made the project a pilot to determine how statistical methods can be used during method development. The tablet dose is intended to dissolve in the body after passing through the stomach. Therefore, a slightly basic media is used to test for the amount of drug released in 60 minutes. The team is utilizing the USP Dissolution Apparatus 1, an automated dissolution bath that includes six wells with baskets used to manipulate the tablets during testing. A high-performance liquid chromatography (HPLC) instrument will be used to detect the amount of drug present in a small sample collected from the sample media. The technical term for the small sample is an aliquot.

The team has not used structured, multivariate experimentation to develop similar methods in the past. Each member of the team obtained a JMP license and was able to go through a training course to get them started on using JMP to add value. The statistician for research and development is coaching and mentoring the team through the project. The statistician worked with the team to break down the method into controls, inputs with varied levels intended for study, and potential noise factors.

Experimental Planning

The analytical team took time to list each and every factor that is involved in creating dissolution results. An excerpt of the list is shown in Figure 16.1.

Figure 16.1: Controls List

Untitled - Notepad
File Edit Format View Help

Controls

Media Dispensing Unit (automatic) .... Instrument ID X234Z1 used for all runs
Media Carboy.. Equipment ID E54S231 used for all runs
Timer.. Equipment ID E56D101 used for all runs
Volumetric Flask .... 900ml, model FVOL900Z mfg by Scigay Limited used for all runs
Graduated Cylinder.. model GC201F mfg by Scigay Limited used for all runs
Sampling Unit... Automatic Samptech 8200 Autosampler used for all runs
Sample Filters... Filltron 0.45micron nylon used for all runs
Reagents.. per LABMETHOD202A used for all runs
Chromatographic System
    Mode.. LC
    Detectors
        Assay............................ UV 254nm
        ID Test A....................... Diode array UV 200-400 nm
        Column.......................... 4.6mm x 25 cm, 5 micron packing L1
        Flow Rate...................... 2.0 ml/min
        Injection Volume............ 10micron
    System Suitability
        Sample.......................... Standard Solution

The team assigns a person to each major group of controls so that all can be verified as in place before the set of runs are executed. If any changes are made to the controls, the lead of the experiments is notified, and the change is noted on the controls list. Once the experiments begin, no changes can be made to the controls unless the subject matter experts and statistical mentor evaluate the change to be sure that runs are not affected.

The team compiles and discusses a list of noise factors as part of the planning to ensure that potential effects on the runs have been mitigated. Noise includes anything other than a control or factor that can change and possibly affect the results. The laboratory environment could be a noise factor since changes in ambient temperature, relative humidity, and room air pressure could influence results. If there are controls for noise factors, they can be added to the controls list for verification. The statistics mentor might decide to use the amount of noise as a random factor of the model if it can be quantified, and check for influence. In this example, the noise is not quantifiable, so no additional blocks or random factors are to be included in the model. If the results of the experiments include odd results that are not related to the input variable changes, noise might need to be revisited.

The last list that is compiled includes all of the inputs that will be varied to specific levels. In JMP terminology, such inputs are called factors. The list is determined by subject matter experts with help from a statistics mentor. Seven method inputs have been chosen for the study based on principal scientific and experience. All seven of the inputs can be easily changed between the runs, making the task of model creation easy.

Methods with inputs that are difficult to change need to be studied by using a split plot design. Split plot designs are not covered in this chapter. There are several excellent books on the subject, as well as the set of JMP books that is included with the software license to provide guidance for planning purposes. Keep in mind that such designs require more resources to execute than does this example.

Levels of each input need to be challenged to be as wide as possible, yet be a good reflection of the real method in order to provide the best chance for detecting significance. Inputs to be studied are shown in Table 16.1.

Table 16.1: Method Inputs List

Input	Variable	Description	Low Level	High Level
1	bathID	Identity of the baths used	A	B
2	vibration mat	The use of a vibration isolation mat under the bath	no	yes
3	age of standard	The elapsed time in days from the creation of the standard to its use	1	3
4	basket height	The distance in millimeters from the bottom of the well to the bottom side of the basket	2.3	2.7
5	shaft RPM	The speed of the shaft that drives the basket in revolutions per minute	48	52
6	age of sample	The elapsed time in days from the creation of the sample to its use	1	3
7	media pH	The actual pH of the media measured by a device sensitive to 0.001 units	7.0	7.4

Design Creation Using the Definitive Screening Design (DSD)

All of the important planning elements are complete, and it is time to determine an experimental model that can balance the amount of information needed with available resources. Callie has limited resources available, yet needs a model that can extract as much information as possible. Different model designs were evaluated by using the model diagnostics and comparison techniques described in chapter 9. The team determined that the definitive screening design (DSD) is an excellent choice since more than six inputs are to be considered. The team members rely on their scientific knowledge and experience to choose appropriate inputs for study. The team believes that at most there will be only a couple of inputs that have an effect on the release at 60 minutes output. The DSD will provide information about the effect of the individual inputs (main effects), combined effects between pairs of inputs (interactions), and the curved terms (quadratic effects). Select *DOE* ▶ *Definitive Screening* ▶ *Definitive Screening Design* to get to the *Definitive Screening Design* dialog box shown in Figure 16.2.

Figure 16.2: Definitive Screening Design Model Inputs

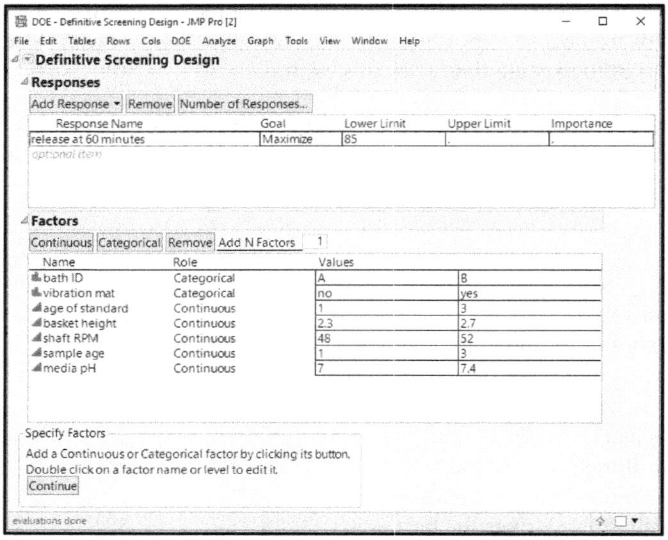

The output and inputs must be set up to initiate the experimental design. Click *Add Response* and select *release at 60 minutes* as the response set. Specify *Maximize* as the *Goal* and 85 for *Lower Limit*. Add the two categorical inputs and five continuous inputs with the levels shown in Figure 16.2. Click *Continue* to select the options shown in Figure 16.3. It is good practice to use the red triangle menu located next to the *Definitive Screening Design* header and select *Save Factors* to create a factors data file to save in case you need it for later use.

Figure 16.3: Setting Design Options

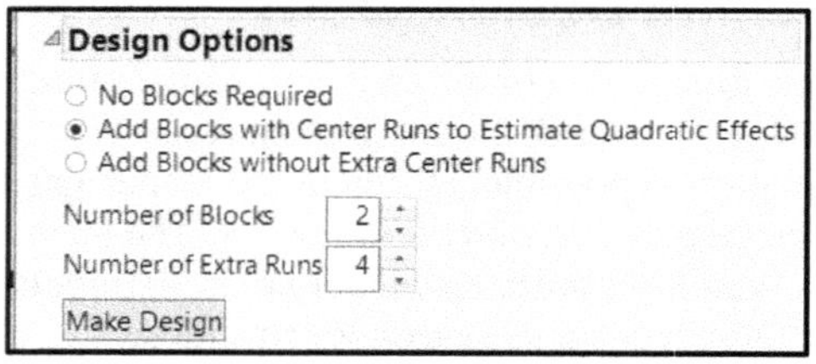

The basic DSD is an excellent design, but you can extract more robust information with the addition of only a few more runs. The model options enable you to add blocks with extra runs so that the quadratic effects can be estimated. Since most analytical methods involve a great deal of chemistry, quadratic effects are a good possibility since effects might show a rate of change as inputs change. An output trend with a rate of change is a non-linear response. Select *Add Blocks with Center Runs to Estimate Quadratic Effects* to create a robust design. Click *Make Design*. The design is added to the output, as shown in Figure 16.4.

Figure 16.4: DSD Design Output

			vibration	age of	basket			
Run	Block	bath ID	mat	standard	height	shaft RPM	sample age	media pH
1	1	B	yes	2	2.7	52	3	7.4
2	1	A	no	2	2.3	48	1	7
3	2	A	yes	3	2.5	48	1	7
4	2	B	no	1	2.5	52	3	7.4
5	1	B	no	3	2.3	50	1	7.4
6	1	A	yes	1	2.7	50	3	7
7	2	B	yes	3	2.3	48	2	7.4
8	2	A	no	1	2.7	52	2	7
9	1	A	no	3	2.3	52	3	7.2
10	1	B	yes	1	2.7	48	1	7.2
11	2	B	yes	3	2.3	52	3	7
12	2	A	no	1	2.7	48	1	7.4
13	1	B	yes	3	2.7	48	3	7
14	1	A	no	1	2.3	52	1	7.4
15	2	A	no	3	2.7	48	3	7.4
16	2	B	yes	1	2.3	52	1	7
17	1	B	no	3	2.7	52	1	7
18	1	A	yes	1	2.3	48	3	7.4
19	2	A	yes	3	2.7	52	1	7.4
20	2	B	no	1	2.3	48	3	7
21	1	A	no	2	2.5	50	2	7.2
22	1	B	yes	2	2.5	50	2	7.2

The DSD includes 22 runs with blocks and center points to study the seven inputs. Notice that the first two inputs involve group levels instead of numeric values since each is a categorical input. The ability to easily incorporate both categorical and numeric variables is a great feature of a DSD. Keep in mind that the design shown in the output does not include randomized runs. The *Design Evaluation* header located underneath the design provides important details about the robustness of the model. A few evaluation sections are shown in Figure 16.5.

Figure 16.5: Design Evaluation

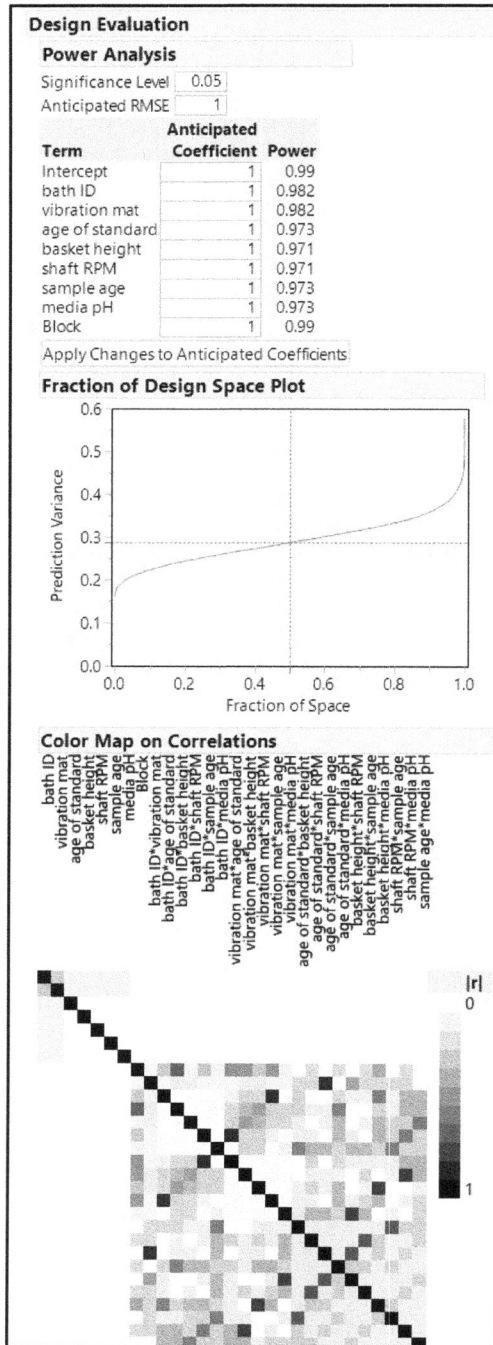

Design Evaluation

Power Analysis

Significance Level   0.05
Anticipated RMSE   1

Term	Anticipated Coefficient	Power
Intercept	1	0.99
bath ID	1	0.982
vibration mat	1	0.982
age of standard	1	0.973
basket height	1	0.971
shaft RPM	1	0.971
sample age	1	0.973
media pH	1	0.973
Block	1	0.99

Apply Changes to Anticipated Coefficients

Fraction of Design Space Plot

Color Map on Correlations

The estimated power of the DSD with 22 runs is excellent for the inputs at 97% to 98%. No data is available from previous multivariate experiments to be used to get a more precise estimate of power by entering the anticipated RMSE and input coefficients. The Fraction of Design Space plot illustrates a trend of prediction variance that increases little from the center point of the design space and out to nearly the edges. The inner 50% of the design space includes under 0.3 units of prediction variance, which is minimal. The Color Map on Correlations illustrates the minimal correlation of 10% between the first two categorical inputs and the five continuous outputs. There is no correlation between the individual inputs and the interactions between inputs. There is varying correlation among interactions, which is not of great initial concern for the goal of the study. If an interaction has a high amount of significant influence on the dissolution output from the analysis of the model, the team might need to consider augmenting the design with additional runs to mitigate correlation. The evaluation of the design provides evidence of likely robustness for results of the analysis.

With reasonable design evaluation results obtained, the design is to be randomized into a JMP data table. Prior to the randomization, use the red triangle menu next to the *Definitive Screening Design* header, select *Set Random Seed*, and enter 2017. Click *Make Table* to create the data collection plan shown in Figure 16.6.

Figure 16.6: DSD Data Collection Plan

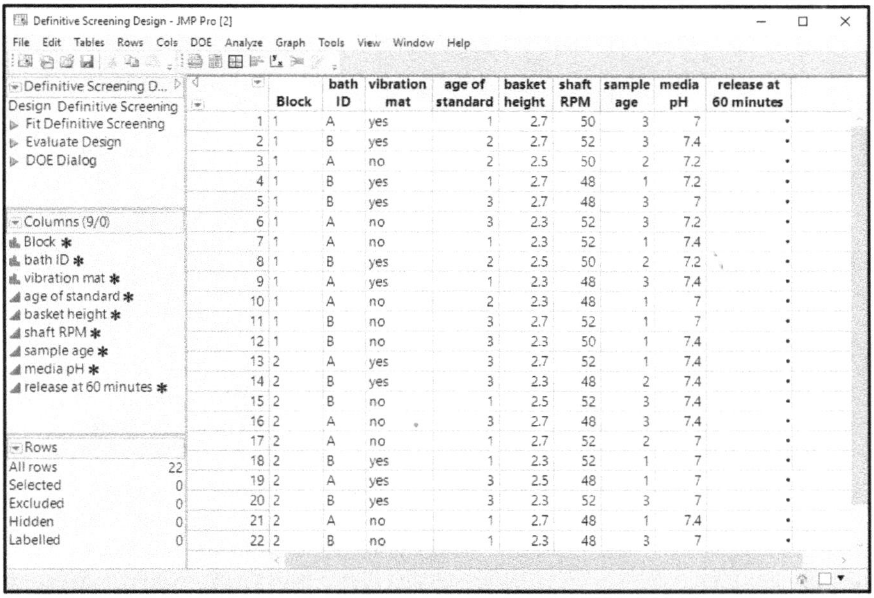

	Block	bath ID	vibration mat	age of standard	basket height	shaft RPM	sample age	media pH	release at 60 minutes
1	1	A	yes	1	2.7	50	3	7	•
2	1	B	yes	2	2.7	52	3	7.4	•
3	1	A	no	2	2.5	50	2	7.2	•
4	1	B	yes	1	2.7	48	1	7.2	•
5	1	B	yes	3	2.7	48	3	7	•
6	1	A	no	3	2.3	52	3	7.2	•
7	1	A	no	1	2.3	52	1	7.4	•
8	1	B	yes	2	2.5	50	2	7.2	•
9	1	A	yes	1	2.3	48	3	7.4	•
10	1	A	no	2	2.3	48	1	7	•
11	1	B	no	3	2.7	52	1	7	•
12	1	B	no	3	2.3	50	1	7.4	•
13	2	A	yes	3	2.7	52	1	7.4	•
14	2	B	yes	3	2.3	48	2	7.4	•
15	2	B	no	1	2.5	52	3	7.4	•
16	2	A	no	3	2.7	48	3	7.4	•
17	2	A	no	1	2.7	52	2	7	•
18	2	B	yes	1	2.3	52	1	7	•
19	2	A	yes	3	2.5	48	1	7	•
20	2	B	yes	3	2.3	52	3	7	•
21	2	A	no	1	2.7	48	1	7.4	•
22	2	B	no	1	2.3	48	3	7	•

The table is created and Callie is delighted that so much information can be extracted with only 22 runs. The risk is that if many of the inputs are found to be significant, the DSD might not provide the best results. Since it is not expected that many inputs will have an effect on the release at 60 minutes, the risk of using the DSD model is well justified.

Model Analysis of the DSD

The set of 22 experiments was run in the exact random order indicated by the DSD model data collection plan. Results for the release at 60 minutes is added to the original data table, and the DSD scripts included in the design table are available for efficient analyses. The first step is to screen the seven inputs to determine whether any are important. The fit definitive screening script created for JMP 13 is an excellent start to the analyses of the model. Open the file *Disso Method DSD.jmp* and run the *Fit Definitive Screening* script for the first level of analysis shown in Figure 16.7.

Figure 16.7: DSD Model Screening Results

Fit Definitive Screening for release at 60 minutes

Stage 1 - Main Effect Estimates

Term	Estimate	Std Error	t Ratio	Prob>|t|
vibration mat	2.5357	0.4204	6.0317	0.0003*
sample age	-5.276	0.4624	-11.41	<.0001*
media pH	4.4056	0.4624	9.5276	<.0001*

Statistic	Value
RMSE	1.9518
DF	8

Stage 2 - Even Order Effect Estimates

Term	Estimate	Std Error	t Ratio	Prob>|t|
Intercept	78.765	0.3094	254.6	<.0001*
Block[1]	0.4788	0.3094	1.5475	0.1561

Statistic	Value
RMSE	1.4451
DF	9

Combined Model Parameter Estimates

Term	Estimate	Std Error	t Ratio	Prob>|t|
Intercept	78.765	0.3645	216.11	<.0001*
Block[1]	0.4788	0.3645	1.3136	0.2064
vibration mat	2.5357	0.3667	6.9153	<.0001*
sample age	-5.276	0.4033	-13.08	<.0001*
media pH	4.4056	0.4033	10.923	<.0001*

Statistic	Value
RMSE	1.7024
DF	17

Make Model | Run Model

The screening analysis output includes initial results for the model, which lists the important inputs of the dissolution testing method. Stage 1 shows the individual inputs that have influence on the measured output of drug release at 60 minutes. The vibration mat, sample age, and media pH have significant influence on the output; with p-values that are all less than the default significance level (α=0.05). The stage 2 estimates list significant interactions and the effect from the blocking variable among the inputs. The stage 2 estimates have no significant interactions listed and indicate no significant effect from the blocks. There is evidence that the intercept is significant; however, this only means that it is not zero. A non zero intercept for this example is expected and not particularly noteworthy. The lack of significance for the blocks indicates that the runs of the successive halves of the 22 experimental runs are not significantly affecting the output. Screening has given you a gross idea of which of the seven inputs might have influence on drug release at 30 minutes. Fit screening includes plots to provide visualization of the trends detected in the model, as shown in Figure 16.8.

Figure 16.8: DSD Model Screening Plots

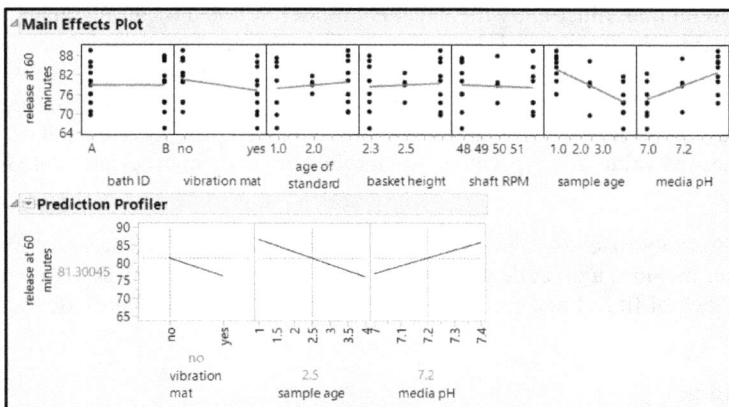

Main Effects plots display the 22 individual runs as black dots, with a blue line that shows the average trends. Inputs that have no slope have no relationship to the drug release at 60 minutes. The sloped trend lines for sample age, media pH, and vibration mat illustrate the significance of the inputs.

The Prediction Profiler provides a rich, dynamic visualization of the input trends alone. Exploration of how changes in the three significant inputs affect the release at 60 minutes is held off for now. It is time to fully analyze the model.

Two options are available for analyzing the model: making the model manually or using the *Run Model* button to automatically include only the inputs of importance. This example uses the automated run model feature for simplicity. Click *Run Model* to get the analysis output shown in Figure 16.9.

Figure 16.9: DSD Model Prediction Plot and Effect Summary

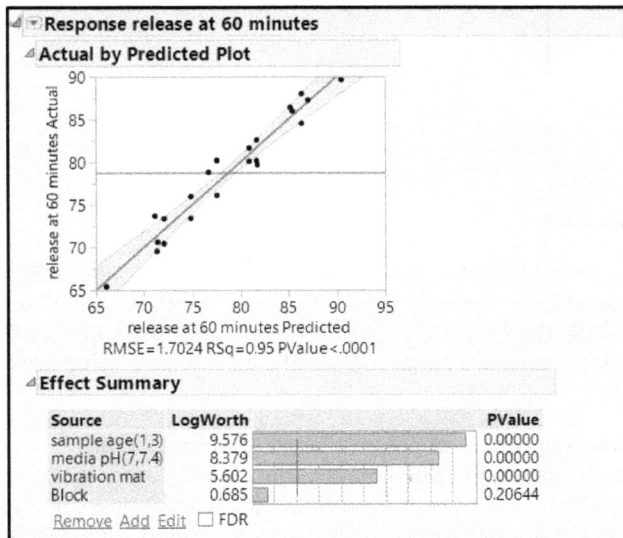

The Actual by Predicted plot provides the view of the overall model formed by the three significant inputs. Points are located very close to the trend line, illustrating the excellent model fit noted by an R square of 0.95. The variation in the model noted by the RMSE of 1.7024 is minimal, evidenced by the narrow confidence interval shown in red shading about the model prediction line. The model is highly significant (p-value <0.0001) in influencing changes in the release at 60 minutes. The Effect Summary provides the same information about the important inputs as did the model screening, with a slightly different set of statistics. The LogWorth is a transformed value of the p-values so that they are at an appropriate scale for plotting.

One aspect of analysis that has not been approached is diagnostics to ensure that appropriate assumptions have been met. The full model output includes a great deal of information to evaluate for the determination of the robustness of the model. The lack of fit test and residual plots shown in Figure 16.10 provide the first line of diagnostics.

Figure 16.10: DSD Model Diagnostics

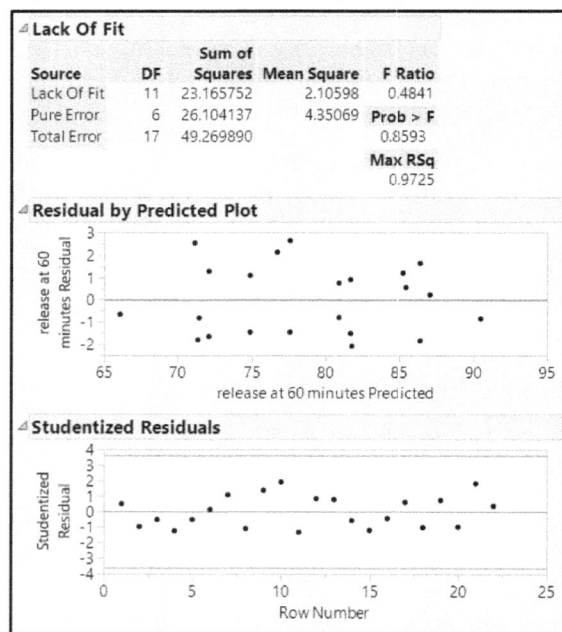

The Actual by Predicted plot in Figure 16.9 contains visual evidence of how closely actual values align with the model prediction line. The model fit provides a summary to qualify the fit, but there might be a limited number of observations that do not fit well. The lack of fit test provides the sensitivity needed for further digging. The resulting probability of 0.8593 indicates almost no significant evidence that points have a poor fit.

The Residual by Predicted plot should have a random pattern of points in order to meet modeling assumptions. Coned patterns or curved patterns are especially troublesome. The plot illustrates the random pattern and meets assumptions. The Studentized Residuals by row are within the statistical limits of randomness (red lines) with no trend, which meets assumptions. The diagnostics are useful because they remove doubt that the trends detected are not robust.

There is one last option to consider for making the best model possible and reducing the random error. Since the data is to be used to support an analytical method, precision in modeling must be a high priority. The Box-Cox Transformations plot shown in Figure 16.11 illustrates the set of all possible lambda transformations for the output that can optimize the assumption of normality and constant variance.

Figure 16.11: Box-Cox Transformations

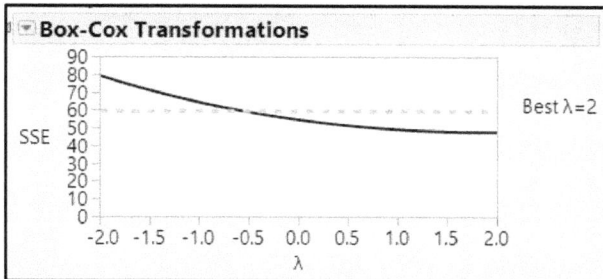

The lowest model error (SSE) can be obtained with a lambda transformation of 2 because it is the lowest point of the blue SSE curve. The red line indicates the average SSE for the space. You can make a new model with the transformed results to determine whether analysis results can be improved. Utilize the red triangle menu next to the *Box-Cox Transformations* header and select *Refit with Transformed*. JMP chose the best lambda value 2 as the default. Keep the default and click *OK* to get the transformed model output shown in Figure 16.12.

Figure 16.12: DSD Model Diagnostics (Transformed)

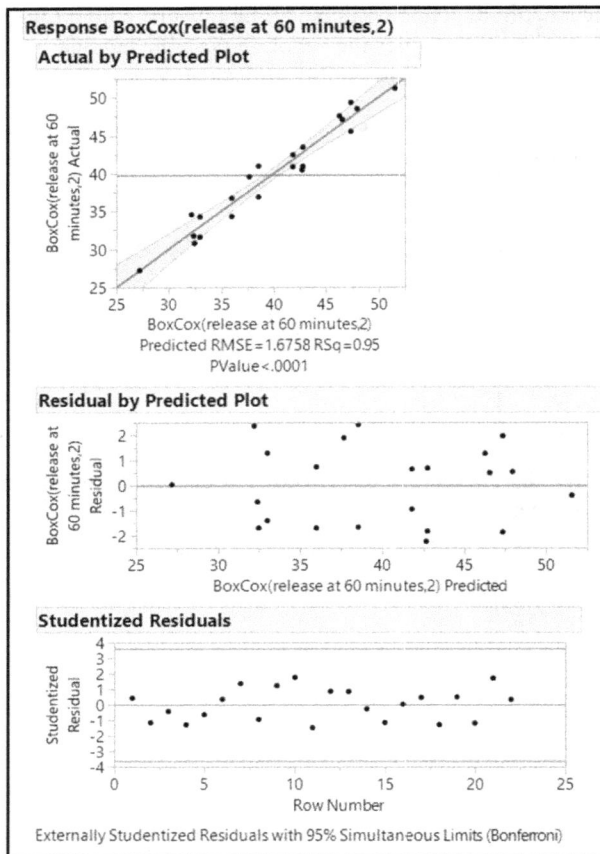

The model error for the transformed model (RMSE = 1.6758) is only slightly less than the initial model (RMSE = 1.7024). The model fit and the residual plots show almost no change. This is no surprise since the previous diagnostics indicated that each assumption is robustly met. Therefore, there is no need to complicate the modeling with transformations.

Making Estimates from the Model

The model for the dissolution method is robust and the important predictors have been identified. It is time to use the model to make estimations and find the optimum levels for the important predictors. There is great value in knowing that there are four other inputs that have no significant influence on the output measurements. The method can include the full range of levels studied with no risk of effecting a change in the measured outputs. The influence of the important predictors is shown in the Parameter Estimates table in Figure 16.13.

Figure 16.13: Parameter Estimates

Term	Estimate	Std Error	t Ratio	Prob>|t|
Intercept	80.523563	0.388472	207.28	<.0001*
Block[1]	0.4787527	0.364466	1.31	0.2064
vibration mat[no]	2.5356848	0.36668	6.92	<.0001*
sample age(1,3)	-3.517605	0.268885	-13.08	<.0001*
media pH(7,7.4)	4.4056072	0.403327	10.92	<.0001*

The intercept of 80.5 is significantly different from 0, which is not all that interesting. There is no significance of the blocked runs, so all 22 runs can be pooled for making estimates. When a vibration mat is added underneath the bath, the dissolution values drop by 2.5 percentage points. As the sample age increases by one day, the dissolution values drop by 3.5 percentage points. Increasing the media pH by 0.2 units would increase the dissolution value measured by 4.4 percentage points.

Recall that the goal is to obtain the most accurate and precise signal from the method as possible. If a vibration mat is not used, the signal is increased, but this is not a good effect. The USP method guidance clearly indicates that all known sources of influence from vibration or agitation of the media must be controlled. The increased results from vibration are not acceptable, and the team must use a mat to isolate vibration in order to reduce bias in results.

The laboratory acceptance criteria for the age of sample preparations is up to three days. Modeling has indicated that the dissolution results are likely to drop as the age of the sample increases. More exploration of this important input is needed to ensure that the method has appropriate instructions.

The estimates indicate that the dissolution output can change up to 8.8 percentage points for the range of levels studied, which is a big shift. The team should determine how closely the actual pH of the media can be controlled in order to scratch this input off the list of contributions to measurement uncertainty.

All three influential inputs can be manipulated dynamically with the Prediction Profiler to try different levels and to determine the changes in the dissolution values. This example uses static screen shots of the tool. However, it is highly advised that the team takes advantage of this powerful tool in JMP.

Figure 16.14: DSD Model Prediction Profiler

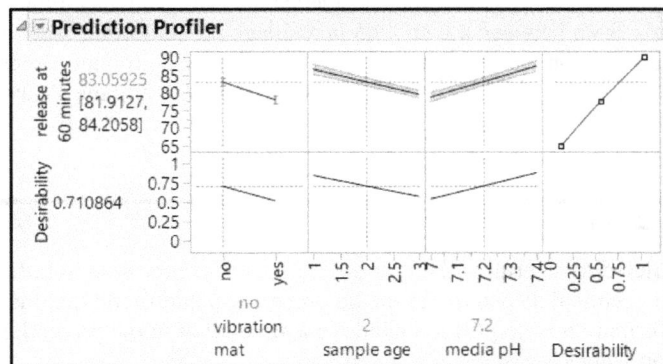

The Prediction Profiler is the default tool that appears with the model analysis. The first thing to do is lock in the use of the vibration mat to determine how changes in the other two inputs affect results. Press the

Ctrl key and right-click in the white space of the *vibration mat* plot, to get the *Factor Settings* options. Select the *Lock Factor Setting* check box, and click *OK*.

It is clear that the newest sample age and highest pH of media provide the highest amounts of release at 60 minutes. The team is concerned that the results change by almost four percentage points for each day of sample age, which translates to a difference of over 10 percentage points over the acceptable three days of storage. Visualization of the effect will be important for the report of the modeling results.

1. Make sure that the *media pH* is set to 7.2.
2. Move the vertical red slider line for *sample age* to 1.
3. Use the red triangle menu next to the *Prediction Profiler* header and select *Factor Settings* ▶ *Remember Settings*.
4. Enter "New Sample" as the name, and click *OK*.
5. Move the vertical red slider line for *sample age* to 3.
6. Use the red triangle menu next to the *Prediction Profiler* header and select *Factor Settings* ▶ *Remember Settings*.
7. Enter "3 Day Old Sample" as the name, and click *OK* to get the output shown in Figure 16.15.

Figure 16.15: DSD Model Prediction Profiler Sample Remembered Settings

◢ **Remembered Settings**

Setting	Block	vibration mat	sample age	media pH	release at 60 minutes	release at 60 minutes Lower CI	release at 60 minutes Upper CI	Desirability
○ New Sample		yes	1	7.2	81.505484	80.17098	82.839987	0.650690
● 3 Day Old Sample		yes	3	7.2	74.470274	73.323722	75.616825	0.391650

◢ **Differences**

Setting A	- Setting B	release at 60 minutes	Desirability
3 Day Old Sample	New Sample	-7.03521	-0.25904

The JMP journal with the output can be used to help the stakeholders of the project see that the predicted dissolution results are likely to drop seven percentage points with the oldest acceptable sample. This is an average estimate, good for only the subject set of experiments. The confidence interval estimates indicate that the population of results is likely to drop from between 4.6 and 9.5 percentage points for the 3-day-old sample prep, dissolution method run with a vibration mat, and media set to 7.2 pH. The confidence interval of the difference is calculated by subtracting the 1-day lower CI from the 3-day upper CI, and the 1-day upper CI from the 3-day lower CI.

Using Simulations to Estimate Practical Results

An evaluation of the media prep was completed to determine the best possible level of precision in hitting the target pH. The team tried an automated chemical dispenser for media preparation and found that the media is within a range of 0.1 pH. This information is helpful to include in a simulation of the prediction profiler and assess the impact on dissolution values measured.

1. Use the red triangle menu next to the *Prediction Profiler* header and select *Simulator*.
2. Make sure the *media pH* is set to the 7.2 target.

3. Click *Fixed* under the *media pH* plot, and select *Random*.

4. Keep the default normal distribution (many options are available).

5. Enter 0.0167 as the *SD* (standard deviation), which is the range divided by 6.

6. Keep the default value 5000 in the *N Runs* box.

7. Under the *Simulate Table* header, click *Make Table*.

8. When the simulated table appears, select *Analyze* ▶ *Distribution*.

9. Move *release at 60 minutes* to the *Y, Response* box, and click *OK*.

10. Utilize the red triangle menu next to the *release at 60 minutes* header and select *Tolerance Interval*.

11. Keep the defaults and click *OK*.

12. Utilize the red triangle menu next to the *Distributions* header and select *Stack* to get the output shown in Figure 16.16.

Figure 16.16: DSD Model Prediction Profiler Sample Remembered Settings

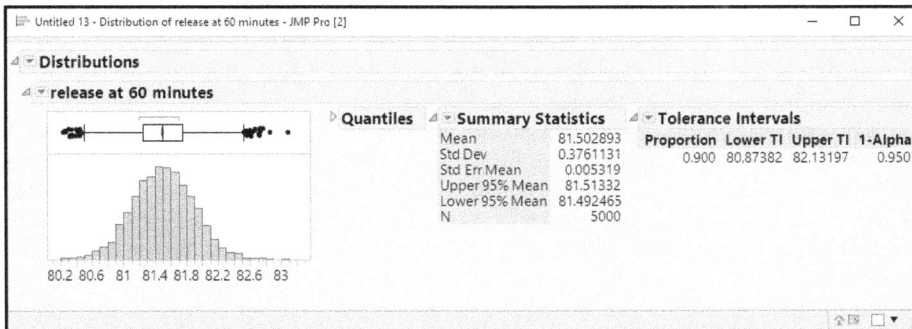

The distribution of simulated results, given the method variability in media pH, show a small amount of variability. The tolerance interval provides estimates for 90% of the individual results given that you utilize a 95% confidence interval to estimate the population average. The project stakeholders need to discuss whether it is acceptable to have a 1.2 percentage point range for measurement uncertainty due to an expected variability in actual media pH.

Practical Conclusions

Callie now has the information she needs to bring to the project stakeholders so that they can make decisions about the amount of measurement uncertainty that can be tolerated from the age of the sample and the actual pH of the media. Informed decisions can balance the costs of added precision with the amount of uncertainty added from variability in the factors.

This chapter is focused on early structured experimentation with the goal of detecting method inputs that have significant influence on results. Definitive screening designs work extremely well for experimentation since analytical methods use multiple steps, which include several inputs for study. Design of experiments has not been typically used for method development within the pharmaceutical industry. It is very likely that analytical teams will learn that inputs previously thought to be robust actually have significant evidence of influence. This is especially true for interactions between inputs, which have not previously been included in method diagnostics. The value added by using DOE extends throughout the lifecycle of a product since more robust methods require fewer resources to investigate unexpected results. Control of influential inputs can also reduce complexity in method procedures as engineered controls of inputs are adopted due to model analysis results.

The next step of analytical method development includes the study of variance from method inputs. Response surface studies with replicates offer insight into the variability in the response that comes from the inputs. Measurement systems analysis is also a great option for the study of variance components that make up the analytical results. In either case, the initial screening of inputs focuses the team on important inputs and reduces resources.

Meaningful updates can be made from the information and a strong justification for the changes documented from the results of the experiments. The goals of the proposed General USP Chapters <1220> and <1210> have been met. The structured experimentation executed is the purest form of the application of Quality by Design principles.

Exercises

E16.1—An analytical R&D method of development involves the assay of a low dose tablet that has had a previous history of low values. The team did not utilize structured, multivariate experimentation to develop the method. The problem of low assay persists and questions regarding the robustness of the method are mounting. A set of experiments is planned to fully vet the method, and leadership has requested that an appropriate DOE is executed. All potential contributions to variation in the assay output have been reviewed using principal science and experience. Six variables have been established, and multiple controls and noise factors have been identified. The goal of the experimentation is to screen for the inputs that have significant influence on the results. The scientists have prior evidence to suggest that only a few of the inputs migh have influence on the assay results. Leadership has allocated resources to allow for up to 22 runs to be completed for the set of experiments. The factors for the method are stored in *Assay factors.jmp*.

1. Create experimental designs using *DOE ▶ Custom Design* for the default D-Optimal screening design and *DOE ▶ Definitive Screening ▶ Definitive Screening Design*. Be sure to include the individual inputs and two-way interactions for all designs.
2. Use *DOE ▶ Diagnostics ▶ Compare Designs* to determine the advantages of each design.
3. How would you summarize the information to suggest the best design choice for moving forward to the project stakeholders?

E16.2—A design has been chosen for the problem presented in E16.2, and the data was collected in a JMP data table. Run the experimental analysis to screen for important inputs. The team is interested in any method input that significantly explains at least 2% of the change in results to be considered practically relevant.

1. Open *DSD for assay method 18R.jmp*. Select *Analyze* ▶ *Distribution* to visualize the *% extracted* results before you run the model analysis. How much variation is there in the overall results before the analysis divides them by significant inputs?

2. Use the green arrow next to *Fit Definitive Screening* to run the script for the analysis.

3. The team is very concerned that significant interactions are present and are influencing results. Use the *Stage 2- Even Order Effects* analysis results to determine whether the concern is valid.

4. Click *Run Model* to get the detailed analysis. The assay specification is 90% to 110%, and the team is most interested in 2% or more influence from method inputs to be considered as practically relevant. Is there a cause for concern based on the analysis of the method experimentation regarding an input that might need to be controlled?

5. The proposed method utilizes the A type of solvent. The B type of solvent was tried as a possible alternative. What do the results suggest regarding the optimum type of solvent?

6. Summarize the results into a presentation for leadership. Be sure to explain whether the information from the analysis indicates that additional studies into analytical methods are warranted.

Chapter 17: Exploring Stability Studies with JMP

Overview

Pharmaceutical products have rigorous requirements. The product must resist environmental exposure so that all critical quality attributes (CQAs) meet specifications through the expiration date. Maintaining a stabile formulation can be one of the most challenging aspects of drug development since many drug substances are sensitive to exposure to light, temperature, or humidity. The label claim cannot be adjusted for the effects of environmental exposure over time. It is extremely important for development teams to explore the trends of how the product changes during storage. Stability studies quantify trends in CQAs over time. The studies are completed during product development and continue throughout the commercial lifespan of a product. Specialized software is available to analyze stability data. However, the applications are not easy to use and tend to provide information limited to evidence of meeting the regulatory requirements for expiration dating. JMP includes a set of stability analysis tools that are very easy to use. The stability analysis provides high-quality graphs, tables, and analysis options that set JMP apart from other software solutions.

The Problem: Transdermal Patch Stability

This chapter involves a development project for a transdermal patch that delivers a drug used for cardiac care. The example utilizes data collected from project batches that are stored in accelerated conditions of high temperature and humidity. The tools in this chapter work just as well for real-time stability projects that are stored in ambient conditions.

Bryce is a Senior Scientist in charge of developing a drug product that is dosed by a transdermal patch that is used for the treatment of angina. The patch is made with a polymer that ensures a controlled dose to the patient over a 24-hour period. The team has two alternate products that use different polymers and has seen very good results for both in physical properties and analytical test results. An assessment is needed to ensure that the patch will be able to fully release the drug at the 24-hour point of the dosing regimen for the entire length of product life. Commercial needs dictate that the patch must have a 24-month expiration date to ensure it is competitive with other like products that are on the market. The specification for % drug release is between 90% and 110%.

The development team is utilizing accelerated stability studies to get results quickly so that a final product formula can be chosen. Prior studies indicate that one week for patches in the accelerated conditions is equivalent to three months in ambient conditions. The information previously provided to the team is limited to results data, stability model estimates, and shelf-life conclusions. Bryce recently discovered that

JMP includes stability studies, which happen to meet the International Council for Harmonisation of Technical Requirements for Pharmaceuticals for Human Use (ICH) guidelines. He is hoping to extract additional useful information from the JMP results to look at stability from a QbD perspective.

Summarizing Stability Data

The stability data includes all the individual tablet test values for each of the weekly pulls from the accelerated stability chamber. There is work to be done on the data since pharmaceutical stability studies typically deal with average trends over time. The *Tables* functions are used to first stack the results and then create a summary table. Open *Accelerated stability table 24 hrs raw data.jmp*, shown in Figure 17.1.

Figure 17.1: Raw Stability Data

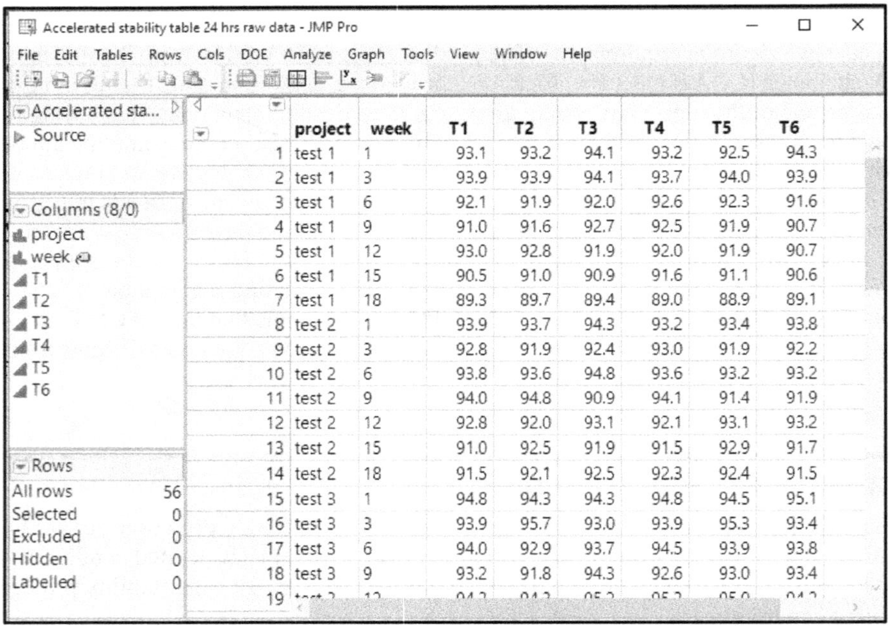

The data is in an unstacked form and must be changed to a stacked and summarized data table for stability analysis. Use the *Tables* platform to get the final desired format. Select *Tables ▸ Stack* from the main menu, and select columns T1–T6. Move the selected variables to the *Stack Columns* box in the *Stack* dialog box, as shown in Figure 17.2.

Figure 17.2: Stack Dialog Box

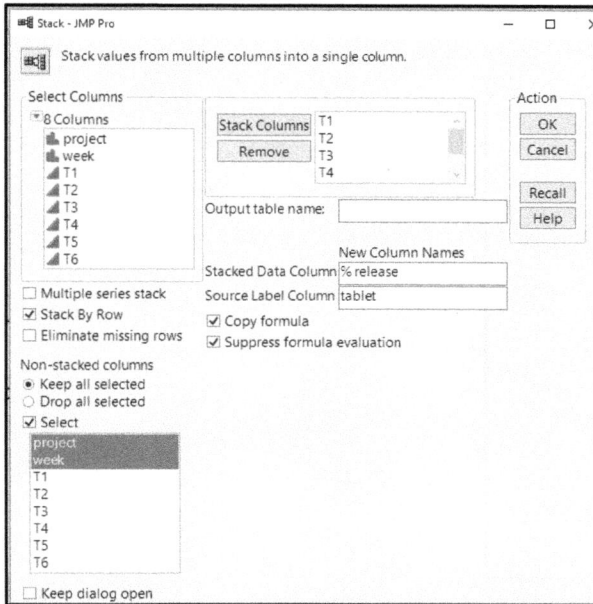

Leave the *Stack by Row* check box selected. Under Non-*stacked columns*, select the *Select* check box. Press the Ctrl key and select *week* and *project*. Under *New Column Names*, enter "% release" in the *Data* box and "tablet" in the *Label* box. Click *OK* to get the *Untitled ##* data table shown in Figure 17.3. Keep in mind that the number represented by ## in your results will be different from the number in the title in the figure.

Figure 17.3: Raw Dataset Stacked

The data needs to be summarized into means of *% drug release* by *project* for each *week*. With *Untitled ##.jmp* open, select *Tables* ▶ *Summary* to get the dialog box shown in Figure 17.4.

Figure 17.4: Tables Summary Dialog Box

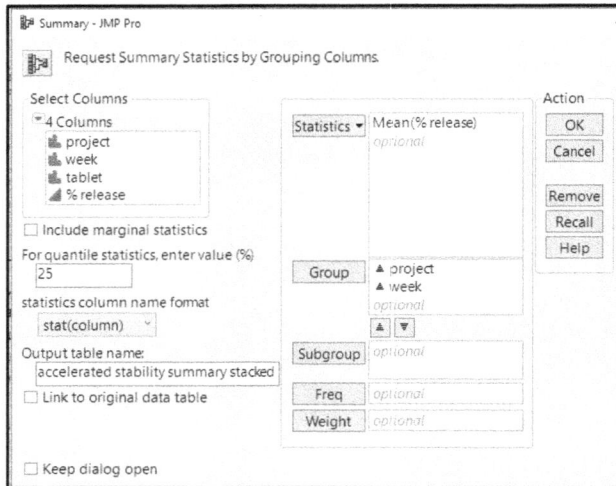

Select *% release*. Click *Statistics* and select *Mean* option; *Mean (% release)* appears in the box. Select *project* and *week*, and then click the *Group* button; the pyramid shape indicates that the groups will be sorted from low to high, which is desired. Enter "accelerated stability summary stacked" as the *Output Table Name*, and deselect the *Link to original data table* check box. Click *OK* to get the table shown in Figure 17.5. Save the data table as *accelerated stability summary stacked.jmp* to an accessible location.

Figure 17.5: Accelerated Stability Summary Stacked Data Set

Adding Initial Results and Formatting for Stability Studies

The data is now in the format needed to run stability analysis. The data is missing key information: the initial *%drug release* values for the project batches. The initial release values are often in a separate data source because they come from the quality assurance group. Quality assurance in this example is separate from the group that compiles stability studies. A data file of initial results is preformatted to add to the stacked summary of stability data. Open the file *initial 24hr results.jmp*, shown in Figure 17.6.

Figure 17.6: Initial 24 Hour Results Data Set

		sample	test 1	test 2	test 3	test 4	test 5	test 6	test 7	test 8
1	1	94.0	94.1	94.4	93.7	87.8	89.5	85.3	88.8
2	2	93.5	94.5	94.5	92.9	87.0	89.5	86.4	89.3
3	3	93.1	94.0	94.7	93.5	88.1	89.6	85.9	89.4
4	4	94.1	94.3	95.8	93.5	89.5	89.7	86.0	89.5
5	5	94.2	93.9	95.1	94.8	88.3	88.8	85.8	89.3
6	6	93.7	94.4	95.3	93.9	88.4	89.1	86.5	88.7

Select *Tables* ▶ *Stack* to get the dialog box shown in Figure 17.7.

Figure 17.7: Stack Dialog Box

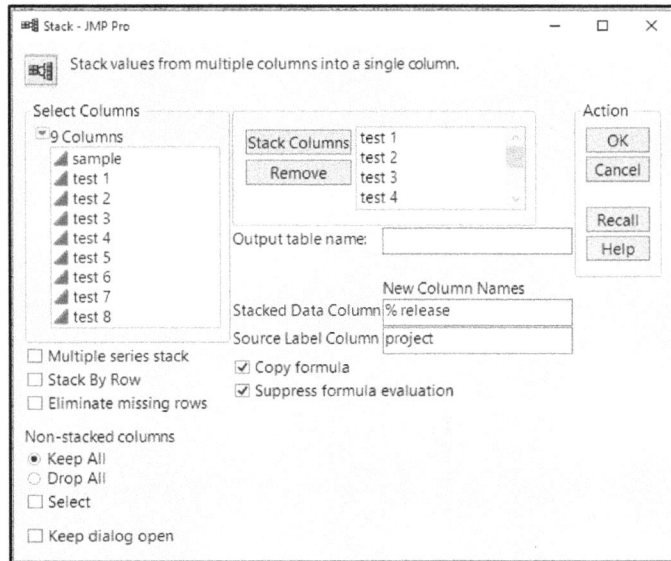

Select *test 1* through *test 8* and click *Stack Columns*. Deselect the *Stack by Row* check box. Enter "% release" in *Stacked Data Column*, enter "tablet" in *Source Label Column*, and then click *OK* to get the data table shown in Figure 17.8.

Figure 17.8: Stacked Initial 24-Hour % Release Data

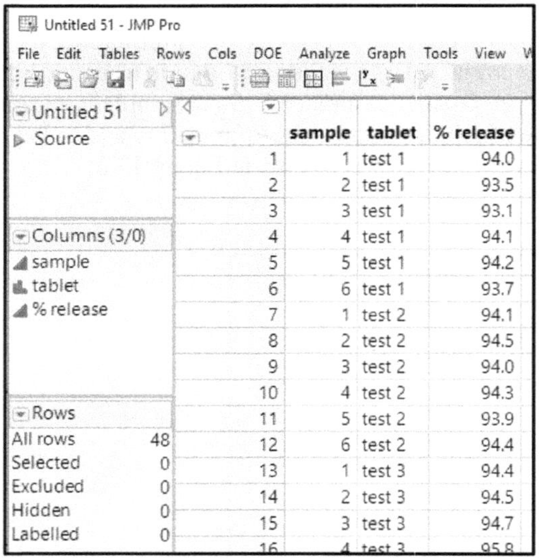

The initial 24-hour data is now summarized to match up with the stability data and combine into one set of data. Select *Tables* ▶ *Summary* to get the dialog box shown in Figure 17.9.

Figure 17.9: Table Summary Dialog Box

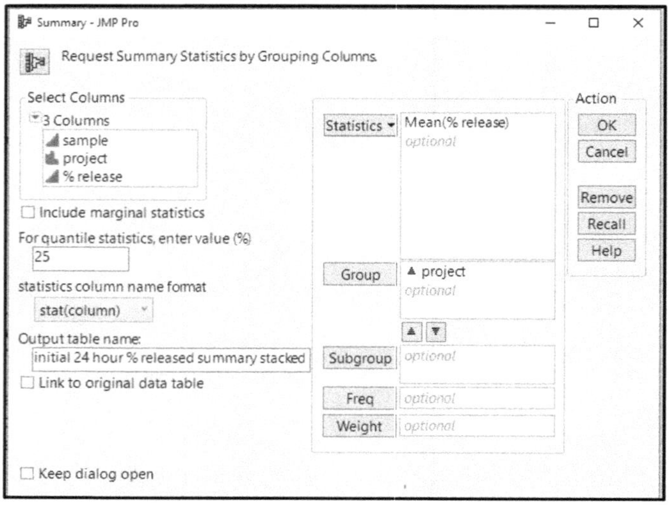

Select *% release*. Click *Statistics* and select *Mean*; *Mean (% release)* appears in the box. Select *project*, and then click the *Group* button; the pyramid shape indicates that the groups will be sorted from low to high, which is desired. Enter "initial 24 hour % released summary stacked" as the *Output Table Name*, and deselect the *Link to original data table* check box. Click *OK* to get the table shown in Figure 17.10. Save the data table as *initial 24 hour % released summary stacked.jmp* to an accessible location.

Figure 17.10: Initial 24-Hour Results Summary Stacked Data

initial 24 hour % released summary stacked - JMP Pro			
File Edit Tables Rows Cols DOE Analyze Graph Tools View Window Help			
	project	N Rows	Mean(% release)
1	test 1	6	93.78
2	test 2	6	94.22
3	test 3	6	94.98
4	test 4	6	93.73
5	test 5	6	88.19
6	test 6	6	89.38
7	test 7	6	85.99
8	test 8	6	89.20

initial 24 hour % r...
▷ Source

▼ Columns (3/0)
🔓 project
◢ N Rows
◢ Mean(% release)

Now you must add a column for week and make the table format match the stability data table. Select *Cols* ▶ *New Columns* to get the dialog box shown in Figure 17.11.

Figure 17.11: New Column Dialog Box

New Column - JMP Pro	— □ ×
Add columns to 'initial 24 hour % released summary stacked'	OK
Column Name: week	Cancel
☐ Lock	Apply
Data Type: Numeric	Next
Modeling Type: Nominal	Help
Format: Best ▾ Width 12	
☐ Use thousands separator (,)	
Initialize Data: Constant	
0	
Number of columns to add: 1	
After last column	
Column Properties ▾	

Enter "week" for Column Name, change Modeling Type to Nominal, and for *Initialize Data* select the *Constant* option to make all values 0. Click *OK* to add the column to the data.

The last thing to do for this table is to put the columns into the same order as in the accelerated stability data table. Highlight the *week* column and select *Cols* ▶ *Reorder Columns* ▶ *Move Selected Columns* to move *week* to be after *project* in the table, as shown in Figure 17.12. Save the updated data table file.

Figure 17.12: Initial 24-Hour % Released Data Table in Matching Format

		project	week	N Rows	Mean(% release)
	1	test 1	0	6	93.78
	2	test 2	0	6	94.22
	3	test 3	0	6	94.98
	4	test 4	0	6	93.73
	5	test 5	0	6	88.19
	6	test 6	0	6	89.38
	7	test 7	0	6	85.99
	8	test 8	0	6	89.20

The accelerated stability data needs to include initial results. Make sure that both data tables are open and that *accelerated stability summary stacked.jmp* is the active window. Select *Tables* ▶ *Concatenate* to get the dialog box shown in Figure 17.13.

Figure 17.13: Table Concatenate Dialog Box

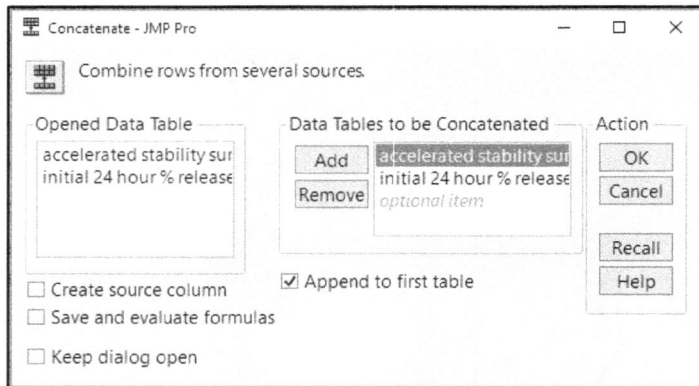

In the *Concatenate* dialog box, select *initial 24hr %released summary stacked* in the *Open Data Table* box, and click *Add*. Select the *Append to first table* check box, and then click *OK* to execute. Select *File* ▶ *Save* or press the *Ctrl S* keys to save the data file. Close the *initial 24hr %released summary stacked.jmp* file to mitigate confusion.

There are two remaining tasks needed to get the table ready for stability analysis. You must add the *polymer* variable, and the data needs to be sorted by *project* and *week*. Click on the *project* column header to select it. Select *Cols* ▶ *Recode* to open the *Recode* dialog box shown in Figure 17.14.

Figure 17.14: Recode Dialog Box

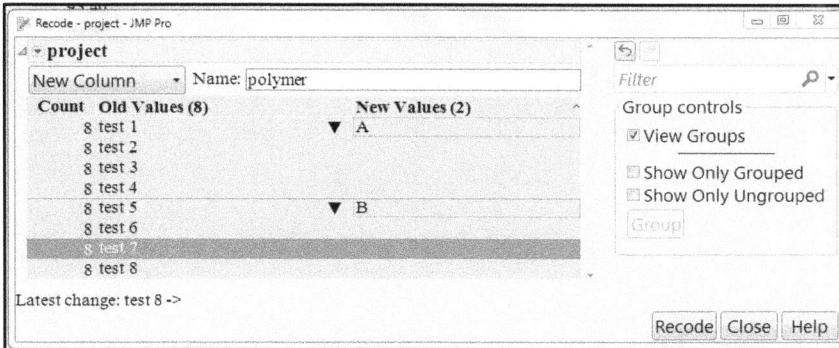

By default, JMP 14 creates a new column with the recoded data, which is desired. Enter "polymer" as the column name. Change test 1, test 2, test 3, and test 4 by entering A for each box. Change test 5, test 6, test 7, and test 8 by entering B. The recode function will automatically group the tests into an A and B group as shown in Figure 17.14. Click *Recode* to add the polymer column to the data table, shown in Figure 17.15.

Figure 17.15: Accelerated Stability Data Set with Polymer

	project	polymer	week	N Rows	Mean(% release)
1	test 1	A	1	6	93.40
2	test 1	A	12	6	92.04
3	test 1	A	15	6	90.94
4	test 1	A	18	6	89.24
5	test 1	A	3	6	93.89
6	test 1	A	6	6	92.08
7	test 1	A	9	6	91.74
8	test 2	A	1	6	93.76
9	test 2	A	12	6	92.73
10	test 2	A	15	6	91.90
11	test 2	A	18	6	92.07
12	test 2	A	3	6	92.36

Sorting the data table is not required in order to execute the stability study, but sorting cleans up the information in the table. Select *Tables* ▶ *Sort* to get the dialog box shown in Figure 17.16.

Figure 17.16: Sort Dialog Box

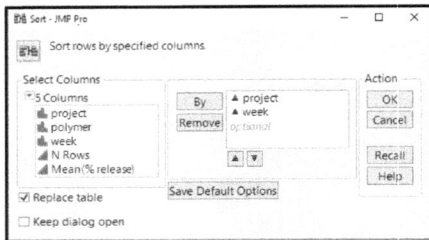

In the *Sort* dialog box, move *project* to the *By* box first, and then move *week* to be underneath *project*. Select the *Replace Table* check box, and click *OK* to execute the sort. Save the table to ensure that your formatting work is not lost.

The steps required to upload the stability data, summarize it, add the initial results, sort the table, and format the result into a stacked data table take time to complete. There are ways to speed up the process through scripting, but the tools in JMP enable you to efficiently get data ready for analysis.

Running Stability Analysis

Make sure that the file *accelerated stability summary stacked.jmp* is open. The data is organized into a stacked data set that sorted by week in the table shown in Figure 17.17. There are 64 observations that come from eight projects with eight weekly pulls from the stability chamber. It is good practice to double-check that the total rows match what you expect before you waste time on analyzing an incomplete set of data.

Figure 17.17: Summary Data Set for Stability Analysis

A complete set of reliability and survival tools are in the *Analyze* menu. Degradation studies are appropriate since the % drug release is likely to decline over time with exposure to environmental conditions. The degradation studies include a special set of customized tools for stability studies that comply with the ICH guidelines utilized by the pharmaceutical industry.

The first thing to do with the data is to study overall stability for all eight projects and evaluate the trends. Select *Analyze* ▶ *Reliability and Survival* ▶ *Degradation.* Click the *Stability Test* tab to get the dialog box shown in Figure 17.18.

Figure 17.18: Degradation Data Analysis Dialog Box

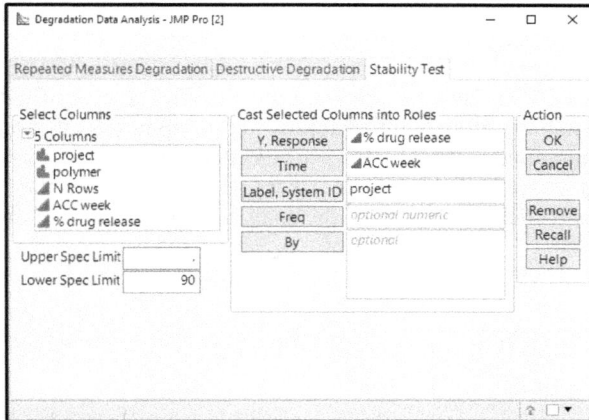

Move *% drug release* to *Y,Response*, *ACC week* to *Time*, and *project* to *Label, System ID*. Recall that the label claim for % drug release is 90% to 110%. Enter 90 as the *Lower Spec Limit*. Because there is no prior history of increases in *% drug release* over time, there is no *Upper Spec Limit*. You can enter 110 in that field, but it serves no purpose for this example. Click *OK* to get the output shown in Figure 17.19.

Figure 17.19: Stability Analysis Initial Output

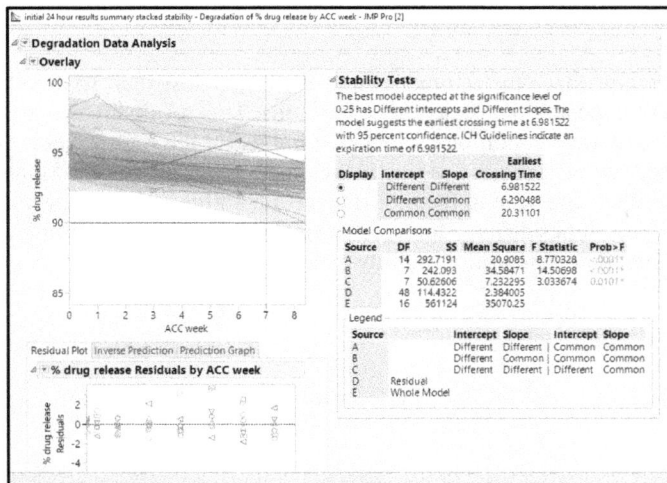

The stability trends for the pooled data show a large amount of variation present among both the intercepts and the slopes for the eight projects. The excessive variability dashes the hope that the team will be able to interchangeably utilize either polymer, which would have provided a commercial advantage for the procurement team. According to the ICH Q1E guideline, the differences exceed what is acceptable for the pooling of the models, which includes common intercepts, slopes, or both. The most conservative model creates eight individual stability trends and uses the worst case to estimate the expected shelf life. Shelf life is determined by the point at which the confidence interval, illustrated with a shaded area about the model line, crosses the 90% drug release lower limit. One of the projects defines the shelf life to be a bit less than seven months, which is far too short for the needs of the product team. You can identify the offending batch by using the red triangle menu next to the **Degradation Data Analysis** heading to show a legend. Because the team plans to dig deeper into the data for the polymer groups, there is no need to identify the worst-case project at this point.

Most analysis in JMP can be completed by using a BY variable to split analyses. Stability studies offer this option as well, and you will complete individual stability models for each polymer group next so that you can make comparisons. Use the red triangle menu option next to the *Degradation Data Analysis* header to select *Redo* ▸ *Relaunch Analysis*. The *Degradation Data Analysis* that you used to launch the analysis appears. Move *polymer* to the *By* box, and click *OK* to launch the analysis. The output is shown in Figure 17.20.

Figure 17.20: Stability Analysis for Polymer A

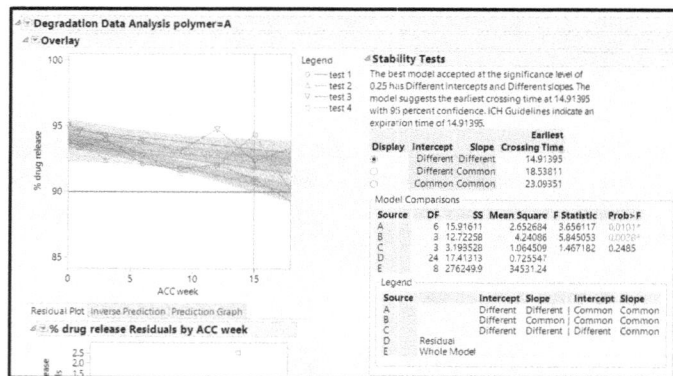

The stability trend for polymer A is encouraging since the estimated shelf life of just less than 15 months exceeds the goal of no less than 12 weeks of exposure to accelerated conditions. This result equates to an estimated shelf-life of 30 months in ambient storage conditions. The linear trends for the four projects do not pool since individual models are utilized for each project, which is the most conservative of estimates. The results for the test 4 seem to reflect the shelf life seen in the plot. The results of the stability analysis for polymer A indicates that the intercepts and slopes differ. The worst-case test indicates the expiration time of just under 15 months. Figure 17.21 includes the initial stability analysis for polymer B.

Figure 17.21: Stability Analysis for Polymer B

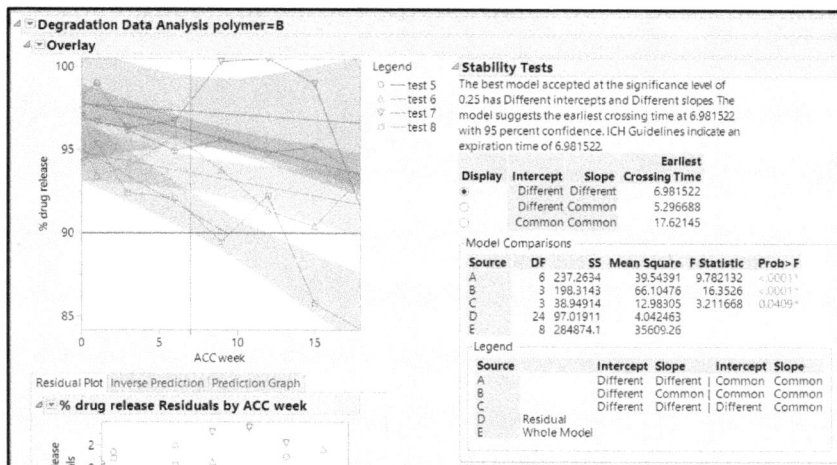

The stability trends for the polymer B projects also do not pool. The worst-case model for test 8 defines the shelf-life of just under seven weeks of accelerated stability exposure. Recall that one week of time in accelerated conditions represents three months of exposure to ambient conditions. The team can predict that the product would last only 21 months before the % drug release might not meet the 90% minimum. The stability plot also illustrates shaded confidence intervals that are much wider than the trends of polymer A projects. It seems that polymer A is preferred to polymer B. However, more digging into the model details will ensure that all required evidence is interpreted, and the best decision is reached.

Expanding the analysis with additional detail adds to the interpretation. The trend plots of *% drug release* by *ACC* must be modified to have similar scaling so that a robust visualization of the data can take place. The *polymer=A* plot scale for *ACC weeks* is the widest of the two; it spans from 0 weeks to 18 weeks. The *polymer=B* plot *% drug release* scale is the widest of the two; it spans between 84 and 101. Right-click on each of the scales for each of the plots to adjust the scales so that both plots reflect the extremes of 0 weeks to approximately 16 weeks, and % drug release of just under 85% to just over 100%.

Several models are available for stability analysis and the platform uses the ICH guidelines to select the most appropriate model. The data from the projects of polymer A include initial results that are significantly different, as well as slopes that differ significantly. The most appropriate analyses include individual models for each of the four project batches. All possible models are outlined in the JMP output and are located underneath the stability plots. Open the Model 4 outline to look at the estimates for the four linear models shown in Figure 17.22.

Figure 17.22: Stability Model Estimates for Polymer A

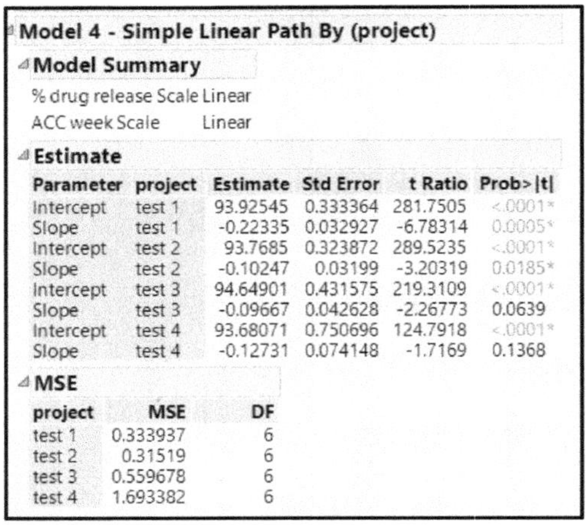

Model 4 - Simple Linear Path By (project)

Model Summary

% drug release Scale	Linear
ACC week Scale	Linear

Estimate

Parameter	project	Estimate	Std Error	t Ratio	Prob>|t|
Intercept	test 1	93.92545	0.333364	281.7505	<.0001*
Slope	test 1	-0.22335	0.032927	-6.78314	0.0005*
Intercept	test 2	93.7685	0.323872	289.5235	<.0001*
Slope	test 2	-0.10247	0.03199	-3.20319	0.0185*
Intercept	test 3	94.64901	0.431575	219.3109	<.0001*
Slope	test 3	-0.09667	0.042628	-2.26773	0.0639
Intercept	test 4	93.68071	0.750696	124.7918	<.0001*
Slope	test 4	-0.12731	0.074148	-1.7169	0.1368

MSE

project	MSE	DF
test 1	0.333937	6
test 2	0.31519	6
test 3	0.559678	6
test 4	1.693382	6

The estimate table provides detail for the intercepts, slopes, and the statistics that are used to indicate the significant differences of the model parameters. Another important table includes the mean standard errors (MSE). MSE is a measure of spread for the various weekly pulls throughout the analysis timespan. The test 4 project batch includes much more variability among pulls than the other three project batches. It is good practice for the team to review the results for test 4 with the analytical team to ensure that there are no errors in the data used for this study. Six degrees of freedom are used because sever observations were made for the span of stability storage (1 wk, 3 wks, 6 wks, 9 wks, 12 wks, 15 wks, 18 wks). If the DF is not the value for the number of pulls minus one, there might be issues with missing data that need to be explored.

You will use the model estimates in a later section, so you should make them into a JMP data set for ease of use. Right-click on the *Estimate* table and select *Make into Data Table*, as shown in Figure 17.23.

Figure 17.23: Stability Model Estimates for Polymer A Converted to Data Table

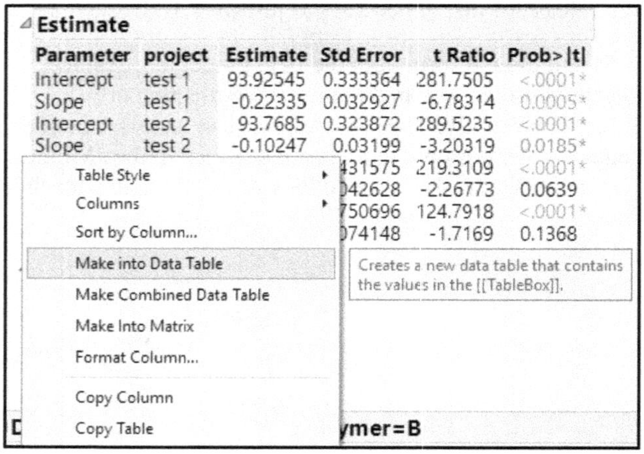

Estimate

Parameter	project	Estimate	Std Error	t Ratio	Prob>|t|
Intercept	test 1	93.92545	0.333364	281.7505	<.0001*
Slope	test 1	-0.22335	0.032927	-6.78314	0.0005*
Intercept	test 2	93.7685	0.323872	289.5235	<.0001*
Slope	test 2	-0.10247	0.03199	-3.20319	0.0185*

Table Style ▸	431575 219.3109 <.0001*
Columns ▸	042628 -2.26773 0.0639
Sort by Column...	750696 124.7918 <.0001*
	074148 -1.7169 0.1368
Make into Data Table	Creates a new data table that contains the values in the [[TableBox]].
Make Combined Data Table	
Make Into Matrix	
Format Column...	
Copy Column	
Copy Table	ymer=B

The data table of model estimates is shown in Figure 17.24. Save the table as "Polymer A Stability Estimates."

Figure 17.24: Table of Stability Model Estimates for Polymer A

	Parameter	project	Estimate	Std Error	t Ratio	Prob>|t|
1	Intercept	test 1	93.925445123	0.3333639021	281.75049704	<.0001
2	Slope	test 1	-0.223350647	0.0329273393	-6.783136815	0.0005
3	Intercept	test 2	93.768500069	0.3238717813	289.52352596	<.0001
4	Slope	test 2	-0.102469455	0.0319897744	-3.203194042	0.0185
5	Intercept	test 3	94.649012691	0.4315745974	219.31089844	<.0001
6	Slope	test 3	-0.096668607	0.0426279003	-2.267730909	0.0639
7	Intercept	test 4	93.680708313	0.7506959117	124.79181897	<.0001
8	Slope	test 4	-0.127305304	0.074148457	-1.71689755	0.1368

The table of model estimates is in the stability analysis output for polymer B, shown in Figure 17.25.

Figure 17.25: Stability Model Estimates for Polymer B

Model 4 - Simple Linear Path By (project)

Model Summary

% drug release Scale Linear
ACC week Scale Linear

Estimate

Parameter	project	Estimate	Std Error	t Ratio	Prob>|t|
Intercept	test 5	97.8202	0.579994	168.6574	<.0001*
Slope	test 5	-0.24177	0.057288	-4.22022	0.0056*
Intercept	test 6	94.79696	0.892364	106.2313	<.0001*
Slope	test 6	-0.17378	0.088141	-1.97164	0.0961
Intercept	test 7	97.70487	1.78387	54.77129	<.0001*
Slope	test 7	-0.06172	0.176198	-0.3503	0.7381
Intercept	test 8	95.02974	1.032619	92.02786	<.0001*
Slope	test 8	-0.54193	0.101995	-5.31334	0.0018*

MSE

project	MSE	DF
test 5	1.010819	6
test 6	2.392824	6
test 7	9.5621	6
test 8	3.20411	6

The slopes of the four project batches do not differ radically from the polymer A project batches, with only test 8 as an exception. Test 8 had a drop in % drug release that is more than double the amount of drop for any other project batch regardless of polymer type. Reviewing the MSD paints a picture of the most interesting differences between the polymer groups. The variability within the project batches of polymer B is larger than the polymer A group by several degrees of magnitude. The high variability is reflected in the confidence interval areas about the model lines that are much wider for the polymer B group. It is good practice to go back to the analytical team to review the source data to be sure there are no errors in the data set.

The heightened variability in the project batches of the polymer B group make the bad results of shortened shelf life worse. It might be possible to identify issues with the test batch that defines the shortened shelf life, but the variability creates unacceptable risk for the development team. There is little value to digging deeper into the linear models for the polymer B group at this point. The team decides to shift focus to fully explore the trends in the polymer A group.

Stability—Linear Model Diagnostics

ICH guidelines recommend the use of linear models in the analysis of stability trends. The first level of diagnostics for analysis models is to visualize the residuals to look for non-random patterns. Recall that residuals are the difference between actual values and the predicted values from the model. Figure 17.26 includes residuals for the projects made with polymer A.

Figure 17.26: Residual Plot for Polymer A

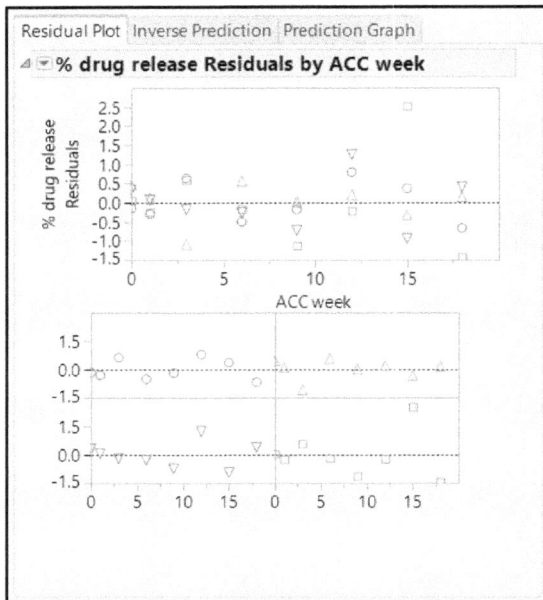

The overall plot and the plots by project batch show random patterns of residuals. A point from test 4 at 15 weeks was underestimated by the model and can be considered extreme. One extreme point does not make a trend and is not much to worry about; the linear models have an adequate fit to the data. The next plot to review is the inverse prediction plot shown in Figure 17.27.

Figure 17.27: Inverse Prediction Plot for Polymer A

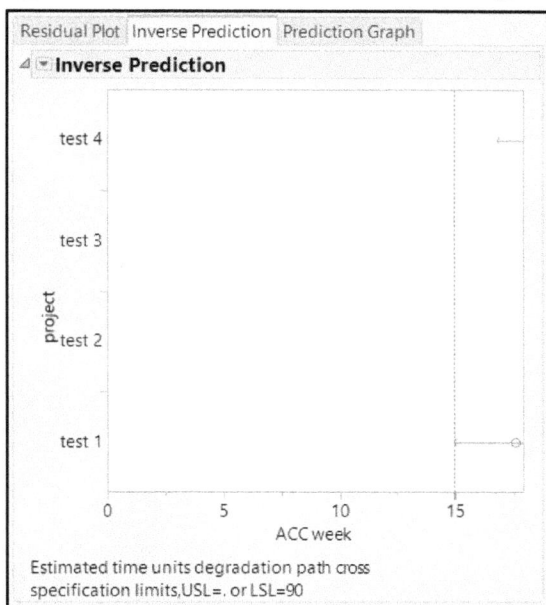

The inverse prediction plot indicates which of the projects have the shortest shelf life. The plot clearly shows that the test 1 project batch intersects the lower limit at 15 weeks. The project batch test 4 is next with an estimated shelf life of just over 16 weeks. Right-click on the X axis scale to adjust the tick marks. You can also use plot sizing to adjust the scale so that you can see all the project batches.

You can include a specific prediction on the % drug release for any week that you specify. Click the *Prediction Graph* tab to get the dialog box shown in Figure 17.28.

Figure 17.28: Prediction Plot Dialog Box

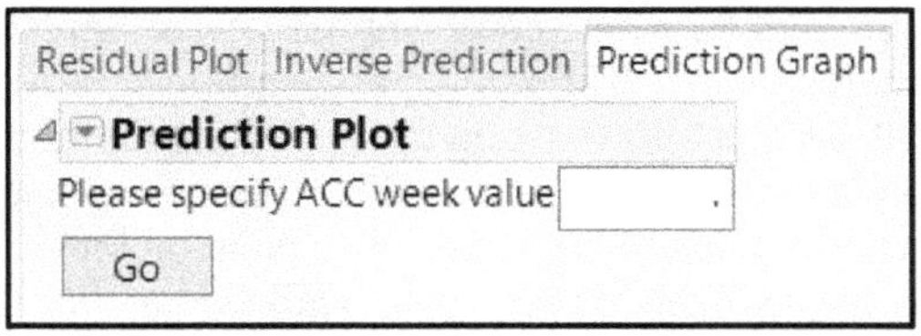

Enter 12 to get predictions at the 12-week date of storage in the accelerated conditions, and click *Go* to get the plot shown in Figure 17.29.

Figure 17.29: Prediction Plot for Polymer A

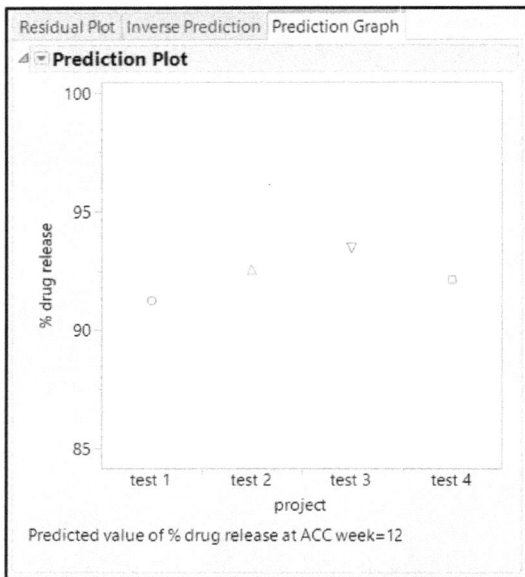

The prediction plot indicates that all four project batches meet the 90% lower limit. The predictions of % drug release are 91% and 93% at 12 weeks of accelerated stability exposure. The stability analysis provides clear evidence of the superior results of the polymer A group. Bryce is impressed with the amount of information provided by JMP, coupled with plots that clearly illustrate trends. The detailed information is shared with leadership to justify the direction of formulating the commercial product with polymer A.

The analyses of stability trends with JMP was executed more quickly than Bryce expected. He knows that the quality team needs to use the results to come up with internal alert limits and internal release limits. Internal limits provide added confidence that future commercial batches of product chosen for stability testing during the life cycle of the product will meet the 90% lower specification limit. Internal limits can add to the commercial risk of the product if they are set too conservatively. Extremely conservative internal limits incorrectly identify batches at risk when they are actually likely to meet the label claim on the expiration date. Bryce intends to use the information from the JMP analysis to create realistic internal limits for the quality team to consider.

Using Stability Estimates to Calculate Internal Limits

The stability model estimates for polymer A were saved as the data file *Polymer A Stability Estimates.jmp*. Open this data file so that you can organize it into the most useable format. Select *Tables*
▶ *Split* to get the dialog box shown in Figure 17.31.

Figure 17.31: Split Table Dialog Box

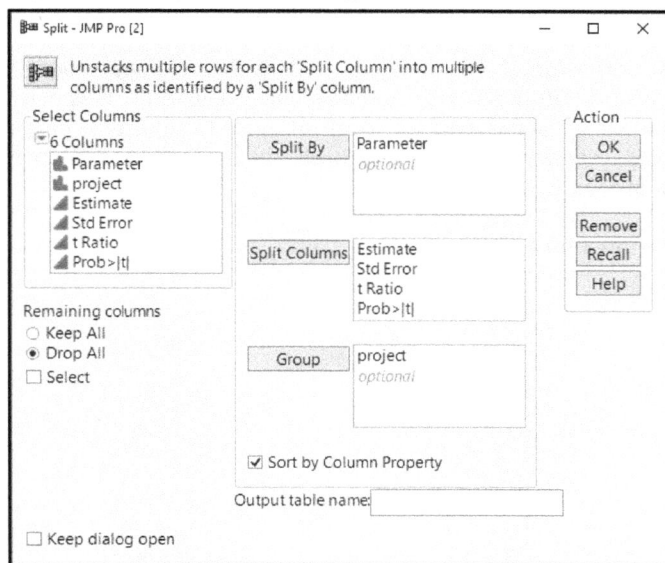

Move *Parameter* to the *Split By* box, and move *Estimate, Std Error, t Ratio,* and *Prob>|t|* to the *Split Columns* box. Move *project* to the *Group* box, and click *OK* to get the new data table shown in Figure 17.32.

Figure 17.32: Stability Model Estimates for Polymer A

	project	Estimate Intercept	Estimate Slope	Std Error Intercept	Std Error Slope	t Ratio Intercept	t Ratio Slope	Prob>|t| Intercept	Prob>|t| Slope
1	test 1	93.925445...	-0.22335...	0.3333639...	0.032927...	281.7504...	-6.7831...	<.0001	0.0005
2	test 2	93.768500...	-0.10246...	0.3238717...	0.031989...	289.5235...	-3.2031...	<.0001	0.0185
3	test 3	94.649012...	-0.09666...	0.4315745...	0.042627...	219.3108...	-2.2677...	<.0001	0.0639
4	test 4	93.680708...	-0.12730...	0.7506959...	0.074148...	124.7918...	-1.7168...	<.0001	0.1368

The project that defines the shelf-life limit is test 1, which happens to have the most extreme slope, losing 0.22% of drug release for every week of storage in accelerated conditions. The ICH guidelines for stability tend to be extremely conservative when a campaign of projects does not have intercepts and slopes pools. In this case, it is not recommended to apply extremely conservative confidence levels since the estimates are very likely to be unrealistic. In many cases, accelerated stability tends to yield more extreme results than the trends observed in ambient conditions, which adds justification for less conservative estimation techniques. Bryce's team chooses an 80% level of confidence for an infinite sample size for the calculation of a reasonable internal release limit and internal alert limit.

The z estimate for an 80% confidence level modifier is 1.3 and is used to calculate the internal limits. The standard error for the worst-case slope (test 1) is 0.0329. Therefore, a conservative estimate for the worst possible degradation is 0.22+(1.3*0.0329) =0.26. Working backward from the drug release low limit of 90%, at the 12-week accelerated estimated expiration date, you can see that the product must yield an initial

% release of at least 90+(12*0.26) =93.12. The initial release limit for internal quality use with the transdermal patch made with polymer A is % drug release of no less than 93%.

Leadership might want an even more conservative initial alert limit to prompt quality and operations of batches that are at risk for stability issues. Alert limits are used for trending changes in test results that might indicate added product risk. The problem is that the standard errors differ for the four projects. Rather than chase down the different intercept estimates for the four projects, the pooled model intercept estimate is used for the calculations shown in Figure 17.33.

Figure 17.33: Pooled Stability Model Estimates for Polymer A

Model 1 - Simple Linear Path

⊿**Model Summary**

% drug release Scale Linear	
ACC week Scale	Linear
SSE	17.41313
Nparm	8
DF	24
RSquare	0.692371
MSE	0.725547

⊿**Estimate**

Parameter	Estimate	Std Error	t Ratio	Prob>|t|
Intercept[project=test 1]	93.92545	0.491382	191.1454	<.0001*
Slope[project=test 1]	-0.22335	0.048535	-4.60182	0.0001*
Intercept[project=test 2]	93.7685	0.491382	190.826	<.0001*
Slope[project=test 2]	-0.10247	0.048535	-2.11124	0.0454*
Intercept[project=test 3]	94.64901	0.491382	192.618	<.0001*
Slope[project=test 3]	-0.09667	0.048535	-1.99172	0.0579
Intercept[project=test 4]	93.68071	0.491382	190.6474	<.0001*
Slope[project=test 4]	-0.12731	0.048535	-2.62294	0.0149*

Utilization of the standard error to add half of the margin of error of an 80% confidence interval for the mean (intercept) provides a conservative estimate for the alert limit. The lower half of the margin of error is calculated as 1.3*0.49 = 0.64. The most conservative estimate for an initial release limit is 93 + 0.64 = 93.6. The team rounds up to simplify the internal alert limit to no less than 94% release.

All four projects are at least 1% (rounded) above the proposed internal release limit and right on the alert limit. This is not necessarily a cause of alarm for the development team and leadership. As real-time stability projects mature to the point of having a reasonable group of projects with stability results beyond the 24-month expiration date, teams should review the stability trends. Internal limits can easily be adjusted by the quality team without the need to submit documentation to regulatory agencies. The changes will be captured in new revisions of the product release specifications if questions arise during the life cycle of the product. It is best practice for pharmaceutical manufacturers to plan to continue stability studies on at least the first five stability projects for a period that is at least one year beyond the expiration date. This ensures that the statistical trends and estimations made at the time of expiration include a reasonable amount of statistical error.

Practical Conclusions

Bryce can communicate a great deal of information about how the transdermal patches change in the % drug release over time. The JMP degradation platform with stability analysis options is easy to use and quickly summarizes trends, based on ICH guidelines. The best estimates of stability trends result from analyses that include multiple test batches over a time period that is as long as possible. This example included eight batches with accelerated exposure times that went significantly beyond the shelf-life goal for the transdermal product. Therefore, results from the example are robust.

The high-quality graphics and summary tables enabled the development team to robustly justify selecting polymer A for commercial product manufacturing. Not only is the polymer A product highly likely to meet the commercial expiration date of 24 months, the minimal variability among batch results make the predictions very robust.

The predictions made from the stability modeling add significant practical value. The quality team used the predictions to create a set of internal limits used to evaluate the initial release of commercial product batches. The internal limits help ensure that the % drug release meets the 90% lower specification throughout the life of the batch.

Exercises

E17.1—A new drug formula for a capsule product has been produced for several months. Data is available from a long-term study of several packaging configurations that are stored in a real-time stability chamber (25 degrees C/ 60% relative humidity). The product includes two different active ingredients and is produced in a regular strength (400 mg) and maximum strength (800 mg) formula. The marketing team inquired about the potential for placing more configurations into the commercial product mix. The requirement for assay of both actives is to 95% to 105% for the life of the product. The data includes the following grouping variables:

Dose:	400 mg, 800 mg
Active:	A, B
Container:	bottle (plastic), blister (strip with 20 capsules)
Size:	20count, 50count, 80count, 200count. 20count_film100 (blister), 20count_film200 (blister), 20 count_film300 (blister)
Desiccant:	none, low (1unit), high (2 units)

1. Open *two strength assay RT stability data.jmp*.
2. Select the *container, size*, and *desiccant* columns. Select *Cols ▶ Utilities ▶ Combine Columns* to create a new column with the combined information as a labeling group for stability study. Name the new column "configuration" and click OK to add it to the JMP data sheet.

3. The blisters were not marketed due to poor stability study results from the initial accelerated stability studies.

   ○ Filter the data to select and include the blister configurations.

   ○ Use the techniques noted in the chapter to analyze the assay stability trend for blisters.

   ○ Be sure to include *dose* and *active* in the *by* window.

   ○ Is there evidence that the stability for assay will be adequate for marketed product to maintain a 24-month expiration date?

4. Change the filter to run stability analysis on assay data for the bottles. Is there evidence that the stability for assay will be adequate for marketed product to maintain a 24-month expiration date?

5. Utilizing filtering and By grouping to investigate whether the stability trends differ by desiccant group.

6. How would you summarize this information to the leadership of the marketing team?

E17.2—You analyzed the assay stability for the capsule product with two active ingredients in the previous exercise. However, that was only part of the story. You also need to evaluate the product for dissolution since it is formulated for extended release. Keep in mind that dissolution testing involves multiple capsules tested at each pull; filtering the data and using By groups are likely to have significant effects on the stability trends.

1. Open *two strength disso RT stability data.jmp*.

2. The data includes the added grouping of the four hourly pulls for measuring the % released in the dissolution media (1, 2, 6, and 12 hours). The analyses require that the data be filtered by hour since the specifications for *% released* are specific for each pull:

1 hour:	20 to 45%
2 hours:	35 to 60%
6 hours:	60 to 85%
12 hours:	no less than 80%

3. Be sure to include *dose* and *active* in the *By* window.

4. Explore use of the Data Filter by *container* to isolate the trends for bottles and blisters. Once you have filtered the data by selecting either container type, use the red triangle options for the stability analysis to pick Redo. Evaluate the trend for the bottles and the blisters to determine whether the estimated shelf life for the various configurations is likely to exceed the planned 24-month expiration date.

5. How would you summarize the results of the stability analysis to present to the marketing team?

References

Carver, Robert H. *Practical Data Analysis with JMP®*. 2nd ed. Cary, NC: SAS Institute, 2014.

Gabrosek, John, and Paul Stephenson. *Introductory Applied Statistics: A Variable Approach*. 2nd ed. Allendale, MI: Grand Valley State University, 2016.

International Council for Harmonisation of Technical Requirements for Pharmaceuticals for Human Use (ICH). Geneva, Switzerland: Pharmaceutical Development Q8(R2). (ICH Q8(R2)), 2009.

International Council for Harmonisation of Technical Requirements for Pharmaceuticals for Human Use (ICH). Geneva, Switzerland: Quality Risk Management Q9. (ICH Q9), 2005.

Juran, Joseph M., and Joseph A. DeFeo. *Juran's Quality Handbook: The Complete Guide to Performance Excellence*. 7th Edition. New York, NY: McGraw-Hill Education, 2016.

United States Pharmacopeia and National Formulary (USP <1220>). Rockville, MD: United States Pharmacopeal Convention. 2016

United States Pharmacopeia and National Formulary (USP <1210>). Rockville, MD: United States Pharmacopeal Convention. 2015

Wheeler, Donald J. EMP III (Evaluating the Measurement Process): Using Imperfect Data. Knoxville, TN: SPC Press, 2006.

www.ingramcontent.com/pod-product-compliance
Lightning Source LLC
Chambersburg PA
CBHW081037220326
41598CB00038B/6907